DATE DUE			
NOV 23 1981			

WITHDRAWN

STUDIES IN NEW ENGLAND GEOLOGY

The Geological Society of America, Inc.
Memoir 146

Studies in New England Geology

A Memoir in Honor of C. Wroe Wolfe

Edited by
PAUL C. LYONS
AND
ARTHUR H. BROWNLOW

1976

*The Memoir series was originally made possible
through the bequest of
Richard Alexander Fullerton Penrose, Jr.*

*Copyright 1976 by The Geological Society of America, Inc.
Copyright is not claimed on any material prepared by
U.S. Government employees within the scope of their employment.
Library of Congress Catalog Card Number 75-30494
I.S.B.N. 0-8137-1146-0*

Published by
THE GEOLOGICAL SOCIETY OF AMERICA, INC.
3300 Penrose Place
Boulder, Colorado 80301

Printed in the United States of America

Contents

Preface vi
Acknowledgments vii
Introduction to C. Wroe Wolfe . . . .*Paul C. Lyons and Arthur H. Brownlow* ix

Part I. Geology of eastern Massachusetts

Introduction . . . .*Paul C. Lyons and Arthur H. Brownlow*		3
Geology of the Boston basin . . . . *Marland P. Billings*		5
Petrography and geochemistry of the Nashoba Formation, east-central Massachusetts . . . .*Adel Abu-moustafa and James W. Skehan, S.J.*		31
Petrology, chemistry, and age of the Rattlesnake pluton and implications for other alkalic granite plutons of southern New England . . . .*Paul C. Lyons and Harold W. Krueger*		71
Ayer Crystalline Complex at Ayer, Harvard, and Clinton, Massachusetts . . . . *Richard Z. Gore*		103
Stratigraphy and petrography of the volcanic flows of the Blue Hills area, Massachusetts . . . .*Uldis Kaktins*		125
Fossil plants of Pennsylvanian age from northwestern Narragansett basin . . . . *John Oleksyshyn*		143
Early Pennsylvanian age of the Norfolk basin, southeastern Massachusetts, based on plant megafossils . . . . . . *Paul C. Lyons, Bruce Tiffney, and Barry Cameron*		181

Part II. Geology of northern New England

Introduction . . . .*Paul C. Lyons and Arthur H. Brownlow* 201
Chronology and styles of multiple deformation, plutonism, and polymetamorphism in the Merrimack synclinorium of western Maine . . . . *Robert H. Moench and Robert E. Zartman* 203
Prehnite-pumpellyite facies metamorphism in central Aroostook County, Maine . . . . *Dorothy A. Richter and David C. Roy* 239
Stratigraphic relationships on the southeast limb of the Merrimack synclinorium in central and west-central Maine . . . .*Kost A. Pankiwskyj, Allan Ludman, John R. Griffin, and W.B.N. Berry* 263
Structural evolution of the White Mountain magma series . . *Carleton A. Chapman* 281
Gravity models and mode of emplacement of the New Hampshire Plutonic Series . . . . . . *Dennis L. Nielson, Russell G. Clark, John B. Lyons, Evan J. Englund, and David J. Borns* 301
Nickeliferous pyrrhotite deposits, Knox County, southeastern Maine . . . . *George D. Rainville and Won C. Park* 319

Preface

The Wolfe Volume was written to honor C. Wroe Wolfe on his retirement from Boston University. Dr. Wolfe founded the Geology Department at Boston University and guided the thinking of Boston University geology students for over thirty years. In addition, he has been a regular contributor to studies of the geology of New England. This memoir pays tribute to him as teacher, colleague, researcher, and geologic thinker.

Professor Wolfe and his students have made their greatest contributions to the geology of northern New England and eastern Massachusetts. In the early 1950s Dr. Wolfe and his Ph.D. candidates, in connection with Boston University's summer geologic field camp, started a program of detailed bedrock geologic mapping in west-central Maine. Later during the 1960s, he and his students concentrated their efforts in eastern Massachusetts. Dr. Wolfe guided numerous geologic field trips in this area throughout his teaching career at Boston University. Because of Dr. Wolfe's interest and work in these two regions, the Wolfe Volume is divided into two sections: (1) Geology of Northern New England and (2) Geology of Eastern Massachusetts.

Many of the authors of papers in the Wolfe Volume are former students of Professor Wolfe: Moench, Lyons (P. C.), Richter, Rainville, Abu-moustafa, Kaktins, Gore, and Tiffney. Other authors are colleagues, old friends, or admirers: Skehan, Cameron, Park, Oleksyshyn, Billings, Chapman, Lyons (J. B.), and Pankiwskyj.

Acknowledgments

We are very much indebted to Edwin B. Eckel, who, as Executive Secretary of the Geological Society of America, encouraged us to proceed with the Wolfe Volume in spite of our utopian dream to get it published before the Wolfe Retirement Banquet in 1974. We did, however, manage to get a commitment to publish the collection of papers by that time. We give our sincere thanks to the staff of the Geological Society of America, who quickly answered the numerous questions put to them from editors and authors and helped us through the seemingly endless technical aspects of preparing a GSA Memoir.

We are grateful to Arthur M. Hussey II, coordinator of the section on northern New England, and to James W. Skehan, S.J., coordinator of the section on eastern Massachusetts, who did an excellent job of securing formal reviewers for the papers and gave suggestions for improving papers in their sections. Marland Billings gave us great encouragement by his willingness to do a major paper on the Boston basin; in addition, he gave valuable service as an informal reviewer of several manuscripts. Cornelius S. Hurlbut, Jr., Won C. Park, and Anthony N. Mariano were very helpful in the early planning for the volume. We thank John Raabe, who was instrumental in getting the Pankiwskyj paper.

We owe a great deal of gratitude to all the contributors to the Wolfe Volume, who met the almost impossible deadlines and eagerly modified their papers in light of tough reviews. We also thank all the reviewers, most of whom are acknowledged in the papers, who gave rapid reviews and helped us meet our tight deadlines.

John W. Stuart, Curator of the Department of Geology at Boston University, was the principal compiler of the Wolfe bibliography, and for this assistance and various pleasantly undertaken courier missions, we are especially grateful. Lillian D. Paralikis was a constant supporter of the volume since its mental birth and played numerous roles as co-compiler of the Wolfe bibliography, and, at times, liaison person between authors and editors. Alice Wolfe provided valuable biographical data on her husband, particularly for his early years, and we acknowledge with great appreciation this information, which has been incorporated into the "Introduction to C. Wroe Wolfe."

Of course, and lastly, we owe our greatest debt to Dr. Wolfe, who inspired this undertaking and without whom there would not have been a Wolfe Volume. We want him to know that this undertaking was more a labor of love than a chore.

C. Wroe Wolfe

Introduction to C. Wroe Wolfe

Caleb Wroe Wolfe has been many things: farmer, crystallographer, family man, gemologist, preacher, world traveler, storyteller, clandestine poet, inventor, educator, and geologist. For 33 years his name has been intimately associated with the Geology Department of Boston University.

He was born the second youngest of six children in Washington, D.C., on October 22, 1908. His early childhood was spent in Minnesota. His mother died when he was seven years old, and in the next years his family moved often as his father, who was in the real estate business, shifted his real estate interests. During high school he worked on a farm in Roberts, Wisconsin. He then entered River Falls State College in River Falls, Wisconsin, and continued to work on the farm to earn his way through college. His major, both in high school and in college, was agriculture. His days on the farm were long and hard, with chores including milking the cows and delivering milk; during the bitterly cold Wisconsin winters, he would sometimes go to bed with his clothes on because of lack of heat in the farmhouse. One of Wroe Wolfe's favorite stories is about burying an old horse in hardpan on the farm, a job at which he was not entirely successful, as his calculation of depth of burial was off by a "horse's leg."

In the summer of 1930, after his graduation from college, he left Wisconsin with $23 in his pocket and a shoebox full of sandwiches, fruit, and cookies. He hitchhiked his way to Boston to enter the School of Theology at Boston University, with the intention of becoming a Methodist minister. After a year of study, he found that his own theology was too unorthodox to fit into any formal church scheme, although in later years he sometimes gave the sermon at the local church in Maine during the weeks of the field course he gave there. Wroe Wolfe was later to be a rebel in his geological views as well, which distinguished him among his contemporaries.

Jobs were scarce during the depression years, but Wolfe managed to get a position as a physical education instructor at a tuberculosis sanitarium in Reading, Massachusetts. After two years, he decided to enter Harvard University as a M.A. candidate in geology, the only subject he had really enjoyed in college. A year and a half later, in 1935, he received his M.A. degree and again found jobs hard to come by. This time he found his way into teaching sports (tennis, golf, and swimming) under the WPA. Within a year or two he decided to re-enter Harvard to study for his Ph.D. degree, but the entrance examination in German proved to be a temporary obstacle, as he was ill-prepared by his major in agriculture. However, through independent study he managed to pass the examination on the second attempt.

He was a research assistant at Harvard from 1937 to 1941, mainly working under the great crystallographer Charles Palache, and to a lesser extent under the mineralogist Harry Berman. During that time he was associated with Clifford Frondel and Cornelius S. Hurlbut, Jr., as co-workers, and later he worked with them for a time on a revision of *Dana's System of Mineralogy*. These four years represented a very fruitful period for Wolfe; it was marked by the publication of 17 papers in crystallography and mineralogy, almost all of them appearing in the *American Mineralogist.* Wolfe did his thesis work on the triclinic hydrated phosphates, a mineral group no one wanted to investigate because of the difficulty of study. Wolfe, never one to let a challenge pass by, successfully classified this mineral group. He received his Ph.D. degree in 1940 and published the results of his study in two parts in 1941. These papers and others that appeared in the 1940s were to establish Wolfe as a leading crystallographer. His stature was later enhanced by his invention of a new two-circle goniometer (1948) and by publication of his *Manual for Geometrical Crystallography* (1953).

In the fall of 1941, Harry Berman convinced Ralph Taylor, then dean of the College of Liberal Arts at Boston University and a fellow member of the newly formed Boston Mineral Club, to hire Wolfe as a member of the geography and geology department at Boston University. During the early 1940s, Wolfe was the field assistant of Marland P. Billings during the search for lithium in the pegmatites of New England. In 1943 Wolfe started an adult education course in Earth science through the Harvard Extension Program, a course he has continued to teach to this day. In 1946 he founded a separate geology department at Boston University and was its chairman from 1946 to 1963. During its early days, Wolfe *was* the department, teaching some 14 courses ranging from paleontology to economic geology. It was his view that his students should be broadly prepared in the field of geology. This teaching experience was to give Wroe Wolfe one of the broadest geological perspectives of any contemporary geologist. He likes to consider himself a generalist, a vanishing breed of geologist in this age of overspecialization.

In 1948 Professor Wolfe started a geology field camp in Maine, and during the 1950s and early 1960s his graduate students got their field training in the geology of west-central Maine. In 1949 he started a Ph.D. program in geology at Boston University, and he and his Ph.D. students laid the foundation for the study of the stratigraphy and general geology of west-central Maine. His first two doctoral students, Wolfgang Swarzenski and Robert H. Moench, went on to successful careers with the U.S. Geological Survey.

Professor Wolfe's students have many pleasant memories of their field days in Maine. Oh those peanut butter and raisin sandwiches! Wolfe had a reputation for climbing mountains like a mountain goat, almost always arriving at the summit before anyone else and coming down sometimes at a gallop. The volleyball games at the field camp, the stops for his favorite root beer (the only beer that Wolfe ever came near), and the field trips to Canada and the pegmatite mines were memorable. His sense of hard work, thrift, sacrifice, and fair play and his love for the outdoors and teaching are indelible qualities of C. Wroe Wolfe.

During the 1960s, the interest of Wolfe and his students turned toward the geology of the Boston area, where Wolfe had introduced countless numbers of students to the wonders of geology. Here his students worked on everything from the origin of the Salem gabbro-diorite to a study of the volcanic rocks at Nantasket.

Wroe Wolfe has an uncanny ability to get to the heart of a geologic problem by questioning the basic assumptions that underlie explanations. This gift, his great breadth of knowledge, and his ability to communicate effectively with people from all walks of life have distinguished him as one of the great twentieth-century American teachers of geology and Earth science. In recognition of this, Professor Wolfe was awarded the 1974 Neil A. Miner award by the National Association of Geology Teachers for outstanding teaching in the Earth sciences. It is as a classroom teacher that Wolfe has probably made his greatest contributions. He has regularly carried as much as twice the average daytime teaching load and has also taught two or three courses each semester at night. Thousands of Boston University undergraduates have taken his courses, and those of 30 years ago still remember him vividly. For many years Dr. Wolfe taught a course in gemology which was attended by jewelers and amateur collectors from the Boston area. He has always been a popular and available speaker for mineral clubs and other organizations. During the 1950s he frequently lectured on geology on FM radio and offered a 15-hour course on television. His lectures were described in one Boston University course and teacher evaluation book as "interesting, fun, clear, and organized." The fact that his courses are also challenging is explained by Wolfe to complaining students with this statement: "Rocks are a hard subject anyway."

Dr. Wolfe is married to the former Alice Caras and is the father of six children, four by previous marriages. He has traveled widely with his family and with his students; he has been to Alaska, Mexico, the Pacific, Ontario, and Iceland, and has made several excursions by automobile across Europe. During a Fulbright scholarship in 1963 and 1964, he traveled to and lectured in Paris, Iran, India, and East Pakistan. While in Dacca, Professor Wolfe wrote a collection of 11 poems (unpublished) on his observations and reflections on Dacca's culture, scenery, and mighty mountain ranges.

The following poem from this collection reveals the spirit of Wolfe the geologist:

The Himalayas

Rock!
Jagged, sculptured rock!
Where air is thin,
Too thin to breathe.
Where mind and blood congeal
And sky is just a step away.

There you stand,
Mighty monuments—ramparts rising—
Born from out a Tethys sea
You reach for Heaven from Earth.

Young you are
As mountains go.
Only yesterday the waters crossed your face;
But deep below the waters your strength was growing.

Then—
Quake by quake, fault by fault
You rose—
To claim the sky.

And now 'tis yours.
Drink deep of that ethereal blue,
For as your winter snows
Melt and fade away,
So will your form and power decay,
And you will sink to prepare the way
For a loftier, nobler range of yet another day.

Wolfe claims he has seen more mountain ranges than most geologists, and always has his own views on their origin. His attention turned toward mountain-building hypotheses in the late 1940s, and in 1949 he published his blister hypothesis, which maintained that radioactivity was the source of the internal heat of the Earth, an idea which now has virtually total support. In 1956 he published his idea on pleated folding to explain the vertical structure of west-central Maine; this idea was later supported by the detailed work of Moench (1970). Wolfe has always felt that a compressional tectonic model did not give an adequate explanation for the structure of that area of Maine. From the days of the granite controversy to the present, he has been an articulate spokesman of the transformationist school of granite origin. Today he is a healthy skeptic of plate tectonic theory because he says it offers no explanation for vertical motions of the Earth's crust. Another fruitful idea that Wolfe has nurtured is isostateism (as opposed to isostasy; 1966, 1968, 1973), which he uses to explain anything from Pleistocene glacial rebound to the uplift of the Himalayas and the Colorado Plateau. Wolfe has never been found wanting for an explanation of a geological phenomenon, whether it be the origin of diatomite deposits in Maine or the origin of plutonic breccia at Marblehead, Massachusetts.

Wolfe was president of the Boston Mineral Club from 1945 to 1946 and later was president of the now-defunct Boston Geological Society during the 1950s, also serving as its secretary-treasurer. He also served on the American Geological Institute Visual Aid Committee during the 1950s. He was a consultant in crystallography for Lincoln Laboratories from 1961 to 1964. Wolfe has authored or co-authored a number of books, including *This Earth of Ours—Past and Present* (1950, the book used in his introductory geology course), *Manual for Geometrical Crystallography* (1953), *Earth Science* (1953), and *Earth and Space Science* (1966). He is a fellow of the Geological Society of America and of the American Mineralogical Society.

Wroe Wolfe was honored in April, 1974, at a surprise banquet attended by nearly 300 of his friends, colleagues, and students, which marked his entrance into the ranks of professor's emeriti at Boston University. At the banquet it was announced that a new mineral was being named after him: Wroewolfeite, a beautiful blue hydrated copper sulfate mineral from Loudville, Massachusetts, described by Pete J. Dunn and Roland C. Rouse (1975, *Mineralogical Magazine*, v. 40, p. 1-5). He had previously been honored by the new mineral name Wolfeite, for an iron phosphate mineral described by Frondel (1949, *American Mineralogist*, v. 34, p. 692-705). Wolfe and Harry Berman co-discovered the mineral bellingerite from Chile (1940).

Wroe Wolfe's friends, colleagues, and former students now look toward a period when he will be more available to spend time in the field and to share his ideas on the origin of geologic phenomena.

We are indebted to Alice Wolfe, *American Men and Women of Science*, and to Wolfe himself for information in preparing this biography.

Paul C. Lyons
Arthur H. Brownlow

SELECTED BIBLIOGRAPHY OF C. WROE WOLFE

1937 Re-orientation of römerite: Am. Mineralogist, v. 22, p. 736–741.
1938 Calculation of angles for parahilgardite: Am. Mineralogist, v. 23, p. 767.
—— Note on römerite: Am. Mineralogist, v. 23, p. 468.
—— (and Richmond, W. E.) Crystallography of lanarkite: Am. Mineralogist, v. 23, p. 799–804.
—— Cannizzarite and bismuthinite from Vulcano: Am. Mineralogist, v. 23, p. 790–798.
1939 (and Berman, H.) Crystallography of aramayoite: Mineralog. Mag., v. 25, p. 466–473.
—— Symmetry and unit cell of hopeite [abs.]: Am. Mineralogist, v. 24, p. 194–195.
1940 Crystallography of aramayoite [abs.]: Am. Mineralogist, v. 25, p. 153.
—— Classification of minerals of the type $A_3(XO_4)_2 \cdot nH_2O$: Am. Mineralogist, v. 25, no. 11, p. 738–753.
—— Classification of minerals of the type $A_3(XO_4)_2 \cdot nH_2O$: Am. Mineralogist, v. 25, no. 12, p. 787–809.
—— (and Berman, H.) Bellingerite, a new mineral from Chuquicamata, Chile: Am. Mineralogist, v. 25, p. 505–512.
—— (and Richmond, W. E.) Crystallography of dolerophanite [abs.]: Am. Mineralogist, v. 25, p. 212.
—— (and Richmond, W. E.) Crystallography of dolerophanite: Am. Mineralogist, v. 25, p. 606–610.
1941 Crystallographic procedures: Am. Mineralogist, v. 26, p. 55–91.
—— A check on unit cell constants derived from 1-layer-line Weissenberg pictures: Am. Mineralogist, v. 26, p. 134.
—— X-ray data on diaboleite: Am. Mineralogist, v. 26, p. 610.
—— The unit cell of dickinsonite: Am. Mineralogist, v. 26, p. 338–342.
1943 (and Palache, C., and Richmond, W. E.) On amblygonite (Maine): Am. Mineralogist, v. 28, p. 39–53.
1944 Hexagonal zone, symbols and transformation formulae: Am. Mineralogist, v. 29, p. 49–54.
—— (and Billings, M. P.) Spodumene deposits in Leominster-Sterling area, Mass.: Massachusetts Dept. Public Works Inf. Circ. 3, 9 p. (with geologic map).
1945 Crystallography of cristobalite from Ellora Caves, India: Am. Mineralogist, v. 30, p. 536–537.
1947 (and Heinrich, E. W.) Triplite crystals from Colorado: Am. Mineralogist, v. 32, p. 518–526.
1948 (and Heinrich, E. W.) Triplite crystals from Colorado (additional note): Am. Mineralogist, v. 33, p. 92.
—— A new two-circle goniometer: Am. Mineralogist, v. 33, p. 739–743.
—— The past and future at Nantasket Beach, Mass.: Earth Sci. Digest, v. 2, no. 9, p. 19–23.
—— The blister hypothesis and the orogenic cycle [abs.]: Geol. Soc. America Bull., v. 59, p. 1364.
1949 The blister hypothesis: Sci. American, v. 180, no. 6, p. 16–21.
—— The blister hypothesis and the orogenic cycle: New York Acad. Sci. Trans., ser. 2, v. 11, no. 6, p. 188–195.
—— (and Franklin, V.) Refractive indices of high index liquids by the prism method on the two-circle goniometer: Am. Mineralogist, v. 34, p. 893–895.
—— The blister hypothesis and geological problems: Earth Sci. Digest, v. 3, no. 2, p. 3–11.
—— Surface geology at the border of an ice sheet: Earth Sci. Digest, v. 3, no. 9, p. 3–8.
—— Ludlamite from the Palermo Mine, North Groton, New Hampshire: Am. Mineralogist, v. 34, p. 94–97.
1950 Secondary earth science education through the earth science institute: Earth Sci. Digest, v. 4, no. 3, p. 14–15.
—— This Earth of ours—Past and present: Revere, Massachusetts, Earth Sci. Pub. Co., 374 p.

1950 (and Ramsdell, L. S.) The unit cell of malachite: Am. Mineralogist, v. 34, p. 119-121.
1951 Blister hypothesis and the petrogenic cycle [abs.]: Geol. Soc. America Bull., v. 62, no. 12, pt. 2, p. 1941.
—— (and Caras, A.) Unit cell of schairerite: Am. Mineralogist, v. 36, p. 912-915.
—— Blister hypothesis and the petrogenic cycle [abs.]: Am. Mineralogist, v. 37, p. 303-304.
1952 Review of Porter, M. W., and Spiller, R. C., The Barker Index of Crystals: A Method for the Identification of Crystalline Substances, Vol. 1: Am. Mineralogist, v. 37, p. 875-877.
—— Rock furnaces: Explosives Engineer, v. 30, no. 1, p. 7-11.
—— Outstanding pegmatites of Maine and New Hampshire *in* Billings, M. P., and others, eds., Guidebook for field trips in New England: Geol. Soc. America Ann. Mtg., p. 73-101.
1953 Manual for geometrical crystallography: Ann Arbor, Michigan, Edwards Bros., 263 p.
—— (and Fletcher, G. L.), Earth science, (3rd ed.): Boston, Massachusetts, D. C. Heath and Co., 556 p. (includes lab manual).
—— Review of Hurlbut, C. S., Jr., Dana's manual of mineralogy (16th ed.): Am. Mineralogist, v. 38, p. 423-425.
—— Recorded in rocks: Bostonia, v. 26, no. 2, p. 23-26.
1954 Using local resources in teaching earth science: Metro. Detroit Sci. Rev., v. 15, no. 1, p. 28-30.
1955 Crystallography of jadeite crystals from near Cloverdale, California: Am. Mineralogist, v. 40, p. 248-260.
—— Ontario-Quebec: A mineral collector's paradise: Earth Sci., v. 8, no. 3, p. 13-15.
—— Ontario-Quebec: A mineral collector's paradise II: Earth Sci., v. 8, no. 4, p. 16-22.
—— (and Block, M. L., and Baker, L.C.W.) The dimeric nature and crystallographic unit cell of ammonium 6-molybdochromiate: Am. Chem. Soc. Jour., 77, p. 2200.
—— Adult education in geology: Boston Univ. Graduate Jour., v. 4, no. 4, p. 53-55.
1956 Blister hypothesis and geomorphology [abs.]: Geol. Soc. America Bull., v. 67, no. 12, pt. 2, p. 1745.
—— Boston University's summer field camp—A Doorway to geological research: Boston Univ. Graduate Jour., v. 4, no. 9, p. 166-167.
—— The interrelationships of geology and chemistry: Scientia, Boston Univ. Chemia Soc., v. 8, no. 4, p. 3.
—— Underground storage of hydrocarbons: Interstate Oil Compact Comm., p. 45-47.
—— Pleated folding in northwestern Maine [abs.]: Geol. Soc. America Bull., v. 67, no. 12, pt. 2, p. 1745.
1957 (and Swarzenski, W. V.) The tectonic significance of the erosion surfaces in northwestern Maine: Internat. Geol. Cong., 20th, Mexico [D. F.] 1956, sec. 5, p. 491-500.
—— The blister hypothesis and the origin of mineral deposits: Earth Sci., v. 10, no. 1, p. 17-22.
1959 Geology on TV: Geotimes, v. 3, no. 7, p. 10-11.
—— Polarizing adapters for the Wolfe goniometer: Am. Mineralogist, v. 44, p. 182-184.
1960 (and Vilks, I.) Pseudomorphes after datolite, prehnite and apophyllite from East Granby, Connecticut: Am. Mineralogist, v. 45, p. 443-447.
—— (and Swarzenski, W. V.) The tectonic significance of the erosion surfaces in northwestern Maine: Zeitschr Geomorphologie, Band 4, Heft 1, p. 53-68.
—— On being a scholar: Boston Univ. Graduate Jour., v. 9, no. 1, p. 19-22.
—— Crystal synthesis by refrigeration: Am. Mineralogist, v. 45, p. 1211-1220.
—— Stratigraphy and general geology, Rangeley to Phillips, Maine, *in* Field trips in west-central Maine: New England Intercollegiate Geol. Conf., 52nd Ann. Mtg. Guidebook, p. 9-17.
—— Review of Hurlbut, C. S., Jr., Dana's Manual of Mineralogy (17th ed.): Am. Mineralogist, v. 45, p. 750-751.
—— Volume change forces in geologic processes: Copenhagen, 21st Internat. Geol. Cong., Rept., pt. 18, p. 287-294.

1961 Religion in higher education—Approach to "religious experience": Zion Herald, p. 139-144.
1964 Research in Pakistan: Geotimes, v. 9, no. 4, p. 6.
—— (and Latif, M. A., and Hoque, M.) Isostateism versus isostasy: A study in vertical tectonics [abs.]: Am. Geophys. Union, 4th Western Nat. Mtg., Trans., p. 638.
—— (and Boutilier, R.) Re-evaluation of the Salem gabbro-diorite complex, Mass. [abs.]: Geol. Soc. America Spec. Paper 76, p. 18.
—— (and Furlong, I. E.) Pluton genesis as related to geomorphology in northwest-central Maine [abs.]: Geol. Soc. America Spec. Paper 82, p. 368.
1966 (and Battan, L. J., Fleming, R. H., Hawkins, G. S., and Skornick, H.) Earth and space science: Boston, Massachusetts, D. C. Heath and Co., 630 p.
—— Energy, time, and physical morphology, *in* Planetology and space mission planning Sec. 1, Environments: New York Acad. Sci. Annals, v. 140, p. 16-34.
1967 Is moral law "natural" law?: Zion Herald, v. 145, no. 5, p. 4-7.
—— Marginal folding by displacement: Earth Sci., v. 20, p. 220-227.
1968 Marginal folding by displacement [abs.]: Geol. Soc. America Spec. Paper 101, p. 283-284.
—— Crustal deformation under glacial loading and unloading by shifting Moho [abs.]: Geol. Soc. America Spec. Paper 115, p. 303.
—— (and Farnsworth, R. L.) Analysis of Massachusetts erosion surfaces by the band width method [abs.]: Geol. Soc. America Spec. Paper 115, p. 261.
1969 Traverse information scanning: New York Acad. Sci. Annals, v. 163, p. 236-239.
—— Secondary relief features as clues to planetary evolution: New York Acad. Sci. Annals, v. 163, p. 81-89.
1971 (and Kessler-Richardson, C. A., and Park, W. C.) Experiments on movement and precipitation of salt solutions in unconsolidated sediments [abs.]: 8th Internat. Sedimentological Cong., Heidelberg, p. 50.
—— (and Lyons, P. C.) Correlation of granites by soda/potash ratios: Geol. Soc. America Bull., v. 82, p. 2023-2026.
—— (and Gheith, M.) Origin of two granite dikes in east New Portland, Maine: Geol. Soc. America, Abs. with Programs, v. 3, no. 1, p. 62-63.
—— (and Keidel, F. A., Montgomery, A., and Christian, R. P.) Calcian ancylite from Pennsylvania: New data: Mineral Record, v. 2, no. 1, p. 18-25, 36.
1972 Surface topography of the inner planets as related to planetary origins: New York Acad. Sci. Annals, v. 187, p. 82-87.
1973 Isostateism—An alternative to isostasy, *in* Recent researches in geology: Delhi, India, Hindustan Pub. Corp., p. 8-24.
—— (and Estelle, S.) Magma simulation with sugar solutions: Geol. Soc. America, Abs. with Programs, v. 5, no. 2, p. 156.
1974 (and Koza, D. M.) Ordered lithologic and textural variations in the Andover granite: Geol. Soc. America, Abs. with Programs, v. 5, no. 1, p. 46.
—— (and Raman, S. V.) Petrogenic study of granites and associated rocks at Clifton, Marblehead, Mass: Geol. Soc. America, Abs. with Programs, v. 5, no. 1, p. 65.
—— Geochemical differentiation and mineral deposits: Dhanbad, India, Coal and Mining Jour., The New Sketch, Republic Day Spec. No., Jan., p. 11-19.
—— (and Chasen, E.) Crystal habit as related to position and orientation in growth solution: Am. Mineralogist, v. 59, p. 1105-1112.
—— (and Estelle, S.) Magma simulation with sugar solutions: Earth Digest, v. 27, no. 4, p. 203-209.
1975 Isotherm migration and pegmatite evolution: Dhanbad, India, Coal and Mining Jour., The new sketch, Republic Day Spec. No., Jan., p. 24-31.
—— Pseudoapparent crustal shortening: Delhi, India, Hindustan Pub. Corp., Chayanica Geologica, v. 1, no. 1, p. 25-43.
—— (and Tekverk, R. W.) Stratigraphy of glacial clays in Norridgewock, Maine: Geol. Soc. America, Abs. with Programs, v. 7, p. 123.

PART I
GEOLOGY OF EASTERN MASSACHUSETTS
JAMES W. SKEHAN, S.J.
Co-ordinator

Introduction

This section includes seven papers: three on the late Paleozoic sedimentary basins of southeastern Massachusetts, one on the alkalic granite of eastern Massachusetts and Rhode Island and another on a chemically related volcanic sequence in the Blue Hills area, one on the Nashoba Formation, and one on the areally associated Ayer Granodiorite. All the papers present new data and interpretations on various aspects of the geology of eastern Massachusetts.

The paper by Marland Billings on the geology of the Boston basin is a major synthesis based mainly on recent tunnel work. This is the most important paper to appear on the Boston basin since his 1929 paper, "Structural Geology of the Eastern Part of the Boston Basin." Billings presents maps and cross sections of the basin and gives new interpretations of its structure and stratigraphy, as well as of the structural evolution of the Blue Hills complex.

Two papers deal with recent biostratigraphic work on the Pennsylvanian basins of southeastern Massachusetts. John Oleksyshyn reports an early Alleghenyan (Westphalian C) megaflora from the Rhode Island Formation in the northern part of the Narragansett basin. This is the most important contribution to appear on the paleobotany of the basin since A. S. Knox's 1944 paper, "A Carboniferous Flora from the Wamsutta Formation of Southeastern Massachusetts." A *Neuropteris obliqua–Lonchopteris* floral zone in the Norfolk basin is reported by Lyons and others. This work shows that the Pondville Conglomerate of the Norfolk basin correlates with the Upper Pottsville of the central Appalachians. It provides the best evidence to date that the Norfolk basin is of Pennsylvanian age and, furthermore, refines this age to Early Pennsylvanian.

Lyons and Krueger give new petrographic, chemical, and radiometric data on the rocks of the Rattlesnake pluton of Sharon, Massachusetts. This pluton is part of a belt of alkalic granite plutons that extends from Cape Ann, Massachusetts, to Rhode Island. Important chemical data are reported for the feldspar and riebeckite. The difficulty of making correlations based on mineralogic and petrographic similarity is made clear in this paper. Radiometric data are presented to contest age findings by R. E. Zartman and R. F. Marvin ("Radiometric Age (Late Ordovician) of the Quincy, Cape Ann, and Peabody Granites from Eastern Massachusetts," 1971).

A detailed stratigraphy for an ash-flow sequence in the Blue Hills area is reported by Kaktins. In the older literature, these volcanic rocks were referred to as "aporhyolite," but no important stratigraphic, petrographic, or chemical work has been done until now. Kaktins indicates that there are six volcanic-stratigraphic units of a probable peralkaline magmatic origin.

Abu-moustafa and Skehan report a wealth of new modal and petrochemical data on the Nashoba Formation, as revealed in the Wachusett-Marlborough Tunnel.

The authors distinguish seven lithologic units and relate their origin to a eugeosynclinal environment dominated by volcanic activity. The authors maintain that the dominantly volcanic pile was metamorphosed under moderate pressures at near-magmatic temperatures to a gneiss, schist, and amphibolite sequence.

New data on the petrography and field geology of the Ayer Granodiorite, as well as x-ray data on its potash feldspars, are presented in a paper by Gore. Bulk composition of potash feldspar samples provides a basis for subdivision of the Ayer into a younger porphyritic Clinton facies and an older gneissic Devens–Long Pond facies. The Clinton facies intrudes the Worcester and Oakdale Formations, which overlie the Nashoba Formation in the Wachusett-Marlborough Tunnel. The author suggests a Devonian age for this facies.

PAUL C. LYONS
ARTHUR H. BROWNLOW

Geology of the Boston Basin

MARLAND P. BILLINGS
Geological Museum
Harvard University
Oxford Street
Cambridge, Massachusetts 02138

ABSTRACT

The Boston basin is one of several late Paleozoic nonmarine sedimentary basins that developed in eastern New England subsequent to the Acadian revolution. Most of the sedimentary rocks in these basins are known to be Pennsylvanian in age; those in the Boston basin are presumably of this age. The principal map units—except for the Blue Hills and Nahant—are the Precambrian basement, the Mattapan and Lynn Volcanic Complexes (Mississippian?), and the Boston Bay Group (Pennsylvanian?). The Boston Bay Group consists of the Cambridge Argillite and the Roxbury Conglomerate. The Roxbury Conglomerate in turn is subdivided, from bottom to top, into the Brookline, Dorchester, and Squantum Members.

During the past 25 years, a series of bedrock tunnels, driven for water supply and drainage purposes, have added greatly to our knowledge. The tunnels, 3 to 3.5 m in diameter, are at a depth of 30 to 90 m below the surface. The total length of these tunnels is 39.57 km; the Dorchester Tunnel, under construction, is another 10.19 km long.

New observations and interpretations are as follows: (1) The maximum thickness of the Boston Bay Group is 5,700 m. (2) The Boston Bay Group thins to the south. (3) The Roxbury Conglomerate, with a maximum thickness of 1,310 m, is a southerly facies of the lower part of the Cambridge Argillite. (4) The Cambridge Argillite reaches a maximum thickness of 5,700 m in the northern part of the basin. (5) The sedimentary rocks were derived from a highland to the south.

The most important new results that bear on structure are that (1) the Northern border fault, where exposed in a tunnel, dips 55°N; (2) the Charles River syncline, exposed in two tunnels 10.5 km apart, plunges 19° in a direction N84°E; and (3) many minor folds and faults complicate the structure.

Although no tunnel crosses the Blue Hills, a new interpretation of the structure is presented. The volcanic complex of that area was erupted onto flat-lying Cambrian sedimentary rocks. The Quincy Granite and Blue Hill Granite Porphyry were injected into the horizontal Cambrian strata and volcanic complex. After a period of uplift and erosion, the Pennsylvanian strata of the Norfolk basin were deposited. All the rocks were then folded into a syncline, the vertical north limb of which is now the Blue Hills. The Blue Hills were then thrust northward over the Boston basin.

INTRODUCTION

History

This paper is concerned with the bedrock geology of the Boston basin and, to a much lesser extent, with that of the Blue Hills area to the south. The main features of the bedrock geology have been well known for one hundred years, and much was learned in the last quarter of the nineteenth century (LaForge, 1932). The principal contributor was W. O. Crosby, who in 1880 laid the foundation for the stratigraphy and structure of the area. His superb program of detailed mapping (Crosby, 1893, 1894, 1900) was unfortunately terminated by his commitment to engineering geology, a subject in which he was the American pioneer (Shrock, 1972).

The knowledge available in the early part of this century has been admirably summarized by LaForge (1932), whose bedrock map distinguished areas with no outcrop from those with outcrops. During the period 1933–1964, while I was in charge of the course in field geology at Harvard University, detailed maps were prepared of selected areas, notably Mattapan, Newton-Brookline, Blue Hills, and Nantasket-Hingham (Billings and others, 1939).

A new era began shortly after World War II, when a series of bedrock tunnels were driven under the supervision of the Construction Division of the Metropolitan District Commission. These tunnels, 3.0 to 3.7 m in diameter and built for water supply, drainage, and sewerage, are at a depth of 30 to 90 m beneath the surface. They total 39.57 km in length; the Dorchester Tunnel, under construction, adds an additional 10.19 km. The geology of all these tunnels was mapped, some in great detail. The results of some of these studies have been published (Rahm, 1962; Billings and Tierney, 1964; Billings and Rahm, 1966; Tierney and others, 1968). This paper will be largely concerned with the integration of these studies into a coherent picture. The Dorchester Bay Tunnel, driven in the 1880s, is 2.11 km long (Clarke, 1888). The West Roxbury Tunnel, driven in the 1960s, is 3.81 km long and has been mapped by the author, but the essential analysis has not been completed.

The origin of the locality names used for stratigraphic and plutonic units are adequately described in Wilmarth (1936) and need not be repeated here.

Frustrations

The bedrock geology of the Boston basin presents numerous difficulties. The lack of outcrops in some areas impedes definitive conclusions on the stratigraphy and structure; thus, alternative interpretations are often possible. For example, in an earlier publication (Billings, 1929), a large transverse fault is shown in the Weymouth area. In the present paper, a curving thrust fault is shown, but critical outcrops are lacking. Of course, tunnels give invaluable data in areas that lack outcrops. Drill holes into bedrock would also help, but generally such data are not published and drillers' logs are unreliable as sources of scientific data. The scarcity of fossils prevents their use in correlation. Moreover, many of the major

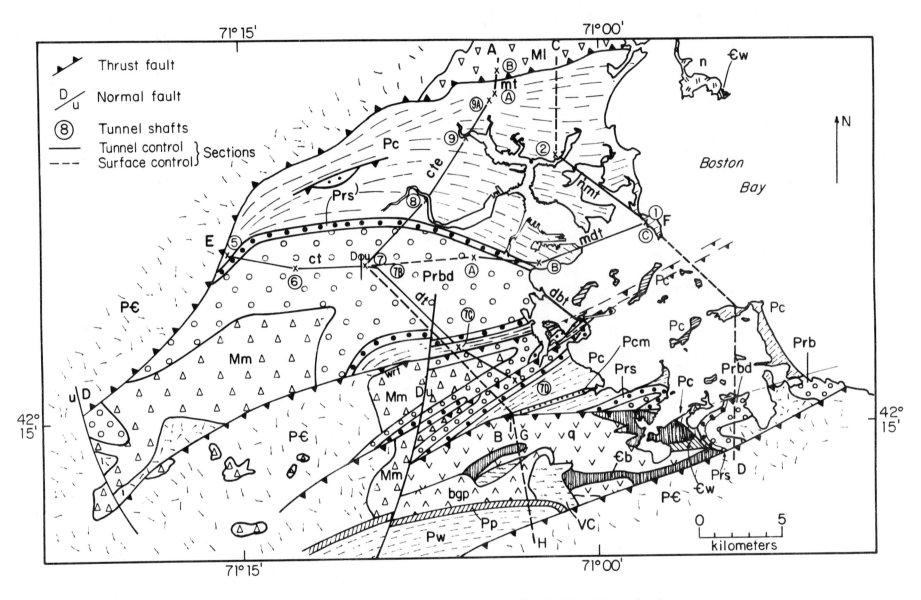

Figure 1. Geological sketch map of Boston basin and Blue Hills. See Figure 2 for explanation of symbols for formations. In Figure 1 Brookline and Dorchester Members are combined as Prdb. Tunnels: mt, Malden Tunnel; cte, City Tunnel Extension; ct, City Tunnel; nmt, North Metropolitan Relief Tunnel; dbt, Dorchester Bay Tunnel; wrt, West Roxbury Tunnel; dt, Dorchester Tunnel, under construction.

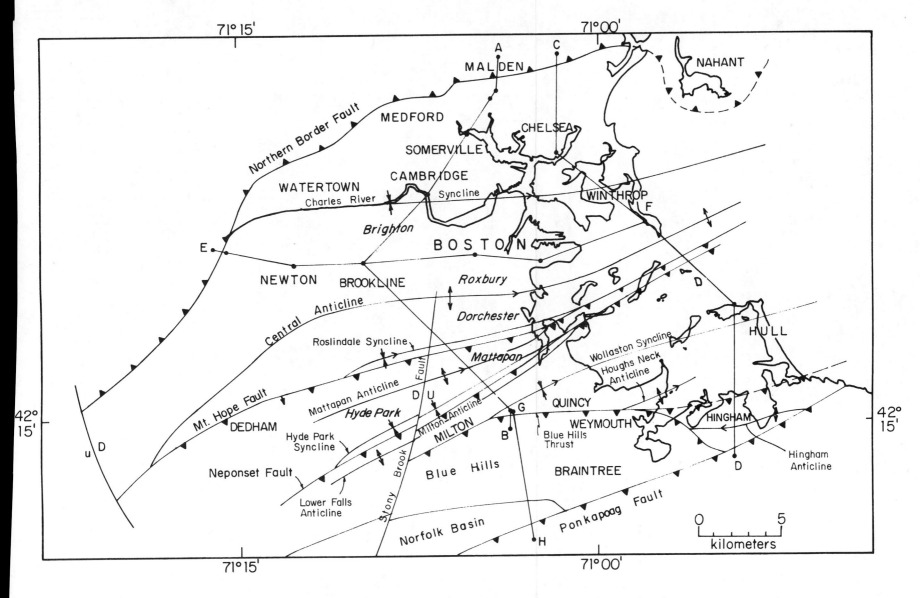

Figure 3. Tectonic map of Boston basin and Blue Hills. Also shows cities and towns referred to in text and location of structure sections.

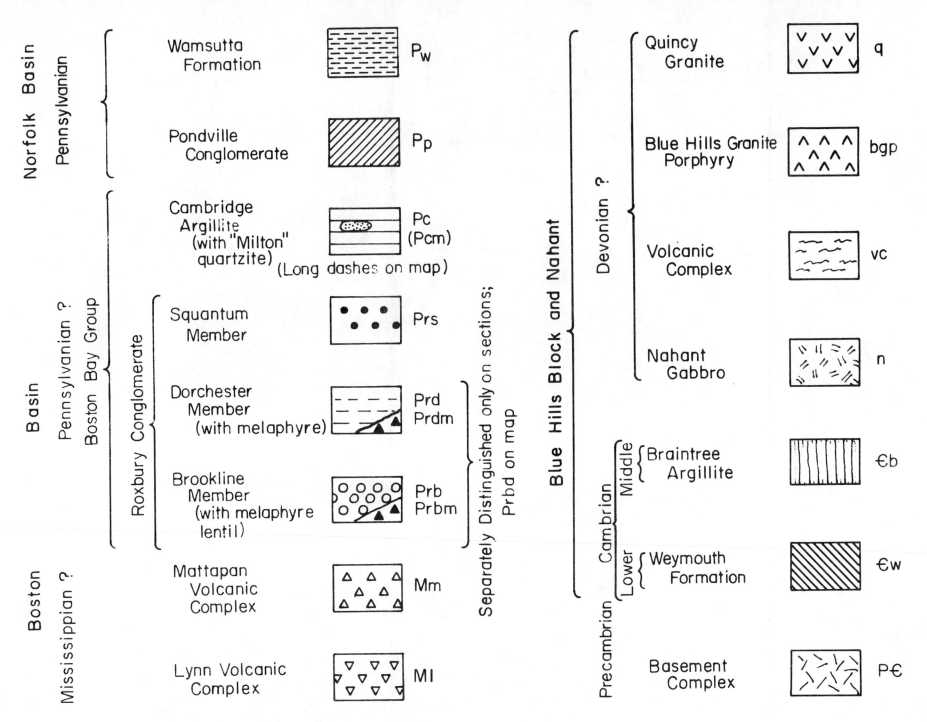

Figure 2. Explanation for Figures 1, 4, 5, 6, 7, and 13.

map units cannot be dated with any precision. The reader should also realize that the scale of the maps and diagrams in this article prevents the representation of many details observed in the field.

BOSTON BASIN

The Boston basin, like the Norfolk and Narragansett basins to the south, is a structural depression filled by late Paleozoic rocks younger than those in surrounding areas. The Boston basin is also a topographic basin. The northern border is clearly defined by an escarpment along a thrust fault. The southern border is marked by several en echelon thrust faults and locally by escarpments. The western boundary is not clearly defined topographically or geologically, inasmuch as older rocks outside the basin merge with similar rocks in the cores of easterly-plunging anticlines. The eastern limits are covered by the waters of Boston Harbor and the Atlantic Ocean.

Stratigraphy

General Statement. Three major units in the Boston basin are shown in Figure 1: (1) the Precambrian basement; (2) the Mississippian(?) Mattapan and Lynn Volcanic Complexes; and (3) the Pennsylvanian(?) Boston Bay Group. Figure 2 is the explanation for the symbols used in Figures 1, 4, 5, 6, 7, and 13. In Figure 3, the names of pertinent municipalities are given. Figure 4 shows the stratigraphic relations within the Boston Bay Group.

Many of the values of thickness in Table 1 are given to the nearest metre. Obviously these numbers could be rounded off, but those given to the nearest metre are based on calculations made from the continuous exposures in the tunnels. The data published in the original papers include some beds that are only 1 m thick. Moreover, comparison with the original papers would be difficult if the numbers were rounded off. On the other hand, I have no illusions about the precision of the numbers. Minor faulting and the necessity of averaging progressively changing dips and strikes in making the calculations introduces errors. An analysis shows that the errors in thickness may be as great as 5 percent. Values for distances and thicknesses were given in feet or inches in the original papers, but they have been converted to metric units here.

Precambrian Basement. The Precambrian basement (Fig. 1) underlies large areas northwest and southeast of the Boston basin. Moreover, it is extensively developed in the southwestern part of Figure 1, where it appears in the cores of northeasterly plunging anticlines. A detailed description of these rocks is beyond the scope of the present paper. The interested reader is referred to Emerson (1917), LaForge (1932), and Chute (1966, 1969). The Dedham Quartz Monzonite of eastern Massachusetts is overlain unconformably by fossiliferous Early Cambrian rocks (Dowse, 1950). Many of the rocks of eastern Massachusetts are older than the Dedham.

Mattapan and Lynn Volcanic Complexes. The Mattapan Volcanic Complex occupies three large areas on Figure 1, two to the south of the center of the map, the other to the southwest. Smaller areas are also present. Volcanic rocks in the Blue Hills are considered to be Mattapan by Chute (1969). This may be correct, but they will be described separately in a later section of this paper.

The Mattapan Volcanic Complex is composed of felsite and melaphyre (altered basalt and andesite), some of which is amygdaloidal. The felsite and melaphyre are pyroclastic rocks as well as flows; in some areas the Mattapan forms dikes and small stocks cutting across the Dedham Quartz Monzonite and other Precambrian rocks (Crosby, 1905; Crosby, 1928; Chute, 1966). I have also observed such relations in the West Roxbury Tunnel (Fig. 1). The abundance of such dikes and stocks undoubtedly explains the complex distribution of the Mattapan relative to older units.

The Lynn Volcanic Complex lies north of the Boston basin (Fig. 1) and generally is correlated with the lithologically similar Mattapan Volcanic Complex.

The Lynn and Mattapan Volcanic Complexes are older than the Boston Bay Group, because pebbles and cobbles of the volcanic rocks are common in the Roxbury Conglomerate. Based on the paper by Pollard (1965), they are here classified as Mississippian(?) in age, but the paleontological evidence is not conclusive. They may be the same age as the Newbury Volcanic Complex, which is of Late Silurian or Early Devonian age (Toulmin, 1964, p. A17).

General Statement Concerning the Boston Bay Group. One of the major contributions of the present paper is the description and interpretation of the stratigraphy of the Boston Bay Group (Fig. 4). This study, although utilizing surface exposures, is based primarily on observations made in five tunnels (Main Drainage Tunnel, Rahm, 1962; City Tunnel Extension, Billings and Tierney, 1964; Malden Tunnel, Billings and Rahm, 1966; City Tunnel, Tierney and others, 1968; North Metropolitan Relief Tunnel, Billings, in prep.).

Traditionally the Boston Bay Group has been divided into two formations, the Cambridge Argillite above and the Roxbury Conglomerate below. The latter in turn has been divided into three members, from top to bottom, the Squantum,

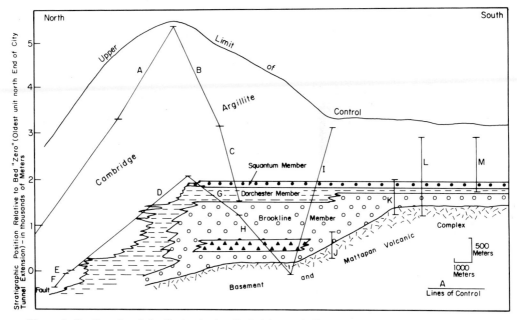

Figure 4. Stratigraphy of the Boston basin. Capital letters indicate nature of control. A. North part of North Metropolitan Relief Tunnel. B. South part of North Metropolitan Relief Tunnel. C. Main Drainage Tunnel. D. North part of City Tunnel Extension. E. Surface geology in Malden. F. Malden Tunnel. G. South part of City Tunnel Extension. H. Surface geology in Brookline. I. Surface geology from Brookline to Dorchester. J. Surface geology, north limb of Mattapan anticline. K. Surface geology, south limb of Mattapan anticline in Hyde Park. L. Surface geology, Dorchester Lower Mills. M. Furnace Brook at Adams Street, Quincy.

TABLE 1. LITHOLOGY OF BOSTON BAY GROUP

		Thickness (m)*	Conglomerate (%)	Sandstone† (%)	Argillite§ (%)	Tillite (%)
Cambridge Argillite	15	369	..	..	100	..
	14	90#	..	..	100	..
	13	2,060	..	3	97	..
	12	1,162	..	2	98	..
Roxbury Conglomerate						
Squantum Member	11	41	100	..	..	..
	10	19	95	5	..	..
	9	92	..	..	4	96
	8	122	20	..	..	80
Dorchester Member	7	187	11	46	43	..
	6	485	16	26	58	..
	5	399	9	8	83	..
	4	405	18	22	60	..
Brookline Member	3	146	52	31	17	..
	2	423**	40	11	49	..
	1	288**	49	27	24	..

*Values for thickness are given in meters to nearest unit for reasons explained in the text.
†Includes argillaceous sandstone.
§Includes some arenaceous argillite.
#Only bottom of formation penetrated.
**Only the upper part of this member was penetrated by the tunnels.
1. City Tunnel Extension, stations 255 + 11 to 333 + 74. 2. City Tunnel, stations 146 + 20 to 287 + 13 (includes tunnel from shaft 7 to 7B). 3. Hingham, surface data. 4. City Tunnel Extension, stations 333 + 74 to 368 + 97. 5. Main Drainage Tunnel, stations 0 + 00 to 119 + 16. 6. City Tunnel, stations 24 + 60 to 146 + 20. 7. Hingham, surface data. 8. Main Drainage Tunnel, stations 119 + 16 to 137 + 94. 9. City Tunnel, stations 11 + 60 to 24 + 60. 10. City Tunnel Extension, stations 368 + 97 to 371 + 08. 11. Hingham, surface data. 12. Main Drainage Tunnel, stations 137 + 94 to 375 + 86. 13. City Tunnel Extension, stations 398 + 98 to 627 + 72. 14. City Tunnel, stations 0 + 00 to 11 + 60. 15. Malden Tunnel, stations 2 + 00 to 24 + 57.

Dorchester, and Brookline Members (Emerson, 1917; LaForge, 1932). These subdivisions of the Roxbury can generally be readily recognized in the southern half of the basin. In the northern part of the basin, there is a major facies change; the Cambridge Argillite in Cambridge, Somerville, Medford, and Malden is the northerly facies of the Roxbury Conglomerate (Fig. 4). Nevertheless, in Boston, Winthrop, and Chelsea there are 3,000 m of argillite above the Roxbury. Kaye (1967) has discussed the kaolinization of the Boston Bay Group.

Roxbury Conglomerate. The Brookline Member consists primarily of the typical Roxbury "puddingstone," but there are interbedded argillite, sandstone, and melaphyre. The total thickness of the Brookline Member in the southern part of the basin is 1,310 m, but the tunnels penetrate only the upper part of the unit. Table 1 (line 1) shows a thickness of 288 m of this member in the City Tunnel Extension; 49 percent is conglomerate, 27 percent is sandstone, and 24 percent is argillite. The City Tunnel penetrates a thickness of 423 m, including an exceptionally thick section of argillite (Table 1, line 2). The data for Hingham (Table 1, line 3) are based on surface exposures of the upper part of the member. In natural exposures the conglomerate forms a much higher proportion of outcrops, because the weaker argillite and sandstone are more readily eroded.

The matrix of the typical conglomerate is a gray feldspathic sandstone. The pebbles and cobbles are well rounded, normally range in size from 1 to 15 cm, but locally may be as much as 30 cm in diameter. The clasts are chiefly quartzite, quartz monzonite, granite, and felsite, with smaller amounts of melaphyre and argillite. Bedding is uncommon.

The sandstone is fine to medium grained, and some is so feldspathic that it could be called arkose. Some displays beds that range in thickness from a few millimeters to 1 cm. Others that are 10 m or more thick show no bedding. The colors are white, gray, pink, and red. A few oscillatory ripple marks were observed.

In the Boston area, the term argillite has been applied to rocks that some may prefer to call mudstone or siltstone. The argillite is generally laminated (2 mm to 1 cm) but lacks the papery (2 mm) or platy (2 mm to 1 cm) splitting property characteristic of shale. The argillite in the Brookline Member tends to split into flags (1 to 5 cm thick) or slabs (5 to 60 cm thick) parallel to the bedding. The colors are pink or red; light-gray, dark-gray, and green-gray shades that are typical of the Cambridge Argillite are rare.

Melaphyre is associated with the Brookline Member. Some rocks are intrusive, others are flows. A large body in the western part of Brookline is in places a well-bedded tuff and breccia; but melaphyre in other places appears to be composed of flows.

The rocks in the Dorchester Member are similar to those in the Brookline Member, but the percentages are different. Table 1 lists the thickness of the Dorchester Member in three of the tunnels and the percentages of the main varieties of rocks. Table 2 is a sample to show how the various rocks are interbedded. The argillitic rocks are white, pink, red, and purplish gray; some are gray and greenish gray, similar to those in the Cambridge Argillite. In the western part of the Main Drainage Tunnel, the pelitic rocks are sufficiently fissile to be called shale. The sandstones are pink, red, and white; all samples are feldspathic, some containing so much feldspar that they are arkose. The conglomeratic rocks are similar to those in the Brookline Member.

The top of the Dorchester Member is defined rather readily by the base of the Squantum Member, which in most places contains the distinctive tillite. Locally, as in the City Tunnel Extension and at Hingham, tillite is absent but conglomerate at the appropriate stratigraphic position is assigned to the Squantum. The base of the Dorchester is defined as the horizon above which conglomerate is less than 20 percent of the rock, with sandstone and argillite constituting the balance.

The Squantum Member contains an unusual rock that many geologists consider tillite (Sayles, 1914, 1916; Sayles and LaForge, 1910; Rehmer and Hepburn, 1974), although others dispute or question this interpretation (Dott, 1961; Pollard, 1965; Caldwell, 1964). The most characteristic variety has a massive dark-gray, purple, or green-gray sandy to argillaceous matrix. In the tunnels the subrounded to angular clasts range from 5 to 60 cm in diameter. In the City Tunnel, a block of argillite 6 m long was incorporated in the tillite. In the Main Drainage Tunnel, pebbly shale and sandstone constitute part of the Squantum. The clasts are quartz monzonite, quartzite, felsite, and melaphyre. Cleavage is prominent in the tillite, notably in the City Tunnel. Lines 8 to 11 in Table 1 show the composition of the Squantum Member in three of the tunnels and in Hingham.

In the City Tunnel Extension, the Squantum Member consists entirely of conglomerate. The well-rounded clasts average 3 to 8 cm in diameter; some are 15 to 20 cm. A few thin quartzite and sandstone beds are interbedded.

Two poorly preserved tree trunks were found in the Brookline Member of the Roxbury Conglomerate many years ago by Burr and Burke (1900). The identification and significance of these specimens has been reviewed by Pollard (1965) and Lyons and others (1976). They indicate that the formation may range from Late Devonian through Permian in age (Rahm, 1962, p. 329).

Dating of the Boston Bay Group based on the supposed absence of clasts of the Quincy Granite is quite irrelevant. The oft-repeated statement that clasts of this granite are not found in the Roxbury Conglomerate must be reviewed. Moreover, as described below, the Quincy and related rocks have been thrust and uplifted into their present position.

Cambridge Argillite. This formation is composed almost exclusively of gray argillite, in which the beds range in thickness from 1 mm to 8 cm. The shades of gray differ in intensity; the grains in the light-gray rocks are silt or fine sand, whereas in the darker layers they are clay or fine silt. Much of the argillite shows a rhythmic layering, resulting from the alternation of lighter and darker beds 1 to 8 cm thick. Graded bedding is scarce. Occasional sandstone units show minute cross-bedding. Some beds of argillite are 1 m thick, but these larger beds consist of laminae a few millimetres thick. The argillite is slightly calcareous. The fractures parallel to the bedding are 15 to 120 cm apart; because of joints making a high angle with the bedding, the rocks break up into parallelopipeda. Table 1, lines 12 through 15, show the composition of the Cambridge Argillite.

"Milton" Quartzite. This unit is confined to the southern margin of the basin. The name was proposed by Billings (1929) but was preoccupied (Wilmarth, 1936). Rather than propose a new name, it seems best to retain the old name but to place it in quotes. The best exposures are in Milton just north of the Quincy Granite (Fig. 1). They extend from Randolph Avenue to East Milton in a belt 1.6 km long. The typical "Milton" Quartzite is a white coarse sericitic quartzite. When the Southeast Expressway was being built as a depressed highway through East Milton, the great trench through the glacial sand plain exposed a section

TABLE 2. SAMPLE OF LITHOLOGY OF PART OF BOSTON BAY GROUP*

			Thickness (m)
Cambridge Argillite	11	Argillite; gray slabby	18
	10	Sandstone; red, fine grained	38
	9	Argillite and sandy argillite; gray, buff, red, and purplish-red; a 1-m bed of gray grit at top.	8
Roxbury Conglomerate			
Squantum Member	8	Conglomerate. Pebbles well rounded 2 to 8 cm in diameter, some 15 to 20 cm. A few beds of quartzite and sandstone, each 5 to 8 cm thick	19
Dorchester Member	7	Argillite, some quartzite, and conglomerate. Argillite red, pink, gray, and greenish gray. One conglomerate bed 1.5 m thick	16
	6	White shale with some buff and green quartzite beds 5 mm to 20 cm thick	12
	5	Sandstone; pink, fine grained. Bed of quartzite 1 m thick at top	20
	4	Argillite; gray and greenish gray. One bed of fine grained sandstone 1 m thick	39
	3	Argillite, pink and red.	4
	2	Conglomerate. Pebbles rounded, 2 to 5 cm in diameter, maximum 12 cm; mostly quartzite and granite	68
	1	Pink, sandy argillite, red sandstone, and pink argillite.	50

*From City Tunnel Extension, stations 351 + 74 to 398 + 98 (stations in feet).

of bed rock 228 m long. The section trends N25°W. Because the beds dip 72° to 90°N, the breadth of outcrop is only slightly greater than the thickness. The measured section is presented in Table 3.

The south end of the section, units 1 through 4, 32.3 m thick, is typical "Milton" Quartzite. Above this are units 5 through 9, 31.7 m thick, consisting of red sericitic sandstone and red and green shale. The upper part of the section, 156.3 m thick (units 10 through 26), is composed of gray argillite typical of the Cambridge Argillite. The lower part of the section appears to be conformable with the gray argillite. I conclude, therefore, that the "Milton" is a member of the Cambridge. There is no evidence of a large fault between the typical Cambridge rocks and the "Milton" Quartzite. If the "Milton" is projected 1.6 km to the east-northeast, it lies 915 m northwest of the outcrops of the Squantum Member at Furnace Brook Parkway. If the dips are 70°NW and there is no folding, the "Milton" lies 820 m stratigraphically above the Squantum Member.

Major Structural Features

General Statement. The major structural features of the Boston basin trend east-northeast, and most of the folds plunge in that direction. Hence the Precambrian basement rocks and the Mattapan Volcanic Complex are exposed to the southwest, whereas the youngest units of the Boston Bay Group are found in Boston Harbor.

The principal tectonic units are given in Figure 3. They are assigned to the following categories:

1. Anticlines: Central, Mattapan, Milton, and Hingham. The Houghs Neck anticline is in an area of poor outcrops, and other interpretations are possible.

2. Synclines: Charles River, Roslindale, Hyde Park, and Wollaston. The Roslindale and Hyde Park synclines consist only of the northern limbs, because the south limbs are eliminated by thrust faulting. The Wollaston syncline is poorly exposed.

3. Northerly dipping thrusts: Northern border fault.

4. High-angle thrusts: Mount Hope, Neponset, and Blue Hills. The Mount Hope and Blue Hills thrusts are known to be essentially vertical. It is believed that originally they dipped south but have been rotated into their present position. The dip of the Ponkapoag fault is unknown.

5. Transverse faults: Stony Brook fault.

Western North-South Section of Boston Basin (AB). The location of section AB (Fig. 5) is shown on Figure 1. It extends across the basin from the Lynn Volcanic Complex on the north to the Blue Hills on the south. This section, 24.30 km long, includes the Malden Tunnel (1.60 km) and the City Tunnel Extension (11.43 km). The rest of the section is based on surface geology.

At the north end of the section, the Lynn Volcanic Complex is thrust over the Cambridge Argillite along a fault exposed in the Malden Tunnel; the fault dips 55°N. The contact is sharp, with no evidence of brecciation. Data are unavailable to calculate the net slip, but it is probably large because this fault can be traced for at least 40 km. The south end of the Malden Tunnel, the connection between the two tunnels, and the northern part of the City Tunnel Extension are in the north limb of the Charles River syncline. In general the strata dip south, but occasional reversals indicate subsidiary anticlines and synclines. The hinge lies just south of the Charles River in this structure section. This northern limb is composed largely of the Cambridge Argillite, which contains a few red sandstones. The south limb is composed of the Cambridge Argillite and the Roxbury Conglomerate. Only 288 m of the Brookline Member is exposed in the tunnel.

The hinge of the Central anticline lies 4 km southeast of the southwest end of the City Tunnel Extension. This is a broad open fold plunging about 15°E. The Brookline, Dorchester, and Squantum Members and the Cambridge Argillite are exposed on the southern limb of the anticline. The line of section AB crosses the Stony Brook fault, a steep fault that bears N10°E and along which the west wall has moved 700 m downward relative to the east wall. East of the fault, the Dorchester and Squantum Members and the lower part of the Cambridge Argillite are repeated. The Cambridge in this area forms the north limb of the Roslindale syncline.

There is no southeast limb of the Roslindale syncline, because it is eliminated by the Mount Hope fault, a vertical fault formerly exposed on Middleton Street in Dorchester. At this locality the Roxbury Conglomerate lies directly south of the fault. But 1.6 km to the west the Mattapan Volcanic Complex lies directly south of the fault, and 5 km to the west the Dedham Quartz Monzonite lies directly south of the fault. The stratigraphic throw along the fault is about 3,000 m.

The belt of Roxbury Conglomerate to the south of the Mount Hope fault is about 760 m wide. In an earlier paper (Billings, 1929), the rocks throughout this belt were shown as becoming younger to the south, necessitating a fault between them and the Mattapan further south. This was called the Sally Rock thrust. But

TABLE 3. SECTION IN EAST MILTON*

Lithology and attitude	Breadth (m)	Thickness (m)
26. Gap	15.2	14.6
25. Gray argillite, N. 75° E., 80° N.	3.7	3.4
24. Gap	57.0	55.2
23. Gray argillite	0.9	0.9
22. Gap	10.7	10.4
21. Gray argillite; thin beds fine-grained quartzite	9.1	8.8
20. Gray quartzite with quartz veins	0.6	0.6
19. Gray argillite, N. 62° E., 90°	2.4	2.1
18. Gap	0.9	0.9
17. Gray shale	0.9	0.9
16. Gap	9.7	9.5
15. Gray argillite, N. 80° E., 80° N.	6.1	6.1
14. Gap	22.9	22.2
13. Light-gray argillite, N. 70° E., 75° N.	7.6	7.3
12. Gap	12.2	11.9
11. Light-gray shale	0.9	0.9
10. Gap	0.6	0.6
9. Red shale, N. 66° E., 73° N.	10.7	10.4
8. Red shale†	5.2	4.9
7. Green shale	1.8	1.8
6. Red shale, N. 87° E., 72° N.	14.3	13.7
5. Red, sericitic sandstone	0.9	0.9
4. Gray sericitic quartzite N. 80° E., 73° N.	0.6	0.6
3. Quartzite with a few beds of red shale 2 to 3 cm thick	2.4	2.4
2. Gap	0.3	0.3
1. Slightly sericitic quartzite ("Milton")	30.2	29.0
Totals	227.8	220.3

*Section of Cambridge Argillite including "Milton" Quartzite at intersection of State Highway 135 and Southeast Expressway in East Milton. Outcrops were found in a road excavation and are now completely covered by concrete wall. Section trends N. 25° W. South end of section is north end of bridge abutment on Ledge Hill Road overpass. No. 1 is south end of section. Measured by M. P. Billings and P. R. Brayton, Jr., November 19, 1955.

†Top of this unit is south end of bridge abutment of Route 135 over Southeast Expressway, on east side.

subsequently a shale and sandstone contact indicated that the beds in the southern half of the belt become younger to the north (Billings, 1972, p. 88). This implies that the contact with the Mattapan to the south is a sedimentary contact, as shown in Figure 5.

Still further south the map pattern indicates that the Mattapan Volcanic Complex is in an anticline plunging northeast. This is the Mattapan anticline.

The Neponset River flows northeast in a belt of the Roxbury Conglomerate (Fig. 5). Still further southeast the Mattapan Volcanic Complex is exposed in the Milton anticline. This anticline was formerly well exposed in argillite near the base of the Roxbury Conglomerate (Fig. 1) at Milton Lower Mills along what are now MBTA tracks. The anticline plunges 18° in a direction N67°E. Exposures to the north were excellent and showed that the Roxbury Conglomerate is in a syncline plunging southwest; this, of course, is contrary to the general northeasterly plunge of the folds in the Boston basin.

But along the line of the cross section (Fig. 5), the Squantum Member lies directly north of the Mattapan in the Milton anticline. A fault is indicated. Five kilometres to the southwest, similar relations occur in Hyde Park just east of East Dedham (Chute, 1966). The stratigraphic throw along this fault is at least 550 m, but the throw decreases northeast of Milton Lower Mills.

The Milton anticline brings up the Mattapan Volcanic Complex. To the southeast there are no exposures within several kilometres of the line of section, but exposure of the Cambridge Argillite to the northeast in Quincy and to the southwest in Milton demonstrate the presence of a syncline. As indicated in a previous section, the "Milton" Quartzite is on the southeast limb of the syncline. The Quincy Granite is in fault contact with the "Milton."

Eastern North-South Section of Boston Basin (CD). The location of section CD (Fig. 6) is shown in Figure 1. It extends across the basin from the Lynn Volcanic Complex on the north to the Precambrian basement on the south. This section, 36 km long, includes the North Metropolitan Relief Tunnel (6.33 km long). It is especially significant because, due to the easterly plunge of the folds, higher stratigraphic units of the Boston Bay Group are present than farther west. Moreover, such data are pertinent to planning any future tunnels in the eastern part of Boston Harbor.

East and west of the north end of the section, the Lynn Volcanic Complex is exposed in an escarpment 50 m high. It is assumed that, as in the Malden Tunnel to the west, the Lynn is thrust over the Cambridge Argillite. The structure shown in the segment that extends for 6.5 km south of the thrust is based on a few exposures in which the Cambridge Argillite dips 15° to 40°SE. Moreover, the Cambridge Argillite may be projected along strike from the Malden Tunnel. The strata just south of the Lynn Volcanic Complex are 1,800 m stratigraphically below those at the north end of the North Metropolitan Relief Tunnel. This gives us an additional control on the structure at the north end of section CD.

The North Metropolitan Relief Tunnel extends for 6.33 km on an average bearing of S53°E. from Chelsea to Deer Island. The report on this tunnel will be published shortly (Billings, in prep.). The rocks are gray argillite of the Cambridge Argillite. The structure is relatively simple. The hinge of the Charles River syncline is located 2,900 m south of shaft 2 (Fig. 1, northwest end of tunnel). The strata in the north limb dip at angles ranging from 35° to 60°SE. A minor syncline and a minor anticline lie, respectively, 230 and 110 m northwest of the hinge of the Charles River syncline. The dip of the south limb ranges from 30° to 70°NW.

The location of the hinge of the Charles River syncline is also known in the City Tunnel Extension. Moreover, the stratigraphic position of the strata in the hinges in both tunnels is known. It is therefore possible to calculate that the average bearing of the hinge between the two tunnels, a distance of 10.5 km, is N84°E.; it changes from N90°E in the City Tunnel Extension to N60°E in the North Metropolitan Relief Tunnel. The average plunge is 19° in a direction N84°E.

That portion of the section extending for 5 km southeast of shaft 1 (Fig. 1) has very little control. In about half of this distance, however, data are projected to the section from islands in Boston Harbor. A short way north of the shore in Hull, the section assumes a southerly trend. The open anticline in Hull is based on natural exposures. For 3 km south of this anticline, no data are available. In the southern third of Hingham Bay, data are projected from exposures on some of the islands.

The southernmost 3 km are based on good surface data (Billings and others, 1939; Crosby, 1893, 1894). The rocks belong to the Dedham Quartz Monzonite (part of the Precambrian basement), Roxbury Conglomerate, and Cambridge Argillite. Although the three units of the Roxbury can not be easily recognized, the uppermost conglomerate is unusually coarse and is presumably representative of the Squantum. Whereas the Roxbury is only 408 m thick in the southern end of the section, it is more than 710 m thick at the northern end. The Dedham Quartz Monzonite is exposed in the core of a westerly plunging anticline. Moreover, near the south end of the section, the Dedham is thrust over the Roxbury. West of the entrance of Hingham Harbor, the Roxbury is thrust over the Cambridge.

East-West Section of Boston Basin (EF). The location of section EF (Fig. 7) is shown in Figure 1. This section, 27.4 km long, includes the City Tunnel (8.74 km long) and the Main Drainage Tunnel (11.46 km long). This section is parallel to the regional structure, hence the attitudes of the strata are apparent dips rather than true dips.

The fault at the west end is shown as a thrust because of the evidence in the Malden Tunnel. Precambrian basement rocks are thrust over the Cambridge Argillite. Surface exposures over the City Tunnel are good. From west to east, the principal units in the City Tunnel are Cambridge, Squantum, Dorchester, and Brookline. A large mass of melaphyre lies within the Dorchester Member. The two bodies of Dorchester Member shown within the melaphyre are based on surface data; they were not encountered in the tunnel. A small body of Dorchester at tunnel level is too small to show on the scale of section EF. A large normal fault is shown just west of shaft 7. At the surface this fault is expressed as a north-trending scarp 25 m high on the Boston College campus. In the tunnel this fault is occupied by a basalt dike 1 m wide that dips 75°W. The net slip is at least 70 m, west wall down. The argillite beds east of this fault are exceptionally thick, 213 m.

The section between the City Tunnel and the Main Drainage Tunnel has no surface control. That part of the section following the Main Drainage Tunnel is based entirely on data in the tunnel because surface data are not available.

In the western 6 km of Figure 7, the strikes are northeast and the dips are generally 20° to 50°NW. The northwesterly dips are the result of a swing in the regional trend. For 2.5 km west of shaft 7, the dips are 30° to 35°N, but the apparent dip in the section is 0°, because it is parallel to the north limb of the Central anticline. East of shaft 7 and in the Main Drainage Tunnel, the true dips are northeast because of the easterly plunge of the Central anticline.

Dorchester Bay Tunnel. The Dorchester Bay Tunnel (Clarke, 1888) trends N52°W between Squantum and Columbia Point (Fig. 1). The main tunnel extends for 1,856 m between the west and east shafts at a depth of about 43 m below sea level. An incline 259 m long rises at an angle of $9\frac{1}{2}°$ toward the east from the bottom of the east shaft and comes to the surface at Squantum (Fig. 8). A geologic section accompanies the publication by Clarke (1888) but does not appear to be very accurate or precise. The northwest end of the tunnel is in "slate" with a breadth outcrop of 535 m; this is the Dorchester Member. Southeast of this the rock is shown as conglomerate; of course, tillite was not separately distinguished when the tunnel was driven. This unit is considered to be the Squantum Member. The rock in the southeast part of the tunnel is shown as "slate" and has a breadth of outcrop of 1,020 m. Most of this is Cambridge Argillite, but the southeasterly part is probably the Dorchester Member thrust over the Cambridge. The incline is shown as "slate," but this is not consistent with the surface exposures, which are tillite and conglomerate.

Minor Structural Features

General Statement. The minor structural features of the Boston basin include minor folds, slaty cleavage, faults, joints, and dikes. The ensuing discussion is based entirely on observations in the tunnels.

Folds. These folds range in wavelength from 1 m to about 10 m. Such folds may be observed directly, or they may be deduced from differences in the attitudes (strikes and dips) of beds at nearby points of observation (Billings, 1972, p. 75, 570–575).

In the City Tunnel Extension, the observed minor folds on the north limb of the Charles River syncline plunge gently east. The average plunge of 28 such folds is 8° in a direction N87°E (Billings and Tierney, 1964, p. 138). No such minor folds were recorded in the Roxbury Conglomerate on the south limb of the Charles River syncline, but differences in the attitude of the bedding indicate broader folds, several tens of metres in wavelength. About 20 such folds were inferred, most of them plunging on the average 20°NW. These are considered to be a product of deformation subsequent to the main folding.

In the Main Drainage Tunnel, there are three kinds of folds (Rahm, 1962, Fig. 11). The oldest are slump folds formed during sedimentation. Of intermediate age is a relatively large drag fold in the Cambridge Argillite in the eastern part of the tunnel. The anticlinal hinge of this pair plunges 1° in a direction N74°E. The synclinal hinge plunges 6° in a direction N80°E. The two hinges are 40 m apart measured in a direction perpendicular to the axes. This is a congrous drag fold in the southern limb of the Charles River syncline. The plunges of 87 folds, each several tens of meters in wavelength, were calculated from differences in the attitude of the beds. Most of these plunge east, northeast, and north at angles ranging from 10° to 30°; they appear to be congrous drag folds (Hills, 1953). About a third of these folds plunge from 10° to 50°NW and SE; they are "warps" or "buckles" that are later than the main folding.

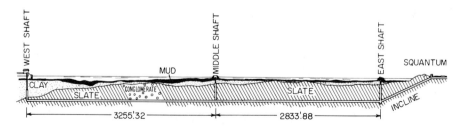

Figure 8. Structure section of Dorchester Bay Tunnel (after Clarke, 1888). Distances in feet.

Cleavage. Cleavage is present in some places in the Boston basin, but it is distinctly limited in its occurrence.

In the City Tunnel Extension, 111 measurements were made on slaty cleavage in the Cambridge Argillite. The average strike is N80°E and the average dip is 70°N but ranges from 60° to 90°N (Billings and Tierney, 1964, p. 130, 146). This is an axial plane cleavage because the folds are essentially symmetrical.

In the Main Drainage Tunnel, cleavage is weak or absent (Rahm, 1962, p. 354). However, in the tillite of the Squantum Member, the fracture cleavage has an average strike of N70°W and dips steeply north. It is cut by a later slip cleavage that deforms the fracture cleavage into small crinkles. Further east in this same tunnel, slaty cleavage is found locally in the Cambridge Argillite. The strike changes from N70°W in the west to N70°E in the east; the dip is steep to the north.

Joints. A statistical analysis of joints is difficult because of the sampling problem (Billings, 1972, p. 147-151). Literally tens of thousands of joints are exposed in the tunnels, and it was obviously impossible to gather all the data. For a complete analysis, it would be necessary to record the size of each joint, the spacing, and the orientation. Consequently the gathering of data is subjective.

The joints are generally planar but locally curved. They range in length from a few centimeters to 30 or more metres and are spaced at intervals of 1 cm to 15 m. Although many of the joints are tight, clean fractures, some have a thin coating of chlorite. Some are slickensided and perhaps should be classified as faults. Generally the attitude was measured on a representative joint in a set consisting of six or more joints. Where two or three sets of joints are closely spaced, the rock breaks into parallelopipeda that are 5 to 50 cm on a side.

Contour diagrams of the poles of the perpendiculars to joint sets are presented in Figures 9 and 10, the former for the City Tunnel, City Tunnel Extension, and the Malden Tunnel and the latter for the Main Drainage Tunnel.

Data on 1,107 joint sets enter into these diagrams. Most of the joints dip steeply, the only exception being some horizontal joints in the Lynn Volcanic Complex in the Malden Tunnel (Fig. 9D). In the City Tunnel, City Tunnel Extension, and Malden Tunnel, the most pronounced strikes of these steep joints are north-northeast. Similar sets are common in the Main Drainage Tunnel, but in addition there is a set striking nearly east.

Faults. The large faults have been discussed above. Here we are concerned with the small faults that have been observed in the tunnels.

The geometry of faulting is a complex subject (Billings, 1972, p. 174-276), and it must be assumed that the reader is familiar with the problem. The true movement, that is, the net slip, could rarely be calculated, primarily because of the lack of data but partly because of the lack of time to obtain them. Hence the ensuing discussion will be based on the apparent movement as seen on the vertical walls of the tunnels. Pertinent data from the different tunnels have been assembled in Table 4 to facilitate comparison.

The line labeled "number" refers to the number of faults that were observed. The total number would exceed this figure, because in places observations were prevented by structural steel, lagging, or metal sheeting.

"Normal" means faults in which the hanging wall appears to have gone down. "Reverse" means faults in which the hanging wall appears to have gone up. Neutral faults are those that dip 90° and hence have no hanging wall or footwall. The apparent movement can be deduced if the disrupted stratum is visible on both sides of the fault; it can also be deduced by drag along the fault or by evidence from slickensides.

Vertical separation is the separation measured along a vertical line. It can be

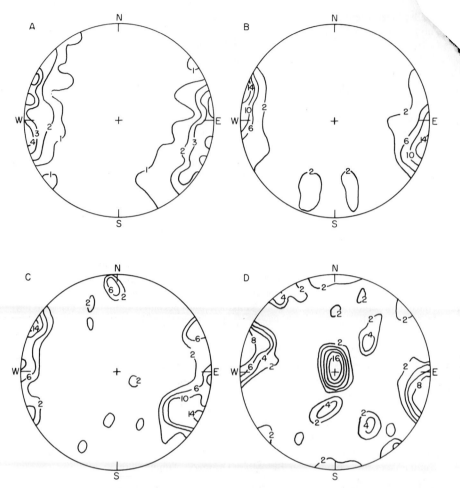

Figure 9. Contour diagrams of joints in City Tunnel, City Tunnel Extension, and Malden Tunnel. Equal-area projection, lower hemisphere, contours of poles of perpendiculars to joint sets. A. City Tunnel, 251 sets of joints, maximum 5 percent. B. City Tunnel Extension, north limb of Charles River syncline, 130 sets of joints, maximum 14 percent. C. Malden Tunnel, Cambridge Argillite, 28 sets of joints, maximum 16 percent. D. Malden Tunnel, Lynn Volcanic Complex, 49 sets of joints, maximum 16 percent.

determined only if the disrupted stratum is present on both sides of the fault. Moreover, if observations are confined to one wall of the tunnel, the value can not exceed the height of the wall (4 m). If movement on the fault is so great that the disrupted stratum is found on only one side of the fault, the vertical separation may exceed the height of the tunnel. Hence two categories are listed: those in which the vertical separation is 4 m or less and can be measured and those in which it exceeds 4 m and cannot be measured.

In some of the tunnels, the term "shear" was used to refer to a planar feature that was generally a few centimetres to 1 m thick; composed of a platy rock, slickensided rock, gouge, or breccia; and along which there was no demonstrable offset, in many instances because of absence of bedding in the adjacent rock.

Contour diagrams showing the orientation of the faults and shears are presented in Figure 11. Figure 11A is for the faults in the Main Drainage Tunnel, Figure

11B for the faults in the City Tunnel, Figure 11C for the faults and shears on the southern limb of the Charles River syncline, Figure 11D for the faults and shears on the northern limb, and Figure 11E for shears in the Lynn Volcanic Complex of the Malden Tunnel.

On the basis of the data in Table 4 and Figure 11, several generalizations may be made. The minor faults are relatively narrow, ranging from tight fractures to zones 1 m wide, characterized in some instances by platy rock, slickensides, gouge, and breccia. In 80 percent of the faults measured, the vertical separation averages about 1 m, but in 20 percent it exceeds 4 m. Sixty-seven percent are "normal," 18 percent are "reverse," and 15 percent are neutral. From Figure 11 it is apparent that most of the faults dip steeply. Although there is considerable spread in the strike, the most common strike is north-northeast; other strikes are north, northwest, and east.

Hypabyssal Intrusives. Data on the hypabyssal intrusions are assembled in Table 5. These intrusions may be classified as dikes or sills, although some bodies are irregular. In Table 5 the rocks are classified as diabase, melaphyre, and granite

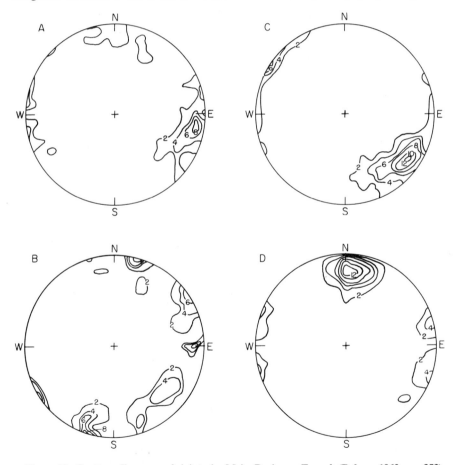

Figure 10. Contour diagrams of joints in Main Drainage Tunnel (Rahm, 1962, p. 353). Equal-area projection, contours of poles of perpendiculars to sets of joints. A. Dorchester Member, shaft A to station 118 + 80, 180 sets. B. Squantum Member, stations 119 + 16 to station 137 + 94, 78 sets. C. Cambridge Argillite, station 137 + 94 to 262 + 55, 215 sets. D. Cambridge Argillite, station 263 + 65 to shaft C, 181 sets.

TABLE 4. MINOR FAULTS AND SHEARS

	1*	2.*	3*	4*
Faults				
Number	138	74	106	..
Width (centimeters)	0.1 to 30	0.1 to 400	0 to 100	..
Displacement				
"Normal"	68	48	70	..
"Reverse"	17	20	14	..
"Neutral"	27	6	8	..
Indeterminate	26	..	14	..
Vertical separation				
Number less than 4 m	54†	52	89	..
Range (cm)	2 to 400	3 to 300	..	..
Average (cm)	80	100	100	..
Number greater than 4 m		22	17	..
Shears				
Number	..	..	114	40
Width	..	..	0.03 to 1.0	0.06 to 1.0

*1. Main Drainage Tunnel. 2. City Tunnel. 3. City Tunnel Extension. 4. Malden Tunnel.
†Vertical separations calculated for only 54 faults in this tunnel.

aplite. The least altered diabase is black to dark-gray, fine-grained to medium-grained, and ophitic; it is composed principally of labradorite and pyroxene, with such accessories as olivine, biotite, magnetite, and pyrite. Alteration minerals, which may be of minor importance or may replace the whole rock, are albite, chlorite, epidote, calcite, kaolinite, and limonite. Some of the rocks listed as diabase in Table 5 may be diorite or some similar rock. Complete petrographic studies have not been made, but those rocks studied in thin section are diabase or altered diabase. The term melaphyre was used in mapping in the tunnels for dark-green to yellow-green fine-grained rocks composed largely of alteration minerals, chiefly albite, chlorite, calcite, and limonite. The rocks are altered basalt or andesite. Distinction between altered diabase and melaphyre was not always readily made in the tunnels. Some melaphyre is extrusive, but in this section we are primarily concerned with intrusive melaphyre.

Granite aplite constitutes only 1 percent of the hypabyssal rocks. They are fine-grained gray highly altered rocks, consisting originally of quartz and feldspar, now composed of quartz, kaolinite, and sericite.

In Table 5 the distance that hypabyssal rocks are exposed on the walls of the tunnels is given. In essence this is a gigantic Rosiwal traverse, 33.24 km long, within the Boston basin. The traverse is 6.33 km less than the total length of the five tunnels (39.57 km) because data are unavailable on hyperbyssal intrusive rocks in the North Metropolitan Relief Tunnel. The hypabyssal rocks average 8.2 percent of the bed rock. If, however, the calculations were based on mafic rocks, the results would differ. This is because of a large body of melaphyre in the City Tunnel that extends for 2,282 m and is 320 m thick. It is partly, if not entirely, extrusive. It is consequently excluded from the calculations, but footnotes indicate how inclusion would modify the results. In other words, mafic rocks, including intrusive and extrusive, average 14.6 percent of the bed rock in the areas occupied by the tunnels.

As shown by Table 5, 356 dikes of diabase and similar rocks were recorded in the tunnels. They average 4.2 m in thickness. Twenty-nine melaphyre dikes, averaging 15 m in thickness, were observed in the City Tunnel Extension. Only three granite aplite dikes, averaging 2.4 m in thickness, were recorded.

The orientation of the dikes is shown in Figure 12. Most of the dikes dip steeply,

60° to 90°. The diabase and related rocks (Fig. 12A, B, D) show two preferred strikes, north-northeast and east. The melaphyre (Fig. 12C) ranges in strike from north through northwest to west.

BLUE HILLS

General Statement

For many years the students in the course in field geology at Harvard University mapped the geology of the Blue Hills under my supervision. An excellent account of the bedrock geology has also been published by Chute (1969). His data have

TABLE 5. HYPABYSSAL ROCKS

	1	2	3	4	5
Quantity in tunnels					
Length exposure*	767	258§	1,783	137	2,945**
Length of tunnel	12,323	9,404	12,299	1,792	35,818
Percentage of tunnel	6.2	2.7#	13.7	7.7	8.2††
Dikes					
Diabase†					
Number	43	101	189	23	356
Range in thickness	0.1 to 46.0	0.2 to 15.0	0.1 to 30.0	0.07 to 4.5	0.1 to 30.0
Average thickness	5.0	6.0	3.3	1.5	4.2
Melaphyre					
Number	..	..	29	..	29
Range in thickness	..	..	0.2 to 134.0	..	0.2 to 134.0
Average thickness	..	..	15.0	..	15.0
Granite Aplite					
Number	..	..	3	..	3
Range in thickness	..	..	0.8 to 5.0	..	0.8 to 5.0
Average thickness	..	..	2.4	..	2.4
Orientation					
Number plotted	30	..	221	15	..
Maxima	N. 10° E., 80° NW.	..	N. 70° W. 90°	large spread	..
Sills					
Diabase†					
Number	28	..	53	..	81
Range in thickness	0.3 to 60	..	0.2 to 22.0	..	0.2 to 22.0
Average thickness	8	..	2.8	..	6.1
Melaphyre					
Number	..	..§	3	..	3
Range in thickness	..	..	5.6 to 30.0	..	5.6 to 30.0
Average thickness	..	..	18.0	..	18.0
Granite Aplite					
Number	..	..	2	..	2
Range in thickness	..	..	2.9 to 3.3	..	2.9 to 3.3
Average thickness	..	..	3.1	..	3.1

*Distances and thicknesses are given in meters.
†May include some diorite and related rocks.
§If the large body of melaphyre, 320 m thick, extending for 2,282 m in the tunnel, is included, this number becomes 2,540 m.
#If the 320-m thick melaphyre is included, this number becomes 27.0.
**If the 320-m thick melaphyre is included, this would be 5,227.
††If the 320-m thick melaphyre is included, this number would be 14.6.
Note: 1. Main Drainage Tunnel; 2. City Tunnel; 3. City Tunnel Extension; 4. Malden Tunnel; 5. total for the four tunnels.

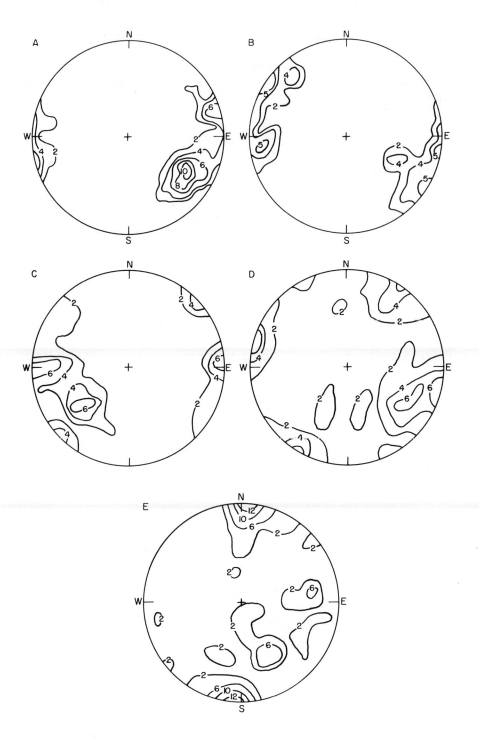

Figure 11. Contour diagrams of faults and shears. Equal-area projection, lower hemisphere, contours of poles of perpendiculars to faults and shears. A. Main Drainage Tunnel, 138 faults, maximum 11 percent. B. City Tunnel, 74 faults, maximum 5 percent. C. City Tunnel Extension, south limb of Charles River syncline, 67 faults and 53 shears, maximum 7 percent. D. City Tunnel Extension, north limb of Charles River syncline, 39 faults and 61 shears, maximum 6 percent. E. Malden Tunnel, 41 shears in Lynn Volcanic Complex, maximum 12 percent.

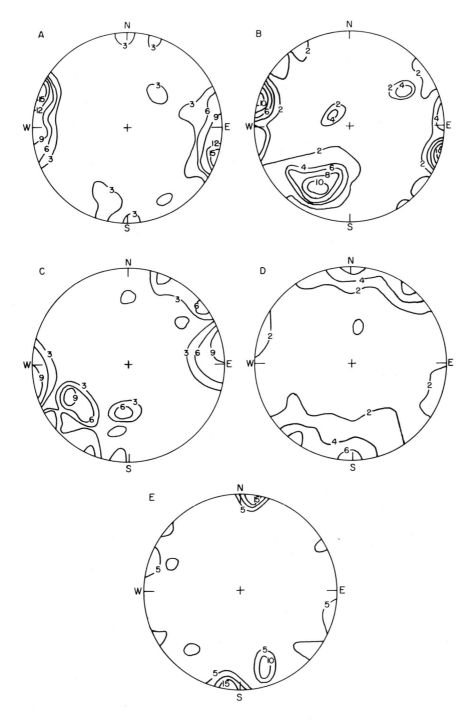

Figure 12. Contour diagrams of dikes in tunnels. Equal-area projection of poles of perpendiculars to walls of dikes, lower hemisphere. A. Main Drainage Tunnel, 30 diabase dikes, maximum 16 percent. B. City Tunnel Extension, south limb of Charles River syncline, 43 diabase and related dikes, maximum 11 percent. C. City Tunnel Extension, south limb of Charles River syncline, 29 melaphyre dikes, maximum 13 percent. D. City Tunnel Extension, north limb of Charles River syncline, 146 diabase and related dikes, 3 granite aplite dikes, maximum 7 percent. E. Malden Tunnel, 15 diabase and related rocks, maximum 19 percent.

been utilized in preparing the ensuing abbreviated descriptions of some of the lithologic units. For more detail the reader is referred to Chute (1969), Kaktins (1976), and Loughlin (1911). A generalized map is shown in Figure 1, and a simplified structure section is shown in Figure 13.

Lithology

The distribution of the Cambrian strata is shown diagrammatically in Figure 1. We await with great anticipation the report by G. Stinson Lord on the stratigraphy and paleontology of the Weymouth-Braintree area. A brief summary has been published (Lord, 1972).

The Early Cambrian Weymouth Formation is a green, gray, and red slate and contains a little gray argillaceous limestone. The Middle Cambrian Braintree Formation is a dark argillite and siltstone. A quartzite in the western part of Milton, which is intruded by the Quincy Granite, has been assigned to the Cambrian by Chute (1969).

Only the three major map units of the comagmatic rocks are shown in Figures 1 and 13: volcanic complex, Blue Hill Granite Porphyry, and Quincy Granite. The volcanic complex consists of red, pink, purple, brown, and gray rhyolitic ash flows, tuffs, breccias and lava flows. The flow structure and occasional bedding dip steeply south (Fig. 13), and the formation appears to be at least 2,400 m thick. Chute (1969) correlated these volcanic rocks with the Mattapan Volcanic Complex. The Blue Hill Granite Porphyry is a dark-gray to blue-gray granite porphyry with 30 to 50 percent of phenocrysts of microperthite, quartz, and riebeckite that are 1 to 10 mm long. The Quincy Granite is a massive gray to blue-gray, medium- to coarse-grained rock composed of microperthite, quartz, riebeckite, and aegirine. Details were given by Warren (1913).

These alkalic rocks are younger than Middle Cambrian and, as will be shown in the next paragraph, are pre-Pennsylvanian. Zartman and Marvin (1971) dated the Quincy Granite as 437 ± 32 m.y. old (Late Ordovician) on the basis of the Pb^{207}/Pb^{206} ratio in zircon. Lyons and Krueger (1976) questioned this particular age assignment and dated similar alkalic rocks as Devonian.

The alkalic rocks are overlain unconformably on the south by the Pennsylvanian Pondville Conglomerate of the Norfolk basin (Chute, 1966, 1969). At the base is the "giant conglomerate," a very coarse conglomerate with cobbles and boulders that are composed of Blue Hill Granite Porphyry and related rocks. The "giant conglomerate" is 90 to 150 m thick. The upper part of the Pondville Conglomerate is a light green-gray granule and pebble conglomerate, 90 to 150 m thick.

The Pennsylvanian Wamsutta Formation, the uppermost unit in the Norfolk basin, is composed of red sandstone, red slate, and light-gray coarse arkosic sandstone (Lyons and others, 1976). This unit is at least 300 m thick.

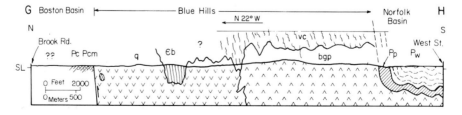

Figure 13. North-south structure section of Blue Hills (GH).

Structure

The structure of the Blue Hills is shown in Figure 13. In the south-central part of Figure 1, the Quincy Granite is in fault contact with the "Milton" Quartzite on the north. Three kilometres further east along this same fault, the Quincy is in contact with the Brookline, Squantum, and Dorchester Members of the Roxbury Conglomerate.

The volcanic complex shown above the topographic surface in Figure 13 is based on exposures 300 to 600 m east of the line of section.

The dips of the Pondville Conglomerate are almost vertical. The deformation that produced these vertical dips is Pennsylvanian or younger and traditionally has been assigned to the Appalachian orogeny. The Braintree Argillite and the volcanic complex of the Blue Hills also dip very steeply. Apparently these steep dips resulted from the Pennsylvanian or post-Pennsylvanian deformation. Thus the volcanic complex and the Braintree Argillite were horizontal when the Pennsylvanian rocks were being deposited. The Blue Hill Granite Porphyry and Quincy Granite were injected as one or more irregular bodies along the unconformity between the Cambrian and Precambrian rocks. The alkalic rocks, including the volcanic complex and the Braintree Argillite, were tipped up on end during the Appalachian revolution.

The Blue Hills block is bounded on the north by a fault. During the Appalachian revolution, a horizontal fracture formed beneath the alkalic complex, and it was thrust northward into contact with the Boston Bay Group. The shortening must have been many kilometres.

Another thrust slice is shown in the southeast corner of Figure 1. Along this thrust the Precambrian basement was driven northward against the Norfolk basin, the Blue Hills block, and the Boston basin in Hingham.

THEORETICAL CONSIDERATIONS

Sedimentation in the Boston Basin

The pertinent factual data on the stratigraphy of the Boston basin are given in Figure 4 and Table 1. Before discussing the conditions of sedimentation, certain important facts and inferences may be given.

1. The Boston basin is one of several isolated basins that developed in eastern North America on the deeply eroded Acadian orogenic belt. Some are marine, other nonmarine.
2. In New England the basins were nonmarine.
3. The Narragansett, Norfolk, and Worcester basins are definitely of Pennsylvanian age. The Boston basin is probably of this age also.
4. The fact that the Brookline Member thins toward the south, from nearly 1,500 m in the central part of the basin to 300 m in Hyde Park and 408 m in Hingham (Billings and others, 1939), indicates that the center of the basin was going down more rapidly than the southern margin. There is no evidence that faulting was associated with this downwarping.
5. The sediments were derived from the south, but the intrusive alkalic rocks of the Blue Hills were not exposed at the time of sedimentation.
6. The sedimentary basin was much wider than the present structural basin.

The conditions of sedimentation will now be considered.

1. During the accumulation of the lowest 1,500 m of the Boston Bay Group,

deposition in the northern part of the basin was in a body of standing water, presumably a lake. In the southern part of the basin, much of the deposition was by streams (Fig. 14A). The water level was constantly shifting due to changes in precipitation, the formation or destruction of temporary dams holding in the lake, or differential subsidence of the floor of the basin. Gravel was deposited in deltas above water level, sand near shore, and mud in deeper water (Fig. 14B). If the water level rose, mud was deposited further south on top of gravel (Fig. 14C). If the water level fell, gravel was deposited over mud (Fig. 14D). Thus the complex relations shown in Table 2 are explained. Of course, other factors were involved. Channeling at times of lowered water level resulted in contacts that transgress bedding. When the source area was low, only sand and mud were available. Sand might be deposited in channels at the same time that mud was being deposited in adjacent flood plains.

2. The location of the source area is a problem. The clasts in the Roxbury Conglomerate are chiefly quartz monzonite, granite, volcanic rocks, and quartzite. The quartz monzonite and granite were derived from the Dedham Quartz Monzonite, which is extensively developed in eastern Massachusetts (Emerson, 1917). Similarly the volcanic rocks were presumably derived from the Mattapan and other volcanic complexes. The quartzite presents a problem. Quartzite pebbles in the Dighton Conglomerate of the Narragansett basin contain Late Cambrian fossils (Shaler and others, 1899) and were presumably derived from exposures that now lay to the south under the waters of the Atlantic Ocean. It is difficult to see how the quartzite clasts in the Boston basin could have been derived from such a source, because in some way they seemingly would have had to bypass the Narragansett and Norfolk basins. It seems probable that such a source was present between the Norfolk and Boston basins but has since been removed by erosion or is still present beneath the Blue Hills thrust sheet.

Tectonics

General Statement. The folding in the Narragansett, Norfolk, and Worcester basins is Pennsylvanian or post-Pennsylvanian (Emerson, 1917; Lyons and others,

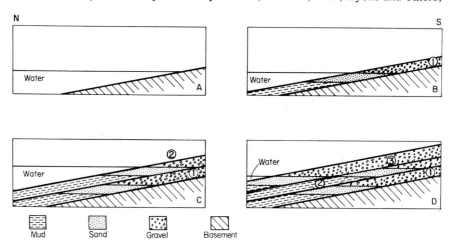

Figure 14. Evolution of Facies changes in Boston basin. A. Prior to sedimentation. B. Stage 1, gravel deposited above water level, sand in shallow water, salt and clay in deeper water. C. Stage 2, water level higher. D. Stage 3, water level lower.

1976; Grew, 1972) and is commonly assigned to the Appalachian revolution. The folding in the Boston basin is undoubtedly of the same age.

Stress Axes. The regional stresses during orogeny were deduced from structural data in the City Tunnel Extension (Billings and Tierney, 1964, p. 146). There it was concluded that the greatest principal stress axis (compression) was perpendicular to the slaty cleavage and that the intermediate principal stress axis was parallel to the fold axes. As shown in Figure 15C, which is slightly modified from the original figure, the greatest principal stress axis plunges 20° in a direction S10°E, the intermediate axis plunges 8° in a direction N77°E, and the least stress axis 68° in a direction N35°W. Toward the east the intermediate principal stress axis, as deduced from data in the North Metropolitan Relief Tunnel, plunges somewhat more steeply.

The major thrusts are interpreted as shear fractures consistent with this stress distribution. Such thrusts would dip northwest or southeast at intermediate angles. The southeast-dipping thrusts were subsequently rotated to steep dips.

Fracture Pattern. The fracture pattern as represented by the joints, minor faults, and dikes appears to be related to this same stress pattern. The maxima shown in Figures 9, 10, 11, and 12 have been plotted in Figure 15A. Many of the maxima are on the periphery of the diagram and thus appear twice on opposite diameters. Four concentrations of maxima are shown in Figure 15A, concentrations A, B, C, and D. The strike of the fracture systems represented by these maxima is

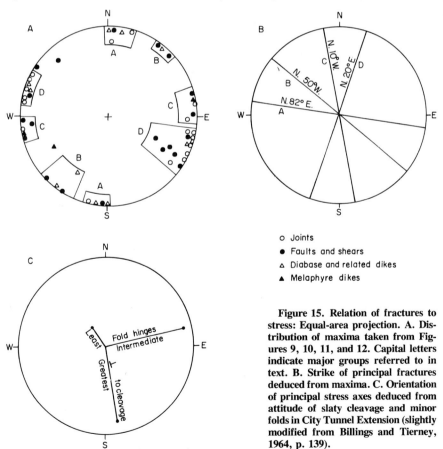

o Joints
● Faults and shears
△ Diabase and related dikes
▲ Melaphyre dikes

Figure 15. Relation of fractures to stress: Equal-area projection. A. Distribution of maxima taken from Figures 9, 10, 11, and 12. Capital letters indicate major groups referred to in text. B. Strike of principal fractures deduced from maxima. C. Orientation of principal stress axes deduced from attitude of slaty cleavage and minor folds in City Tunnel Extension (slightly modified from Billings and Tierney, 1964, p. 139).

shown in Figure 15B. The greatest concentration of maxima represents a fracture system striking N20°E. The other three concentrations are of about equal weight. It is apparent that fracture sets B, C, and D are symmetrically distributed relative to the greatest principal stress axis (Fig. 15C). Set C, striking N10°W, is parallel to the greatest principal stress axis and appears to represent extension fractures. Sets D (N20°E) and B (N50°W) are, respectively, 30° clockwise and 40° anticlockwise from set C. They appear to be the complementary set of shear fractures related to the greatest principal stress axis.

If this analysis is correct, the intermediate principal stress axis is essentially vertical rather than horizontal, as indicated by the fold axes and major thrusts. This suggests that during the deformation, as the rocks became more brittle, the intermediate axis changed from a horizontal position to a vertical position. An analogous situation is found in the Jura Mountains, where the least principal stress axis was at times parallel to the fold axes; the large vertical shear fractures, which became strike-slip faults, formed diagonally to the fold axes.

The direction of the net slip along most of the minor faults in the Boston basin is not known. The faults were classified as normal and reverse merely on the basis of the apparent movement. Theoretically the net slip should have been a strike slip if the same stresses caused both the fractures and the displacement along them. The slickensides along the faults have an average rake of 45°, suggesting some strike-slip component. In conclusion, most of the fractures seem to have been produced by compression. Displacement along them may have resulted from the same stresses as those that produced the fractures but may have also occurred later under different conditions of stress.

ACKNOWLEDGMENTS

I wish to express my deepest gratitude to the late F. Lyle Tierney and David Rahm for their participation in the mapping of the geology of the tunnels. The late Frederick W. Gow, Chief Engineer, and Martin Cosgrove, Deputy Chief Engineer, Construction Division, Metropolitan District Commission, provided superb logistic support during the mapping. I am also indebted to those who have critically read the present manuscript and made many valuable suggestions: Arthur H. Brownlow, John Haller, Paul C. Lyons, Lincoln R. Page, Stephen Richardson, and James W. Skehan, S.J. Figures have been drafted by William Minty of the Cruft Laboratory, Harvard University.

REFERENCES CITED

Billings, M. P., 1929, Structural geology of the eastern part of the Boston basin: Am. Jour. Sci., 5th ser., v. 18, p. 97-137.
——1972, Structural geology (3rd ed.): New York, Prentice Hall, 606 p.
Billings, M. P., and Rahm, D. A., 1966, Geology of the Malden Tunnel, Massachusetts: Boston Soc. Civil Engineers, Jour., v. 53, p. 116-141.
Billings, M. P., and Tierney, F. L., 1964, Geology of the City Tunnel Extension, Greater Boston, Massachusetts: Boston Soc. Civil Engineers Jour., v. 51, p. 111-154.
Billings, M. P., Loomis, F. B., Jr., and Stewart, G. W., 1939, Carboniferous topography in the vicinity of Boston, Massachusetts: Geol. Soc. America Bull., v. 50, p. 1867-1884.
Burr, H. T., and Burke, R. E., 1900, The Occurrence of fossils in the Roxbury Conglomerate: Boston Soc. Nat. History Proc., v. 29, p. 179-184.
Caldwell, D. W., 1964, The Squantum Formation: Paleozoic tillite or tilloid, *in* Skehan, J. W., ed., Guide to field trips in the Boston area and vicinity: New England Intercollegiate Geol. Conf. Guidebook, 56th Ann. Mtg., p. 53-60.
Chute, N. E., 1966, Geology of the Norwood quadrangle, Norfolk and Suffolk Counties, Massachusetts: U.S. Geol. Survey Bull. 1163-B, 78 p.
——1969, Bedrock geologic map of the Blue Hills quadrangle, Norfolk, Suffolk, and Plymouth Counties, Massachusetts: U.S. Geol. Survey Geol. Quad. Map GQ-796.
Clarke, E. C., 1888, Main drainage works of the city of Boston (3rd ed.): Boston, Rockwell and Churchill, 217 p.
Crosby, I. B., 1928, Boston through the ages: The geological story of Greater Boston: Boston, Marshall Jones Co., 166 p.
Crosby, W. O., 1880, Contributions to the geology of eastern Massachusetts: Boston Soc. Nat. History Occasional Paper 3, 286 p.
——1893, Geology of the Boston basin: Nantasket and Cohasset: Boston Soc. Nat. History Occasional Paper 4, v. 1, pt. 1, p. 1-177.
——1894, Geology of the Boston basin: Hingham: Boston Soc. Nat. History Occasional Paper 4, v. 1, pt. 2, p. 179-288.
——1900, Geology of the Boston basin, the Blue Hills Complex: Boston Soc. Nat. History Occasional Paper 4, v. 1, pt. 3, p. 289-563.
——1905, Genetic and structural relations of the igneous rocks of the lower Neponset Valley, Massachusetts: Am. Geologist, v. 36, p. 34-47, 69-83.
Dott, R. H., Jr., 1961, Squantum "tillite," Massachusetts—Evidence of glaciation or subaqueous mass movements?: Geol. Soc. America Bull., v. 72, p. 1289-1305.
Dowse, A. M., 1950, New evidence on the Cambrian contact at Hoppin Hill, North Attleboro, Massachusetts: Am. Jour. Sci., v. 248, p. 95-99.
Emerson, B. K., 1917, Geology of Massachusetts and Rhode Island: U.S. Geol. Survey Bull. 597, 289 p.
Grew, E., 1972, Stratigraphy of the Pennsylvanian and pre-Pennsylvanian rocks of the Worcester area, Massachusetts: Am. Jour. Sci., v. 273, p. 113-129.
Hills, E. S., 1953, Outlines of structural geology: New York, John Wiley & Sons 182 p.
Kaktins, U., 1976, Stratigraphy and petrography of the volcanic flows of the Blue Hills area, Massachusetts, *in* Lyons, P. C., and Brownlow, A. H., eds., Studies in New England geology: Geol. Soc. America Mem. 146, p. 125-142 (this volume).
Kaye, C., 1967, Kaolinization of bedrock of the Boston, Massachusetts, area: U.S. Geol. Survey Prof. Paper 575-C, p. C165-C172.
LaForge, L., 1932, Geology of the Boston area, Massachusetts: U.S. Geol. Survey Bull. 839, 105 p.
Lord, G. S., 1972, The geology of Weymouth: *in* Cassese, L. W., and Belcher, H. C., eds., Weymouth 350th Anniversary: p. 14-17.
Loughlin, G. F., 1911, The structural relations between the Quincy Granite and the adjacent sedimentary formations: Am. Jour. Sci., 4th ser., v. 32, p. 17-32.
Lyons, P. C., and Krueger, H. W., 1976, Petrology, chemistry, and age of the Rattlesnake pluton and implications for other alkalic granite plutons of southern New England,

in Lyons, P. C., and Brownlow, A. H., eds., Studies in New England geology: Geol. Soc. America Mem. 146, p. 71-102 (this volume).

Lyons, P. C., Tiffney, B., and Cameron, B., 1976, Early Pennsylvanian age of the Norfolk basin, southeastern Massachusetts, based on plant megafossils, *in* Lyons, P. C., and Brownlow, A. H., eds., Studies in New England geology: Geol. Soc. America Mem. 146, p. 181-198 (this volume).

Pollard, M., 1965, Age, origin, and structure of the post-Cambrian Boston strata, Massachusetts, Geol. Soc. America, Bull., v. 76, p. 1065-1068.

Rahm, D. A., 1962, Geology of the Main Drainage Tunnel, Boston, Massachusetts: Boston Soc. Civil Engineers Jour. v. 49, p. 319-368.

Rehmer, J. A., and Hepburn, J. C., 1974, Quartz sand surface textural evidence for a glacial origin of the Squantum "tillite," Boston basin, Massachusetts: Geology, v. 2, p. 413-415.

Sayles, R. W., 1914, The Squantum Tillite: Harvard Coll. Mus. Comp. Zoology Bull., v. 66 (Geol. Ser. 10), p. 141-175.

——1916, Banded glacial slates of Permo-Carboniferous age, showing possible seasonal variations in deposition: Nat. Acad. Sci. Proc., v. 2, p. 167-170.

Sayles, R. W., and LaForge, L., 1910, The glacial origin of the Roxbury Conglomerate: Science, new ser., v. 22, p. 723-724.

Shaler, N. S., Woodworth, J. B., and Foerste, A. F., 1899, Geology of the Narragansett basin: U.S. Geol. Survey Mon. 33, 402 p.

Shrock, R. R., 1972, The Geologists Crosby of Boston: Cambridge, Mass., Mass. Inst. Tech., 175 p.

Tierney, F. L., Billings, M. P., and Cassidy, M. M., 1968, Geology of the City Tunnel, Greater Boston, Massachusetts: Boston Soc. Civil Engineers Jour., v. 55, p. 60-96.

Toulmin, P., III, 1964, Bedrock geology of the Salem quandrangle and vicinity, Massachusetts, U.S. Geol. Survey Bull. 1163A, 79 p.

Warren, C. H., 1913, Petrology of the alkali granites and porphyries of Quincy and Blue Hills, Mass., U.S.A.: Am. Acad. Arts and Sci. Proc., v. 49, p. 203-331.

Wilmarth, M. G., 1936, Lexicon of geologic names of the United States: U.S. Geol. Survey Bull. 896, 2,396 p.

Zartman, R. E., and Marvin, R. F., 1971, Radiometric age (Late Ordovician) of the Quincy, Cape Ann, and Peabody Granites from eastern Massachusetts: Geol. Soc. America Bull., v. 82, p. 937-957.

MANUSCRIPT RECEIVED BY THE SOCIETY JUNE 6, 1974
REVISED MANUSCRIPT RECEIVED DECEMBER 11, 1974
MANUSCRIPT ACCEPTED JANUARY 3, 1975

Petrography and Geochemistry of the Nashoba Formation, East-Central Massachusetts

ADEL ABU-MOUSTAFA
Salem State College
Salem, Massachusetts

AND

JAMES W. SKEHAN, S. J.
Weston Observatory, Boston College
Weston, Massachusetts

ABSTRACT

A 3,740-m-thick section of the Nashoba Formation was exposed in the Wachusett-Marlborough Tunnel. We have divided this formation into 30 members, composed of distinctive sequences of 7 lithologic types.

The Nashoba Formation of this report lies east of the Merrimack synclinorium and is part of a folded, faulted, metamorphosed sequence between westward-dipping faults—the Clinton-Newbury fault zone and the Bloody Bluff fault zone. The Nashoba Formation is underlain by and grades into the Marlboro Formation, which is composed of 7 of the same lithologic units as the Nashoba, although each formation is distinctive in terms of the stratigraphic succession of its members. The Nashoba Formation is overlain unconformably by unnamed formations U-1 through U-15.

A total of 155 modal analyses of rocks of units 1 through 7 were made from samples collected during mapping of the tunnel. The chemical compositions of 61 rocks were determined and, together with field and petrographic information, were used to determine source rocks and environmental setting.

The Nashoba Formation was derived from a sequence of volcanogenic sediments, interbedded volcanic rocks, and limey marine sediments. The Nashoba Formation, in contrast to the Marlboro Formation, is alumina-rich as evidenced by the occurrence of sillimanite and has marble and calc-silicate beds. Otherwise, the Nashoba and Marlboro Formations are similar, and it is clear that the major source rocks were in part deeply saprolitized, even lateritized, and in part unweathered volcanic and plutonic rocks, volcanogenic sediments in the average range of composition of dacite.

The quartz-feldspar rocks of units 1, 2, and 6 have modes and consistently variable chemical compositions that indicate mixed sources. The hornblende-rich rocks of unit 4 consist of flows, pyroclastics, and volcanogenic sediments of dominantly basaltic composition. These have the average composition of central basalt.

The environmental setting for the deposition of the Nashoba Formation is inferred to have been a relatively shallow marine basin that received deeply weathered soils, unaltered volcanogenic sediments, and some volcanic rocks, occasionally interbedded with thin limey beds (unit 3), thin quartz and quartz-feldspar sand (unit 5), and distinctive manganiferous-iron chert (unit 7).

Mineral assemblages are referred to the sillimanite-potash feldspar isograd and indicate metamorphism corresponding to the sillimanite-almandine-orthoclase subfacies of the almandine-amphibolite facies. The observed mineral associations indicate that the rocks of this study recrystallized in the range of 625° to 675°C at 5 to 6 kb. It is probable that the more precise value is 625° to 650°C at 6 kb. This pressure, assuming $P_{load} = P_{H_2O}$, implies an unusually thick pile exceeding 20 km in thickness.

INTRODUCTION

Nashoba Formation (Fig. 1) was proposed by Hansen (1956) as the name for a sequence of metamorphic rocks of presumed Paleozoic age that extends northeastward across much of east-central Massachusetts. This metamorphic unit, typically exposed in the town of Bolton, Massachusetts, was previously called the "Bolton Gneiss" by Emerson (1917). The Nashoba Formation has been described (Hansen, 1956) as composed of biotite gneiss with interbedded layers of amphibole schist, hornblende gneiss, and feldspathic quartzite; there are also beds of marble and abundant pegmatite and quartz veins (Hansen, 1956). The biotite gneiss contains varying amounts of pink to gray lenses and rounded aggregates or crystals of microcline, which give the rock in places a conglomeratic appearance. An index to more detailed geologic mapping of the Nashoba Formation and other rock units in nearby areas is presented as Figure 2.

The unusually complete section of the Nashoba Formation of the Wachusett-Marlborough[1] Tunnel (Fig. 2) is an eugeosynclinal sequence about 3,740 m thick. This stratigraphic succession has been divided by Skehan and Abu-moustafa (1976) into 30 members. These members are composed of distinctive sequences of 7 lithologic types or facies referred to herein as lithologic units. Several of these units have recognizable subdivisions. The Nashoba Formation in the tunnel overlies the Marlboro[1] Formation (Fig. 3) into which it grades and is, in turn, overlain unconformably by a series of 15 as yet unnamed[2] rock units (U-1 through U-15).

The Nashoba Formation in the tunnel was carefully sampled during mapping

---

[1]The spelling "Marlborough" is used for the township and "Marlboro" for the city in which the type section is exposed.

[2]Because these formations and members are known only in the subsurface, it is not possible to assign them formal formation or member names. These sections are no longer accessible because the rock tunnel is encased by a concrete lining. They have, therefore, been given the informal names U-1 through U-15, until such time as they can be assigned names from surface exposures.

on a scale of 1:240 while excavation was in progress. These samples were collected by Skehan in order that detailed petrographic, chemical (Tables 1 through 12), and age relationship studies could be made.

The Wachusett-Marlborough Tunnel was constructed in 1960–1962 by the Metropolitan District Commission (MDC) of Boston, Massachusetts, as part of a long-range hydrologic program developed over the past one hundred years for water distribution to the Boston area. The tunnel extends 12.8 km from the Wachusett Reservoir in Clinton to Marlborough (Figs. 1, 2). Its trend is N42°36'48"W; its depth below the surface varies from 54 to 150 m, depending on surface elevation. There are three access shafts (Fig. 3): at Clinton, at a position near the middle of the tunnel in Northborough, and in Marlborough Townships. Preliminary preconstruction mapping of the tunnel was conducted by M. P. Billings of Harvard University.

GEOLOGIC SETTING

The Wachusett-Marlborough Tunnel penetrates a highly folded and faulted, metamorphosed, eugeosynclinal sequence, which is located between two large, westward-dipping fault zones—the Clinton-Newbury fault zone and the Bloody Bluff fault zone (Figs. 1, and 3). The block in which the tunnel is driven lies immediately east of the Merrimack synclinorium, which extends from eastern Connecticut through east-central Massachusetts and southern New Hampshire into west-central Maine. The block is separated from the synclinorium by the Clinton-Newbury fault zone.

Figure 1 shows the Nashoba Formation and other stratigraphic divisions of the tunnel in relation to the general geology of the region. One of the major fault systems of the region, the Clinton-Newbury fault zone, forms the western margin of the map. It should be noted that the block composed of hornblende-actinolite ultramafic rock and the Rattlesnake Hill pluton is shown near the tunnel (designated "ha"; Fig. 1). The nature of the Clinton-Newbury fault zone is in part a network of faults enclosing fault slivers and blocks. The eastern margin of this block is marked by the Rattlesnake Hill fault and the western margin by the Clinton fault (Fig. 3).

LITHOLOGIC UNITS

The entire stratigraphic sequence of the tunnel is made up of 15 distinct lithologic units that have been further divided into subunits. The entire Nashoba Formation exposed in the tunnel is composed of 7 of these units as follows:

Unit 1: Biotite-rich Gneiss and Schist

The dominant lithologic subdivisions are porphyroblastic and nonporphyroblastic varieties.

Unit 1A: Porphyroblastic, Biotite-Feldspar-Quartz Gneiss and Schist. Dominantly gneissic, in places schistose. Mostly well-layered, prophyroblastic, biotite-rich (20 to 60 percent) with variable amounts of garnet, sillimanite, muscovite, and undifferentiated opaque minerals. The megacrysts are microcline, plagioclase, and garnet. Unit 1A can be subdivided on the basis of the porphyroblasts into undifferentiated, microcline, plagioclase, microcline and plagioclase, and garnet units.

Unit 1B: Nonporphyroblastic, Biotite-Feldspar-Quartz Gneiss and Schist. Domi-

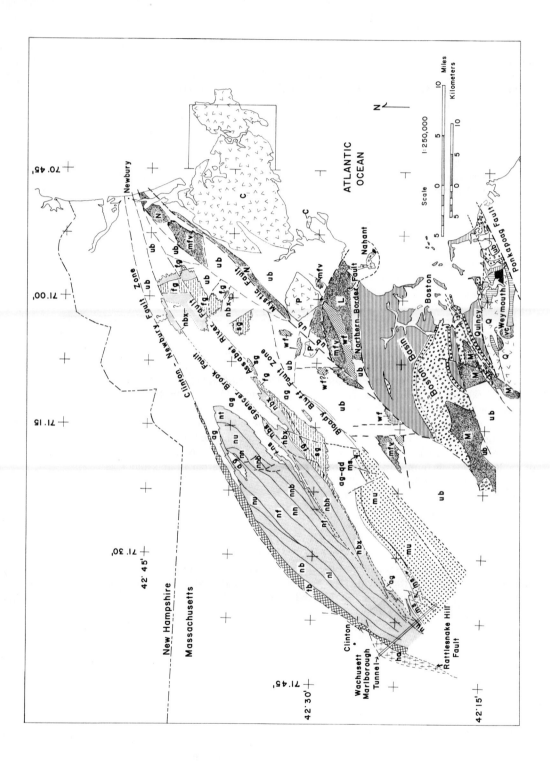

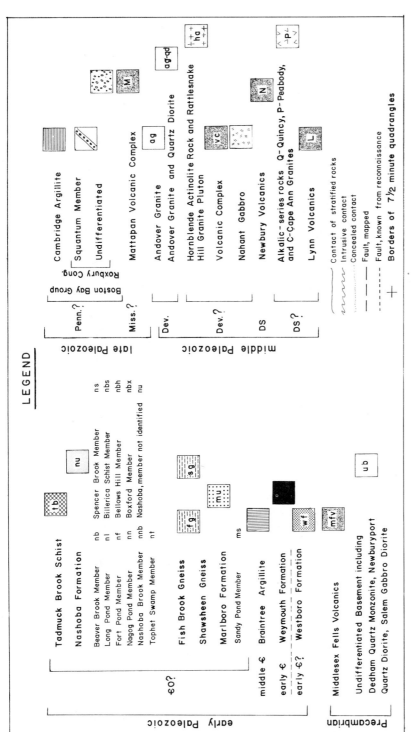

Figure 1. Geologic map of northeastern Massachusetts (modified from Bell and Alvord, 1974, 1976; Billings, 1976; Emerson, 1917; Toulmin, 1964; Zartman and Marvin, 1971). Distribution of major faults and plutons.

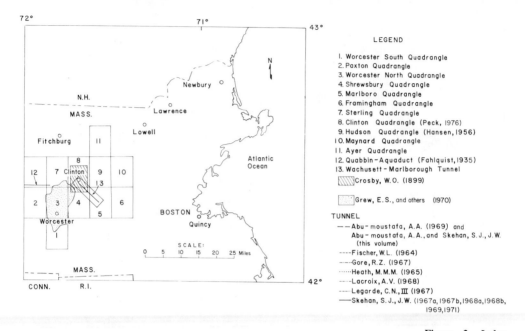

Figure 2. Index to geologic mapping in the vicinity of the Wachusett-Marlborough Tunnel.

nantly gneissic, in places schistose. Mostly well-layered to equigranular, biotite-rich (20 to 60 percent), and contains variable amounts of garnet, sillimanite, muscovite, and opaque minerals.

Unit 2: Quartz-Feldspar-Biotite Gneiss and Schist

Well-foliated to poorly foliated, gneissic to schistose, equigranular to porphyroblastic. Contains less biotite (<20 percent) and more quartz and feldspar than unit 1 (see Fig. 4A and 4B). Variable amounts of chlorite, muscovite, garnet, and opaque minerals. This unit can be subdivided on the basis of the porphyroblasts into undifferentiated, sillimanite and (or) muscovite, plagioclase-rich, microcline-rich, and garnet-rich units.

Unit 3: Biotite-Quartz Granulite, Calc-Silicate Granulite, and Marble

These occur in separate beds or interlayered in zones with other units. There is a wide variation in mineralogical composition.

Unit 3-1: Brown Biotite-Quartz or Biotite-Quartz-Feldspar-Calcite-Calc-Silicate Granulite. Commonly thin-bedded; garnet-rich and associated with biotite- and garnet-rich gneiss of unit 1A or with calc-silicate granulite (unit 3-2).

Unit 3-2: Calc-Silicate Granulite. Greenish-gray to brown; thin-bedded to thick-bedded; fine grained to coarse-grained; in part porphyroblastic; calcite-, plagioclase-, quartz-rich rocks that contain variable amounts of diopside, graphite, serpentine, epidote, biotite, garnet, clinozoisite, pyrophyllite, and talc.

Unit 3-3: Marble. Mostly calcite, white, gray, green, or pink; well-bedded, banded to massive; in places may also contain quartz, biotite, clinozoisite, grossularite, diopside, forsterite, chondrodite, muscovite, graphite, and rhodochrosite.

Unit 4: Hornblende Gneiss, Schist, and Amphibolite

Amphibolite is massive to poorly foliated; gneiss and schist are well-bedded, mostly nonporphyroblastic. These rocks contain variable amounts of biotite, quartz, epidote, garnet, muscovite, sericite, chlorite, opaque minerals, and clinozoisite. Unit 4 is interbedded with all other units and is closely associated with units 3-2, 3-3, and 7. Unit 4 can be subdivided on the basis of rock fabric into undifferentiated gneiss, schist, and equigranular amphibolite units.

Unit 5: Quartzite and Quartz Schist

These rocks are subdivided on the basis of the presence or absence of foliation as follows:

Unit 5-1: Quartzite. Thinly bedded; faintly foliated to massive; pink to brown, gray to black; fine- to coarse-grained, in part pebbly and feldspathic; vitreous or granular.

Unit 5-2: Quartz Schist. Faintly foliated, brown biotite- and (or) muscovite-quartz schist.

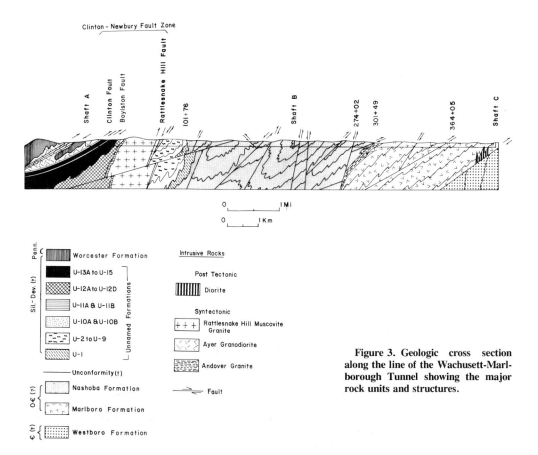

Figure 3. Geologic cross section along the line of the Wachusett-Marlborough Tunnel showing the major rock units and structures.

Unit 6: Muscovite- and (or) Chlorite-rich Schist and Gneiss

Quartz- and plagioclase-rich gneiss and schist with muscovite and (or) chlorite and with or without biotite. These rocks are probably variants of units 1A, 1B, and 2 but are more commonly schistose than gneissose. As much as 12 percent opaque minerals occur; sillimanite and garnet may be abundant and associated with clinozoisite and calcite. Feldspar and garnet megacrysts occur in places.

Unit 7: Coticule

Quartz and garnet rock (coticule, as defined in Renard, 1878, and Clifford, 1960) in thin beds and lenses; cherty to very fine-grained to medium-grained and associated with rocks of units 1A, 2, 3, 4, and 5.

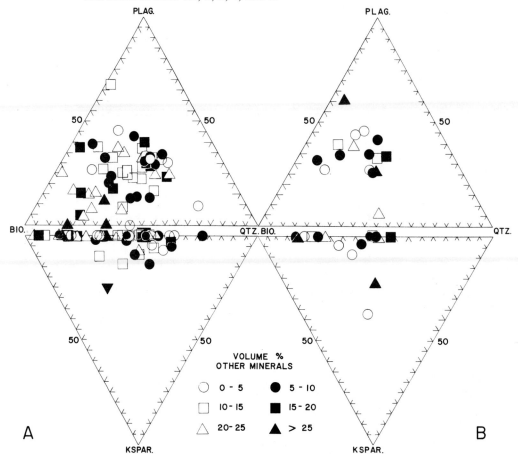

Figure 4A. Summary of modes showing chief mineral components of 58 samples of porphyroblastic biotite-rich gneiss and schist of unit 1A based on the data of Table 1. One hundred percent minus the total percentages of plagioclase, quartz, biotite, and potash feldspar equals the volume percentage of other minerals. *Note:* In 44 samples there is no potash feldspar. Due to extreme clustering, 8 of these samples have not been plotted: one with 5 to 10 volume percent other constituents; three with 10 to 15 volume percent; one with 15 to 20 volume percent; and three with 20 to 25 volume percent. B. Summary of modes showing chief mineral components of 16 samples of nonporphyroblastic biotite-rich gneiss and schist of unit 1B on basis of data in Table 2. One hundred percent minus the total percentages of plagioclase, quartz, biotite and potash feldspar equals the volume percentage of other minerals.

STRATIGRAPHY

The stratigraphic units in and near the tunnel are part of an intensely folded and faulted metamorphic sequence. The upper part of the Worcester Formation has been determined by Grew and others (1970) and by Grew (1973) as Pennsylvanian. This indicates that the lower part of the Worcester Formation, not yet officially renamed, is Pennsylvanian or older. The other Formations in Figure 3 probably range from Cambrian to Devonian in age. Bell and Alvord (1976), however, suggested that the Westboro and Nashoba Formations may be of late Precambrian age. The stratigraphic sequence from youngest to oldest is generally considered to be as follows: Worcester Formation, unnamed formations U-1 through U-15, Nashoba Formation, Marlboro Formation, and Westboro Formation.

Detailed stratigraphic descriptions of those formations exposed in the tunnel are presented by Skehan and Abu-moustafa (1976). The part of the Marlboro Formation that is exposed in the tunnel is at least 2,704 m thick and is composed of 31 members of 8 distinctive lithologic units, 7 of which are composed of the units 1 through 7 described above for the Nashoba Formation; unit 8, epicule rock (epicule is quartz and epidote rock in thin beds or lenses, very fine to cherty or medium-grained, typically associated with coticule; see Skehan and Abu-moustafa, 1976), a cherty to very fine-grained to medium-grained, light to dark green analogue of the coticule rock, is unique to the Marlboro Formation.

The unnamed formations (15 in number) are in apparent stratigraphic succession above the Nashoba and are designated U-1 through U-15 (Fig. 3).

The total measured stratigraphic thickness in the tunnel is 7,834 m. The Marlboro Formation composes 34.6 percent; the Nashoba Formation, 47.7 percent; and the unnamed formations U-1 (oldest) through U-15 (youngest) form 17.7 percent of the measured stratigraphic sequence.

In addition to these units, four plutons (Fig. 3) are exposed in the tunnel. These consist of a posttectonic diorite that was intruded into the Marlboro Formation; the Andover Granite (Hansen, 1956; Skehan, 1968b, 1969; Bell and Alvord, 1976), which was intruded near the boundary of the Nashoba and the Marlboro Formations; the Rattlesnake Hill pluton (Skehan, 1968b, 1969), which occurs only in the unnamed formations; and the Ayer Granodiorite (Jahns, 1952; Gore, 1976), which is intrusive into unnamed formation U-15 in the westernmost part of the tunnel (Fig. 3).

ANALYTICAL METHODS

A total of 155 modal analyses were made from a minimum of 8 linear traverses totaling 20 cm for each thin section; thus, components up to 1 mm in diameter were measured with 1 percent accuracy (Larsen and Miller, 1935). All thin sections were cut perpendicular to the foliation. Standard immersion methods were used for determining the plagioclase feldspar. Thin sections containing untwinned feldspar were tested by the use of standard staining techniques, and a positive reaction indicated potash feldspar. Thus, we designated this feldspar as orthoclase.

The chemical compositions of 61 rocks listed in this paper were determined by the mutual standard method for silicate analysis (described by Dennen and Fowler, 1955).

In preparing specimens, precautions were taken to avoid contamination. The samples were ground in tungsten blenders to avoid contamination with the elements that were being sought by spectrographic analysis. The standards used were G-1, W-1, T-1, and CAAS. The major element concentrations of these standards used were those of Ingamells and Suhr (1963). The analyses were carried out in the Starnd Laboratory of Massachusetts Institute of Technology (MIT). The spectrograph used was a grating-type model 700-000 Marck IV EBERT Convertible.

Spectrochemical determinations were made for 61 samples. C.I.P.W. norms of the metamorphic rocks were computed. These are shown in Tables 1 through 11.

NASHOBA FORMATION

The Nashoba Formation occurs between tunnel stations 101+76 and 301+49, measured from shaft A (Fig. 3). Petrographic investigation was carried out on 350 hand specimens. Modal analyses of 156 samples allowed the chemical compositions to be determined for evaluating metamorphic effects. Spectrographic analyses were made when determinations were not conclusive.

Unit 1: Biotite-rich Gneiss and Schist

Petrography. The biotite-rich gneiss and schist are well layered and in many places have well-developed cleavage. About 90 percent of the specimens studied in this unit are prophyroblastic.

The gneissic rocks have granulose layers rich in quartz and feldspar alternating with biotite-rich, schistose layers (Figs. 4A, 4B, 5). The layers, which range in thickness from less than 1 mm to 2 cm, are in part diffuse and irregular and in part very regular.

The porphyroblasts, which range in grain size from a few millimeters to a few centimeters, are essentially feldspar and pink garnet. There are distinctive rounded to subrounded megacrysts of microcline scattered irregularly throughout the rock. Some of these megacrysts are generally anhedral and appear to be premetamorphic. In other places they appear to grade into thin pegmatitic veinlets and patches scattered along the foliation planes; the quartz in these pegmatite masses is usually smokey, and the enclosing rocks have a migmatitic appearance.

The grain size and texture of these rocks are variable. Some rocks are dense, with little or no foliation in hand specimens; some are very coarse grained; others are medium grained to coarse grained with many megacrysts. The modes of 56 porphyroblastic samples of unit 1A are recorded in Table 1. The modes of 16 nonporphyroblastic samples of unit 1B are given in Table 2. Figures 4A and 4B summarize these data. Figure 4A shows that plagioclase is the dominant feldspar and that the percentage of biotite is high. The percentage of all other constituents is also provided. Figure 4B shows the same basic pattern of distribution as does Figure 4A, which indicates that the major difference between units 1A and 1B is dominantly textural.

Quartz commonly occurs in the felsic layers intergrown with feldspar and is elongated parallel to the micas and the foliation. It has well-developed undulatory extinction. It is a major constituent of most of the rocks, ranging from 2 to 43 percent, and is generally more than 15 percent.

Plagioclase feldspar occurs in the groundmass and as small (5 mm in diameter), rounded to almond-shaped megacrysts (Fig. 6), which apparently shouldered the layers aside as they crystallized. The composition of both the feldspar megacrysts and the feldspar in the groundmass is andesine (An_{30-35}). Unaltered plagioclase shows well-developed albite twinning; some specimens are altered to sericite and (or) epidote. Plagioclase feldspar ranges from 0 to 59 percent.

Figure 5. Photomicrograph of specimen of unit 1A (108+85) showing sillimanite (S) replacing orthoclase feldspar (O) and biotite (B) outlining left limb of fold.

Orthoclase is present in a few specimens as porphyroblasts (2 mm to 5 cm in length); it also occurs in the groundmass. The orthoclase is unaltered and difficult to identify without staining, as no twinning is apparent. Of the 72 thin sections studied, orthoclase occurs in nine; in one it is associated with microcline. Orthoclase ranges from trace amounts to 6 percent (Tables 1 and 2). It contains poikilitic biotite, muscovite, and sillimanite.

Microcline occurs in 12 thin sections as porphyroblasts, usually 5 mm to 5 cm in size, and also in the groundmass. It has cross-hatched twinning, is usually unaltered, and has irregular boundaries.

Biotite is a major constituent of these rocks and ranges from 20 to 56 percent. It occurs as fine-grained to coarse-grained parallel flakes separated by quartz-feldspar layers. In many places, however, a decussate structure is observed. There are two strongly pleochroic varieties, dark-brown and reddish-brown. The latter is usually intergrown with sillimanite. Biotite in some instances is kinked and folded, which indicates that it has been subjected to postmetamorphic deformation. It is intergrown with muscovite, sillimanite, and magnetite.

Sillimanite occurs as porphyroblasts and in the groundmass as needles and long prismatic crystals; it ranges from 0.001 to 3 mm in length. It is intergrown with muscovite, biotite, garnet, and magnetite. Two orientations are seen in most thin sections: parallel and at an angle to the plane of foliation. Thus, both longitudinal and cross sections of sillimanite can be seen in the same thin section. Sillimanite

TABLE 1. MODES OF UNIT 1A

	111+22*	112+27	114+90	115+5	116+78*	117+43	118+92	120+14
Quartz	33.4	31.5	33.9	16.8	30.0	39.3	31.6	24.7
Andesine ($An_{30}-An_{35}$)	27.6	19.3	16.6	24.0	24.3	19.3	24.6	32.3
Orthoclase	1.6	6.3	..	..	..	1.0	tr	..
Microcline	1.0	..	..	..	..	..	..	..
Biotite	27.0	32.0	27.6	46.6	27.0	22.3	32.3	23.3
Muscovite	..	5.6	3.6	tr	6.0	2.6	0.6	4.3
Sericite†	1.6	..	5.3	..	2.6	11.6	3.0	2.3
Epidote	3.6	..	..	8.8	1.0	..	..	..
Clinozoisite	..	..	..	..	..	..	..	0.6
Chlorite	..	..	..	..	2.6	..	..	1.6
Sillimanite	1.0	tr	..	..	0.6	1.6	1.3	tr
Almandine	..	1.0	tr	tr	..	tr	0.3	tr
Apatite	..	tr	..	0.6	0.3	..	tr	..
Zircon	0.3	..	tr	0.2	tr	tr	0.3	tr
Opaque minerals	0.3	4.3	12.0	3.0	5.3	2.3	6.0	8.6
Calcite	1.6	tr	1.0	tr	0.3	..	tr	2.3
Hornblende	1.0	..	..	..	..	..	..	..
Total	100.0	100.0	100.0	100.0	100.0	100.0	100.0	100.0
Grain size in mm	ND	0.01-1.6	0.02-1.55	0.01-1.88	0.02-1.72	0.02-3.6	0.35-0.65	0.01-2.8

	120+95	122+41	125+22	125+22A	127+20*	134+35	134+62	141+15
Quartz	16.8	33.6	26.7	43.2	32.3	24.6	16.7	17.5
Andesine ($An_{30}-An_{35}$)	12.2	7.6	32.0	2.3	23.2	24.3	4.3	3.0
Orthoclase	tr	..	..	..	..	3.7	..	..
Microcline	..	..	..	..	..	..	..	..
Biotite	47.8	48.5	26.6	52.3	26.4	25.0	60.6	58.6
Muscovite	9.0	4.0	6.9	..	9.0	9.6	..	tr
Sericite†	2.0	..	2.0	..	tr	0.3	..	15.6
Epidote	..	..	..	..	0.3	..	..	..
Clinozoisite	4.0	..	tr	0.3	..	..	..	..
Chlorite	..	..	tr	0.6	1.3	..	..	2.3
Sillimanite	..	..	..	..	0.6	5.3	4.2	..
Almandine	..	..	..	..	tr	0.3	3.6	..
Apatite	..	..	..	..	..	..	..	..
Zircon	tr	tr	tr	0.3	0.3	tr	tr	..
Opaque minerals	3.6	0.3	5.6	1.0	4.6	6.6	10.6	tr
Calcite	4.6	6.0	0.2	..	2.0	0.3	..	3.0
Hornblende	..	..	..	..	..	..	..	..
Total	100.0	100.0	100.0	100.0	100.0	100.0	100.0	100.0
Grain size in mm	0.06-1.26	ND	ND	ND	ND	0.03-4.29	0.002-3.6	ND

	143+75*	146+32	148+74	149+50	150+93	151+80	156+57	157+26*
Quartz	35.3	35.0	21.3	19.6	33.2	38.8	24.3	31.3
Andesine ($An_{30}-An_{35}$)	12.3	31.7	..	27.3	19.3	26.6	21.0	28.2
Orthoclase	1.0	tr	..	..	..	4.6	..	..
Microcline	..	..	..	..	..	..	9.6	8.6
Biotite	45.4	30.8	38.5	32.6	25.0	25.6	35.0	26.0
Muscovite	..	..	..	..	2.6	..	..	1.3
Sericite†	2.0	0.4	..	0.3	7.3	1.5	4.3	tr
Epidote	..	..	..	..	..	..	..	..
Clinozoisite	..	..	..	..	..	..	0.6	tr
Chlorite	..	..	..	..	0.7	..	..	..
Sillimanite	..	0.6	29.6	7.3	6.0	1.6	0.3	tr
Almandine	2.0	..	8.3	10.6	3.6	..	0.3	0.6
Apatite	..	..	..	..	..	..	tr	..
Zircon	tr	0.3	0.3	tr	..	tr	tr	1.3
Opaque minerals	2.0	0.6	2.0	2.3	2.3	1.3	0.3	0.3
Calcite	..	tr	..	..	tr	..	..	..
Hornblende	..	..	..	..	..	..	4.3	..
Myrmekite	..	0.6	..	..	..	..	..	2.4
Total	100.0	100.0	100.0	100.0	100.0	100.0	100.0	100.0
Grain size in mm	ND	ND	0.01-3	ND	0.83-4.62	ND	ND	ND

TABLE 1. (Continued)

	161+88*	164+60	169+59*	170+65	173+30	175+08*	176+00	176+63*
Quartz	38.3	9.8	42.6	23.3	14.5	15.7	9.2	12.8
Andesine (An$_{30}$–An$_{35}$)	13.5	34.0	12.3	14.6	11.6	6.6	13.4	12.6
Microcline	4.2	..	..	..	..	17.0	..	..
Biotite	30.0	45.3	37.3	46.0	51.3	35.3	54.0	57.3
Muscovite	1.0	..	..	0.3	..	tr	3.0	tr
Sericite†	1.6	5.6	5.6	1.6	tr	0.3	0.3	..
Epidote	..	0.3	..	..	..	..	..	..
Clinozoisite	..	tr	..	..	..	..	tr	..
Chlorite	0.3	1.0	..	..	..	..	..	..
Sillimanite	6.6	1.0	tr	10.9	8.3	15.2	13.3	9.0
Almandine	0.6	..	..	1.3	12.3	9.3	tr	..
Zircon	0.3	tr	0.3	tr	tr	tr	..	0.3
Opaque minerals	3.0	3.0	1.3	2.0	2.0	..	6.8	8.0
Calcite	0.6	tr	0.6	..	..	0.6	..	..
Total	100.0	100.0	100.0	100.0	100.0	100.0	100.0	100.0
Grain size in mm	ND	ND	ND	0.02–3	0.01–2.54	0.005–07.26	0.005–4.13	0.02–12

	180+49	182+61	186+43	187+64	187+64A	200+26	200+60*	207+59
Quartz	21.6	33.3	66.7	23.3	37.6	11.7	24.6	16.0
Andesine (An$_{30}$–An$_{35}$)	18.6	29.2	8.6	22.3	26.3	..	21.6	29.6
Microcline	1.3	..	..	..	..	..	..	..
Biotite	49.6	29.6	20.3	47.6	27.6	54.0	40.3	46.0
Muscovite	5.0	..	..	tr	tr	2.0	..	2.7
Sericite†	1.6	3.0	1.3	3.6	6.3	10.0	2.6	tr
Clinozoisite	..	..	..	..	..	..	..	..
Chlorite	..	1.3	..	..	tr	..	..	..
Sillimanite	..	..	1.9	0.6	..	4.0	3.3	0.7
Almandine	2.0	1.3	0.3	2.0	0.6	12.4	5.6	tr
Apatite	..	..	..	..	0.3	..	..	..
Zircon	0.3	0.3	..	tr	tr	tr	tr	tr
Opaque minerals	..	2.0	0.9	0.6	tr	5.3	2.0	5.0
Calcite	tr	tr	..	..	1.3	..	..	..
Total	100.0	100.0	100.0	100.0	100.0	100.0	100.0	100.0
Grain size in mm	0.005–1	ND	0.005–1.65	ND	ND	0.01–3	ND	0.01–5

	211+16	215+67	215+70*	225+04	231+16	234+76	244+10	245+86
Quartz	29.4	41.0	43.6	23.1	27.3	16.6	18.3	32.4
Andesine (An$_{30}$–An$_{35}$)	20.3	30.3	28.5	40.3	21.0	26.7	6.4	29.0
Microcline	..	..	..	..	..	6.0	..	..
Biotite	37.3	24.6	23.3	30.0	37.6	38.3	51.0	25.3
Muscovite	0.7	0.3	1.3	0.3	7.3	..	0.9	..
Sericite†	0.3	..	tr	tr	0.3	1.6	3.6	12.3
Clinozoisite	tr	..	..	2.3	..	..	..	tr
Chlorite	..	..	..	..	1.6	..	..	..
Sillimanite	5.0	..	..	..	..	5.6	12.6	tr
Almandine	..	..	..	..	tr	1.6	1.6	tr
Apatite	..	..	..	..	..	..	..	..
Zircon	tr	0.4	tr	tr	..	0.3	1.0	tr
Opaque minerals	7.0	3.4	3.3	3.7	3.3	3.0	4.6	1.0
Calcite	..	..	tr	0.3	1.6	0.3	..	..
Total	100.0	100.0	100.0	100.0	100.0	100.0	100.0	100.0
Grain size in mm	ND	ND	ND	ND	ND	ND	ND	ND

Continued on next page

TABLE 1. *(Continued)*

	247+52	252+52	256+52*	259+01	261+39	262+05	262+05A	270+26
Quartz	34.8	16.4	38.0	30.1	7.8	3.2	2.4	2.4
Andesine ($An_{30}-An_{35}$)	12.3	44.0	31.0	4.6	37.3	31.6	19.8	58.5
Microcline	7.3	..	2.6	..	..	..	..	..
Biotite	39.0	36.0	21.6	43.5	48.0	50.0	55.6	25.6
Muscovite	tr	..	2.6	4.3	..	6.0	3.6	..
Sericite†	..	..	..	..	tr	tr	tr	3.6
Epidote	..	..	tr	..	..	..	..	..
Clinozoisite	tr	tr	0.3	..	..	..	..	..
Chlorite	..	..	..	..	..	tr	..	..
Sillimanite	tr	tr	..	15.2	5.3	5.0	12.3	3.9
Almandine	0.3	1.3	..	..	..	..	..	..
Zircon	0.3	..	0.3	tr	tr	0.6	..	tr
Opaque minerals	4.0	2.0	1.0	2.3	1.6	3.6	6.3	6.0
Calcite	2.0	0.3	2.6	..	..	tr	tr	..
Total	100.0	100.0	100.0	100.0	100.0	100.0	100.0	100.0
Grain size in mm	ND	ND	0.01–5.8	0.007–5	ND	ND	0.002–5.0	0.02–62

Note: ND = not determined.
*Spectrochemical analysis of the specimens are listed in Table 3.
†Including sericitized feldspar.

was recorded in 48 of the 72 thin sections studied and ranges from trace amounts to 30 percent. In some places it encloses biotite and muscovite; in others it embays plagioclase feldspar. In four sections, sillimanite is associated with orthoclase feldspar. Where the relationship is known, sillimanite replaces orthoclase feldspar (Fig. 5) and contains poikilitic orthoclase and biotite; it is also associated with magnetite.

Almandine, in anhedral grains and euhedral fractured crystals, usually occurs as porphyroblasts as much as 5 cm in diameter and is wrapped around by the foliation. It shows a sieve texture of quartz, feldspar, magnetite, and biotite.

Sericite, epidote, and calcite occur as alteration products of plagioclase feldspar. Clinozoisite occurs as small anhedral grains, usually associated with biotite and plagioclase feldspar.

Opaque minerals occur in 95 percent of the samples and range in abundance from trace amounts to 12 percent. In hand specimen, the opaque minerals appear to be magnetite, pyrite, and chalcopyrite. Apatite and zircon occur in minor amounts enclosed in micas and plagioclase feldspar.

Hornblende associated with biotite is a minor constituent in a few thin sections. It is dark green and occurs as anhedral crystals.

Characteristic features of the fine-grained gneissic rocks are rounded to subrounded megacrysts and layering. A few specimens show apparent relict vesicular structures and relict flow banding (Fig. 7).

Chemical Composition and Source Rocks. Twenty spectrochemical analyses and C.I.P.W. norms of unit 1A, porphyroblastic biotite-feldspar-quartz gneiss and schist are recorded in Table 3. Eight analyses and C.I.P.W. norms of unit 1B nonporphyroblastic biotite-feldspar-quartz gneiss and schist, are given in Table 4.

Three samples of unit 1A have between 50 and 60 percent SiO_2 (Fig. 9A; Table 3). These samples (167+42, 175+08, and 176+63), as well as 252+25, are the only ones that contain more than 20 percent Al_2O_3. One of these (175+08) contains 29 percent Al_2O_3, the highest amount of iron oxide, 12.18 percent, and the lowest values of Na_2O and CaO, 0.17 and 0.17 percent, respectively. Another sample

(167+42) contains the lowest value for iron, 4.12 percent, but has a high value for Al_2O_3, 21.70 percent.

Many other samples contain SiO_2 in the range from 62.10 to 71.43 percent. Although the lower values of Al_2O_3 tend to be associated with rocks of higher SiO_2 content, this relationship is highly variable. As a general conclusion, the data of Figure 9A indicate a considerable variability in chemical trends, a condition compatible with the source rock having been subjected to various conditions of leaching and deposition such as would be expected in an environment of intense chemical weathering. The presence of abnormally high K_2O, for example, in 167+42, 175+08, and 176+63 and in other samples as well, would also be consistent with a lateritic source rock in which the clay had adsorbed unusual amounts of K_2O.

TABLE 2. MODES OF UNIT 1B

	112+83*	193+99*	201+34	201+37*	201+37A	223+10	227+44	235+89
Quartz	31.3	29.2	26.4	15.8	12.0	19.6	35.3	28.1
Andesine (An_{30}–An_{35})	27.6	30.3	31.2	31.6	34.0	42.6	3.5	39.0
Microcline	··	··	2.0	··	··	4.6	··	··
Biotite	23.6	30.3	32.0	46.0	41.0	30.3	32.6	27.3
Muscovite	7.4	··	0.5	··	··	tr	9.3	2.6
Sericite†	4.6	5.0	3.6	1.6	0.6	1.0	··	··
Epidote	··	··	··	tr	3.7	··	··	··
Clinozoisite	··	0.6	··	1.0	0.3	tr	··	tr
Chlorite	··	··	··	··	··	··	··	tr
Sillimanite	0.5	··	tr	tr	3.7	tr	12.0	··
Almandine	0.3	··	··	··	0.7	0.6	··	··
Apatite	0.5	··	··	··	··	··	··	tr
Zircon	0.2	tr	tr	··	··	tr	1.3	tr
Opaque minerals	4.0	0.3	3.5	4.0	4.0	1.3	6.0	1.4
Calcite	··	3.3	0.8	··	··	tr	··	1.6
Sphene	··	··	··	··	··	··	··	··
Hornblende	··	1.0	··	··	··	··	··	··
Perthite	··	··	··	··	··	··	··	··
Total	100.0	100.0	100.0	100.0	100.0	100.0	100.0	100.0
Grain size in mm	0.004–1.36	ND	0.01–4	0.005–7.6	0.005–0.34	0.01–4.2	ND	0.01–2.1

	250+71A	250+81B*	264+04	271+56	276+97*	285+14	289+12	290+15
Quartz	17.5	7.7	21.3	3.6	31.7	22.6	15.6	13.5
Andesine (An_{30}–An_{35})	43.0	29.8	14.7	45.1	24.0	18.6	28.8	26.0
Microcline	2.2	··	12.2	··	··	30.0	··	··
Biotite	34.6	56.3	21.0	25.1	34.6	27.0	32.0	56.3
Muscovite	··	··	1.0	3.6	2.0	tr	··	··
Sericite†	··	··	1.6	13.0	2.5	1.2	7.3	0.6
Epidote	0.5	2.0	··	1.7	0.6	··	5.0	··
Clinozoisite	1.0	1.3	··	··	3.3	tr	0.3	tr
Chlorite	··	··	tr	··	··	··	··	··
Sillimanite	··	··	4.2	1.3	··	··	··	··
Almandine	··	··	··	0.3	··	··	··	··
Apatite	··	··	··	··	··	··	··	··
Zircon	tr	tr	tr	··	tr	0.3	0.1	tr
Opaque minerals	0.6	1.6	1.3	2.3	··	0.3	0.3	3.6
Calcite	0.6	tr	··	4.0	0.3	··	tr	··
Sphene	··	··	··	··	··	··	0.6	··
Hornblende	··	1.3	··	··	1.0	··	10.0	··
Perthite	··	··	22.7	··	··	··	··	··
Total	100.0	100.0	100.0	100.0	100.0	100.0	100.0	100.0
Grain size in mm	ND	ND	ND	ND	ND	ND	ND	ND

Note: ND = not determined.
*Spectrochemical analysis of the specimens are listed in Table 4.
†Including sericitized feldspar.

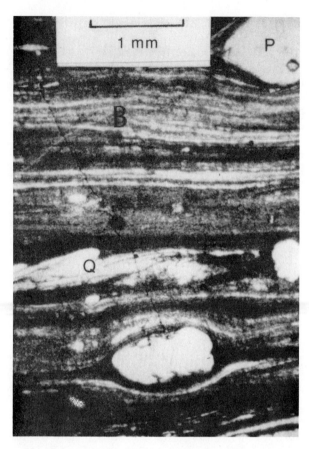

Figure 6. Photomicrograph of specimen of unit 1A (134+62). Thin section shows well-developed layering shouldered aside by crystallization of rounded to subangular plagioclase feldspar porphyroblasts (P); apparently microslump folds in quartz layer (Q), and biotite (B) which is not developed around plagioclase as in other parts of the sample.

We thus conclude that some components have been introduced from saprolitized source areas and added to an essentially unweathered or unleached sediment having, in most components, the approximate composition of Nockold and Allen's (1954) average dacite and in other components that of Daly's (1933) average dacite (Fig. 9A).

A few specimens of the biotite-feldspar-quartz gneiss and schist exhibit structures interpreted as gas cavities and relict flow structure (Fig. 7, specimen 134+62). The rounded to subrounded feldspar in most specimens may be recrystallized amygdaloidal minerals, presumably zeolite. Many of these rocks show layering (Fig. 7), which is interpreted as relict original flow banding or the original layering in pyroclastic materials. Fragmental textures are not observed in these specimens.

Unit 1B (Fig. 9B; Table 4) represents a group of rocks that have been derived from a similar mixed source (probably including rhyolite) as the rocks of unit 1A, which is indicated by the variation in percentages of SiO_2, Al_2O_3, Fe_2O_3, CaO, MgO, and K_2O. Some layers show chemical concentrations that indicate a high proportion of clays and other weathering products, whereas other layers were composed almost wholly of unweathered products.

Unit 2: Quartz-Feldspar-Biotite Gneiss and Schist

Petrography. The quartz-feldspar-biotite gneiss and schist is well-foliated to poorly foliated and commonly exhibits a well-developed cleavage. About 50 percent of

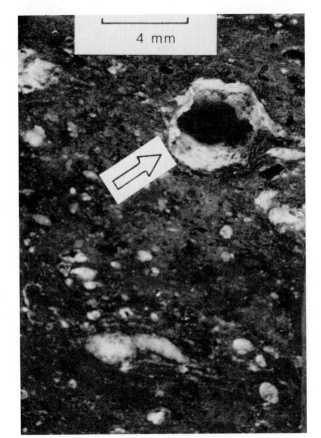

Figure 7. Photograph of a sample of unit 1A (134+62) shows amygdular cavity fillings (arrow) and relict flow banding.

the thin sections are porphyroblastic. The prophyroblasts are chiefly plagioclase, orthoclase, microcline, garnet, sillimanite, and (or) muscovite and range in size from a few millimeters to 5 cm.

The gneissic layering is the result of granulose layers, which are rich in quartz and feldspar, alternating with schistose layers, which are rich in biotite with or without muscovite. The layers range in thickness from less than 1 mm to about 1 cm. The grain size and texture of these rocks are variable. Some rocks are dense with little foliation; others are medium grained to coarse grained with moderate to well-developed foliation. The modes of 32 samples are recorded in Table 5. Figure 8 shows that these quartz-rich rocks, which have biotite in amounts less than 20 percent and greater than 5 percent, contain substantial amounts of both plagioclase and potash feldspar in about two-thirds of the samples.

Quartz commonly occurs in felsic layers intergrown with feldspar in grains elongated parallel to the foliation plane. It is a major constituent of all rocks studied and ranges from 31 to 51 percent. It commonly shows undulatory extinction.

Plagioclase feldspar (An_{30-35}) occurs both in the groundmass and as large porphyroblasts (up to 3 cm long). It is generally unaltered with well-developed albite twinning and shows slight alteration in a few places. Plagioclase is a major constituent of most rocks and ranges from 1 to 57 percent.

Orthoclase feldspar is present in 11 of 32 samples studied; in 9 it is associated with microcline and in 2 it is associated with sillimanite. It occurs usually as large, unaltered prophyroblasts and shows poikilitic sillimanite, biotite, and (or)

TABLE 3. CHEMICAL COMPOSITIONS OF UNIT 1A

	111+22*	111+26	116+78*	118+92*	122+22	127+20*	143+75*	151+10*	157+26*	161+80*
SiO_2	69.10	68.96	64.41	71.43	67.92	66.90	65.10	70.52	68.05	62.87
TiO_2	0.60	0.79	0.82	0.75	0.59	0.64	1.15	0.70	0.53	1.10
Al_2O_3	13.20	12.20	18.14	12.42	14.30	15.70	12.05	13.51	17.45	16.76
Total Fe†	6.45	7.26	6.53	6.77	5.83	5.50	9.40	5.40	4.97	8.38
MnO	0.05	0.06	0.10	0.07	0.06	0.15	0.15	0.08	0.08	0.18
MgO	2.48	3.07	2.18	2.08	2.93	2.67	4.34	2.15	0.70	2.21
CaO	4.35	3.23	2.50	1.68	3.00	2.69	1.74	2.31	2.50	1.59
Na_2O	1.92	1.76	3.72	2.42	2.28	2.68	2.59	2.94	3.49	2.06
K_2O	1.82	2.62	1.60	2.33	3.07	3.05	3.43	2.32	2.22	4.81
Total	99.97	99.95	100.00	99.95	99.98	99.98	99.95	99.93	99.99	99.96
					C.I.P.W. NORMS					
Quartz	35.00	33.57	25.14	38.98	29.69	26.50	25.90	34.46	31.81	21.40
Corundum	0.10	0.37	3.69	1.80	1.00	1.63	1.48	1.20	3.06	3.42
Orthoclase	9.85	14.10	8.84	12.65	16.71	16.90	11.30	12.66	12.41	27.05
Albite	15.82	14.42	31.40	19.98	18.87	22.34	24.60	24.30	29.61	17.57
Anorthite	19.81	14.65	11.64	7.66	13.73	13.65	17.15	10.60	11.73	7.52
Hypersthene	18.44	21.51	17.93	17.70	19.01	17.89	17.60	15.60	10.52	21.20
Ilmenite	0.96	1.24	1.33	1.19	0.93	1.03	1.96	1.10	0.83	1.81
Total	99.98	99.86	99.97	99.96	99.94	99.94	99.99	99.92	99.97	99.97
Johannsen no.	227	227	227	227	227	227	227	227	227	227

	166+62	167+42	169+59*	175+08*	176+63*	200+60*	215+70*	231+66	252+25*	256+52*
SiO_2	67.11	59.69	69.22	50.52	51.30	64.60	64.88	67.12	62.10	70.00
TiO_2	0.68	0.50	0.98	1.36	0.94	0.87	1.52	0.72	1.06	0.34
Al_2O_3	15.98	21.70	14.57	29.00	23.21	17.40	16.22	17.52	20.16	16.00
Total Fe†	4.86	4.12	7.14	12.18	6.17	5.90	7.30	5.25	6.98	4.30
MnO	0.12	0.09	0.05	0.62	0.13	0.10	0.11	0.06	0.10	0.04
MgO	2.60	1.19	2.18	2.17	2.31	3.40	2.54	2.21	1.81	0.51
CaO	3.68	3.91	2.33	0.17	8.13	3.82	2.26	1.93	3.50	3.57
Na_2O	2.78	4.02	2.18	0.17	3.25	2.12	3.15	3.42	3.05	3.60
K_2O	2.19	4.78	1.34	3.80	4.46	1.74	1.52	1.75	1.21	1.70
Total	100.00	100.00	99.99	99.97	99.90	99.95	99.50	99.98	99.97	100.06
					C.I.P.W. NORMS					
Quartz	30.33	5.25	40.64	30.51	0.23	29.77	29.21	31.12	26.80	31.84
Corundum	0.21	1.86	3.35	14.86	5.02	3.24	3.38	3.90	4.85	1.16
Orthoclase	12.31	27.28	7.31	21.51	28.00	9.63	8.42	9.80	6.78	9.72
Albite	23.61	34.87	18.08	1.45	30.00	17.81	26.53	28.80	26.03	31.34
Anorthite	17.41	18.75	10.69	0.86	16.30	17.73	10.42	9.62	16.59	17.18
Hypersthene	14.98	11.14	18.34	28.52	18.80	20.57	19.56	15.60	17.22	8.18
Ilmenite	1.12	0.83	1.57	2.26	1.67	1.24	2.48	1.14	1.74	0.56
Total	99.97	99.98	99.98	99.97	100.02	99.99	100.00	99.98	100.01	99.98
Johannsen no.	227	2211	227	227	2211	227	227	227	227	227

*Modal analysis of the specimens are listed in Table 1.
†As Fe_2O_3.

TABLE 4. CHEMICAL COMPOSITIONS OF UNIT 1B

	112+83*	124+26	143+35	193+99*	201+37*	250+1B*	270+71B	276+97*
SiO_2	67.47	64.84	59.11	65.88	74.00	61.24	61.24	69.69
TiO_2	0.91	0.88	1.03	0.97	0.90	1.35	1.35	0.40
Al_2O_3	15.69	16.63	21.11	14.32	11.60	17.75	17.74	16.70
Total Fe†	6.27	7.31	6.54	7.16	5.02	8.61	8.61	2.63
MnO	0.04	0.04	0.10	0.12	0.09	0.13	0.09	0.03
MgO	1.96	2.74	3.17	3.36	2.23	4.26	4.30	1.15
CaO	2.43	1.82	1.47	4.12	2.47	2.31	2.31	2.50
Na_2O	2.66	2.71	3.15	2.72	2.47	2.72	2.72	3.70
K_2O	2.52	2.99	4.23	1.33	1.20	1.63	1.63	3.20
Total	99.95	99.96	99.91	99.98	99.98	100.00	99.99	100.00
C.I.P.W. NORMS								
Quartz	32.07	27.60	13.22	23.70	44.65	22.00	23.25	29.46
Corundum	2.66	3.62	5.67	0.66	1.03	3.00	4.66	1.55
Orthoclase	13.94	16.60	24.11	7.04	6.48	9.20	8.98	17.86
Albite	22.32	22.80	27.15	23.25	20.35	23.10	22.74	30.97
Anorthite	11.27	8.45	7.00	18.40	11.23	10.80	10.67	11.55
Hypersthene	16.26	19.60	21.13	25.37	14.82	29.70	27.49	7.97
Ilmenite	1.48	1.30	1.72	1.60	1.42	2.20	2.11	0.64
Total	100.00	99.97	100.00	100.02	99.98	100.00	99.90	100.00
Johannsen no.	227	227	227	227	223	227	227	227

*Modal analysis of the specimens are listed in Table 2.
†As Fe_2O_3.

muscovite. Orthoclase ranges from trace amounts to 11 percent.

Microcline occurs in the groundmass and as porphyroblasts. It is unaltered and show excellent cross-hatched twinning. In some places it contains poikilitic biotite and muscovite.

Biotite, usually a major constituent of these rocks (Fig. 8), ranges from 2 to 18 percent. It occurs as fine to coarse flakes in layers or bands associated with muscovite, sillimanite, and magnetite oriented parallel to the foliation plane. In many places dark-brown, strongly pleochroic to pale-brown decussate biotite occurs.

Sillimanite occurs as needles and crystals that range in size from 0.001 to 4 mm; in places it forms large porphyroblasts. Two orientations were noted: parallel or at an angle to the plane of foliation. It is present from trace amounts to 27 percent. Sillimanite is intergrown with biotite, muscovite, and locally with abundant magnetite.

Almandine occurs as small (0.2 to 0.6 mm), moderately fractured euhedral to subhedral crystals associated with biotite, sillimanite, and muscovite. It shows a sieve texture of quartz, chlorite, magnetite, and biotite.

Sericite, epidote, chlorite, and some calcite appear to be alteration products of plagioclase feldspar, biotite, and orthoclase. Zircon, apatite, and sphene occur in minor amounts usually enclosed in biotite.

Magnetite is the major opaque mineral; pyrite and chalcopyrite occur in smaller amounts.

Chemical Composition and Source Rocks. Fourteen spectrochemical analyses and the C.I.P.W. norms of the quartz-feldspar-biotite gneiss and schist are recorded in Table 6. Figure 9C summarizes the chemical data.

In unit 2 (Fig. 9C), SiO_2 has an averaged value of 70.9 percent. Samples 210+78 and 145+76A show the extremes of Al_2O_3 content, 18.76 and 10.47 percent, respectively. There is generally an inverse relationship between SiO_2 and Al_2O_3 content (Fig. 9C).

Total iron oxide is variable and ranges from 2.2 to 7.8 percent. Both high and low values of iron are generally associated with high and low values of MgO and less consistently with high and low values of both SiO_2 and Al_2O_3. K_2O is highly variable with concentrations ranging from 0.88 to 5.8 percent. The TiO_2 content ranges from 0.3 to 1.7 percent; the highest value is associated with high values for Fe_2O_3 and Al_2O_3.

Figure 9C shows the average chemical analysis of 30 graywackes (Tyrrell, 1933, in Pettijohn, 1957, p. 307). The average of these graywackes is generally close to the average of the 14 samples of this study. The average rhyolite of Daly (1933) also provides a useful comparison for silica and alumina. The most notable deviation of samples from the average for graywackes is that Fe_2O_3 is less highly concentrated in unit 2 rocks.

Thus it is inferred that the source rock of unit 2 was graywacke, with some additional rhyolitic or granitic sand, which had concentrated silica and alumina to some degree by laterization.

Unit 3: Biotite-Quartz Granulite, Calc-Silicate Granulite, and Marble

Petrography. The biotite-quartz granulite of unit 3-1, while similar to that of unit 2, is intimately interbedded with the calc-silicate granulite and marble of units 3-2 and 3-3. It occurs as thin layers commonly showing a rhythmic alternation with these beds. Rocks of unit 3-1 also contain small amounts of calcite and (or) calc-silicate minerals.

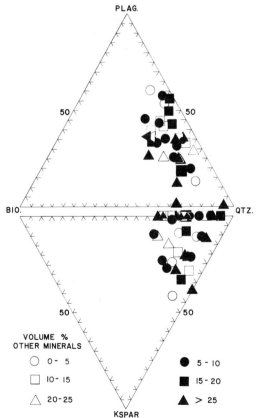

Figure 8. Summary of modes showing ratios of chief mineral components of 32 samples of quartz-feldspar-biotite gneiss and schist of unit 2 on the basis of data in Table 6. One hundred percent minus the total percentages of plagioclase, quartz, biotite, and potash feldspar equals the volume percentage of other minerals. Volume percentages correspond to those of Figures 4A and 4B.

TABLE 5. MODES OF UNIT 2

	113+11	113+61*	116+30	118+95*	121+10	121+34	124+20	124+95
Quartz	32.4	38.1	51.0	34.3	49.7	41.0	38.2	30.6
Andesine (An$_{30}$–An$_{35}$)	31.6	18.3	14.4	14.3	30.6	2.0	39.3	16.9
Orthoclase	10.6	4.0	2.0	..	..	..	..	tr
Microcline	5.3	5.0	3.6	16.3	..	25.6	..	..
Biotite	10.3	13.6	12.3	7.3	7.0	tr	11.3	16.6
Muscovite	8.6	12.4	7.6	2.3	1.3	2.6	tr	17.7
Sericite†	0.3	..	2.0	7.3	0.6	0.6	6.0	tr
Clinozoisite	..	3.6	0.6	tr	1.6	..	tr	..
Chlorite	..	..	0.3	2.3	3.3	..	..	3.6
Sillimanite	..	tr	..	12.3	..	..	2.3	..
Almandine	..	tr	..	tr	2.0	tr	..	tr
Apatite	tr	tr	0.3	..	..	..	tr	..
Zircon	tr	..	0.3	..	tr	tr	tr	tr
Opaque minerals	0.3	2.0	5.0	3.0	2.6	0.6	1.3	9.0
Calcite	0.6	3.0	0.6	0.6	1.3	1.6	1.6	5.6
Perthite	..	..	..	..	..	26.0	..	..
Total	100.0	100.0	100.0	100.0	100.0	100.0	100.0	100.0
Grain size in mm	0.01–2.7	0.02–2.64	0.01–1.16	0.01–4.8	0.7–4.5	ND	0.1–2.81	0.005–1.32

	127+43	133+56*	145+12*	145+76*	147+41	147+89	151+10*	152+95*
Quartz	38.1	42.6	40.5	47.0	36.6	54.1	33.8	52.2
Andesine (An$_{30}$–An$_{35}$)	28.6	7.8	25.0	28.0	23.0	29.6	12.5	14.6
Orthoclase	..	..	..	1.5	..	..	..	6.9
Microcline	..	..	..	..	11.3	..	11.0	9.3
Biotite	18.2	13.0	4.3	18.3	6.3	10.0	6.6	7.6
Muscovite	1.3	2.3	..	1.6	3.3	..	5.6	3.3
Sericite†	5.3	tr	27.3	..	16.6	..	16.0	2.6
Clinozoisite	..	..	..	tr	..	..	..	..
Chlorite	tr	14.5	..	..	2.6	..	3.3	0.6
Sillimanite	3.0	3.2	1.6	..	..	..	1.0	..
Almandine	..	1.6	..	..	..	6.0	..	1.3
Apatite	..	..	..	..	..	0.3	..	..
Zircon	0.3	tr	tr	0.3	tr	..	tr	tr
Opaque minerals	5.2	12.0	0.3	3.3	tr	tr	8.6	1.6
Calcite	tr	3.0	1.0	..	0.3	..	1.6	..
Perthite	..	..	..	..	..	..	..	..
Sphene	..	..	..	..	..	..	..	..
Total	100.0	100.0	100.0	100.0	100.0	100.0	100.0	100.0
Grain size in mm	0.02–5.0	ND	ND	0.03–5.0	ND	ND	0.01–4.0	ND

	153+50	155+19	157+45	158+17	160+94	164+80*	181+86	181+86A
Quartz	65.9	29.6	42.4	30.2	41.7	49.6	38.3	34.3
Andesine (An$_{30}$–An$_{35}$)	10.6	32.9	23.2	25.9	17.9	1.0	35.6	46.0
Orthoclase	4.0	..	..	..	..	..	..	..
Microcline	5.3	16.9	18.6	..	6.4	..	..	..
Biotite	10.0	3.6	4.0	16.0	3.3	18.3	10.6	2.0
Muscovite	tr	8.3	3.6	14.0	3.6	..	..	..
Sericite†	1.3	3.6	4.0	7.3	11.7	..	12.6	16.3
Epidote	..	tr	..	tr	..	..	..	..
Chlorite	tr	1.3	2.6	2.0	1.0	..	0.4	..
Sillimanite	0.6	..	..	1.0	9.2	27.1	1.6	0.6
Almandine	..	tr	tr	..	..	..	0.6	..
Apatite	tr	..	..	..	..	..	..	..
Zircon	tr	..	tr	tr	tr	..	tr	tr
Opaque minerals	2.3	1.6	1.6	3.6	4.6	4.0	0.3	tr
Calcite	..	1.9	tr	..	0.6	..	tr	0.6
Perthite	..	0.3	..	..	..	..	..	..
Sphene	..	..	tr	tr	..	..	..	..
Total	100.0	100.0	100.0	100.0	100.0	100.0	100.0	99.8
Grain size in mm	ND	ND	ND	0.02–5.78	ND	ND	ND	ND

TABLE 5. (Continued)

	188+45	188+90	214+70*	235+00	247+90	248+85	261+08*	276+90*
Quartz	51.0	30.5	55.2	39.3	30.0	37.3	30.8	36.9
Andesine (An$_{30}$–An$_{35}$)	30.0	25.6	24.0	33.2	39.0	35.3	40.6	25.6
Orthoclase	1.2	tr	..	2.0	4.6	..	..	..
Microcline	6.0	5.6	..	14.0	20.0	..	5.0	15.6
Biotite	5.6	15.3	13.5	11.0	4.3	16.3	17.0	14.6
Muscovite	..	1.0	1.8	..	..	1.0	1.3	4.3
Sericite†	..	3.0	0.5	..	..	..	3.3	0.3
Epidote	..	..	..	..	..	..	..	1.6
Clinozoisite	..	..	..	..	..	..	tr	..
Chlorite	..	..	..	..	1.5	..	..	..
Sillimanite	1.4	..	..	..	..	10.1	..	tr
Almandine	3.3	..	..	..	..	..	..	..
Apatite	..	..	..	..	0.3	tr	..	..
Zircon	0.3	tr	tr	..	tr	..	tr	tr
Opaque minerals	1.2	1.0	1.5	0.3	0.3	tr	tr	..
Calcite	..	18.0	3.5	0.2	..	..	2.0	0.7
Sphene	..	..	..	..	..	..	..	0.4
Total	100.0	100.0	100.0	100.0	100.0	100.0	100.0	100.0
Grain size in mm	ND	ND	ND	ND	0.01-3	0.05-6.0	ND	ND

Note: ND = not determined.
*Spectrochemical analysis of the specimens are listed in Table 6.
†Including sericitized feldspar.

The calc-silicate granulite is green, gray, or brown; thin-bedded to thick bedded; fine grained to coarse grained; in part porphyroblastic, well bedded, and banded to massive; and in places contains variable amounts of diopside, clinozoisite, graphite, serpentine, epidote, plagioclase feldspar, biotite, garnet, talc, and pyrophyllite. The marble is mostly calcite—white to gray, green or pink—well-bedded, banded to massive, and in places contains quartz, biotite, epidote, garnet, diopside, forsterite, chondrodite, muscovite, graphite, and rhodochrosite. Figure 10 shows forsterite apparently replacing calcite. The marble unit and the calc-silicate unit may grade into each other within short distances.

The calc-silicate granulite and marble are compact and very fine grained to coarse grained. The modes of 7 samples are recorded in Table 7 and are summarized more generally in Figure 11. Calcite, in amounts that range from 7 to 77 percent, is present as euhedral to anhedral grains; some show lamellar twinning.

The following assemblages are recorded in these rocks:
calcite-forsterite-clinozoisite (195+70 and 195+75);
calcite-quartz-clinozoisite (165+30);
calcite-quartz-grossularite-clinozoisite (139+45);
calcite-quartz-diopside-clinozoisite (261+10);
calcite-quartz-diopside-grossularite-clinozoisite (111+35); and
calcite-diopside-forsterite-grossularite-clinozoisite (161+15).

The calc-silicate granulite samples include 111+35, 139+45, 161+15, 165+30, 261+10, and the marble samples include 195+70 and 195+75.

Source Rocks. The rocks of unit 3 and its subdivisions are characteristically enriched in lime and magnesia. These rocks are closely associated with metabasaltic rocks of units 4, as well as with rocks of units 1, and 2, which are more abundant in the Nashoba Formation than in the Marlboro Formation where unit 4 rocks predominate. Unit 3 was probably derived from lime, lime muds, and siliceous, feldspathic lime muds.

These relationships suggest that the sea water of the volcanic environment was enriched in lime and magnesia. Thus the sources of unit 3-1, enriched to some

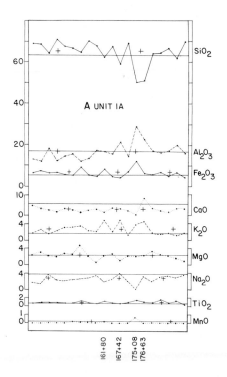
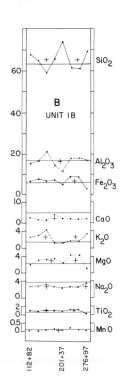
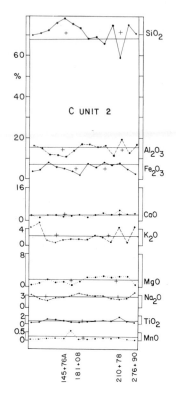
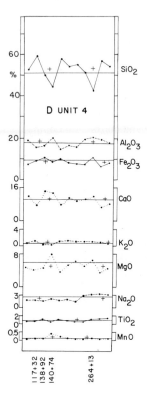
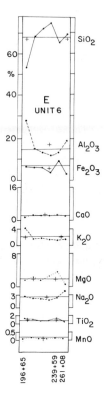

degree in alkalis and lime, may have been volcanogenic sediments and precipitates of calcium and magnesium compounds, probably from admixed calcareous sediments. The protolith of unit 3-2 was richer in lime, magnesia, and silica. The protolith of unit 3-3 was calcite limestone with variable but lesser amounts of silica, iron oxides, and alumina.

Unit 4: Hornblende Gneiss, Schist, and Amphibolite

Petrography. The amphibole schist, gneiss, and amphibolite are greenish black to black in color and in texture range from massive to well foliated. Locally, they contain abundant porphyroblasts.

The gneissic structure, where present, is due to very thin, irregular layers (1 to 1.5 mm) of felsic components alternating with hornblende-rich layers. In some places, the layers are folded, and quartz lenses lie in the plane of foliation. Some gneissic varieties contain layers of coticule a few millimeters to 5 cm thick. Modes of unit 4 rocks are recorded in Table 8, and a more general summary is presented in Figure 12.

The porphyroblasts are euhedral to anhedral and vary in size. They are mostly almandine, hornblende (as much as 5 cm), feldspar, and small pyrite cubes (as much as 3 mm).

Hornblende ranges from 22 to 73 percent; commonly it composes between 45 and 60 percent of the rock. It is euhedral, coarse grained, and occurs rarely in subhedral grains that contain inclusions of anhedral quartz, pyrite cubes, subhedral and anhedral zircon, biotite flakes, anhedral calcite, and in places anhedral clinozoisite. In most of the hornblende-rich rocks, the minerals have a preferred orientation. Samples 114+29, 117+79, and 201+85 are amphibolite. In many specimens, biotite, epidote, and clinozoisite enclose hornblende. In one specimen, the hornblende has lamellar twinning.

The plagioclase is andesine (An_{35-40}) and ranges from 5 to 42 percent. It is usually unaltered and shows excellent albite twinning. Plagioclase porphyroblasts occur in irregular shapes and are completely unaltered.

Quartz shows undulatory extinction and mosaic texture. It is most abundant in veins and, in places, forms inclusions in hornblende and plagioclase. Quartz ranges from 0 to 15 percent.

Biotite is found as small- to medium-size flakes, generally is dark brown, and strongly pleochroic golden or pale yellowish brown. Some red and green varieties occur in a few places. In most samples, biotite is on the borders of hornblende, associated with chlorite, clinozoisite, and magnetite, which probably indicates replacement of hornblende; it also occurs as inclusions in hornblende. The entire biotite content ranges from 0 to 35.5 percent but is generally less than 12 percent.

Figure 9A. Chemical compositions of 20 samples of porphyroblastic biotite-rich gneiss and schist of unit 1A compared with Nockold and Allen's (1954) average dacite and dacite-obsidian (solid line); average of the oxides of 20 samples of this study shown by crosses. Chemical compositions of 8 samples of nonporphyroblastic biotite-rich gneiss and schist of unit 1B compared with Nockold and Allen's (1954) average dacite and dacite-obsidian (solid line); average of the 8 samples of this study is shown by crosses. C. Chemical compositions of 14 samples of quartz-feldspar-biotite gneiss and schist of unit 2 compared with Tyrrell's (Pettijohn, 1957) average graywacke (solid line); average of the 14 samples of this study is shown by crosses. D. Chemical compositions of 11 samples of unit 4 hornblende-rich schist, gneiss, and amphibolite compared with Nockold and Allen's average central basalt. E. Chemical compositions of 6 samples of muscovite- and (or) chlorite-rich rocks of unit 6 compared with Tyrrell's (Pettijohn, 1957) average graywacke.

Muscovite, chlorite, epidote, and calcite are commonly present in minor amounts. Clinozoisite is abundant in two samples (114+39 and 138+92). These rocks grade into calc-silicate granulite.

Almandine, sphene, pyrite, magnetite, and zircon occur in small amounts. Sample 140+74, however, contains 29 percent almandine, and the rock grades into coticule.

TABLE 6. CHEMICAL COMPOSITION OF UNIT 2

	113+61*	118+95*	133+56*	145+76*	145+76A	175+95	181+08
SiO_2	69.84	70.61	71.90	75.45	77.51	74.85	73.07
TiO_2	0.30	0.47	1.24	1.00	0.71	0.55	0.44
Al_2O_3	15.87	14.35	11.62	11.26	10.47	13.35	16.24
Total Fe†	3.33	4.23	7.80	5.63	4.73	3.75	2.16
MnO	0.05	0.06	0.16	0.09	0.10	0.66	0.04
MgO	0.60	0.99	1.86	1.00	1.01	0.34	0.66
CaO	2.30	1.33	2.52	2.31	1.78	2.00	2.60
Na_2O	3.25	2.16	1.74	2.36	2.46	2.94	3.38
K_2O	4.44	5.78	1.18	0.88	1.20	1.56	1.42
Total	99.98	99.98	100.02	99.98	99.97	100.00	100.01
C.I.P.W. NORMS							
Quartz	29.03	31.70	34.56	50.70	52.46	46.48	43.26
Corundum	1.00	1.04	1.80	1.38	1.20	2.05	2.80
Orthoclase	24.94	32.10	6.36	4.73	6.41	8.40	7.76
Albite	27.75	18.20	14.28	19.39	20.11	24.14	28.29
Anorthite	10.86	6.15	22.83	10.48	8.01	9.06	11.73
Hypersthene	5.93	10.00	18.20	11.71	10.70	8.99	5.42
Ilmenite	0.49	0.76	1.97	1.59	1.10	0.86	0.69
Total	100.00	99.95	100.00	99.98	99.99	99.98	99.95
Johannsen no	227	227	227	223	223	223	223

	181+86*	200+91	201+93	208+13	210+78	214+70*	276+90*
SiO_2	67.94	68.40	65.30	74.00	58.70	74.69	70.05
TiO_2	0.74	0.86	0.93	0.77	1.70	0.72	0.27
Al_2O_3	16.17	15.09	15.50	11.00	18.76	12.15	16.15
Total Fe†	7.03	5.73	7.53	6.15	7.21	4.25	2.12
MnO	0.06	0.09	0.12	0.11	0.16	0.06	0.02
MgO	2.18	2.23	2.56	2.35	2.50	2.50	0.38
CaO	1.45	2.37	3.40	2.67	4.70	2.79	2.95
Na_2O	3.00	2.82	2.80	2.20	1.92	2.14	3.66
K_2O	1.37	2.40	1.90	0.88	4.32	0.70	4.40
Total	99.94	99.99	100.04	100.13	99.97	100.00	100.00
C.I.P.W. NORMS							
Quartz	37.61	37.52	26.91	46.00	14.80	49.08	26.06
Corundum	4.44	2.92	1.68	0.97	1.40	1.73	0.13
Orthoclase	7.30	7.67	10.42	4.71	24.90	3.75	23.92
Albite	24.59	23.44	23.26	17.90	16.70	17.42	30.95
Anorthite	6.55	10.89	15.63	12.00	22.52	12.55	13.80
Hypersthene	18.41	16.18	20.60	17.25	16.80	14.33	4.69
Ilmenite	1.09	1.38	1.50	1.21	2.88	1.14	0.45
Total	99.99	100.00	100.00	100.04	100.00	100.00	100.00
Johannsen no.	223	227	227	223	227	223	227

* Modal analysis of the specimens are listed in Table 5.
† As Fe_2O_3.

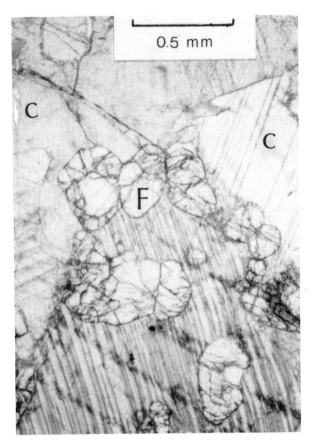

Figure 10. Photomicrograph of marble, unit 3-3 (195+75) showing subhedral crystals of forsterite (F) recrystallized in a calcite (C) groundmass.

Chemical Composition and Source Rocks. Eleven spectrochemical analyses and C.I.P.W. norms of the hornblende gneiss, schist, and amphibolite are recorded in Table 9. The chemical data are summarized in Figure 9D.

The hornblende-rich rocks of unit 4 compose about 50 percent of the Marlboro Formation but constitute only about 8.5 percent of the Nashoba Formation. They constitute a series of rocks whose composition is approximately that of the average of Nockold and Allen's (1954) central basalt (Fig. 9D). The hornblende schist is commonly quartz-rich, whereas the amphibolite and hornblende gneiss contain little or no quartz. These rocks do not show any clear relict textures except for flow(?) banding.

In unit 4, SiO_2 (Fig. 9D) has a mean of 53.29 weight percent. The average value for Al_2O_3 is essentially that of Nockold and Allen's (1954) central basalt (18.04 percent). Alumina generally varies inversely with respect to SiO_2, although this is not uniformly consistent. Sample 264+13, having a very low value of SiO_2, also shows relatively high values of Al_2O_3 (21.30 percent), Fe_2O_3 (10.65 percent), CaO (11.98 percent), and, to a limited degree, MgO (6.40 percent) by comparison with the mean central basalt. The mean value of total iron oxide in unit 4 (8.42 percent) is slightly less than the average value for central basalt of 9.73. The mean percentage values of K_2O (2.30), Na_2O (0.96), MgO (5.09), and CaO (10.19) are close to those of Nockold and Allen's (1954) average central basalts.

Samples 138+92, 140+74, and 264+13 have notably low SiO_2 (Fig. 9D), 50.57, 45.00, and 43.70, respectively. The latter two samples lie in the chemical domain of basalt (Nockold and Allen, 1954, p. 1021). They are sufficiently depleted in MgO as to be excluded from olivine-rich alkali basalt. They are also somewhat enriched in Al_2O_3, which possibly indicates enrichment in clay.

The hornblende schist, gneiss, and amphibolite of unit 4 consist of a variety of protoliths. The schist, dominantly of basaltic composition except that it commonly contains quartz in small to greater amounts, probably consisted of pyroclastics or volcaniclastic rocks with an added component of detrital quartz. The amphibolite was probably formed from basalt. The gneiss has compositions similar to the average of basaltic compositions and therefore may be derived from basalt and (or) basaltic pyroclastics or volcanoclastic rocks.

The low percentage of quartz and calcite and the association sphene-magnetite (Table 8) strongly suggest an igneous origin (Moorhouse, 1959, p. 459). Elemental concentrations of these rocks point strongly to the same conclusion. The predominance of CaO over MgO (Table 8) and the molecular ratio of Al_2O_3 to the sum of Na_2O, K_2O, and CaO, which is less than 1, are indications of igneous origin (Grout, 1932).

The norms of the hornblende gneiss, schist, and amphibolite (Table 9) have a basaltic composition according to Johannsen's classification (1931, 1938).

We therefore interpret the source rocks as having an average composition of central basalt, those found in association with the typical calc-alkali andesite, dacite, and rhyodacite at volcanic centers. The source rocks were probably basaltic flows, pyroclastics, and certain components of volcanogenic sediments derived from the effusive pile. These were deposited in a basin receiving sediments enriched in Al_2O_3 from a saprolitized terrain.

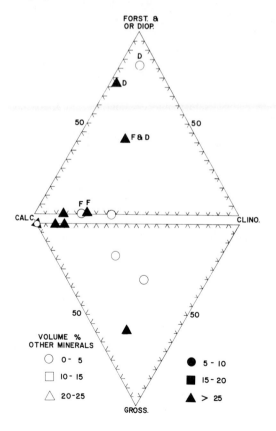

Figure 11. Summary of modes showing ratios of chief mineral components of 7 samples of calc-silicate granulite and marble of unit 3 on basis of data in Table 7. One hundred percent minus the total percentages of calcite, forsterite (F), diopside (D), grossularite, and clinozoisite equals the volume percentage of other minerals. Volume percentages of "other minerals" correspond to those of Figures 4A and 4B.

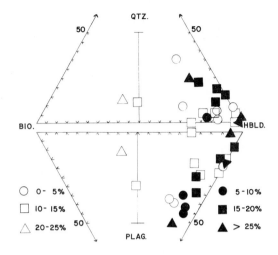

Figure 12. Summary of modes showing ratios of chief mineral components of 20 samples of hornblende gneiss, schist, and amphibolite of unit 4 on basis of data in Table 8. One hundred percent minus the total percentages of quartz, hornblende, plagioclase, and biotite equals the volume percentage of other minerals. Volume percentages of "other minerals" correspond to those of Figures 4A and 4B.

Unit 5: Quartzite and Quartz Schist

Petrography. Much of the quartzite and quartz schist of unit 5-1 is vitreous and made of quartz, which is estimated in hand specimen to range up to 95 percent. Many specimens contain fine-grained to medium-grained feldspar; it ranges from 5 to 30 percent.

More micaceous varieties of unit 5-2 were studied in thin sections. The modes of units 5 and 7 are presented in Table 12.

Their texture varies from granulose to slightly foliated. Quartz is the major constituent of these rocks and averages about 60 percent. It occurs as elongated, undulatory grains in mosaic habit. Andesine occurs as anhedral grains and shows undulatory extinction. Andesine makes up about 20 percent of the rock and shows excellent lamellar twinning, which in places is bent. Biotite occurs in amounts from 3 to about 10 percent. In one section it is associated with muscovite and

TABLE 7. MODES OF UNIT 3

	111+35	139+45	161+15	165+30	195+70	195+75	261+10
Diopside	74.3	..	8.5	..	..	..	43.0
Clinozoisite	9.3	30.4	5.0	3.2	tr	tr	1.6
Calcite	7.4	50.3	7.6	19.3	55.6	76.7	14.6
Biotite	..	tr	..	..	0.6	..	..
Muscovite	..	tr	..	2.0	19.0	..	0.3
Sericite*	..	..	..	..	1.5	..	..
Quartz	1.6	0.7	..	27.0	..	..	5.0
Plagioclase	tr	..	59.6	35.6	..	..	24.0
Chlorite	..	..	..	6.3	..	..	0.6
Sphene	..	1.6	1.1	2.6	..	..	0.6
Pyrite	..	tr	..	2.0	2.0	..	1.0
Opaque minerals	..	..	0.3	2.0	2.3	0.6	0.3
Zircon	..	..	tr	..	tr	..	..
Grossularite	7.4	17.0	17.6	..	..	..	..
Hornblende	..	..	..	..	..	..	9.0
Forsterite	..	..	0.3	..	19.0	22.7	..
Total	100.0	100.0	100.0	100.0	100.0	100.0	100.0
Grain size in mm	0.01–0.7	ND	0.01–0.9	0.03–10	ND	ND	ND

*Including sericitized feldspar.

sericite. It occurs as dark-brown and strongly pleochroic to pale-brown flakes (0.01 to 0.5 mm). Muscovite, which occurs in one section, is visible in hand specimen and grades into sericite. Epidote occurs in minor amounts and is an alteration product of biotite and andesine. Sillimanite, associated with biotite and magnetite, occurs in one thin section. Almandine (0 to 2 percent; up to 4.2 mm in length) occurs in subhedral fractured crystals and contains poikilitic quartz and magnetite with chlorite in fractures. Sillimanite (0 to 0.3 percent) occurs in thin prismatic crystals associated with biotite and magnetite; in one instance it replaces biotite. These rocks contain opaque minerals in small amounts. Calcite occurs as a secondary mineral along fractures.

Chemical Compositions and Source Rocks. One spectrochemical analysis of this type of rock is shown in Table 12.

TABLE 8. MODES OF UNIT 4

	114+29	114+39A	117+79	134+50	138+92*	140+09	140+74*	142+61*
Hornblende	57.2	55.6	71.2	21.7	53.7	58.1	43.2	45.2
Biotite	6.6	2.0	..	4.6	0.3	12.0	..	6.3
Andesine ($An_{35}-An_{40}$)	20.0	10.0	12.3	26.6	2.6	14.3	8.3	16.6
Quartz	4.0	13.3	2.3	8.6	1.0	4.0	..	15.0
Muscovite	..	0.3	4.6	2.0	1.6	..	6.0	..
Sericite†	..	0.3	1.0	5.0	6.3	10.3	6.9	1.3
Chlorite	3.0	1.3	..	..	4.4	..	tr	..
Epidote	7.6	..	..	..	..	..	6.3	..
Clinozoisite	..	14.0	4.6	5.0	25.6	tr	..	tr
Calcite	..	2.6	tr	2.6	3.6	tr	tr	..
Sphene	tr	..	2.0	..	0.6	0.3	..	..
Apatite	tr	..	tr	..	..	..	..	tr
Zircon	tr	tr	tr	tr	..	..	..	tr
Opaque minerals	1.6	0.6	2.0	3.3	0.3	1.0	0.5	4.0
Sillimanite	..	..	..	6.6	..	..	..	..
Almandine	..	..	tr	14.0	..	..	28.8	11.6
Total	100.0	100.0	100.0	100.0	100.0	100.0	100.0	100.0
Texture	g	G	g	P,G	G,P	S	G	G
Grain size in mm	0.06-1.98	0.123-0.83	0.032-73	0.033-0.8	0.001-0.8	0.07-1.85	0.001-2.5	0.07-2.64

	169+24	194+57	201+85*	226+55	252+70*	260+13*	268+95*	271+38*
Hornblende	66.9	65.3	63.4	72.8	46.3	49.2	50.5	41.6
Biotite	21.3	tr	tr	tr	5.3	6.0	10.0	15.0
Andesine ($An_{35}-An_{40}$)	..	18.6	5.4	16.3	41.6	39.3	32.0	33.6
Quartz	..	3.3	11.3	6.3	3.0	1.0	1.3	5.6
Muscovite	..	..	3.4	..	..	..	..	..
Sericite†	..	3.6	4.6	..	..	1.6	..	1.6
Chlorite	4.6	1.0	1.6	..	1.0	..	..	..
Epidote	..	..	9.0	..	tr	..	tr	..
Clinozoisite	..	2.6	tr	..	0.6	tr	2.9	tr
Calcite	tr	0.3	tr	..	0.6	..	..	tr
Sphene	3.6	5.0	..	tr	..	..	..	tr
Apatite	..	..	..	..	..	0.3	..	tr
Zircon	1.3	..	tr	tr	tr	..	..	tr
Opaque minerals	tr	0.3	1.3	4.6	1.6	2.6	3.3	2.6
Sillimanite	..	..	..	..	..	..	..	tr
Almandine	2.3	..	..	..	..	..	..	..
Total	100.0	100.0	100.0	100.0	100.0	100.0	100.0	100.0
Texture	S,P	S	g	G	S	G	P,S	G
Grain size in mm	0.07-2.8	0.03-3.33	0.01-3.3	0.01-2.8	0.1-2.3	0.1-2.3	0.01-4.3	0.01-3.5

Note: g = granular; G = gneissic; P = porphyroblastic; S = schistose.
*Spectrochemical analysis of the specimens are listed in Table 4.
†Including sericitized feldspar.

TABLE 9. CHEMICAL COMPOSITIONS OF UNIT 4

	114+39A*	117+32	138+92*	140+74*	142+61*	194+57	201+85*	252+70*	264+13*	268+99*	271+38*
SiO_2	53.33	59.99	50.57	45.00	58.36	54.80	55.88	52.09	43.70	57.50	55.02
TiO_2	0.83	0.70	1.21	0.59	1.20	0.98	0.87	0.74	1.38	1.50	1.76
Al_2O_3	19.39	16.03	16.86	21.63	15.56	17.13	16.62	20.07	21.30	20.08	18.34
Total Fe†	7.37	8.90	10.45	8.40	9.93	8.05	7.60	7.02	10.65	6.20	8.07
MnO	0.18	0.13	0.14	0.52	0.28	0.22	0.17	0.12	0.18	0.10	0.18
MgO	4.89	4.03	4.63	8.00	3.89	5.14	6.10	4.98	6.40	3.55	4.40
CaO	11.41	7.31	14.16	13.45	7.78	10.45	9.90	10.73	11.98	6.70	8.25
Na_2O	1.74	1.69	1.50	2.08	1.78	1.88	1.66	3.14	3.34	3.35	3.17
K_2O	0.85	1.22	0.45	0.35	1.20	1.35	1.20	1.10	1.07	1.00	0.81
Total	99.99	100.01	99.97	100.02	99.98	100.00	100.03	100.01	100.00	99.98	100.00

C.I.P.W. NORMS

	114+39A*	117+32	138+92*	140+74*	142+61*	194+57	201+85*	252+70*	264+13*	268+99*	271+38*
Quartz	9.89	18.98	2.02	..	15.95	6.00	10.65	..	..	7.22	5.68
Corundum	..	..	..	..	..	..	..	..	..	0.28	..
Orthoclase	4.67	6.67	2.45	1.90	6.52	7.26	5.55	5.49	5.92	8.20	4.44
Albite	8.35	13.97	12.36	5.78	14.74	15.41	11.71	23.77	1.66	29.56	26.39
Anorthite	42.80	30.04	34.85	44.91	28.55	31.33	26.98	31.51	37.26	30.80	31.15
Leucite	..	..	..	5.71	..	..	..	..	18.16	..	..
Nepheline	..	..	..	16.66	..	..	..	..	13.15	..	6.84
Diopside	9.95	2.59	29.66	..	7.06	18.26	11.58	13.34	..	..	..
Wollastonite	..	..	..	..	..	..	..	..	..	..	..
Hypersthene	23.02	26.65	16.73	..	25.26	20.21	32.35	24.18	21.62	21.53	22.65
Olivine	..	..	..	24.09	..	..	..	0.63	..	..	..
Enstatite	..	..	..	..	..	..	..	..	..	..	..
Ilmenite	1.35	1.12	1.93	0.95	1.90	1.55	1.19	1.08	2.23	2.41	2.84
Total	100.03	100.01	100.00	100.00	99.98	100.02	100.01	100.00	100.00	100.00	99.99
Johannsen no.	238	237	2312	2312	237	2311	237	2312	2319	2311	2312

*Modal analysis of the specimens are listed in Table 7.
†As Fe_2O_3.

Quartzite and schistose quartzite of unit 5 were probably derived from quartz sand, quartz and feldspar sand, pebble-bearing sand, chert, and clay-bearing quartz sand and (or) chert. Unit 5 makes up about 6 percent of the Marlboro Formation, about 1 percent of the Nashoba, and 22 percent of the unnamed formations.

The greater part of unit 5 rocks of the Marlboro Formation probably was derived dominantly from chert with lesser quartz sand, relative to the Nashoba Formation. The lesser amounts of chert in the Nashba relates to less intense volcanism in a marine environment by comparison with the Marlboro Formation. The protolith of unit 5 in the unnamed formations is concentrated largely in the basal part of the section. The thicker sections of unit 5, such as U-1, U-6, and U-7, are strand or nearshore deposits, whereas those in which the clay-rich U-15 is more abundant were deeper basin deposits (Skehan and Abu-moustafa, 1976).

Unit 6: Muscovite- and (or) Chlorite-rich Schist and Gneiss

Petrography. The muscovite- and (or) chlorite-rich schist and gneiss are very well foliated with rounded to subrounded megacrysts of plagioclase and (or) microcline and relict fragmental texture. The modes of 16 samples are recorded in Table 10 and are summarized more generally in Figure 13.

Quartz occurs in thin, deformed layers as equidimensional fresh grains in mosaic and as long, deformed, undulatory grains with the long dimension in the foliation planes. The grain size varies from 0.05 to 1 mm. It ranges from 1 to 56 percent.

Plagioclase (An_{30-35}) occurs in the groundmass and as tiny megacrysts (0.6 to 2 mm), highly altered in places to sericite or saussurite; it shows poikilitic quartz, biotite, and magnetite when unaltered. Some altered plagioclase megacrysts are rounded.

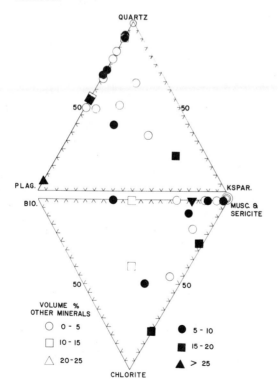

Figure 13. Summary of modes showing ratios of chief mineral components of 18 samples of muscovite- and (or) chlorite-rich schist and gneiss of unit 6 on basis of data in Table 10. One hundred percent minus the total percentages of quartz, plagioclase, potash feldspar, biotite, chlorite, and muscovite + sericite equals the volume percentage of other minerals. Volume percentages of "other minerals" correspond to those of Figures 4A and 4B.

TABLE 10. MODES OF UNIT 6

	112+31	130+05	132+77	134+27	135+21	141+38	146+74	148+08
Quartz	34.5	40.2	32.1	1.4	37.9	21.3	55.8	52.4
Andesine (An_{30}–An_{35})	21.3	4.3	15.0	23.0	14.6	17.3	tr	4.0
Microcline	11.6	..	..	..	..	..	..	..
Biotite	3.3	22.6	7.0	6.0	0.3	14.3	tr	4.0
Muscovite	23.0	20.3	24.0	16.0	20.3	7.6	43.6	33.3
Sericite†	2.0	2.3	9.3	11.0	19.3	7.3	..	..
Epidote	..	tr	tr	..	..	..	0.3	..
Chlorite	..	tr	3.3	..	..	19.0	..	..
Sillimanite	tr	..	..	24.0	..	..	..	..
Almandine	..	..	..	13.0	..	8.3	..	..
Apatite	..	..	..	..	..	..	..	..
Zircon	..	tr	tr	1.0	..	0.3	..	..
Opaque minerals	1.3	4.0	6.3	4.6	4.6	1.6	0.3	1.3
Calcite	3.0	6.3	3.0	..	3.0	3.0	..	5.0
Sphene	..	..	..	..	..	..	..	..
Total	100.0	100.0	100.0	100.0	100.0	100.0	100.0	100.0
Grain size in mm	0.1-2.8	ND	ND	0.04-4.8	0.03-4.29	ND	ND	ND

	155+98	178+80	196+65*	201+10*	203+53*	204+12	206+76	239+59
Quartz	11.7	11.3	26.4	24.7	49.3	54.1	41.3	29.7
Andesine (An_{30}–An_{35})	9.0	9.6	5.6	25.6	13.7	13.7	38.0	29.6
Microcline	34.6	..	tr	12.7	..	16.3	5.1	..
Biotite	..	1.1	6.2	5.7	9.3	..	..	2.6
Muscovite	5.3	17.0	6.3	8.4	25.3	13.0	15.6	17.6
Sericite†	0.6	29.0	42.3	1.7	0.7	tr	..	..
Epidote	..	..	..	..	..	..	..	0.6
Chlorite	21.0	16.0	11.3	15.3	..	1.3	..	16.6
Sillimanite	..	tr	..	..	..	..	..	..
Almandine	..	15.0	..	..	..	..	..	..
Apatite	..	..	..	..	..	..	..	..
Zircon	..	tr	tr	tr	tr	..	..	tr
Opaque minerals	8.6	1.0	1.3	4.6	0.7	0.3	..	3.0
Calcite	6.6	..	0.6	1.3	..	1.3	..	0.3
Sphene	2.6	..	..	..	..	..	..	..
Hornblende	..	..	..	..	1.0	..	..	..
Total	100.0	100.0	100.0	100.0	100.0	100.0	100.0	100.0
Grain size in mm	ND	ND	ND	ND	ND	ND	ND	ND

Note: ND = not determined.
*Spectrochemical analysis of the specimens are listed in Table 12.
†Including sericitized feldspar.

TABLE 11. CHEMICAL COMPOSITIONS OF UNIT 6

	196+65*	201+10*	203+53	205+14	239+59	261+08*
SiO_2	53.27	68.26	71.82	74.90	65.40	69.53
TiO_2	1.17	1.02	0.66	0.83	1.36	0.39
Al_2O_3	28.83	15.59	13.57	12.12	13.62	19.31
Total Fe†	6.54	6.23	6.17	3.88	9.40	2.32
MnO	0.09	0.17	0.09	0.07	0.14	0.05
MgO	1.46	2.29	1.52	2.54	3.54	0.46
CaO	1.85	2.50	2.31	2.22	2.75	2.26
Na_2O	2.68	2.24	2.10	1.98	2.41	4.02
K_2O	4.04	1.67	1.76	1.49	1.35	1.66
Total	99.93	99.97	100.00	100.03	99.97	100.00

*Modal analyses of the specimens are listed in Table 11.
†As Fe_2O_3.

Microcline occurs in eight samples, always as porphyroblasts (0.1 to 2.5 mm) and usually unaltered but with fractures that are apparently the result of postmetamorphic deformation. The microcline shows poor cross-hatching, and the rims are altered to sericite.

Biotite occurs in most samples associated with muscovite and (or) chlorite and sillimanite. It occurs as thin flakes, strongly pleochroic from dark brown to pale brown and is highly chloritized in places.

Muscovite occurs in amounts from 16 to 44 percent in all samples except those that contain chlorite, where it is present in amounts ranging from 5 to 24 percent. In many places, it appears to result from the recrystallization of the sericitized part of the plagioclase feldspar. It occurs in layers with the long direction in the foliation plane. Clusters of muscovite, however, occur in places.

Sericite, epidote, and chlorite appear as alteration products of plagioclase feldspar, biotite, and garnet. In places, chlorite contains many poikilitic almandine crystals that occur as fresh porphyroblasts (0.3 to 7 mm), which may be unfractured to moderately fractured; it occurs in places in major amounts (21 percent in 155+98).

TABLE 12. MODES OF TWO SAMPLES AND CHEMICAL COMPOSITION OF ONE SAMPLE OF UNIT 5 ROCKS: MODES OF TWO SAMPLES OF UNIT 7 ROCKS; CHEMICAL COMPOSITION OF ONE SAMPLE OF COTICULE OF UNIT 7

Modes of unit 5	148+78	201+52
Quartz	56.9	61.1
Andesine ($An_{30}-An_{35}$)	20.3	19.6
Biotite	10.3	3.1
Muscovite	..	9.6
Sericite*	9.3	4.6
Epidote	0.3	0.6
Sillimanite	0.3	..
Almandine	2.0	..
Zircon	tr	..
Opaque minerals	0.6	..
Calcite	..	1.4
Total	100.0	100.0
Grain size in mm	0.01-4.2	0.01-3.2

*Including sericitized feldspar

Modes of unit 7	143+03	227+29
Quartz	37.3	1.3
Almandine*	52.6	80.3
Biotite	3.0	6.3
Epidote	2.6	0.3
Plagioclase	1.6	9.6
Calcite	tr	..
Apatite	0.3	..
Opaque minerals	2.6	1.6
Muscovite	..	0.6
Total	100.0	100.0

*Deep reddish brown, fuses to magnetic globules, n = 1.90, unit cell dimension = 1.64_A^o, total iron = 20.00 ± 2%, and total manganese = 6.28 ± 2% (Fe and Mn percents are determined by atomic absorption method).

Chemical composition of unit 5	148+78*
SiO_2	79.96
TiO_2	0.74
Al_2O_3	8.41
Total Fe*	4.67
MnO	0.10
MgO	0.79
CaO	1.89
Na_2O	2.87
K_2O	0.53
Total	99.96

*As Fe_2O_3.

Chemical composition of coticule of unit 7	227+29
SiO_2	44.26
TiO_2	0.52
Al_2O_3	15.67
Total Fe*	31.60
MgO	0.90
CaO	1.95
Na_2O	2.40
K_2O	2.70
Total	100.00

Note: Total manganese content in the garnet = 8.26 ± 2%.
*As Fe_2O_3.

In this sample, magnetite occurs in major proportions in large anhedral grains associated with garnet, muscovite, and biotite.

Sillimanite occurs in thin fibers oriented along the foliation plane. It is associated with biotite and muscovite and may partly replace them.

Zircon and apatite occur in minor amounts as anhedral crystals, enclosed mostly in biotite and muscovite.

Chemical Composition and Source Rocks. Six spectrochemical analyses of this type rock are recorded in Table 11 and are summarized in Figure 9E.

The schist of unit 6 shows variations in composition similar to that of units 1A, 1B, and 2 and, therefore, is believed to be derived from similar sources. Sample 196+65 shows low SiO_2 (53.27 percent), high Al_2O_3 (28.83 percent), and relatively high K_2O (4.04 percent), which indicates a potash-rich clay source similar in some respects to illite described by Whitehouse and McCarter (1958, Table 5, p. 97).

The SiO_2 and Al_2O_3 have mean values of 67.2 and 17.17 percent, respectively. As regards these oxides, unit 6 in the aggregate compares favorably with Tyrrell's average graywacke (in Pettijohn, 1957). In general lower SiO_2 content relates to higher Al_2O_3 content.

These rocks with an average iron oxide content of 5.76 percent are somewhat depleted in iron by comparison with Tyrrell's average graywacke, which has a value of 7.17 percent.

The rocks of unit 6 are thus similar to rocks of units 1A, 1B, and 2 in that they have components derived from weathered, lateritic source rock as well as components that have been relatively unaltered. The enrichment in sheet silicates in these rocks probably indicates that they were clay-rich sediment.

A comparison of the average dacite of Figures 9A and 9B with the Tyrrell (in Pettijohn, 1957) average graywacke of Figures 9C and 9E shows the variation with rocks of our study. A useful comparison for certain oxides results from superimposing the average for dacite of Figures 9A and 9B on Figures 9C and 9E and vice versa.

Rocks of unit 6 compose nearly 9 percent of the Nashoba Formation but are only a minor component of the Marlboro Formation and the unnamed formations. They may have accumulated on the submarine flanks of island volcanoes or in offshore basins near basalt-andesite-dacite-rhyolite volcanic centers.

Unit 7: Coticule

Petrography. The coticule rocks are composed essentially of manganese-rich almandine and quartz (Fig. 14). These distinctive rocks are gray to pink in color, thinly laminated, and cherty. The garnet occurs in euhedral crystals (0.001 to 1 mm), which may be slightly anisotropic when found in a matrix of sutured quartz and plagioclase feldspar. Two modes of this rock type are shown in Table 12.

Dark-brown biotite, strongly pleochroic to light brown, occurs in prophyroblasts (2 mm) parallel to the layers. Epidote appears as an alteration product of plagioclase feldspar. Muscovite, calcite, magnetite, and apatite occur in minor amounts.

Chemical Composition and Source Rocks. One spectrochemical analysis of this rock is shown in Table 12.

The presence of microbanding (Fig. 14) and the development of persistent layers enclosed between amphibole schist layers probably indicate sedimentary origin. Typically, the coticule is flinty and is dominantly composed of quartz and microscopic manganese-rich almandine garnet. The coticule may have formed as a direct precipitate of aluminum-silicon-iron-manganese carbonate gel.

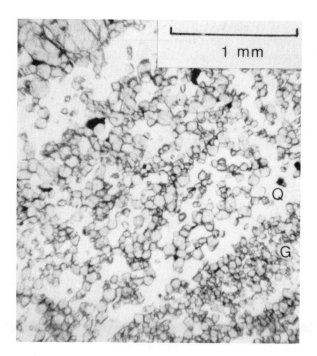

Figure 14. Photomicrograph of coticule, unit 7 (142+61) showing its occurrence as layers of quartz (Q) and garnet (G) in amphibolite indicated by hornblende (H).

The mode of sample 143+03 (Table 12) is more typical of the average coticule, although that of sample 227+29 and its chemical composition underscore how garnet-rich some of the layers may be. The high SiO_2, Al_2O_3, and Fe_2O_3 content and relative depletion in other oxides by comparison suggest, alternatively, that the coticule may possibly have formed by chemical weathering and leaching of volcanic glass, presumably ranging in composition from rhyolite to basalt.

Conditions of Metamorphism. The mineral assemblages in the metamorphic rock sequence of the Nashoba Formation indicate a condition corresponding to the almandine-amphibolite facies of the Barrovian type (see Tables 1, 2, 5, 7, 8, 10, 12). The mineral assemblage, quartz-andesine-biotite-muscovite-sillimanite-orthoclase-almandine, is reported eleven times throughout the samples studied. Texturally, sillimanite appears younger than orthoclase feldspar (Fig. 5). This clearly corresponds to the sillimanite-potash feldspar isograd and therefore indicates a metamorphic condition corresponding to the Barrovian sillimanite-almandine-orthoclase subfacies of the almandine-amphibolite facies (Winkler, 1967). This conclusion is supported by the fact that some of the coarse porphyroblastic varieties of the biotite-feldspar-quartz gneiss and schist are migmatitic and are associated with small to large pegmatite bodies, the presence of which suggests that this grade of metamorphism has been achieved (Thompson and Norton, 1968).

The presence of pegmatite and migmatite probably indicates the beginning of melting. The minimum melting curve of Luth and others (1964) for aqueous granitic systems superimposed on the stability field of Al_2O_5 polymorphs of Richardson and others (1969) intersects the sillimanite field near its most restricted portion. This corresponds to a temperature of about 650°C at a fluid pressure of 5 to 6 kb. Conditions of metamorphism are such that partial melting could have taken place only where the activity of water was approximately equal to 1. The presence of clinozoisite and grossularite in the calc-silicate rocks of unit 3 (Table 7; Fig.

11) requires a rich aqueous fluid phase for crystallization in the higher grades of metamorphism, a condition that prevailed here (Kerrick, 1974).

In some thin sections of the quartz-feldspar-mica gneiss, muscovite is lacking, and the mineral assemblage is quartz-andesine-biotite ± sillimanite ± orthoclase ± almandine ± microcline ± clinozoisite. In others, biotite is lacking, and the mineral assemblage is quartz-andesine-muscovite ± microcline ± sillimanite ± garnet.

The coexistence of sillimanite and clinozoisite in the rocks of this study provides close constraints because clinozoisite, at about 5 kb, is stable only 25° to 35°C into the sillimanite field. This is, of course, only the case if the pistacite (iron) component of the epidote family minerals is less than 25 percent (Liou, 1973), which is probably the case in this instance. Even if the pistacite component of the clinozoisite was greater than 25 percent, which would suggest a higher temperature of crystallization, the presence of plagioclase in the system would have the effect of depressing the temperature. Thus any possible errors in our estimate of the amount of pistacite would give a temperature about equal to or possibly slightly lower than that calculated.

Taking into account the stability field relations of Richardson and others (1969) and those of Holdaway (1971), we found that the rocks of the present study were recrystallized in the range of 625° to 675°C at 5 to 6 kb. It is probable that the more precise value is 625° to 650°C at 6 kb (Daniel Murray, 1975, oral commun.). This pressure, assuming $P_{load} = P_{H_2O}$, implies an unusually thick pile more than 20 km in thickness.

ACKNOWLEDGMENTS

This study was partially supported by National Science Foundation Grant GP 4969.

We are indebted to W. W. Correia and D. L. Guernsey of the Starnd Laboratory of the Massachusetts Institute of Technology for the use of the spectrographic laboratory and for their valuable assistance and advice. We express our gratitude to W. H. Dennen for the use of the Cabot Laboratory of the Massachusetts Institute of Technology and for the preparation of samples for spectrochemical analysis.

Abu-moustafa expresses his appreciation to C. W. Wolfe who supervised the original study embodied in his 1969 Boston University Ph.D. thesis; he also made several improvements in the present paper. M. A. Gheith gave advice and critically read the thesis.

Skehan is grateful to M. P. Billings who suggested the study and made the arrangements with the Metropolitan District Commission (MDC). He acknowledges insights, discussions, and suggestions of M. P. Billings, L. R. Page, Daniel Murray, P. C. Lyons, and, in particular, members of the U.S. Geological Survey and former students. Officials and other employees of the MDC assisted in making the fieldwork possible. The manuscript was reviewed by P. J. Barosh, P. C. Lyons, L. R. Page, and J. H. Peck. We are grateful for their helpful comments.

REFERENCES CITED

Abu-moustafa, A. A., 1969, Petrography and geochemistry of the Nashoba Formation from the Wachusett-Marlborough Tunnel, Massachusetts [Ph.D. thesis]: Boston, Mass., Boston Univ., 223 p.

Abu-moustafa, A. A., and Skehan, J. W., S. J., 1970, Petrography and geochemistry of the Nashoba Formation of the Wachusett-Marlborough Tunnel, Massachusetts: Geol. Soc. America Abs. with Programs v. 2, no. 1, p. 9.

Bell, K. G., 1967, Faults in eastern Massachusetts: Geol. Soc. America Abs. for 1967, Spec. Paper 115, p. 250.

Bell, K. G., and Alvord, D. C., 1974, Geologic sketch map of northeastern Massachusetts: U.S. Geological Survey Open-File Map, no. 74-356, scale 1:250,000.

——1976, Pre-Silurian stratigraphy of northeastern Massachusetts, in Page, L. R., ed., Contributions to the stratigraphy of New England: Geol. Soc. America Mem. 148 (in press).

Billings, M. P., 1976, Geology of the Boston basin, in Lyons, P. C., and Brownlow, A. H., eds., Studies in New England geology: Geol. Soc. America Mem. 146, p. 5-30 (this volume).

Clifford, T. N., 1960, Spessartine and magnesium biotite in coticule-bearing rocks from Mill Hollow, Alstead Township, New Hampshire, U.S.A.—A contribution to the petrology of metamorphosed manganiferous sediments: Neues Jahrb. Mineralogie Abh., Bd. 94, 2 Hälfte, p. 1371-1372.

Crosby, W. O., 1899, Geology of the Wachusett Dam and Wachusett Aqueduct Tunnel of the Metropolitan Water Works in the vicinity of Clinton, Massachusetts: Tech. Quart., v. 12, p. 68-96.

Daly, R. A., 1933, Igneous rocks and the depths of the Earth: New York, McGraw-Hill Book Co., 598 p.

Dennen, W. H., and Fowler, W. C., 1955, Spectrographic analysis by use of mutual standard method: Geol. Soc. America Bull., v. 66, p. 655-662.

Emerson, B. K., 1898, Geology of Old Hampshire County, Massachusetts: U.S. Geol. Survey Monograph 29, 790 p.

——1917, Geology of Massachusetts and Rhode Island: U.S. Geol. Survey Bull. 597, 289 p.

Fahlquist, Frank, 1935, Geology of region in which Quabbin Aqueduct and Quabbin Reservoir are located: Appendix to report of Chief Engineer: Ann. Rept. Metro. District Water Supply Comm., Commonwealth of Massachusetts, Public Doc. No. 147, p. 1-46.

Fischer, W. L., 1964, A magnetic study of the Wachusett-Marlborough Tunnel area [M.S. thesis]: Boston, Mass., Boston College, 74 p.

Gore, R. Z., 1967, An evaluation of the magnetometer in metamorphic terrain [M.S. thesis]: Boston, Mass., Boston College, 69 p.

——1976, Ayer Crystalline Complex at Ayer, Harvard, and Clinton, Massachusetts, in Lyons, P. C., and Brownlow A. H., eds., Studies in New England geology: Geol. Soc. America Mem. 146, p. 103-124 (this volume).

Grew, E. S., 1973, Stratigraphy of the Pennsylvanian and pre-Pennsylvanian rocks of the Worcester Area, Massachusetts: Am. Jour. Sci., v. 273, p. 113-129.

Grew, E. S., Mamay, S. H., and Barghoorn, E. S., 1970, Age of plant fossils, the Worcester coal mine, Worcester, Massachusetts: Am. Jour. Sci., v. 268, p. 113-126.

Grout, F. F., 1932, Petrography and petrology, a textbook (1st ed.): New York, McGraw-Hill Book Co., 522 p.

Hansen, W. R., 1956, Geology and mineral resources of the Hudson and Maynard quadrangles, Massachusetts: U.S. Geol. Survey Bull. 1038, 104 p.

Heath, M. M. M., 1965, Geochronology of the Ayer granite in the Wachusett-Marlborough Tunnel, Clinton, Massachusetts, in Variations in isotopic abundance of strontium, calcium and argon and related topics: U.S. Atomic Energy Commission Contract AT (30-1)-1381, Mass. Inst. Tech., p. 15-19.

Holdaway, M. J., 1971, Stability of andalusite and the aluminum silicate phase diagram: Am. Jour. Sci., v. 271, p. 97-131.

Ingamells, C. O., and Suhr. N. H., 1963, Chemical and spectrochemical analysis of standard silicate samples: Geochim. et Cosmochim. Acta, v. 27, p. 897-910.

Jahns, R. H., 1952, Stratigraphy and sequence of igneous rocks in the Lowell-Ayer region, Massachusetts: Geol. Soc. America, Guidebook for field trips in New England, 1952 Ann. Mtg., Boston, Mass., p. 108-112.

Johannsen, A., 1931, A quantitative mineralogical classification of igneous rocks [rev. ed.]: Jour. Geology, v. 28, p. 36-60, 159-177, 210-232.

——1938, A descriptive petrography of the igneous rocks, Vol. 1: Introduction, textures, classification and glossary: Chicago, Ill., Univ. Chicago Press, 318 p.

Kerrick, D. M., 1974, Review of metamorphic mixed volatile (H_2O-CO_2) equilibria: Am. Mineralogist, v. 59, no. 7-8, p. 729-762.

Lacroix, A. V., 1968, Structure and contact relationships of the Marlboro Formation, Marlboro, Massachusetts [M.S. thesis]: Boston, Mass., Boston College, 83 p.

Larsen, E., and Miller, F., 1935, The rosiwal method and the modal analysis of rocks: Mineralog. Soc. America Jour., v. 20, p. 260-273.

Legarde, C. N., III, 1967, The petrography, structural geology and tectonic history of a portion of the Wachusett-Marlborough Tunnel area, Massachusetts [M.S. thesis]: Boston, Mass., Boston College, 100 p.

Liou, J. G., 1973, Synthesis and stability relations of Epidote: Jour. Petrology, v. 14, no. 3, p. 381-413.

Luth, W. C., Jahns, R. H., and Tuttle, O. F., 1964, The granite system at pressures of 4 to 7 kilobars: Jour. Geophys. Research, v. 69, no. 4, p. 759-773.

Moorhouse, W. W., 1959, The study of rocks in thin sections: New York, Harper and Bros., 514 p.

Nockold, S. R., and Allen, R., 1954, Average chemical composition of some igneous rocks: Geol. Soc. America Bull., v. 65, p. 1007-1032.

Peck, J. H., 1976, Silurian and Devonian stratigraphy in the Clinton quadrangle, central Massachusetts, in Page, L. R., ed., Contributions to the stratigraphy of New England: Geol. Soc. America Mem. 148 (in press).

Pettijohn, F. J., 1957, Sedimentary rocks: New York, Harper and Bros., 519 p.

Renard, A., 1878, Sur la structure et la composition mineralogique du coticule: Acad. Royale Belgique Mem. Sav. Etr., tome XLI.

Richardson, S. W., Gilbert, M. C., and Bell, P. M., 1969, Experimental determination of kyanite-andalusite and andalusite-sillimanite equilibria: The aluminum silicate triple point: Am. Jour. Sci., v. 267, p. 259-272.

Skehan, J. W., S. J., 1967a, Geology of the Wachusett-Marlborough Tunnel, east central Massachusetts, in Farquhar, O. C., ed., Economic geology in Massachusetts: Massachusetts Univ. Graduate School Pub., p. 237-244.

——1967b, Evidence of easterly tectonic transport in the Appalachians of eastern Massachusetts [abs.]: Geol. Soc. America Abs. for 1967, Spec. Paper 115, p. 56.

——1968a, Recognition of faults of regional importance by mapping of joints [abs.]: Geol. Soc. America Abs. for 1968, Spec. Paper 121, p. 376.

——1968b, Fracture tectonics of southeastern New England as illustrated by the Wachusett-Marlborough Tunnel, east central Massachusetts, in Zen, E-an, ed., Studies of Appalachian geology, northern and maritime: New York, John Wiley & Sons, Inc., p. 281-290.

——1969, Tectonic framework of southern New England and eastern New York, in Kay, Marshall, ed., North Atlantic—Geology and continental drift: Am. Assoc. Petroleum Geologists Mem. 12, p. 793-814.

Skehan, J. W., S. J., and Abu-moustafa, A., 1976, Stratigraphic analysis of rocks exposed in the Wachusett-Malborough Tunnel, east-central Massachusetts, in Page, L. R., ed., Contributions to the stratigraphy of New England: Geol. Soc. America Mem. 148 (in press).

Thompson, J. B., Jr., and Norton, S. A., 1968, Paleozoic regional metamorphism in New England and adjacent areas, in Zen, E-an, ed., Studies of Appalachian geology, northern

and maritime: New York, John Wiley & Sons, Inc., p. 319-326.
Toulmin, Priestley, III, 1964, Bedrock geology of the Salem quadrangle: U.S. Geol. Survey Bull. 1163A, 79 p.
Whitehouse, U. G., and McCarter, R. S., 1958, Diagenetic modification of clay mineral types in artificial sea water, in Swineford, Ada, ed., Clays and clay minerals: Fifth Natl. Conf., Clays and Clay Minerals, Natl. Acad. Sci—Natl. Research Council Pub. 566, p. 81-119.
Winkler, H.G.F., 1967, Petrogenesis of metamorphic rocks (2nd ed.): (Trans. by N. D. Chaterjee and E. Froese) New York, Springer-Verlag, 237 p.
Zartman, R. E., and Marvin, R. F., 1971, Radiometric age (Late Ordovician) of the Quincy, Cape Anne, and Peabody Granites from eastern Massachusetts: Geol. Soc. America Bull., v. 82, p. 937-958.

MANUSCRIPT RECEIVED BY THE SOCIETY JULY 8, 1974
REVISED MANUSCRIPT RECEIVED FEBRUARY 10, 1975
MANUSCRIPT ACCEPTED APRIL 2, 1975

Printed in U.S.A.

Geological Society of America
Memoir 146
© 1976

Petrology, Chemistry, and Age of the Rattlesnake Pluton and Implications for Other Alkalic Granite Plutons of Southern New England

Paul C. Lyons*
Division of Science, College of Basic Studies
Boston University
Boston, Massachusetts 02215

AND

Harold W. Krueger
Geochron Laboratories Division
Krueger Enterprises, Inc.
Cambridge, Massachusetts 02139

ABSTRACT

The Rattlesnake pluton of Sharon, Massachusetts, is an elliptical alkalic granite stock composed principally of coarse biotite granite and fine riebeckite granite; minor amounts of granite porphyry, riebeckite-biotite granite, and coarse riebeckite granite; and traces of trachyte and ferrohastingsite granite.

Major-element chemical analyses are reported for the coarse biotite granite, fine riebeckite granite, trachyte, biotite from the coarse biotite granite, and four riebeckite samples (including one from Rhode Island). Sodium and potassium analyses are reported for 11 samples of alkali feldspar, including three from the Quincy pluton.

Feldspar from the Rattlesnake pluton shows a trend in Or enrichment, reflecting a decrease in temperature of crystallization that roughly follows the evolutionary order of the various rock types. Feldspar phenocrysts of the granite porphyry are chemically related to feldspar of the associated coarse biotite granite; however, the groundmass feldspar is more sodic, indicating a markedly different crystallization history, most likely related to a water-vapor pressure-quench episode. In contrast to the fine riebeckite granite, the quartz veins and pegmatite have higher concentrations of Na, Li, Fe, and F, which indicate enrichment in those elements in the late-stage fluids.

---

*Present address: 18 Northwood Drive, Walpole, Massachusetts 02081.

Modal and chemical analyses of the principal granites fall close to the ternary minimum in the system $Ab-Or-SiO_2-H_2O$ and indicate maximum crystallization temperatures of 790° to 800° C. The trachyte probably crystallized at a temperature around 925° C and at a water-vapor pressure of about 500 kg/cm². Most granite bodies of the Rattlesnake pluton crystallized at temperatures between 660° and 800° C and at water-vapor pressures probably close to 1,000 kg/cm². Minor oxide content in the rocks of the pluton suggests their consanguinity with other alkalic granite plutons of southern New England.

Granite magma forming the pluton presumably underwent differentiation associated with transfer of Na, Li, Fe, and F toward the roof of the magma chamber. Two stages of magma upwelling associated with blister collapse and cauldron subsidence are hypothesized. The first stage resulted in the upwelling of a peralkaline magma that formed the fine riebeckite granite. This was followed by vapor-state transfer of Na, Li, Fe, and Si (probably as fluorides) and of H_2O, which produced a pegmatite zone in the fine riebeckite granite and in the country rocks. The residual magma thus became peraluminous by loss of Na and later moved upward during another stage of collapse, crystallizing as coarse biotite granite.

K-Ar dating of the granite at Rattlesnake Hill indicates a Middle Devonian age (about 370 m.y. B.P.), which is in agreement with radiometric dates for the Peabody Granite and with widespread igneous activity thoughout New England during Acadian time. This date, however, is lower than radiometric dates for the Cape Ann and Quincy Granites, which fall around 400 to 450 m.y. B.P.; this variance suggests two or more periods of alkalic magma intrusion in eastern Massachusetts.

INTRODUCTION

The Rattlesnake pluton, which comprises the various granites of the Rattlesnake Hill area of Sharon, Massachusetts, is part of a north-northeast-trending belt of hypersolvus granite plutons that extend from Cape Ann, Massachusetts, through the Quincy area, to the northeastern corner of Rhode Island (Fig. 1). The granite plutons to the north are more calcic and typically contain ferrohastingsite, whereas those to the south are more sodic and typically contain riebeckite or riebeckite and aegirine.

The Rattlesnake pluton is a small elliptical stock (Fig. 2) that occupies an area of about 22 km² and is exposed principally in the Mansfield quadrangle of southeastern Massachusetts. There are two major granite bodies in the pluton: a fine-grained riebeckite granite and a coarse-grained biotite granite. In 1913, Whitehead recognized in these granites a "fine" and a "coarse" type. He noted that the coarse biotite granite grades into the fine riebeckite granite and that these granites are younger than the surrounding Dedham quartz monzonite (Weymouth granite of Whitehead, 1913; Dedham Granodiorite of Emerson, 1917; see Fig. 2).

Whitehead's work was largely reconnaissance, and his area of investigation was confined to the immediate area around Rattlesnake Hill. Chute (1950) recognized a small extension of the fine riebeckite granite in the Brockton quadrangle. The entire pluton, including the units described in this paper, was mapped by Lyons (1969). Sodium/potash ratios for feldspar of the fine riebeckite granite were used to correlate it with the related riebeckite-aegirine granite of Quincy, Massachusetts (Lyons and Wolfe, 1971). Some aspects of the mineral relations and stability in the major granites were reported by Lyons (1971a).

The age of the alkalic granite of the Rattlesnake pluton cannot be firmly established by geologic relations. The granitic rocks are clearly intrusive into the Dedham

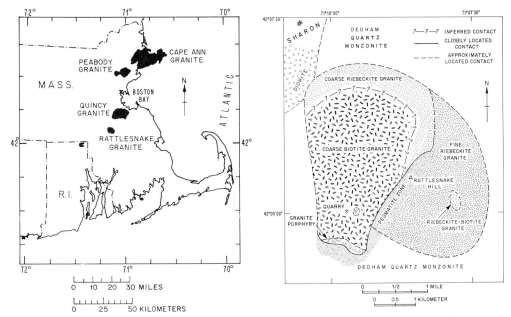

Figure 1. Areal distribution of alkalic granite plutons of southern New England.

Figure 2. Geologic map of Rattlesnake pluton.

quartz monzonite, but no geologic relations with other rocks in the area have been observed. The age of the Dedham quartz monzonite has long been subject to dispute, but it may form part of an extensive late Precambrian basement complex in southeastern New England (Zartman and Naylor, 1972). The nearest outcrops of fossiliferous Pennsylvanian sedimentary rocks occur about 11 km southwest of the Rattlesnake pluton (Lyons, 1969), where they presumably lie unconformably on the Dedham quartz monzonite. Field relations do not establish whether the Rattlesnake pluton was emplaced or exposed by erosion before sedimentation during Pennsylvanian time.

Stratigraphic relations of the other alkalic granite bodies of eastern Massachusetts—the Quincy, Cape Ann, and Peabody Granites—have been reviewed by Zartman and Marvin (1971). A maximum age of Middle Cambrian and a probable minimum age of Pennsylvanian was established for rocks of the Quincy pluton (Crosby, 1900; Chute, 1969).

Radiometric age determinations appear to be the best way of resolving the uncertainty in the age of the alkalic granite bodies of eastern Massachusetts. Zartman and Marvin (1971) used several radiometric methods to try to obtain definitive ages for the Cape Ann, Peabody, and Quincy Granites. They reached the conclusion that all three plutons were Late Ordovician in age (450 ± 25 m.y. B.P.), although their work indicated great complexities in the apparent age patterns presented by different methods of age measurement.

Preliminary age data for granite of the Rattlesnake pluton were reported by Krueger and Lyons (1972). Since then, we have used the K-Ar method to date several more samples representing different rock types from this pluton in order to compare the results with those reported for other alkalic granite bodies of eastern Massachusetts. We have also collected and analyzed riebeckite samples from a riebeckite-aegirine granite occurring in northeastern Rhode Island (see Fig. 1; Quinn and others, 1949; Quinn and Moore, 1968; Quinn, 1971).

ALKALIC GRANITE PLUTONS OF SOUTHERN NEW ENGLAND

The Peabody Granite pluton is the most uniform in grain size of the five plutons examined (Fig. 1) and consists almost entirely of coarse-grained ferrohastingsite granite (Toulmin, 1964). The spatially associated Cape Ann pluton has three major units: one resembling the Peabody Granite; another, the Rockport granite, which is similar in grain size and color to the Quincy Granite; and a third, ferrohastingsite syenite. The Quincy pluton has three closely related granites: the normal medium-grained riebeckite-aegirine granite (Quincy Granite); a mineralogically similar fine granite; and a riebeckite granite porphyry (the Blue Hill Granite Porphyry) of the Blue Hills area. It also has several minor phases consisting principally of rhyolite and rhomb-porphyry (Warren, 1913). Riebeckite-aegirine granite and riebeckite granite porphyry are recognized in Rhode Island (Quinn and others, 1949).

The major common properties of these granite plutons are the generally massive nature of the granites (except in Rhode Island), which has made them famous as quarry rocks, the similarity of the perthitic alkali feldspars, the dominance of amphiboles (both ferrohastingsite and riebeckite) over biotite, and the slight chemical alteration of the feldspar as compared to feldspar of the older Dedham quartz monzonite and associated rocks (Emerson, 1917).

There are subtle differences in the mineralogic and chemical nature of these granites. Aegirine has not been recognized at Rattlesnake Hill, although it is present in the other plutons in the southern part of the belt. Fayalite is present in the Cape Ann Granite but is absent in the Peabody Granite (Toulmin, 1964). Buma and others (1971) demonstrated trace-element differences among the Cape Ann, Peabody, and Quincy Granites; an example is the Eu depletion in the Quincy Granite feldspar relative to that of the Peabody Granite, which suggests higher oxygen fugacity during the crystallization of the Quincy feldspar. Riebeckite granite is absent and rare in the Peabody and Cape Ann plutons, respectively. Preliminary cathodoluminescence studies on the feldspar of the alkalic granites of southern New England indicate differences in the intensity of red luminescence, which suggests varying concentrations of Fe^{+3} and Ti^{+4} (A. N. Mariano, 1973, oral commun. to Lyons).

The Cape Ann Granite, in contrast to the Peabody Granite, is pegmatitic; the Quincy Granite is not very pegmatitic relative to the fine riebeckite granite at Rattlesnake Hill. Pegmatites are very rare in the coarse biotite granite of Rattlesnake Hill and have not been reported in the riebeckite-aegirine granite of Rhode Island (Quinn and others, 1949; Quinn, 1971). These variations in pegmatite content probably reflect different amounts of water and other fluxes in the various granite magmas.

The differences in the amphilboles, the presence or absence of pyroxenes, and the various chemical and petrographic differences in these granites justify separate designations. The term "Quincy Granite" should be restricted to the granite of Quincy, Massachusetts (Toulmin, 1964; Lyons, 1972). Some confusion has resulted from using this term for granite bodies from different plutons (see Toulmin, 1964), and such usage is now clearly inappropriate.

METHODS OF STUDY

Mapping of the Rattlesnake pluton was done on a scale of 1:24,000 using pace-and-compass techniques in more remote areas. Approximately 200 rock samples were collected, and about half of these were stained using a feldspar staining technique described by Lyons (1971b). About 50 thin sections were examined.

Modal analyses were done on the finer rocks using standard point-counting methods; coarser rocks were counted on stained slabs using dot patterns of 45 points/cm² to increase the surface area of the count.

Feldspars were separated and analyzed by procedures given in Lyons and Wolfe (1971). Amphiboles were hand picked, concentrated by heavy liquids and magnetic separation, and analyzed by the Japan Analytical Chemistry Research Institute. Biotite was processed in the same way.

About 20 thin sections were examined by A. N. Mariano using cathodoluminescence techniques; this was particularly helpful in detecting apatite, calcite, fluorite, and zircon.

The freshest samples available were selected for K-Ar age determinations. Whenever possible, relatively pure concentrates of mica or amphiboles, particularly riebeckite, were prepared. The rocks and localities sampled and the concentrates analyzed are described in Appendix 1.

Argon analyses were performed by standard isotope-dilution procedures, using a continuous bulb spike; the samples were analyzed in a MS-10 mass spectrometer. Potassium analyses were performed by flame photometry after dissolution of the sample by either HF digestion or lithium metaborate fusion. The precision of potassium and argon analyses, in the range of concentration found in the samples reported here, is better than ±1.5 and ±2.0 percent, respectively, as judged from statistical analyses of replicate measurements on unknown samples. The overall accuracy of the K-Ar ages, including calibration and decay-constant uncertainties, should be about ±4.0 percent or better. Analyses of several interlaboratory standards indicate that both precision and accuracy are at this level or better. The constants used in the age calculations are $\lambda_\beta = 4.72 \times 10^{-10}$/yr; $\lambda_e = 0.585 \times 10^{-10}$/yr; and $K^{40}/K = 1.22 \times 10^{-4}$ g/g. Complete analytical data for all new K-Ar age determinations are given in Appendix 2.

FIELD RELATIONS AND PETROGRAPHY

Coarse Biotite Granite

Coarse biotite granite was mapped by Lyons (1969) as the Massapoag Lake granite. Fine-grained biotite granite is associated as irregular veinlike masses in the coarse biotite granite; some of these masses have medial zones of coarse biotite granite and micrographic margins. Granite porphyry is also found associated with the coarse biotite granite in the southern part of the pluton as is coarse-grained riebeckite granite in the northern part (Fig. 2).The granite is easily identified in the field by its coarse-grained texture and its "rotten-rock character. This granite is well exposed in the south, but only scattered exposures can be observed in the northern part of the pluton, owing to thick deposits of glacial till. An unusually fresh exposure of this granite can be seen in an old Welch quarry located southwest of Rattlesnake Hill (Fig. 2). The fresh rock is light to medium gray; on weathered surfaces it is yellowish brown. The feldspar grains are gray and average about 15 mm in length. Light- to dark-gray quartz occurs as anhedral grains averaging about 5 mm in length. Biotite is not very obvious in hand specimen because of intense limonitization, but it occurs in grains as long as 8 mm.

The feldspar takes an excellent stain even on badly weathered specimens (Lyons, 1969, Fig. 21). Discontinuous trains of quartz wind around the feldspar, suggesting later crystallization of quartz. The margins of the perthite grains interdigitate in a manner similar to those observed in the coarse sodic granite of southwestern

Greenland (Parsons, 1972, Fig. 12), probably indicating resorption. A small amount of exsolved albite can be observed around the margins of some perthite grains, but apparently less so than around the feldspar of the Peabody and Cape Ann Granites. Some feldspars are zoned with blebs of quartz (less than 0.5 mm in diameter) oriented parallel to the margin of the grains. In such zoned feldspar, the quartz is found near the margin of the grain, but in some perthite grains there is an additional inner zone.

In thin section the perthite is seen as a regular intergrowth of microcline and untwinned albite. Biotite is partly altered to chlorite. Euhedral apatite in one thin section (Fig. 3) cuts across chlorite and biotite, indicating its later crystallization; apatite, in turn, is cut by magnetite concentrated along the cleavage planes of biotite, thus indicating the paragenesis of these four minerals. Anhedral fluorite is also found in the biotite and is probably an alteration product. Traces of calcite were observed under cathodoluminescence around the margins of biotite, indicating alteration associated with chloritization of biotite. Cathodoluminescence studies show fluorite in all the major alkalic granites of southern New England; it appears as a secondary mineral. Slight sericitic alteration is commonly observed in the microperthite. In shear zones the microperthite is strained, fractured, and heavily sericitized; secondary anhedral zircon(?) has replaced feldspar and is in turn cut by veins of secondary biotite. Modes of the coarse biotite granite are given in Table 1.

The coarse biotite granite grades into fine-grained (1 to 2 mm) biotite-riebeckite granite near its contact with the fine riebeckite granite. This transition occurs over a distance of about 30 m. A dike of the coarse biotite granite intrudes the Dedham quartz monzonite country rock near the marginal granite porphyry zone shown in Figure 2.

The biotite granite at Rattlesnake Hill has no exact counterpart in eastern Massachusetts and Rhode Island. Most of the alkalic granite in southern New England is amphibole granite, in contrast to other alkalic granite terranes where biotite granite is usually dominant (Billings, 1956; Jacobson and others, 1958).

Granite Porphyry

The term "granite porphyry" is used here for a rock of overall granite composition of at least 25 percent or more megacrysts of alkali feldspar and quartz set in a fine to very fine phaneritic groundmass of a similar mineralogic composition. Granite porphyry is associated with the coarse biotite granite near the southern part of the Rattlesnake pluton (Fig. 2). It occurs as inclusions up to several hundreds of metres across and as a "chilled" contact zone between quartz monzonite (Barefoot Hill quartz monzonite of Lyons, 1969) and the coarse biotite granite, which indicates a contact phase of the coarse biotite granite.

The granite porphyry is 40 to 60 percent phenocrysts of rounded quartz grains and rectangular crystals of microperthite set in a granitic groundmass. The phenocrysts of feldspar and quartz appear similar in color and size to those of the coarse biotite granite.

In thin section the feldspar phenocrysts are found to be microcline-albite microperthite like the feldspar of the coarse biotite granite. Staining has revealed zoning similar to that described for feldspar of the coarse biotite granite; however, the zoning is more common and distinct in the granite porphyry (Lyons, 1969, Fig. 22), perhaps indicating rapid changes in the physical and chemical conditions of crystallization. Some of the zones appear as micrographic intergrowths. The groundmass also contains small amounts of albite, magnetite, and sericite, and

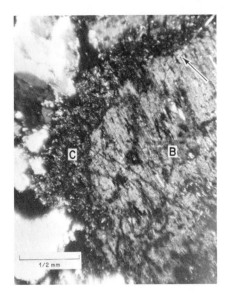 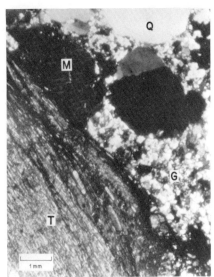

Figure 3. Photomicrograph of coarse biotite granite (433A) showing large grain of biotite (B) with marginal chlorite (C), apatite (arrow), and magnetite alteration (along cleavage traces).

Figure 4. Photomicrograph of granite porphyry-trachyte contact (85B). Granite porphyry shows microperthite (M) and quartz (Q) phenocrysts set in granitic groundmass (G). Trachyte (T) shows euhedral anorthoclase laths with trachytic texture separated by bands of iron chlorite(?).

biotite that is slightly altered to geothite along its margins. Modal analyses of the granite porphyry are given in Table 1.

Granite porphyry was found in contact with trachyte in a boulder located near the southeastern end of a small pond, 350 m N50°W of the summit of Rattlesnake Hill (Fig. 2). The contact between the two rocks is rather sharp (Fig. 4) and is characterized by brecciation of the granite porphyry, euhedral anorthoclase crystals paralleling the contact, microperthite megacrysts in the trachyte, and a concentration of zircon crystals on the granite porphyry side of the contact. The evidence as a whole favors an interpretation of the trachyte as a dike, but it could be a xenolith of a volcanic flow.

Trachyte

The trachyte occurs only in a small boulder in an area mapped as coarse biotite granite, very close to the contact with the fine riebeckite granite. It is an aphanitic rock with a dark-gray color on fresh surfaces and a specific gravity of 2.66.

In thin section the trachyte is composed almost entirely of lath-shaped alkali feldspar arranged in a parallel or subparallel manner (trachytic texture); some feldspar crystals are euhedral, twinned phenocrysts as long as 3 mm (Fig. 4). The rock has medium-gray to black laminae of iron-chlorite(?) that parallel the trachytic texture and the contact. Chemical analysis of the trachyte (Table 2) indicates that the feldspar is anorthoclase ($Or_{35.5} Ab_{63.5} An_{1.0}$). This feldspar does not cathodoluminesce like the microperthite of the pluton. On stained surfaces, the granite porphyry shows areas presumed to represent albite within a few millimetres of the contact with the trachyte, probably indicating sodium metasomatism.

TABLE 1. MODAL ANALYSES OF GRANITES OF THE RATTLESNAKE PLUTON, SHARON, MASSACHUSETTS

Rock name	Coarse biotite granite										Granite porphyry				Coarse riebeckite granite	
Field no.	63	75B	70	98	309	369	406B	433A	479	4H	387A	387A§	387C	431	69	154C
Microperthite	70.8	59.0	71.3	63.2	62.2	66.0	51.7	61.7	61.9	67.0	25.2	57.8	45.5	42.3	64.1	67.4
Albite	0.0	0.0	0.0	0.0	0.0	0.0	0.0	0.0	2.1	0.0	0.0	0.1	0.0	0.0	0.0	0.0
Quartz	27.3	36.8	26.1	34.9	31.8	31.3	45.8	33.8	29.8	28.1	12.0	37.9	7.9	12.5	31.3	26.9
Groundmass	0.0	0.0	0.0	0.0	0.0	0.0	0.0	0.0	0.0	0.0	62.3	100.0	45.2	43.4	0.0	0.0
Riebeckite	0.0	0.0	0.0	0.0	0.0	0.0	0.0	0.0	0.0	0.0	0.0	0.0	0.0	0.0	3.7#	3.0†
Grunerite	0.0	0.0	0.0	0.0	0.0	0.0	0.0	0.0	0.0	0.0	0.0	0.0	0.0	0.0	0.0	0.0
Biotite	1.8*	4.0*	2.6*	1.1*	5.8	2.1	2.5†	3.0	4.5	0.0	0.5	4.1	1.4†	1.8	0.1	0.4
Augite	0.0	0.0	0.0	0.0	0.0	0.0	0.0	0.0	0.0	4.9	0.0	0.0	0.0	0.0	0.0	0.1?
Magnetite	0.1	0.1	X	0.7	tr.	0.6	X	0.8	0.3	X	tr.	0.1	...	...	0.8	0.2
Zircon	tr.	0.1	...	0.1	0.1	tr.	...	0.2	tr.	X	tr.	tr.	...	...	tr.	tr.
Sphene	0.0	0.0	0.0	0.0	0.0	0.0	0.0	0.0	0.0	...	0.0	0.0	...	...	0.0	0.0
Chlorite	X	X	X	X	0.0	0.0	...	0.4	1.4	...	X	0.0	...	...	X?	2.0#
Apatite	0.0	0.0	...	tr.	0.0	0.0	...	0.1	0.0	...	...	0.0	...	...	0.0	0.0
Calcite	0.0	0.0	...	...	0.0	0.0	...	tr.	0.0	...	...	0.0	...	...	0.0	0.0
Fluorite	0.0	0.0	...	...	0.0	0.0	...	tr.	0.0	...	...	0.0	...	...	tr.	0.0
Sericite	tr.	tr.	X	tr.	0.1	tr.	X	...	tr.	X	...	tr.	...	...	tr.	tr.
Points counted	1,000	1,000	1,000	1,000	1,900	1,400	1,100	1,600	1,700	1,800	1,000	1,000	1,000	1,000	1,000	1,000

* Includes chlorite alteration
† Includes other dark minerals, including magnetite and chlorite
§ Composition of groundmass only
# Mainly or partly altered to goethite
X Present, amount not determined

TABLE 1. (continued)

Rock name	Fine riebeckite granite								Quartz syenite	Riebeckite-biotite granite			Riebeckite granite pegmatite					
Field no.	74A	74G	122C	305	307B	307A	383A		109B	39	116A		122DY	146D2	121C	118C	56Y	109
Microperthite	64.3	70.0	66.8	67.8	72.4	57.6	52.9		80.6	69.2	75.7		X	46.7	48.8	X	60.7	45.1
Albite	0.0	0.0	0.0	3.4	0.0	1.9	0.0		0.5	0.0	0.0		..	0.0	0.0	..	0.0	0.0
Quartz	30.8	24.1	30.7	26.6	23.4	35.5	40.5		9.9	27.5	19.3		..	38.7	26.6	X	29.4	43.1
Riebeckite	4.8§§	5.6	1.4	1.6§§	3.8	4.8§§	6.0		8.3	0.8	3.1		..	14.6††	24.5**	X	9.9**	11.8**
Grunerite	0.0	0.0	0.0	0.0	0.0	0.0	0.0		0.0	0.0	0.0		X	X	0.0	..	..	..
Biotite	0.0	0.0	0.0	0.0	0.0	0.0	0.0		0.0	1.9**	1.2**		..	X	0.0	..	..	..
Augite	tr.?	0.0	0.0	0.0	0.0	0.0	0.2		0.0	0.0	0.0		..	..	tr.?	..	..	..
Magnetite	0.0	0.2	0.9	0.6	0.4	0.1	0.2		0.5	0.6	0.6		X	X	X	X	..	..
Zircon	0.1	0.1	0.2	tr.	tr.	0.1	0.2		0.1	tr.	0.1		..	X	0.1	..	..	..
Sphene	0.0	0.0	0.0	0.0	0.0	0.0	0.0		0.0	0.0	0.0		..	..	tr.?	..	..	..
Chlorite	0.0	0.0	0.0	0.0	0.0	0.0	0.0		0.0	0.0	0.0		X	X	0.0	..	..	..
Apatite	tr.	tr.	0.0	0.0	0.0	0.0	0.0		tr.	0.0	0.0		X	..	0.0	..	..	..
Calcite	0.0	0.0	0.0	0.0	0.0	0.0	0.0		tr.	0.0	0.0		X	..	0.0	..	..	..
Fluorite	0.0	0.0	0.0	0.0	0.0	0.0	0.0		0.1	tr.	0.0		X	..	0.0	..	..	..
Epidote	0.0	0.0	0.0	0.0	0.0	0.0	0.0		0.0	0.0	tr.		X?	..	0.0	..	..	..
Sericite	tr.	tr.	tr.	tr.	tr.	tr.	tr.		tr.	tr.	tr.		..	..	tr.	..	..	..
Points counted	1,000	1,000	1,000	1,000	1,000	1,000	1,000		1,000	1,000	1,000		..	7,000	2,800	..	1,000	2,400

** Includes minor amounts of magnetite, chlorite, and other dark minerals
†† Altered to magnetite, zircon, fluorite, quartz, etc.
§§ Altered to goethite, anatase, etc.
X Present, amount not determined

Chemical analysis of the trachyte (Table 2) indicates that it is a peraluminous rock, similar in composition to quartz syenite and syenite of other alkalic granite areas (Parsons, 1972; Jacobson and others, 1958; Warren and McKinstry, 1924).

Coarse Riebeckite Granite

Coarse-grained riebeckite granite appears as an arcuate marginal zone on the northern part of the Rattlesnake pluton (Fig. 2). It is known principally from a study of glacial boulders, and because of this its inner contact with the coarse biotite granite can only be inferred. The coarse riebeckite granite was mapped as part of the Massapoag Lake granite (Lyons, 1969), but recent mapping indicates it is a separate, closely related unit.

Two granite pegmatite pods as much as 70 cm across with black amphibole (probably riebeckite) prisms were found in this unit near a contact with the fine riebeckite granite. Quartz veins are not particularly common in either of the coarse granites; only one 70-cm pod of quartz was observed near a contact with the fine riebeckite granite.

Texturally, the rock is similar to the coarse biotite granite, but it is distinguished by the presence of black amphibole grains as long as 15 mm, orange weathering tones, and less intense weathering due to the relative resistance of the amphibole. The specific gravity of the rock is 2.67 ± 0.01. The feldspar stains nicely, and the fresh amphibole can be distinguished from a small amount of altered amphibole that appears pitted on the surface. In thin section this alteration is made up of euhedral magnetite, quartz, zircon, and brownish-orange goethite.

In thin section the feldspar is similar to that of the coarse biotite granite, but no zoning was observed. The black amphibole is riebeckite, optically similar to that described below. Modes are given in Table 1.

Fine Riebeckite Granite

This is the fine-grained granite of Whitehead (1913) and the Rattlesnake Hill granite of Lyons (1969). The common variety is a nonporphyritic, fine-grained (0.5 to 2 mm) riebeckite granite. Its texture ranges from very fine grained (0.25 mm) to pegmatitic (25 to 75 mm). Several special varieties are recognizable, but only one, a fine riebeckite-biotite granite, was separately mapped (Fig. 2). In addition, there is a leucogranite with less than 1 percent dark minerals (in some places associated with riebeckite-quartz veins), fine graphic granite associated with pegmatite, spongy porphyritic ("spotted") riebeckite granite, and various types of subporphyritic granites that grade into quartz syenite.

The granite is light gray on fresh surfaces and weathers to pink or orange tones, indicating the presence of riebeckite with little or no biotite. Brownish weathering tones indicate the presence of biotite and its alteration to hydrous iron oxides such as goethite.

The feldspar grains are usually subhedral and range from 0.25 mm to 6 mm in length, averaging about 1 to 2 mm. Quartz grains have a similar range in size and average about 0.5 to 1 mm. The riebeckite usually occurs as grains as long as a few millimetres, but it ranges to 10 mm in the spongy variety; it also occurs as shreds, prisms, needles, or in poikilitic masses with inclusions of alkali feldspar. The specific gravity of the fine riebeckite granite averages 2.62.

In thin section the feldspar is microcline-albite microperthite similar to feldspar of other granites of the pluton. Some of the albite has exsolved and appears twinned along the margins of microperthite grains (Fig. 5). The riebeckite often appears dark, almost black, but the edges of the grains always show the bright blue of riebeckite. The specific gravity of the riebeckite is 3.39 ± 0.01. The pleochroic formula of the riebeckite is X = dark blue to bluish black, Y = grayish black, and Z = greenish yellow; the smaller grains show more vivid colors and lack the blackish tones. The riebeckite in several thin sections appears completely altered to iron hydroxides and fine-grained anatase. Some riebeckite has been altered completely to fine-grained euhedral magnetite and biotite. Alteration of the riebeckite is apparently more pronounced in the pegmatite, perhaps reflecting greater concentrations of fluxes and deuteric alteration. Fluorite, apatite, and calcite were detected by cathodoluminescence and appear as secondary minerals.

The quartz syenite phase of the Rattlesnake granite has relatively less quartz (about 10 percent) and much more riebeckite (Table 1). The specific gravity of the rock is 2.70. There appear to be two generations of riebeckite in this rock: one occurs as large, irregular crystals, which are in places altered to biotite, fluorite, magnetite, and zircon; the other occurs as pale-blue acicular sprays (Fig. 6), which appear to follow healed fractures probably suggesting introduction by late-stage vapors.

Near the contact zone with the coarse biotite granite, the fine riebeckite granite has microperthite with light-bluish cores surrounded by bright-red halos, probably reflecting Ti^{+4} and $Fe^{\pm 3}$ activation, respectively, under cathodoluminescence (A. N. Mariano, 1973, oral commun.). Similar zoning was found in the feldspar phenocrysts of the Blue Hill Granite Porphyry and in other feldspars of the alkalic granites of southern New England. Near this same contact, the Rattlesnake granite has very fine grained cognate riebeckite granite inclusions (Lyons, 1969, Figs. 23, 24).

Measurements of the orientation of 100 joints in this granite indicate two prominent, almost vertical joint sets that strike due north and N70°W. Light-gray quartz veins as wide as several centimetres cut the riebeckite granite. Their genesis appears

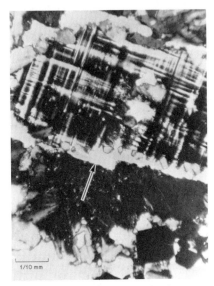

Figure 5. Photomicrograph of fine riebeckite granite showing atypical view of microperthite composed of microcline with gridiron structure and exsolved twinned albite (arrow).

Figure 6. Photomicrograph of quartz syenite phase of fine riebeckite granite showing sprays of secondary riebeckite surrounded by quartz (Q) and microperthite (M) with sericitic alteration.

to be related to the emplacement of the pegmatite dikes as indicated by parallelism of strike. Several slickensided fault-line scarps can be traced as many as 250 m along a general N40°W direction. Riebeckite crystals in the fine riebeckite granite have a steeply inclined orientation very close to contacts with the country rocks, most likely indicating emplacement along nearly vertical fracture planes.

Quartz diorite, a common associate of the Dedham quartz monzonite, was found as a xenolith in this granite. It is cut by a small dike of fine-grained ferrohastingsite granite, probably reflecting the influence of the country rock on the chemical composition of the amphibole.

Riebeckite-Biotite Granite

Riebeckite-biotite granite is found in the transition zone between the coarse biotite granite and the fine riebeckite granite and in a small area in the eastern part of the Rattlesnake pluton (Fig. 2). It is extensively weathered, and the dark minerals are very difficult to recognize in the field. In thin section the typical riebeckite of the fine riebeckite granite as well as vivid brownish-orange aggregates of goethite representing biotite alteration can be observed. Biotite associated with magnetite and zircon has apparently replaced some of the riebeckite, which is found as an optically continuous residuum in the biotite. Traces of fluorite are also associated with altered riebeckite. Purple fluorite, in places associated with galena (both discovered by an amateur collector, W. O. Hocking, Jr.), occurs as veinlets as wide as 3 mm that strike close to N60°W and N10°E and dip near the vertical, suggesting emplacement along the major joint sets mentioned above. Modes are given in Table 1.

Pegmatite in Fine Riebeckite Granite

Riebeckite granite pegmatite is intimately associated with the fine riebeckite granite along the western margin of the pluton (Fig. 2). It occurs as dikes, pods, and radial and irregular masses in the fine riebeckite granite. The pegmatite usually has riebeckite grains (some as long as 10 cm) that also occur in the associated quartz veins. Granite aplite is also associated with the pegmatite in a few places. One large (6 m) podlike pegmatite is underlain by an amphibolite xenolith that is cut through, presumably, by Dedham quartz monzonite; the quartz monzonite is exposed about 30 m away and has similar xenoliths. The pegmatite dikes, as wide as 60 cm, commonly pinch and swell, more rarely branch, and frequently are zoned. The dikes comprise a N70°W- and N70°E-striking dike system. Details of zoning and other features of the pegmatite are described in Lyons (1969).

The average mode of the pegmatite is alkali feldspar, 56.5 percent; quartz, 31.1 percent; and riebeckite, magnetite, and so forth, 12.4 percent. Individual modes are given in Table 1. The modes of the pegmatite indicate more riebeckite than in the fine riebeckite granite and probably a concentration of iron in the pegmatitic fluids.

A nonpleochroic, uncolored amphibole with first- and second-order interference colors and polysynthetic twinning was found in the riebeckite concentrate from the pegmatite. This amphibole is almost certainly grunerite, but this has not been confirmed by chemical analysis. Grunerite was previously discovered in the pegmatite of the Cape Ann Granite (Bowen and Schairer, 1935). The relation between the grunerite and riebeckite of the Rattlesnake pluton is not known, but at Cape Ann, grunerite is an alteration product of fayalite (Warren and McKinstry, 1924).

Altered riebeckite appears as a very fine aggregate of euhedral magnetite, quartz, fluorite, biotite, iron-chlorite(?), and untwinned albite(?). The feldspar is microcline-albite microperthite; the grid twinning of the microcline is unusually clear. Aplite of one thin section shows fresh riebeckite in contrast to altered riebeckite of pegmatite of the same section, probably indicating a lack of water or other fluxes in the aplite and their availability in the pegmatite.

ROCK CHEMISTRY

Major-element chemical analyses of the fine riebeckite granite, coarse biotite granite, trachyte of the Rattlesnake pluton, and comparable rocks are given in Table 2. Molar ratios of $(Na_2O + K_2O)/Al_2O_3$ for the various rocks are also given in Table 2.

The Quincy Granite is clearly peralkaline. The fine riebeckite granite must have a $(Na_2O + K_2O)/Al_2O_3$ ratio greater than 1.00 in order to be peralkaline as is demanded by its mineralogic composition, so its original ratio must have been slightly higher than the 1.00. The Peabody and Cape Ann Granites (Table 2) are

TABLE 2. MAJOR-ELEMENT ANALYSES OF SELECTED ROCKS OF THE RATTLESNAKE HILL AREA, SHARON, MASSACHUSETTS, COMPARED WITH CHEMICALLY SIMILAR ROCKS OF OTHER AREAS

	1	2	3	4	5	6	7	8	9
SiO_2	74.80	73.58	63.40	74.86	74.44	76.15	76.37	72.08	72.56
ZrO_2	..	..	..	0.20	..	0.04	..	0.12	..
Al_2O_3	12.02	12.69	19.32	11.61	11.40	12.48	12.15	13.08	13.21
Fe_2O_3	1.61	1.53	1.10	2.29	2.86	0.50	1.65	0.53	0.78
FeO	1.13	0.68	2.59	1.25	0.66	0.73	1.06	2.81	2.14
MgO	0.12	0.10	0.26	0.05	0.10	0.19	0.10	0.11	0.12
CaO	0.01	0.52	0.15	0.41	0.25	0.51	0.17	1.07	0.89
Na_2O	4.41	3.95	6.60	4.30	4.64	4.06	3.64	4.09	3.82
K_2O	4.43	4.70	5.30	4.64	4.59	4.43	4.68	5.31	5.53
H_2O+	0.79	0.81	1.53	0.31	0.39	0.33	0.08	0.20	0.25
H_2O-	0.43	0.66	0.17	0.04	0.19	0.20	0.13	0.08	0.04
TiO_2	0.11	0.12	0.15	0.20	0.21	0.06	0.18	0.35	0.25
MnO	0.03	0.02	0.04	0.02	0.16	0.01	0.07	0.09	0.05
Li_2O	0.06	0.04	0.01	..	..	0.08	..	..	..
P_2O_5	0.01	0.01	0.01	tr.	0.03	tr.	..	0.07	tr.
CO_2	..	..	..	..	0.05	0.02	..	0.04	..
Cl	..	..	..	..	0.02	0.01	..	..	..
F	0.05	0.29	0.04	..	0.08	0.35	..	..	..
	100.01	99.70	100.67	100.18	100.07	100.15	100.28	100.03	99.64
less O = F	0.02	0.12	0.02	0.00	0.03	0.15	0.00	0.00	0.00
Total	99.99	99.58	100.65	100.18	100.04	100.00	100.28	100.03	99.64
$\frac{Na_2O + K_2O*}{Al_2O_3}$	1.00	0.91	0.86	1.04	1.11	0.92	0.91	0.96	0.93

* Molar ratio.

Column:
1 Fine riebeckite granite (74K), Sharon, Massachusetts, Analyst: T. Asari.
2 Coarse biotite granite (433A), Sharon, Massachusetts, Analyst: T. Asari.
3 Trachyte (85B), Sharon, Massachusetts, Analyst: T. Asari.
4 Quincy Granite, Quincy, Massachusetts, ave. of 3 analyses, Warren (1913).
5 Crystallized comendite tuff, Nevada, ave. of 13 analyses, Noble (1970).
6 Biotite microperthite granite, Liruei, Nigeria (Jacobson and others, 1958).
7 Aporhyolite, Wampatuck Hill, Blue Hills, Massachusetts, Warren (1913).
8 Peabody Granite, ave. of 2 analyses, Tuttle and Bowen (1958), Clapp (1921).
9 Cape Ann Granite, Babson Farm quarry, Cape Ann, Massachusetts, Warren and McKinstry (1924).

metaluminous (that is, $Al_2O_3/(Na_2O + K_2O + CaO) < 1.00$). The coarse biotite granite is slightly peraluminous (that is, $Al_2O_3/(Na_2O + K_2O + CaO) > 1.00$); however, the trachyte is very peraluminous, which reflects the presence of alteration minerals, particularly clay minerals and chlorite.

The chemical composition of the Peabody and Cape Ann Granites are remarkably similar. There appear to be no major differences in their analyses except, perhaps, in the oxidation state of the iron.

The fine riebeckite granite and the Quincy Granite are also similar in composition. Again, one of the important differences is in the oxidation state of the iron—the iron in the Quincy Granite appearing more oxidized. Another important difference between these two peralkaline granites is in the amount of CaO: the Quincy Granite is more calcic than the fine riebeckite granite, but it is much less so than either the Peabody or Cape Ann Granites.

Both the fine riebeckite granite and the Quincy Granite are richer in SiO_2 and Na_2O and poorer in CaO, Al_2O_3, and K_2O than either the Peabody or the Cape Ann Granites. The consanguinity of the alkalic granites of eastern Massachusetts is partly indicated by the similar quantities of minor oxides, particularly MgO, TiO_2, MnO, and P_2O_5.

The chemical composition of the coarse biotite granite is very similar to that of the aporhyolite (ash-flow tuff; see Kaktins, 1976) of the Quincy pluton (Warren, 1913), indicating that this is the volcanic counterpart of the coarse biotite granite and part of the Quincy magma series as previously suggested by Warren (1913). Volcanic rocks similar to the Quincy Granite are the comendites of Nevada (Table 2) and Liruei, Nigeria.

MINERAL CHEMISTRY

Chemical compositions of alkali feldspar, riebeckite, and biotite (annite) of the Rattlesnake pluton are given, respectively, in Tables 3, 4, and 6. Comparative data for minerals of other alkalic granites are also given. The chemistry of these minerals will now be discussed in light of the petrologic relations.

The close chemical affinity between the alkali feldspars of the Rattlesnake granite and the Quincy Granite is clearly shown in Table 3 by their identical Na_2O/K_2O ratios, which correspond to microperthite of bulk composition $Ab_{53}Or_{47}$. This composition closely agrees with the bulk feldspar composition of $Ab_{55}Or_{45}$ (Parsons, 1972) for the sodic granite of the Puklen complex, southwestern Greenland. The feldspar of the Peabody and Cape Ann Granites is also similar, indicating similar crystallization histories. Toulmin (1960) discussed the composition of feldspar from these ferrohastingsite granites and indicated their chemical similarity. Chemical and optical data on such alkali feldspars (Toulmin, 1960, 1964; Warren, 1913; Lyons and Wolfe, 1971; Fabriès and Rocci, 1965) indicate less than 5 percent of the An component; thus, they are binary feldspars as defined by Barth (1969). The rarity of more than 3 percent An in the microperthite (Toulmin, 1964) indicates that the albite member is low albite. Parsons (1972) demonstrated by x-ray analysis that the sodium phase of chemically similar microperthite samples of the Puklen complex is low albite.

Feldspar of the coarse biotite granite consistently shows less Ab than feldspar of the fine riebeckite granite (Table 3). It is interesting to note that the felspar of the riebeckite granite pegmatite of the Rattlesnake Hill area shows Or enrichment as compared to feldspar of the fine riebeckite and coarse biotite granites, indicating their different crystallization histories.

TABLE 3. PARTIAL CHEMICAL DATA FOR ALKALI FELDSPARS OF THE RATTLESNAKE HILL AREA, SHARON, MASSACHUSETTS, COMPARED WITH DATA FOR RELATED ALKALI FELDSPARS*

Granite type	Field No.	Rock designation and locality	1 Na_2O	2 K_2O	3 R	4 Ab	5 Or
Riebeckite granite†	123G	Riebeckite-aegirine granite, Quincy, Mass.	5.32	6.86	0.78	52.6	47.4
	413C1	Riebeckite-aegirine granite, Quincy, Mass.	5.06	6.42	0.79	53.0	47.0
	305	Fine riebeckite granite, Sharon, Mass.	6.10	7.65	0.80	53.4	46.6
	139B	Fine riebeckite granite, Sharon, Mass.	5.81	7.47	0.78	52.7	47.3
	74A	Fine riebeckite granite, Sharon, Mass.	5.43	7.05	0.77	52.4	47.6
Biotite granite	98	Coarse biotite granite, Sharon, Mass.	5.05	6.99	0.72	50.8	49.2
	309	Coarse biotite granite, Sharon, Mass.	4.85	7.89	0.61	47.2	52.8
	387B	Coarse biotite granite, Sharon, Mass.	5.12	8.33	0.61	47.3	52.7
Riebeckite granite pegmatite	121C	Riebeckite granite pegmatite, Sharon, Mass.	3.23	10.34	0.31	30.9	69.1
	146D2	Riebeckite granite pegmatite, Sharon, Mass.	3.93	9.46	0.42	37.3	62.7
Granite porphyry	123A	Granite porphyry, Blue Hills, Mass.#	5.45	7.48	0.73	51.1	48.9
	123A	Granite porphyry, Blue Hills, Mass.**	4.60	8.07	0.57	44.9	55.1
	123B	Granite porphyry, Blue Hills, Mass.#	5.61	7.30	0.77	52.4	47.6
	387C	Granite porphyry, Sharon, Mass. #	5.08	7.57	0.67	49.0	51.0
	387C	Granite porphyry, Sharon, Mass.**	3.86	4.58	0.84	54.7	45.3
Riebeckite-biotite granite	116A	Riebeckite-biotite granite, Sharon, Mass.	4.32	8.74	0.49	41.4	58.6
Ferrohastingsite granite§	..	Peabody Granite, Peabody area, Mass.	5.58	7.14	0.78	52.8	47.2
	..	Cape Ann Granite, Cape Ann, Mass.	6.05	5.25	0.87	55.4	44.6

* Analyst: P. C. Lyons, except as noted.
† From Lyons and Wolfe (1971), Analyst: P. C. Lyons.
§ From Toulmin (1964), Analysts: J. Ito and P.J. Byler.
\# Feldspar from phenocrysts.
** Feldspar from groundmass.

Column:
1 Weight percent Na_2O in sample feldspar.
2 Weight percent K_2O in sample feldspar.
3 Na_2O/K_2O (weight percent).
4 Weight percent albite (corrected to 100% feldspar).
5 Weight percent orthoclase (corrected to 100% feldspar)

TABLE 4. CHEMICAL COMPOSITIONS OF RIEBECKITES OF THE RATTLESNAKE HILL AREA, SHARON, MASSACHUSETTS, COMPARED WITH COMPOSITIONS OF RIEBECKITES OF OTHER AREAS

	1	2	3	4	5	6	7	8	9	10
SiO_2	48.50	50.62	47.06	49.65	48.47	50.01	49.20	45.85	48.47	46.98
Al_2O_3	3.74	0.68	2.36	1.34	1.75	2.07	2.28	2.20	0.59	1.29
TiO_2	0.80	1.28	1.53	..	1.06	1.39	0.94	3.00	0.74	1.49
Fe_2O_3	16.13	14.51	12.91	17.66	16.38	16.51	16.39	13.70	11.89	11.93
FeO	18.97	21.43	20.92	19.55	19.20	17.90	17.21	21.60	22.93	23.38
MgO	0.32	0.10	0.30	..	0.10	0.14	0.09	0.48	0.05	0.13
MnO	0.59	1.15	0.76	..	0.76	0.55	0.72	0.75	0.37	0.24
CaO	1.29	1.28	3.76	3.16	0.87	0.27	0.15	1.66	0.88	1.91
Na_2O	5.91	6.15	5.96	7.61	6.70	6.23	7.48	6.40	8.88	8.90
K_2O	0.55	1.10	1.22	..	1.42	1.22	1.65	1.07	1.60	2.74
Li_2O	0.56	..	0.38	..	0.96	0.75	1.16	..	0.74	..
H_2O+	2.23	1.30	1.64	1.67	1.55	1.83	1.85	2.52	1.00	1.10
H_2O-	0.41	0.10	0.28	..	0.20	0.43	0.27	0.28	0.08	0.00
F	0.74	0.20	1.16	..	1.45	1.08	1.49	..	2.20	..
	100.74	99.90	100.24	100.64	100.87	100.38	100.88	99.51	100.42	100.09
less O = F	0.31	0.09	0.49	0.00	0.57	0.45	0.60	0.00	0.92	0.00
Total	100.43	99.81	99.75	100.64	100.30	99.93	100.28	99.51	99.50	100.09

Column:
1 Riebeckite-aegirine granite, Cumberland, Rhode Island (123F).
2 Pegmatite, Fallon quarry, Quincy, Massachusetts (Warren and Palache, 1911).
3 Riebeckite-aegirine granite, Quincy, Massachusetts (Lyons, 1972).
4 Granite, Cape Ann, Massachusetts (Warren and Palache, 1911).
5 Quartz vein, Rattlesnake Hill, Sharon, Massachusetts (118).
6 Riebeckite granite, Rattlesnake Hill, Sharon, Massachusetts (74K).
7 Riebeckite granite pegmatite, Rattlesnake Hill, Sharon, Massachusetts (118C).
8 Riebeckite granite, Tarraouadji, Nigeria (Fabriès and Rocci, 1965).
9 Riebeckite granite, Amo complex, Nigeria (Borley, 1963).
10 Riebeckite granite, Mill Mountain, New Hampshire (Chapman and Williams, 1935).

New analyses: Analyst, T. Asari.
Methods of analysis and standard deviations: Gravimetric: SiO_2 (0.55%); Al_2O_3 (2.20%); MgO (3.40%); CaO (2.16%); H_2O+ (21.3%); H_2O- (25.0%). Volumetric: Fe_2O_3 (6.13%); FeO (2.09%). Colorimetric: TiO_2 (5.85%); MnO (..); F (15.0%). Flame photometric: Na_2O (4.45%); K_2O (10.4%); Li_2O (10.0%).

The relations between the Ab and Or content of the feldspar of the phenocrysts and groundmass of the granite porphyry of the Rattlesnake Hill area and those of the Blue Hill Granite Porphyry of the Quincy area were determined (Table 3). The feldspar phenocrysts of both granite porphyries are practically identical in Ab and Or content to their associated granite, indicating similar crystallization conditions. The average bulk composition of feldspar of the Quincy Granite is $Ab_{53}Or_{47}$, whereas the feldspar phenocrysts of the Blue Hill Granite Porphyry have the average composition $Ab_{52}Or_{48}$. The average bulk composition of feldspar in the coarse biotite granite is $Ab_{48}Or_{52}$, whereas the feldspar phenocrysts in areally associated granite porphyry have a composition of $Ab_{49}Or_{51}$. The groundmass feldspar of the granite porphyry of the Rattlesnake Hill area is enriched in Ab relative to the feldspar phenocrysts; the reverse is true for the Blue Hill Granite Porphyry. It is clear that the feldspar of the groundmass and phenocrysts of both granite porphyries crystallized under markedly different conditions. The difference in trend in groundmass feldspar crystallization of the two granite porphyries is probably related to the fact that the Blue Hill Granite Porphyry is peralkaline and the granite porphyry of the Rattlesnake pluton is peraluminous.

The chemical composition of feldspar of the trachyte based on whole-rock data (Table 2) is about $Ab_{64}Or_{36}$; it is the most sodic feldspar of the Rattlesnake pluton. Its composition indicates anorthoclase and high crystallization temperatures.

Major-element analyses of pegmatite, riebeckite of the fine granite, and a quartz vein of the Rattlesnake Hill area are shown in Table 4; empirical formulas are given in Table 5. The chemical composition of riebeckite samples from Quincy, Massachusetts, Cumberland, Rhode Island, the White Mountains of New Hampshire, and Nigeria are also shown for comparative purposes. A detailed discussion of the chemistry of riebeckite of the alkalic plutons of eastern Massachusetts and Rhode Island will be treated elsewhere (Lyons, unpub. manuscript). It will suffice here to point out some of the important aspects of their chemistry, particularly as related to petrology.

First, the three riebeckite units from Sharon, Massachusetts, are enriched in fluorine, more so than the riebeckite from Quincy, but less so in general than those of Nigeria (Borley, 1963). The Nigerian riebeckite rocks, particularly those associated with albite granite, are greatly enriched in fluorine (Borley, 1963). The Nigerian sodic and other peralkaline granite bodies have a variety of fluorine-containing minerals including fluorite, pyrochlore, cryolite, and topaz. There is a good possibility that tin mineralization in this province is related to the presence of fluorine, which probably carried Sn, Na, Li, Si, Fe, and other elements as fluorides in the vapor state. The direct correlation of sodium with fluorine in the Rattlesnake pluton indicates that Na was probably carried as NaF.

Second, the riebeckite of the pegmatite is enriched in the alkali metals (Na, K, and Li) as compared to riebeckite of the fine riebeckite granite; this indicates their concentration in the pegmatitic fluids. Riebeckite of the quartz vein shows close chemical affinities to riebeckite of the pegmatite, which was expected from the field association of the two rocks. Neither the riebeckite of the Rattlesnake pluton pegmatite nor that of the Quincy Granite pegmatite (Tables 4 and 5) shows any significant differences in the oxidation of the iron, as compared to riebeckite of the host granite (Table 4). This indicates that similar oxygen fugacities prevailed during the crystallization of riebeckite of the pegmatite and that of the main granite within the same pluton. The Na, K, and Li are concentrated in both the riebeckite of the pegmatite and quartz vein of the Rattlesnake pluton, and at the same time, both are enriched in fluorine as compared to riebeckite of the fine riebeckite granite. This is a good indication of the availability of fluorine and its probable role as

TABLE 5. EMPIRICAL FORMULAS OF RIEBECKITES OF TABLE 4

	1	2	3	4	5	6	7	8	9	10
Si	7.43	7.84	7.41	7.65	7.52	7.66	7.54	7.41	7.67	7.62
Al^{IV}	0.57	0.14	0.44	0.23	0.31	0.34	0.41	0.43	0.11	0.25
$Fe^{+3\,IV}$	0.00	0.02	0.15	0.12	0.17	0.00	0.05	0.16	0.22	0.13
Al^{VI}	0.10	0.00	0.00	0.00	0.00	0.03	0.00	0.00	0.00	0.00
$Fe^{+3\,VI}$	1.85	1.67	1.38	1.83	1.72	1.90	1.84	1.51	1.20	2.00
Ti^{+4}	0.09	0.15	0.18	0.12	0.12	0.16	0.11	0.37	0.09	0.18
Fe^{+2}	2.42	2.77	2.76	2.43	2.49	2.30	2.20	2.91	3.03	3.20
Mn	0.07	0.15	0.10	(0.10)	0.10	0.07	0.09	0.09	0.05	0.03
Mg	0.08	0.02	0.07	(0.04)	0.02	0.03	0.02	0.11	0.01	0.03
Li	0.34	(0.44)	0.24	(0.67)	0.60	0.46	0.72	..	0.43	..
Na^{VI}	0.00	0.00	0.27	0.00	0:00	0.00	0.00	0.00	0.11	0.71
Ca	0.21	0.21	0.63	0.51	0.14	0.05	0.04	0.29	0.15	0.33
Na^{VIII}	1.75	1.84	1.55	2.17	2.02	1.86	2.22	2.00	2.61	2.81
K	0.11	0.22	0.25	(0.24)	0.28	0.24	0.34	0.23	0.32	0.57
OH	2.28	1.34	1.73	1.74	1.60	1.88	1.89	1.55	1.05	1.18
F	0.36	0.12	0.58	(0.58)	0.71	0.53	0.72	..	1.10	..

Note: Values in parentheses are estimates based upon data for closely related riebeckites.

an agent of transport, particularly during the late stages of crystallization of peralkaline granite magmas.

The composition and empirical formula of biotite of the coarse granite, as compared to biotite of other biotite microperthite granites, is shown in Table 6. Biotite from the coarse granite plots (Mg-Fe as ordinate, Si_6Al_2-Si_5Al_3 as abscissa) in the same area as the annite of Nigeria (Fabriès and Rocci, 1965, Fig.3). The annite of the coarse granite has some chlorite in the concentrate; thus, the amount of Mg-component in this biotite probably should be interpreted as a maximum. Practically all of the biotite shown in Table 6, with the exception of that from microperthite granite of the White Mountains, New Hampshire, shows small or insignificant amounts of the Mg-component in solid solution. The chemical composition of annite from the type locality at Cape Ann, Massachusetts, is also shown in Table 6. Judging by the similar Fe^{+3}/Fe^{+2} ratios of biotite of the Cape Ann Granite and coarse biotite granite of the Rattlesnake pluton, the oxidizing conditions that prevailed during their crystallization were probably rather similar. In light of the presumably higher oxidizing conditions that prevailed during the crystallization of riebeckite in the Rattlesnake pluton, this evidence indicates a drop in oxygen fugacities, most likely at a later time and at slightly lower temperatures, during the crystallization of the coarse biotite granite.

PETROGENESIS

In discussing granite petrogenesis in the broad sense, it is helpful to follow Wolfe's fourfold genetic classification of granite (C. W. Wolfe, 1973, oral commun.): (1) allochthonous transformed granite, (2) allochthonous magmatic granite, (3) autochthonous transformed granite, and (4) autochthonous magmatic granite. Allochthonous granite, as opposed to autochthonous granite, originates from magma, solutions, or vapors that have been generated in one place and have crystallized in or been introduced into another place.

The granite units of the Rattlesnake pluton are probably allochthonous, as

TABLE 6. CHEMICAL COMPOSITION AND EMPIRICAL FORMULA OF BIOTITE (ANNITE) FROM THE RATTLESNAKE HILL AREA, SHARON, MASSACHUSETTS, COMPARED WITH SIMILAR BIOTITES OF OTHER AREAS

	Chemical compositions						Empirical formulas				
	1	2	3	4	5		1	2	3	4	5
SiO_2	35.84	32.03	37.40	37.38	33.48	Si	5.73	4.95	5.99	6.03	4.95
Al_2O_3	14.21	11.92	13.92	11.89	13.64	Al^{IV}	2.27	2.19	2.01	1.97	2.32
TiO_2	2.70	3.42	2.00	1.84	2.94	$Fe^{+3\,IV}$	0.00	0.86	0.00	0.00	0.34
Fe_2O_3	6.12	8.00	5.42	4.38	8.00	Ti^{IV}	0.00	0.00	0.00	0.00	0.32
FeO	24.47	30.41	24.34	28.65	23.54	Al^{VI}	0.38	0.00	0.61	0.00	0.00
MnO	0.29	0.21	0.50	0.41	1.02	$Fe^{+3\,VI}$	0.73	0.08	0.65	0.53	0.00
MgO	2.91	0.06	0.40	0.22	4.97	Ti^{VI}	0.32	0.44	0.24	0.22	0.00
CaO	0.99	0.23	1.72	0.16	0.56	Fe^{+2}	3.24	3.96	3.25	3.86	2.84
Li_2O	0.26	tr.	..	0.77	..	Mg	0.69	0.01	0.10	0.05	1.07
Na_2O	0.19	1.54	0.65	0.39	0.53	Mn	0.04	0.03	0.07	0.06	0.12
K_2O	6.86	8.46	7.90	8.78	7.80	Li	0.17	0.00	..	0.50	..
P_2O_5	0.18	..	0.01	..	..	Ca	0.17	0.03	0.29	0.01	0.09
H_2O+	3.44	4.19	5.33	1.84	2.63	Na	0.06	0.46	0.19	0.12	0.15
H_2O-	0.28		0.32	0.67		K	1.39	1.68	1.62	1.81	1.43
F	2.32	tr.	0.00	4.36	0.95						
	101.06	100.47	99.91	101.74	100.06						
less O = F	0.98	0.00	..	1.86	0.40						
Total	100.08	100.47	99.91	99.88	99.66						

Column:
1 Coarse biotite granite, Sharon, Massachusetts (433A). Analyst: T. Asari
2 Ferrohastingsite microperthite granite (Cape Ann Granite), Rockport, Massachusetts (Clark, 1903).
3 Biotite microperthite granite, Tarraouadji Massif, Nigeria (Fabriès and Rocci, 1965).
4 Biotite microperthite granite, Liruei Complex, Nigeria (Jacobson and others, 1958).
5 Biotite microperthite granite, Percy quadrangle, White Mountains, New Hampshire (Chapman and Williams, 1935).

evidenced by a "chilled border" of granite porphyry with the country rocks (quartz monzonites), soaking of the country rocks with riebeckite granite pegmatite ("contact halo" of Wolfe), emplacement of granite dikes in the country rocks, the presence of country rocks (quartz diorite and amphibolite) in the fine riebeckite granite, and the parallel alignment of riebeckite crystals in the fine riebeckite granite at contacts with the Dedham quartz monzonite.

The magmatic origin of the granite units of the Rattlesnake pluton is indicated by the following:

1. Close chemical similarity of the biotite and riebeckite granites to rocks known to be of volcanic origin (Table 2). For example the coarse biotite granite is probably the plutonic equivalent of the rhyolite of the Blue Hills area. In New Hampshire and Nigeria, comparable granite units are known to have volcanic equivalents in the same complex (Billings, 1956; Jacobson and others, 1958), thus indicating a common source magma.

2. The nature of the alkali feldspar, which indicates crystallization at magmatic temperatures above 660°C in the system $Ab-Or-H_2O$ of Tuttle and Bowen (1958).

3. Comparison of the fine riebeckite granite with the riebeckite-aegirine granite of Quincy, Massachusetts (Lyons and Wolfe, 1971), which has a mineral assemblage (quartz, aenigmatite, and acmite) known from phase-equilibria studies to be stable only at magmatic temperatures (Ernst, 1962).

In summary, field, mineralogic, and phase-equilibria data indicate that the alkalic and other granites of the Rattlesnake pluton fall into the category of allochthonous magmatic granite in the Wolfe scheme.

The presence of peralkaline and "subalkalic" (meaning lacking alkali minerals such as riebeckite and aegirine) ash flows and tuffs in the same volcanic center of volcanic rocks of Miocene age in the Great Basin, Nevada (Noble, 1968), indicates the close association of peralkaline and subalkalic magmas in time and space. Plutonic analogues associated with ring complexes can be found in the Pliny Range of the White Mountains, New Hampshire (Billings, 1956), and in the Tunk Lake pluton (Karner, 1963) of the Maine coastal plutons (Chapman, 1968b). The latter pluton shows an annular pattern of rocks from an outer ring of pyroxene granite, through rings of hornblende and biotite granite, to a core of subalkalic granite; this pattern is quite consistent with the cooling of magma from the margin toward the center of the pluton, although C. W. Wolfe (1974, oral commun.) interprets this as owing to different degrees of transformation.

The genesis of subalkalic granite on a large scale such as the Dedham quartz monzonite of eastern Massachusetts may be related to an early stage of collapse of heated crustal sectors ("blisters") associated with the folding, overthrusting, and metamorphism of geosynclinal rocks as outlined in Wolfe's blister hypothesis (Wolfe, 1949). The Dedham quartz monzonite is clearly an allochthonous subalkalic granite whose magmatic or transformed nature has not been resolved; however, the associated quartz diorite is probably a transformed rock, as indicated by its development near contacts with mafic rocks. The Dedham quartz monzonite was probably emplaced during this stage of blister collapse.

It is quite possible that the genesis of many alkalic granite magmas is associated with a later stage of blister collapse during which subalkalic granitic rocks subsided to give rise by selective fusion to alkalic granite magmas at depth. Some of the magma probably reached the surface along ring fractures to form volcanic rocks (rhyolite and trachyte) at or near the surface. A second substage is believed to have occurred after release of heat by volcanism to form peripheral ring fractures at depth followed by the intrusion of alkalic granitic and syenitic magmas, the plutonic equivalents of the volcanic rocks. These two substages correspond to the volcanic and granitic cycles of Jacobson and others (1958). All of this activity could well have occurred during the very late stages of the history of a blister as outlined by Wolfe (1949).

The north-northeast alignment of the alkalic granite plutons of southern New England indicates strong structural control on their emplacement, which is a feature shared with other alkalic granite terranes. Chapman (1968b) suggested that the Maine coastal plutons and the White Mountain alkalic granite plutons are distributed at the nodes of two sets of deep-seated fracture systems. This distribution is also a reasonable one for the alkalic granite plutons of southern New England, where the north-northeast pattern is a rather strong one and the east-west pattern is indicated by the east-west elongation of the plutons (see Fig. 1). This nodal pattern appears to be a good match for that of the Maine coastal plutons (Chapman, 1968b, Fig. 29-1). Page (1968) included both the alkalic granite plutons of southern New England and the Maine coastal plutons in his "Late Devonian Plutonic Series," although his "Late Devonian" age is questionable in light of the Middle Devonian and older radiometric ages discussed below.

As noted above, the alkalic granite of southern New England is peculiar in the absence or rarity of ring dikes and in the high proportion of amphibole granite as compared to biotite granite, which is typically the dominant granite of other alkalic granite provinces. Billings (1976) noted that the Quincy pluton has been tipped up on end (see also Kaktins, 1976); thus, ring dikes would appear as sheets, and this might explain the absence of ring dikes in the other plutons as well.

Jacobson and others (1958) mentioned the profound influence that depth of erosion has on the surface form of hypersolvus granite. They maintained that many circular or elliptical Nigerian granite bodies would probably appear as ring dikes at greater erosional depth. This would indicate that the alkalic granite plutons of southern New England are high-level plutons that extend downward to form ring dikes. Quinn and others (1949) indicated that the riebeckite-aegirine granite and granite porphyry of the Cumberland area, northeastern Rhode Island (Fig. 1), are partly sheetlike structures. Gravity studies indicate that the Cape Ann Granite is a shallow-dipping, sheetlike body (Joyner, 1958; Toulmin, 1964). The shallowness of these bodies and the rarity or absence of volcanic rocks from some of the alkalic granite plutons of eastern Massachusetts (that is, Rattlesnake and Peabody plutons) may indicate that they are relatively lower level plutons.

The dominance of amphibole alkalic granite in southern New England may reflect differences in crystallization temperatures, although phase-equilibria studies of biotite (annite) by Eugster and Wones (1962) and riebeckite-arfvedsonite solid solutions (Ernst, 1962) indicate a common maximum temperature of stability of approximately 700°C in the presence of quartz. These same studies indicated that riebeckite is stable at higher oxygen fugacities when compared to annite at the same temperature. Thus, the common replacement of riebeckite by biotite in the granite of the Rattlesnake pluton may suggest a drop in oxygen fugacity of the magma.

The presence of hypersolvus feldspar in the Rattlesnake pluton indicates temperatures of crystallization greater than 660°C in the system $Ab-Or-H_2O$ of Tuttle and Bowen (1958). Assuming a water pressure of 1,000 kg/cm^2 and neglecting the effect of HF, we find that the initial temperature of crystallization of the riebeckite granite was probably around 800°C, following data for this system. With a similar water pressure, the initial temperature of crystallization of the coarse biotite granite was probably slightly lower (about 790°C), based on data for the same system (Tuttle and Bowen, 1958). The chemistry of the trachyte (Table 2) and the composition of its feldspar ($Ab_{63.5}Or_{35.5}An_{1.0}$) indicate crystallization close to the binary minimum in the system $Ab-Or-H_2O$ of Tuttle and Bowen (1958), crystallization at approximately 925°C, and water pressure of about 500 kg/cm^2.

As mentioned above, the feldspar of the granite porphyry of the Rattlesnake Hill indicates two stages of crystallization. One possible mechanism for the crystallization of the groundmass is the sudden loss of water-vapor pressure (a pressure-quench mechanism) as indicated in Tuttle and Bowen (1958) and discussed by Parsons (1972). Such a loss in pressure would probably cause rapid crystallization of the remaining magma, which had previously crystallized feldspar and quartz. The result would be the granite porphyry texture with the rapidly crystallized magma forming the groundmass. Assuming an initial water pressure of 1,200 kg/cm^2 and the data of Tuttle and Bowen (1958) for the system $Ab-Or-SiO_2-H_2O$, a loss of water pressure of 200 kg/cm^2 in this system would shift the ternary minimum so that a feldspar of composition $Ab_{55}Or_{45}$, the approximate composition of the groundmass feldspar of the granite porphyry of the Rattlesnake pluton, would be the final feldspar to crystallize. This assumes isothermal crystallization and zoning in the phenocryst feldspar, which should have sodic rims of bulk composition $Ab_{55}Or_{45}$ and relatively Or-rich cores.

One possible mechanism for the loss of water pressure might be the fracturing associated with cauldron subsidence. Thus, the granite porphyry is not envisioned as a chilled marginal rock in the classical sense of Crosby (1900) but as a rock that crystallized under two different water pressures, the change brought about by fracturing associated with blister collapse and cauldron subsidence.

Na and Li appear to be relatively concentrated in the fine riebeckite granite as compared to the coarse biotite granite. The same elements are relatively enriched in the pegmatite phase as compared to the fine riebeckite granite. This can be explained by upward movement of Na and Li as NaF and LiF vapor within the magma chamber. Fe was probably also carried upward as indicated by large amounts of riebeckite in the pegmatite bodies. Thus, a sodium-lithium-fluorine-rich zone is envisioned in the roof region of the magma chamber. Deep fracturing of the roof rocks led to upwelling of the sodic magma along nearly vertical ring fractures to form the fine riebeckite granite; some piecemeal stoping probably occurred during this process, as evidenced by xenoliths of country rock in the granite. Higher oxygen fugacities prevailed during the crystallization of the fine riebeckite granite, probably owing to loss of hydrogen from the magma by HF moving upward along ring fractures. The remaining parent magma was relatively enriched in Al, Ca, and K and later moved upward along a second fracture zone, associated with a second stage of cauldron subsidence, to form the coarse biotite granite. Before it moved upward, however, NaF, LiF, FeF_3, SiF_4, H_2O, and other components moved upward along the hanging wall (C. W. Wolfe [1973, oral commun.] mentioned the common association of pegmatite with the hanging wall of faults) of a steeply dipping fault system, associated with the fracture system of the fine riebeckite granite, to form the pegmatite zone at the western margin of the fine riebeckite granite (Fig. 2) and in the Dedham quartz monzonite country rocks. Chapman (1976) gives a detailed account of how alkalic granite might be generated and later emplaced in such fracture systems.

The genesis of the ferrohastingsite granite may be related to the same source magma, and their somewhat different natures are explainable by two major factors: the presence of lower oxygen fugacities associated with deficiency of F favored the crystallization of ferrohastingsite over riebeckite (Lyons, 1972), and more calcic country rocks differentially contaminated the magma during assimilation.

RADIOMETRIC AGE DETERMINATIONS

Radiometric Dating of Rattlesnake Pluton

Most of the rock types occurring in the Rattlesnake alkalic granite pluton were sampled for K-Ar analysis (Table 7). Particular attention was paid to the riebeckite-bearing rocks, because amphiboles are resistant to argon loss that might result from minor thermal events (Hart, 1961, 1964), such as the Permian disturbance of K-Ar ages described by Zartman and others (1970) that affected most of southern New England. All of the six riebeckite samples analyzed appeared quite fresh except sample A-3044, which shows alteration even in hand specimen, probably accounting for its slightly younger apparent age. The five fresh riebeckite samples gave very consistent K-Ar ages, averaging 366 ± 9 m.y.

The analyzed samples were selected partly because of their great variation in grain size. The samples ranged from a coarse-grained quartz-riebeckite vein to a fine-grained riebeckite granite. No correlation of age with grain size is apparent. The riebeckite concentrates have a substantial variation in potassium content. Excluding the partly altered sample A-3044 there is no correlation between K content and the measured ages. No variation of age with field position is apparent.

The concordance of the K-Ar ages on the five fresh riebeckite samples from Rattlesnake Hill, despite variations in grain size, K content, and field position, lends considerable strength to the supposition that the average age of 366 ± 9 m.y. is meaningful. If the ages were simply minimum ages resulting from some disturbance, much more variation under such conditions would be expected.

Furthermore, no similar radiometric dates have been obtained from nearby rocks, and no other igneous or metamorphic events approximating this age are known in the area. The 366-m.y. age is, however, in conflict with the most recent conclusions regarding the age of the alkalic granite bodies of eastern Massachusetts (Zartman and Marvin, 1971).

The two biotites and the trachyte from the Rattlesnake Hill pluton yielded substantially younger apparent ages than the riebeckites (Table 7). The significance of the biotite ages is not clear. They may be minimum ages reflecting the Permian disturbance of K-Ar mica ages as observed to the west of this area (Faul and others, 1963; Zartman and others, 1970). An alternative explanation of the 254-m.y. age of the granite porphyry is that the biotite in this sample is chloritized and badly weathered, and the age may have been lowered by this alteration.

The trachyte was analyzed as a whole-rock sample because of its extremely fine grained nature. Because most of the potassium in this rock is in the form of alkali feldspar, the apparent age of 229 m.y. requires considerable interpretation. Because plutonic alkali feldspar tends to lose about 35 percent of its radiogenic argon (Goldich and others, 1957), correction for such argon loss is necessary. The trachyte therefore gives a corrected age of 342 m.y., which is in reasonable agreement with the riebeckite ages (considering the approximate nature of the correction).

Radiometric Dating of Riebeckite-Aegirine Granite of Northeastern Rhode Island

A small body of riebeckite-aegirine granite with mineralogic characteristics very similar to those of the fine riebeckite granite at Rattlesnake Hill and the Quincy Granite has been mapped (Fig. 1) in the extreme northeast corner of Rhode Island (Quinn and others, 1949; Quinn, 1971). Two samples of this granite and one sample of a quartz-riebeckite-fluorite vein cutting the granite were collected and analyzed by the K-Ar method. The calculated ages are given in Table 8, along with some previously published K-Ar ages from nearby rocks.

The K-Ar ages of riebeckite from all three samples are much younger than the riebeckite ages determined from the alkalic granite of Rattlesnake Hill (see Table 7) and the other alkalic granites of eastern Massachusetts (Zartman and Marvin, 1971). About 35 km to the south, the Westerly and Narragansett Pier Granites have intruded and metamorphosed the Pennsylvanian sedimentary rocks of the Narragansett basin (Nichols, 1956). These granites and the metamorphosed Pennsylvanian sedimentary rocks have been dated by Hurley and others (1960), who found K-Ar ages of 230 to 260 m.y., but the higher grades of metamorphism do not extend very far to the north (Quinn, 1971). It is unlikely that this metamorphic zone caused the apparent riebeckite ages, because they appear to be even younger than the metamorphism. For the same reason it is unlikely that the disturbance of K-Ar mica ages (Zartman and others, 1970) caused these young dates, even though that zone of disturbance comes close to the riebeckite-aegirine granite.

One of the riebeckite samples gives an apparent age of only 188 m.y. after repeated analyses. This is equivalent to K-Ar ages of several of the plutons in the White Mountains magma series in New Hampshire (Foland and others, 1971; H. W. Krueger, unpub. data) and for Triassic lava rocks in the Connecticut River valley (Reesman and others, 1973). There is a small area a few kilometres west of the riebeckite-aegirine granite that has yielded several K-Ar ages of about 200 m.y. on biotite from the Scituate Granite Gneiss (see Table 7). The Scituate Granite Gneiss and other rocks farther south in Rhode Island, all pre-Mississippian in

TABLE 7. K-Ar AGES ON MINERALS FROM THE IGNEOUS ROCKS AT RATTLESNAKE HILL, SHARON, MASSACHUSETTS

Sample no.	Field no.	Rock unit	Mineral analyzed	Grain size of material dated (mm)	% K	K-Ar age (m.y.)
A-1089	118	Coarse quartz-riebeckite vein	Riebeckite	5 to 30	1.362	371
A-1090	118C	Riebeckite granite pegmatite	Riebeckite	3 to 100	1.408	375
A-1091	74K	Fine riebeckite granite	Riebeckite	0.5 to 2	1.178	352
A-2009	71-20	Fine riebeckite granite	Riebeckite	1 to 3	1.311	363
A-2010	71-21	Fine riebeckite granite	Riebeckite	0.2 to 1	1.162	370
A-3044	69	Coarse riebeckite granite	Riebeckite	1 to 5	0.608	343
B-1092	433A	Coarse biotite granite	Biotite	0.5 to 8	5.308	308
B-3043	431	Granite porphyry	Altered biotite	0.5 to 2	0.985	254
R-3042	85B	Trachyte	Whole rock	<0.1 to 3	3.808	229 (342)*

* Corrected for assumed 35% loss of argon from feldspar (Goldich and others, 1957).

TABLE 8. K-Ar AGES ON ALKALINE GRANITES AND OTHER ROCKS FROM NORTHEASTERN RHODE ISLAND

Sample no.	Mineral analyzed and rock unit	Age, (m.y.)	Ref.*
A-2031	Riebeckite from riebeckite-aegirine granite	236	1
A-2033	Same	188	1
A-2032	Riebeckite from quartz-riebeckite-fluorite vein in granite	226	1
9	Biotite from vein in riebeckite-aegirine granite	230	2
8	Biotite from Esmond Granite	230	2
4	Biotite from Scituate Granite Gneiss	200	2
5	Same	215	2
6	Same	200	2
7	Same	200	2

* References:
1. This work, data in Appendixes 1 and 2
2. Harakal (1966) in Zartman and others (1970).

age according to Quinn (1971), generally give biotite K-Ar ages of 230 to 250 m.y. (Hurley and others, 1960; Zartman and others, 1970). These facts suggest that a small "hot spot" may have existed in northeastern Rhode Island, probably during the Triassic Period, well after the remainder of the rocks in central New England had cooled to the point that radiogenic argon was retained in the mica.

Evidence of substantial post-Pennsylvanian faulting, as well as a major silicified zone of unknown age but possibly related to the faulting, is present at Diamond Hill, within 3 km of all of our sampling sites. Quinn (1971) mapped these features, and Rodgers (1970, p. 111) speculated that they might be of Triassic age. Lyons and Faul (1968) referred to a possible "Palisade Disturbance" in this area. One or more of these features may have altered the K-Ar ages found for the riebeckite-aegirine granite of northeastern Rhode Island. In view of these complexities, we believe that no definitive radiometric age for these rocks can be obtained by the K-Ar method, and that K-Ar age comparisons with the alkalic granite of eastern Massachusetts are not useful.

Discussion of Radiometric Ages

K-Ar age determinations on riebeckite from rocks of the Rattlesnake pluton are so consistent that we believe the average age of 366 ± 9 m.y. is significant, even though it differs from the age of 450 ± 25 m.y. proposed by Zartman and Marvin (1971) for the other alkalic granites of eastern Massachusetts. These granite bodies have been correlated on the basis of chemical and mineralogic similarity by other investigators. It is important to resolve the apparent age discrepancy, and we have therefore re-examined all radiometric data relating to the age of the alkalic granite bodies of eastern Massachusetts.

The Late Ordovician age (450 m.y.) of Zartman and Marvin (1971) was based largely on a single concordia plot of zircon data obtained from three samples, one from each of the Peabody, Cape Ann, and Quincy Granites. Zartman and Marvin (p. 947) evaluated" the other geochronological results with respect to this assumed age of intrusion," and arrived at the conclusion that ages by all other methods, including Rb-Sr whole-rock ages and K-Ar hornblende ages, do not represent the true age. They invoked two distinct episodes of disturbance to explain younger ages: a late Paleozoic low-temperature metasomatism south of Boston and an episode in Devonian time of heating north of Boston.

Radiometric ages for the Peabody Granite are reported by Zartman and others (1970) and Zartman and Marvin (1971). Of the K-Ar results, one biotite age probably should be rejected, because it comes from a sample collected close to a diabase dike mapped by Toulmin (1964) and dated at 254 m.y. (see Zartman and Marvin, 1971, p. 950). The other K-Ar ages average 370 m.y., the Rb-Sr whole-rock isochron gives 367 m.y., and only the U-Th-Pb data on two samples of zircon give an older age. Subsequent unpublished work by Zartman (1974, written commun.) has revealed the presence of two generations of zircon within the Peabody Granite, an older variety associated with cognate(?) xenoliths, and a younger one that crystallized with the bulk of the granite. Significantly different ages are recorded by these two populations of zircon, suggesting that intrusion of the alkalic magma series probably spanned a greater interval of time than previously recognized. According to Zartman, the new results are compatible with an approximate 370-m.y. emplacement age for most of the Peabody Granite.

Radiometric data from the Cape Ann Granite are found in Zartman and others (1970), Zartman and Marvin (1971), and Fairbairn and others (1968). Three complicating factors must be accounted for in any interpretation of these data. First, the Peabody Granite lies only about 3 km from the western margin of the Cape Ann pluton, and Toulmin (1964, p. A63) stated that "field evidence supports the interpretation that the Peabody Granite is somewhat younger than at least some of the rocks of the Cape Ann pluton." Secondly, the Cape Ann pluton is a composite intrusion emplaced in several stages according to Toulmin (1964). He noted (p. A41) that the most westerly unit, the Cherry Hill Granite, intrudes the Wenham Monzonite, which in turn lies adjacent to the main Cape Ann Granite. Finally, there are a number of basic dikes intruding the Cape Ann Granite that may have altered the ages of some dated samples.

With the above factors in mind, we find that there are four K-Ar hornblende ages from the northwestern part of the Cape Ann pluton that average 359 m.y., very similar to the probable age of the nearby Peabody pluton. Two of these hornblende ages are from the Wenham Monzonite, but it is not clear whether they represent the age of the monzonitic intrusion or simply ages that have been reset by the intrusion of the Peabody Granite. The southern and eastern portions of the Cape Ann pluton, away from the Peabody Granite, give K-Ar ages averaging 398 m.y. Two Rb-Sr whole-rock isochrons based primarily on samples from this part of the pluton give ages of 414 m.y. and 435 m.y. The zircon used to represent the Cape Ann Granite in the concordia plot of Zartman and Marvin (1971) was obtained from the Rockport quarry near the eastern extremity of Cape Ann. The nearby Flat Ledge quarry has provided the only two K-Ar ages that are substantially younger than 400 m.y. There are several diabase dikes intruding the biotite granite—one dated at 334 m.y.—at this locality. We can only conclude that the bulk of the Cape Ann Granite was emplaced 400 to 450 m.y. ago and that the northwestern portion is either younger (370 m.y.) or was reheated at the time the Peabody Granite was emplaced.

The radiometric data from the Quincy Granite (Zartman and Marvin, 1971; Bottino and others, 1970) are also complex. Five K-Ar ages on riebeckite from the Quincy Granite average 441 m.y., but Zartman and Marvin (1971, p. 948) pointed out that these riebeckite concentrates contain several percent of aegirine. They analyzed one aegirine concentrate, and we have analyzed a very pure aegirine from the Fallon and Ballou quarry in Quincy (see Appendix 2). Both analyses show excess radiogenic argon incorporated in the pyroxene. Zartman and Marvin calculated a 7-m.y. correction for their one analyzed sample, but they have used the riebeckite of highest K content for this analysis and computation. It is very possible that even larger corrections should be applied to some of these riebeckite analyses. The zircon age of the Quincy Granite on the concordia plot of Zartman and Marvin was subject to considerable uncertainty due to a large common lead correction, and singularly it could be interpreted as being anywhere between 400 and 450 m.y. old (R. E. Zartman, 1974, written commun.). Two Rb-Sr whole-rock isochrons are available, giving apparent ages of 313 and 365 m.y. The Rb-Sr data for the Quincy Granite are probably not valid, even though it is tempting to give some meaning to the 365-m.y. isochron in view of the radiometric data from the Rattlesnake and Peabody plutons. Zartman and Marvin (1971) and Bottino and others (1970) have dismissed the Rb-Sr data from the Quincy pluton on the grounds that their samples were apparently open systems with regard to Sr and thus violated the assumptions of the Rb-Sr isochron model. This is not too surprising in view of the low calcium content of the Quincy Granite. If radiogenic Sr^{87} was mobilized in a rock with such low calcium content, it might readily escape rather than redistribute into other minerals. This might be an unrecognized danger in applying Rb-Sr methods to low-calcium rocks. In any case, the extant radiometric data point to a probable age between 400 and 450 m.y. for the Quincy Granite.

It is a common practice to give less significance to K-Ar ages whenever there is a conflict between K-Ar and other methods of age determination. This is frequently done without adequate documentation of evidence that the K-Ar ages might have been altered. Because the K-Ar dating system responds to physical and chemical parameters, as do the other methods, it seems imperative that some reasonably documented mechanism of K-Ar age alteration must be postulated before K-Ar ages can be dismissed.

Our approach to interpreting the radiometric data from samples of alkalic granite of eastern Massachusetts has been a positive one. Whereas a bimodal interpretation of the ages of these granite bodies accounts for virtually all of the radiometric data and is not in conflict with observed field relations, it does present one problem. Petrologic, mineralogic, chemical, and spatial relations of the Peabody and Cape Ann plutons suggest one age; the same is true for the Quincy and Rattlesnake plutons. It is difficult to reconcile two or more magmatic episodes, 30 to 80 m.y. apart, with the petrochemical considerations and with any simple model of genesis and emplacement for the alkalic granite plutons of southern New England. A very similar conflict exists between petrochemical and radiometric data from the White Mountains magma series of New Hampshire, another alkalic plutonic suite (Foland and others, 1971; H. W. Krueger, unpub. data).

Chapman (1968a) considered the possibility that the alkalic granite plutons of eastern Massachusetts and Rhode Island are part of a northeast-trending belt of plutons that includes the mildly alkalic intrusive rocks along the southeastern coast of Maine (Chapman, 1968b). The latter rocks have been dated by the K-Ar method (Faul and others, 1963): four plutons give an average age of 360 ± 8 m.y., and four other plutons average 393 ± 6 m.y. This bimodal age distribution approximates that of the alkalic granite plutons of eastern Massachusetts.

SUMMARY AND CONCLUSIONS

The Rattlesnake pluton is composed of two major hypersolvus granite units: a coarse biotite granite that has a granite porphyry phase and a fine riebeckite granite that ranges into quartz syenite, riebeckite-biotite granite, and riebeckite granite pegmatite. Coarse riebeckite granite, ferrohastingsite granite, and sodium-rich trachyte also occur, the latter two rarely.

Field relations indicate that the granite porphyry is a contact phase of the coarse biotite granite. The relative ages of the two major granite bodies are not clear, but the fine riebeckite granite is interpreted to be slightly older.

Feldspars of the Rattlesnake pluton vary from $Ab_{64}Or_{36}$ in the trachyte to $AB_{34}Or_{66}$ in the pegmatite. This trend in Or enrichment roughly follows the order of crystallization of granite in the pluton. The groundmass feldspar of the granite porphyry has a composition $AB_{55}Or_{45}$ in contrast to a feldspar phenocryst composition of $Ab_{49}Or_{51}$, the latter being very close to the composition of the feldspar of the coarse biotite granite. This indicates two stages of feldspar crystallization for the granite prophyry, probably due to a sudden loss of water pressure in the magma.

Pegmatite of the Rattlesnake pluton shows enrichment in Na, Li, Fe, and F, indicating the concentration of these elements in the late-stage fluids.

Riebeckites in the various alkalic granite plutons in the southern part of the belt are similar, but the Quincy riebeckite is richer in Ca and Fe^{+2}, which is consistent with a somewhat higher temperature of crystallization.

Rock genesis in the pluton is compatible with Wolfe's scheme for blister collapse, which can also explain the older Dedham quartz monzonite, the country rock surrounding the Rattlesnake pluton. A geosynclinal stage is envisioned for the emplacement of the Dedham quartz monzonite, and a posttectonic stage—a second stage of blister collapse—is envisioned for the emplacement of the alkalic granite. This stage begins with the eruption of volcanic material (the trachyte may belong to this substage) followed by a cauldron-subsidence substage during which the granite bodies of the Rattlesnake pluton were emplaced. During this time, magmatic differentiation occurred owing to the upward movement of Na, Li, and Fe as fluoride vapors to form a sodium-lithium-iron-fluorine-rich roof zone that gave rise to the peralkaline fine riebeckite granite followed by the peraluminous biotite granite, in response to two major episodes of roof fracturing.

Temperatures from 660° to 925°C are indicated for crystallization of the rocks, with the temperatures declining from the volcanic substage to the granitic substage. Water-vapor pressure was probably as low as 500 kg/cm² during crystallization of the rocks.

Whole-rock and amphibole compositions of the alkalic granite plutons of southern New England relate reasonably well to a model that assumes contamination by chemically different country rocks during magmatic assimilation and varying oxygen fugacities related to the release of hydrogen from the magma, probably as HF. This would explain the presence of ferrohastingsite granite in the northern part of the belt and riebeckite granite in the southern part.

It is impossible to separate the Cape Ann Granite from the Peabody Granite on the basis of whole-rock and amphibole chemical analysis. The Cape Ann Granite is obviously more enriched in pegmatite and is slightly different than the Peabody Granite.

Biotite of the Cape Ann Granite is chemically similar to that of the coarse biotite granite, and their Fe^{+3}/Fe^{+2} ratios indicate that similar oxygen fugacities prevailed during their crystallization. The riebeckites probably crystallized at somewhat higher oxygen fugacities.

The feldspar composition of the Cape Ann, Peabody, Quincy, and Rattlesnake granite plutons closely approach a bulk composition of $Ab_{53}Or_{47}$, indicating a close parallel in crystallization histories. The minor-element content of these granites also indicates their consanguinity, as does the spatial distribution of the granite plutons. These plutons appear at nodal points of two sets of assumed deep-seated Devonian fracture systems; one set is quite pronounced, is north-northeast-trending, and is parallel to a similar fracture system in the Maine coastal plutons of Devonian age.

K-Ar ages for riebeckites of the Rattlesnake pluton are concordant at 366 ± 9 m.y. B.P. (Middle Devonian), probably the true age of emplacement of the granite.

The riebeckite of the riebeckite-aegirine granite of northeastern Rhode Island and spatially associated older rocks gives K-Ar dates of about 200 m.y., clearly indicating an alteration of the true ages of the rocks.

A re-evaluation of the radiometric data from other alkalic granites in eastern Massachusetts indicates that the Peabody Granite is probably about the same age as the granite bodies of the Rattlesnake pluton, and the Quincy and Cape Ann Granites are 400 to 450 m.y. old.

Two different ages, at least 35 to 40 m.y. apart, for the alkalic granite plutons of southern New England are inconsistent with interpretations based on the chemical, petrologic, and spatial data that indicate a common age.

ACKNOWLEDGMENTS

The gracious help of W. H. Pinson, Jr., is acknowledged with appreciation for making possible the atomic absorption work. The paper was improved by the constructive criticisms of C. W. Wolfe, M. P. Billings, Ian Parsons, Priestley Toulmin III, A. H. Brownlow, W. C. Park, R. E. Zartman, P. M. Hurley, R. S. Naylor, C. A. Chapman, and J. C. Hepburn. We are especially grateful to R. E. Zartman for revealing his recent findings on the age of the Peabody Granite and for sharing his views on the interpretation of the radiometric data. We thank A. N. Mariano who took the photomicrographs and did the work on cathodoluminescence. Boston University grants 123-CBS and 153-CBS were used to support this study.

I (P. C. L.) thank C. Wroe Wolfe who inspired my work in geology. Wolfe's ideas on transformation and petrogenesis greatly helped me to think more clearly and to be skeptical of the magmatic origin of most granites; however, we share a view of the magmatic origin of the alkalic granites of southern New England.

APPENDIX 1. SAMPLE DESCRIPTIONS, SAMPLE LOCATIONS, AND DESCRIPTIONS OF CONCENTRATES ANALYZED BY THE K-AR METHOD

Laboratory sample no.	Field no.	Description of sample	Location of sample	Description of analyzed material
			Rattlesnake pluton samples	
A-1089	118	Quartz-riebeckite vein with coarse blades of riebeckite (up to 30 mm)	Rattlesnake Hill, Sharon, Mass. (lat. 42°05.5'N., long. 71°08.5'W.)	Fresh riebeckite concentrate, -20/+100 mesh; greater than 99% pure with traces of quartz
A-1090	118C	Riebeckite granite pegmatite, very coarse, with large blades of riebeckite (up to 100 mm)	South side of Rattlesnake Hill, Sharon, Mass. (lat. 42°05.5'N., long. 71°08.5'W.)	Fresh riebeckite concentrate, -40/+100 mesh; greater than 99% pure with traces of quartz and feldspar
A-1091	74K	Fine riebeckite granite forming the main phase of the pluton; riebeckite grains 0.5 to 2 mm	Southwest of Rattlesnake Hill, Sharon, Mass. (lat. 42°05.2'N., long. 71°08.9'W.)	Fresh riebeckite concentrate, -40/+200 mesh; 98% riebeckite, 1% biotite, 1% feldspar
B-1092	433A	Coarse biotite granite (Massapoag Lake granite) with biotite 0.5 to 8 mm	1 mile SW of Rattlesnake Hill, Sharon, Mass. (lat. 42°05.0'N., long. 71°09.7'W.)	Biotite concentrate, -40/+100 mesh; 98% slightly chloritized biotite, 1% quartz, 1% feldspar
A-2009	71-20	Fine riebeckite granite; riebeckite 1 to 3 mm	1 mile ENE of Rattlesnake Hill, Stoughton, Mass. (lat. 42°05.5'N., long. 71°08.0'W.)	Riebeckite concentrate, -60/+200 mesh; 96% riebeckite, 4% quartz and feldspar, no biotite observed
A-2010	71-21	Fine riebeckite granite; riebeckite 0.2 to 1 mm	1 mile ENE of Rattlesnake Hill, Stoughton, Mass. (lat. 42°05.5'N., long. 71°08.0'W.)	Riebeckite concentrate, -100/+200 mesh; 95% riebeckite, 5% quartz and feldspar, no biotite seen
R-3042	85B	Very fine grained trachyte with some feldspars up to 3 mm in length	1 mile NE of Rattlesnake Hill, Stoughton, Mass. (lat. 42°05'26"N., long. 71°08'14"W.)	Whole rock, crushed to -80/+200 mesh
B-3043	431	Coarse-grained granite porphyry with biotite 0.5 to 2 mm	About 1 mile SW of Rattlesnake Hill, Sharon, Mass. (lat. 42°05'1"N., long. 71°09'29"W.)	Biotite concentrate, -80/+200 mesh; 99% partly chloritized biotite, traces of quartz and feldspar
A-3044	69	Coarse-grained riebeckite granite with 1 to 5 mm grains of riebeckite	Rattlesnake Hill, Sharon, Mass. (lat. 42°6'43"N., long. 71°9'53"W.)	Riebeckite concentrate, -80/+200 mesh; 98% partly altered riebeckite, 2% quartz and feldspar
			Riebeckite-aegirine granite of northeastern Rhode Island	
A-2031	72-1	Medium-grained riebeckite-aegirine granite	Near E summit of Beacon Pole Hill, Cumberland, R.I. (lat. 41°59.7'N., long. 71°26.8'W.)	Riebeckite concentrate, -40/+200 mesh; 95% riebeckite, 5% aegirine and epidote, trace of quartz, no biotite
A-2032	72-2	Coarse-grained quartz-riebeckite-fluorite vein cutting riebeckite-aegirine granite	W end of Beacon Pole Hill, Cumberland, R.I. (lat. 41°59.7'N., long. 71°27.2'W.)	Riebeckite concentrate, -40/+200 mesh; 96% riebeckite, 2% quartz, 2% iron oxides, trace of aegirine (?)
A-2033	72-3	Medium-grained riebeckite-aegirine granite	W. Wrentham road, Cumberland, R.I. (lat. 41°59.6'N., long. 71°27.6'W.)	Riebeckite concentrate, -40/+200 mesh; 98% riebeckite, 2% quartz and feldspar, aegirine and biotite not seen
			Quincy Granite	
P-0066	65-3	Porous aggregate of coarse quartz, feldspar, and aegirine crystals with minor parisite.	Fallon and Ballou quarry, Quincy, Mass. (lat. 42°15'N., long. 71°1'W.)	Aegirine concentrate, -60/+200 mesh; 99.9% aegirine, traces of feldspar and quartz

APPENDIX 2. ANALYTICAL DATA FOR PREVIOUSLY UNPUBLISHED K-AR AGE DETERMINATIONS

Laboratory sample no.	Field no.	Rock type	Material analyzed	% K	Radiogenic Ar^{40} (ppm)	$\frac{Rad.\ Ar^{40}}{Total\ Ar^{40}}$	Age (m.y.)
P-0066	65-3	Quincy granite, pegmatite	Aegirine	0.058 0.059	0.00555	0.299	1005 ± 50
A-1089	118	Quartz-riebeckite vein	Riebeckite	1.351 1.373	0.0398	0.717	371 ± 14
A-1090	118C	Riebeckite granite pegmatite	Riebeckite	1.409 1.406	0.0417	0.786	375 ± 14
A-1091	74K	Fine riebeckite granite	Riebeckite	1.206 1.150	0.0325 0.0325	0.712 0.763	352 ± 11
B-1092	433A	Coarse biotite granite	Biotite	5.308	0.1270	0.804	309 ± 10
A-2009	71-20	Fine riebeckite granite	Riebeckite	1.311	0.0374	0.762	363 ± 15
A-2010	71-21	Fine riebeckite granite	Riebeckite	1.162	0.0339	0.614	370 ± 16
A-2031	72-1	Riebeckite-aegirine granite	Riebeckite	0.467	0.00838	0.526	236 ± 12
A-2032	72-2	Quartz-riebeckite-fluorite vein	Riebeckite	0.487	0.00833	0.523	226 ± 12
A-2033	72-3	Riebeckite-aegirine granite	Riebeckite	0.823	0.01111 0.01214 0.01153	0.584 0.581 0.537	188 ± 9
R-3042	85B	Trachyte	Whole rock	3.813 3.804	0.06624	0.604	229 ± 11
B-3043	431	Granite porphyry	Altered biotite	0.977 0.993	0.01915	0.350	254 ± 14
A-3044	69	Coarse riebeckite granite	Riebeckite	0.609 0.607	0.01644 0.01621	0.456 0.498	343 ± 14

Analysts: H.W. Krueger and R.H. Reesman

REFERENCES CITED

Barth, T.F.W., 1969, Feldspars: New York, John Wiley & Sons, Inc., 261 p.
Billings, M. P., 1956, The geology of New Hampshire. Pt. II, Bedrock geology: Concord, New Hampshire State Planning and Development Comm., 203 p.
——1976, Geology of the Boston basin, *in* Lyons, P. C., and Brownlow, A. H., eds., Studies in New England geology: Geol. Soc. America Mem. 146, p. 5-30 (this volume).
Borley, G. D., 1963, Amphiboles from the younger granites of Nigeria. Pt. I, Chemical classification: Mineralog. Mag., v. 33, p. 358-376.
Bottino, M. L., Fullagar, P. D., Fairbairn, H. W., Pinson, W. H., Jr., and Hurley, P. M., 1970, The Blue Hills igneous complex, Massachusetts, whole-rock Rb-Sr open systems: Geol. Soc. America Bull., v. 81, p. 3739-3746.
Bowen, N. L., and Schairer, J. R., 1935, Grunerite from Rockport, Massachusetts and a series of synthetic fluor-amphiboles: Am. Mineralogist, v. 20, p. 543-551.
Buma, Grant, Frey, F. A., and Wones, D. R., 1971, New England granites: Trace element evidence regarding their origin and differentiation: Contr. Mineralogy and Petrology, v. 31, p. 300-320.
Chapman, C. A., 1968a, Intersecting belts of post-tectonic "alkaline" intrusions in New England: Illinois Acad. Sci. Trans., v. 61, p. 46-52.
——1968b, A comparison of the Maine Coastal plutons and the magmatic central complexes of New Hampshire; *in* Zen, E-An, and others, eds., Studies of Appalachian geology—Northern and Maritime: New York, Interscience Pubs., Inc., p. 385-396.
——1976, Structural evolution of the White Mountain magma series, *in* Lyons, P. C., and Brownlow, A. H., eds., Studies in New England geology: Geol. Soc. America Mem. 146, p. 281-300 (this volume).
Chapman, R. W., and Williams, C. R., 1935, Evolution of the White Mountain magma series: Am. Mineralogist, v. 20, p. 502-530.
Chute, N. E., 1950, Bedrock geology of the Brockton quadrangle, Massachusetts: U.S. Geol. Survey Geol. Quad., Map 5.
——1969, Bedrock geologic map of the Blue Hills quadrangle, Norfolk, Suffolk, and Plymouth Counties, Massachusetts: U.S. Geol. Survey Geol. Quad., Map 796.
Clapp, C. H., 1921, Geology of the igneous rocks of Essex County, Massachusetts: U.S. Geol. Survey Bull. 704, 132 p.
Clarke, F. W., 1903, Mineral analyses from the laboratories of the United States Geological Survey 1880 to 1903: U. S. Geol. Survey Bull. 220, p. 77.
Crosby, W. O., 1900, Geology of the Boston Basin, the Blue Hills complex: Boston Soc. Nat. History Occasion. Papers 4, v. 1, pt. 3, p. 289-563.
Emerson, B. K., 1917, Geology of Massachusetts and Rhode Island: U.S. Geol. Survey Bull. 597, 289 p.
Ernst, W. G., 1962, Synthesis, stability relations, and occurrence of riebeckite and riebeckite-arfvedsonite solid solutions: Jour. Geology, v. 70, p. 689-736.
Eugster, H. P., and Wones, D. R., 1962, Stability relations of the ferruginous biotite, annite: Jour. Petrology, v. 3, p. 82-125.
Fabriès, Par J., and Rocci, G., 1965, Le massif grantique de Tarraouadji (République du Niger). Étude et signification pétrogénétique des principaux minéraux: Soc. Française Minéralogie et Cristallographie Bull., v. 88, p. 319-340.
Fairbairn, H. W., Bottino, M. L., Handford, L. S., Hurley, P. M., Heath, M. M., and Pinson, W. H., Jr., 1968, Radiometric ages of igneous rocks in northeastern Massachusetts: Geol. Soc. America, Abs. for 1967, Spec. Paper 115, p. 260-261.
Faul, H., Stern, T. W., Thomas, H. H., and Elmore, P.L.D., 1963, Ages of intrusion and metamorphism in the northern Appalachians: Am. Jour. Sci., v. 261, p. 1-19.
Foland, K. A., Quinn, A. W., and Giletti, B. J., 1971, K-Ar and Rb-Sr Jurassic and Cretaceous ages for intrusives of the White Mountain magma series, northern New England: Am. Jour. Sci., v. 270, p. 321-330.
Goldich, S. S., Baadsgaard, H., and Nier, A. O., 1957, Investigations in A^{40}/K^{40} dating: Am. Geophys. Union Trans., v. 38, p. 547-551.

Harakal, J. E., 1966, Potassium-argon ages of the Scituate granite, north-central Rhode Island [M.A. thesis]: Providence, R.I., Brown Univ., 32 p.

Hart, S. R., 1961, The use of hornblendes and pyroxenes for K-Ar dating: Jour. Geophys. Research, v. 66, p. 2995-3001.

——1964, The petrology and isotopic-mineral age relations of a contact zone in the Front Range, Colorado: Jour. Geology, v. 72, p. 493-525.

Hurley, P. M., Fairbairn, H. W., Pinson, W. H., and Faure, G., 1960, K-Ar and Rb-Sr minimum ages for the Pennsylvanian section in the Narragansett Basin: Geochim. et Cosmochim. Acta, v. 18, p. 247-258.

Jacobson, R.R.E., MacLeod, W. N., and Black, R., 1958, Ring-complexes in the younger granite province of northern Nigeria: Geol. Soc. London Mem. 1, 71 p.

Joyner, W. B., 1958, Gravity in New Hampshire and adjoining areas [Ph.D. thesis]: Cambridge, Mass., Harvard Univ., 126 p.

Kaktins, Uldis, 1976, Stratigraphy and petrography of the volcanic flows of the Blue Hills area, Massachusetts, *in* Lyons, P. C., and Brownlow, A. H., eds., Studies in New England geology: Geol. Soc. America Mem. 146, p. 125-142 (this volume).

Karner, F. R., 1963, Petrology of the Tunk Lake Granite pluton, southeastern Maine [Ph.D. thesis]: Urbana, Univ. Illinois, 94 p.

Krueger, H. W., and Lyons, P. C. 1972, K-Ar dating of the Rattlesnake Hill granites of southeastern Massachusetts: Geol. Soc. America, Abs. with Programs (Northeastern Sec.), v. 4, no. 1, p. 26.

Lyons, J. B., and Faul, H., 1968, Isotope geochronology of the northern Appalachians, *in* Zen, E-An, and others, eds., Studies of Appalachian geology, Northern and Maritime: New York, Interscience Pubs., Inc., p. 305-318.

Lyons, P. C., 1969, Bedrock geology of the Mansfield quadrangle, Massachusetts [Ph.D. dissert.]: Boston, Mass., Boston Univ., 283 p.

——1971a, Mineral relations and stability in the alkaline granites of the Rattlesnake Hill pluton: Geol. Soc. America, Abs. with Programs (Northeastern Sec.), v. 3, no. 1, p. 44.

——1971b, Staining of feldspars on rock-slab surfaces for modal analysis: Mineralog. Mag., v. 38, p. 518-519.

——1972, Significance of riebeckite and ferrohastingsite in micropherthite granites: Am. Mineralogist, v. 57, p. 1404-1412.

Lyons, P. C., and Wolfe, C. W., 1971, Correlation of granites by soda/potash ratios: Geol. Soc. America Bull., v. 82, p. 2023-2026.

Nichols, D. R., 1956, Bedrock geology of the Narragansett Pier quadrangle, Rhode Island: U.S. Geol. Survey Quad., Map GQ-91.

Noble, D. C., 1968, Kane Springs Wash volanic center, Lincoln County, Nevada, *in* Eckel, E. B., ed., Nevada test site—Studies in geology and hydrology: Geol. Soc. America Mem. 110, p. 109-116.

——1970, Loss of sodium from crystallized comendite welded tuffs of the Miocene Grand Canyon member of the Belted Range tuff, Nevada: Geol. Soc. America Bull., v. 81, p. 2677-2688.

Page, L. R., 1968, Devonian plutonic rocks in New England, *in* Zen, E-An, and others, eds., Studies of Appalachian geology, Northern and Maritime: New York, Interscience Pubs., Inc., p. 371-383.

Parsons, Ian, 1972, Petrology of the Puklen syenite-alkali granite complex, Nunassuit, South Greenland: Medd. Grönland, v. 195, no. 3, 73 p.

Quinn, A. W., 1971, Bedrock geology of Rhode Island: U.S. Geol. Survey Bull. 1295, 68 p.

Quinn, A. W., and Moore, G. E., Jr., 1968, Sedimentation, tectonism, and plutonism of the Narragansett Basin region, *in* Zen, E-An, and others, eds., Studies of Appalachian Geology, Northern and Maritime: New York, Interscience Pubs., Inc., p. 269-279.

Quinn, A. W., Ray, R. G., and Seymour, W. L., 1949, Bedrock geology of the Pawtucket quadrangle, Rhode Island-Massachusetts: U.S. Geol. Survey Geol. Quad., Map GQ-1.

Reesman, R. H., Filbert, C. R., and Krueger, H. W., 1973, Potassium-argon dating of the upper Triassic lavas of the Connecticut valley, New England: Geol. Soc. America, Abs. with Programs (Northeastern Sec.), v. 5, no. 2, p. 211.

Rodgers, John, 1970, The tectonics of the Appalachians: New York, Interscience Pubs., Inc., 271 p.

Toulmin, Priestley, III, 1960, Composition of feldspars and crystallization history of the granite-syenite complex near Salem, Essex County, Massachusetts: Internat. Geol. Cong., 21st, Copenhagen 1960, Rept., pt. 13, p. 275–286.

——1964, Bedrock geology of the Salem quadrangle and vicinity, Massachusetts: U.S. Geol. Survey Bull. 1163-A, 79 p.

Tuttle, O. F., and Bowen, N. L., 1958, Origin of granite in the light of experimental studies in the system $NaAlSi_3O_8$-$KAlSi_3O_8$-SiO_2-H_2O: Geol. Soc. America Mem. 74, 153 p.

Warren, C. H., 1913, Petrology of the alkali-granites and porphyries of Quincy and the Blue Hills, Massachusetts, U.S.A.: Am. Acad. Arts Sci. Proc., v. 49, p. 203–331.

Warren, C. H., and McKinstry, H. E., 1924, The granites and pegmatites of Cape Ann, Massachusetts: Am. Acad. Arts Sci. Proc., v. 59, p. 315–357.

Warren, C. H., and Palache, C., 1911, The pegmatites of the riebeckite-aegirite granite of Quincy, Mass., U.S.A.: Their structure, minerals, and origin: Am. Acad. Arts Sci. Proc., v. 47, p. 126–168.

Whitehead, W. L., 1913, The geology of the Rattlesnake Hill granite of Sharon, Massachusetts: Massachusetts Inst. Tech. Mining Dept. Thesis (Mining Thesis Case), 35 p.

Wolfe, C. W., 1949, The Blister hypothesis and the orogenic cycle: New York Acad. Sci. Trans., v. 11, p. 188–195.

Zartman, R. E., and Marvin, R. F., 1971, Radiometric age (Late Ordovician) of the Quincy, Cape Ann, and Peabody Granites from eastern Massachusetts: Geol. Soc. America Bull., v. 82, p. 937–958.

Zartman, R. E., and Naylor, R. S., 1972, Structural implications of some U-Th-Pb zircon isotopic ages of igneous rocks in eastern Massachusetts: Geol. Soc. America, Abs. with Programs (Northeastern Sec.), v. 4, no. 1, p. 54–55.

Zartman, R. E., Hurley, P. M., Krueger, H. W., and Giletti, B. J., 1970, A Permian disturbance of K-Ar radiometric ages in New England—Its occurrence and cause: Geol. Soc. America Bull., v. 81, p. 3359–3373.

MANUSCRIPT RECEIVED BY THE SOCIETY MARCH 20, 1974
REVISED MANUSCRIPT RECEIVED JULY 30, 1974
MANUSCRIPT ACCEPTED AUGUST 26, 1974

Printed in U.S.A.

Ayer Crystalline Complex at Ayer, Harvard, and Clinton, Massachusetts

RICHARD Z. GORE
Department of Earth and Environmental Sciences
University of Lowell (South Campus)
Lowell, Massachusetts 01854

ABSTRACT

The Ayer Granodiorite (Jahns, 1952) in the vicinity of Clinton, Harvard, and Ayer, Massachusetts, is divided into two units. The youngest unit, the Clinton facies, is composed of slightly to moderately foliated porphyritic quartz monzonite—the rock that dominates both Emerson's (1917) and Billings' (1956) descriptions. The second unit, the Devens-Long Pond facies, primarily contains feldspathic gneiss varying from porphyroblastic to equigranular varieties. These rocks range in composition from quartz monzonite to quartz diorite.

K-feldspar unit cell analysis indicates that there is more than a 99 percent probability that the microperthite of the Clinton facies comes from a different population than the microperthite of the Devens-Long Pond facies. Twenty homogenized megacryst samples of the Clinton facies have a mean composition of Or 79 ± 5 percent, whereas the mean composition of three homogenized porphyroblastic feldspar samples from the coarsest gneiss of the Devens-Long Pond facies is Or 91.5 ± 2 percent. The similarity in all samples of the unit cell volume of the K-rich phase of the perthite and the maximum microcline structural state suggest that both facies of the "Ayer Crystalline Complex" (named herein) experienced a similar final thermal history.

The Clinton facies is an igneous intrusion that has undergone one, and possibly two, postconsolidation events that produced both cataclastic and recrystallization textural features. The age of the Clinton facies is uncertain but is probably Devonian. The Clinton facies intrudes the Oakdale Quartzite, Straw Hollow Diorite, and Worcester Phyllite but is in fault contact with the Brimfield Schist and the Harvard Conglomerate.

The heterogeneous nature of the Devens-Long Pond facies on mesoscopic and macroscopic scales suggests that their protoliths were part of a volcanic-sedimentary sequence probably of Ordovician age or older. The Devens-Long Pond facies contains some granitoid rocks of uncertain origin and is intruded by the Chelmsford granite. The relationship of the Clinton facies to the Devens-Long Pond facies is uncertain. Portions of the Devens-Long Pond facies may be separated from the presumably younger Oakdale Quartzite by an unconformity.

The preintrusive or synintrusive blastesis of the Devens-Long Pond facies couples with the postconsolidational foliation of the Clinton facies and Chelmsford granite to suggest that the Acadian orogeny was characterized by two distinct pulses in this region.

INTRODUCTION

The name "Ayer Granite," as originally used in east-central Massachusetts by Emerson (1917), included almost all the leucocratic plutonic rocks on the east side of the Merrimack synclinorium (Fig. 1). Currier (1952), Jahns (1952), Hansen (1956), and Grew (1970) locally subdivided this heterogeneous formation into several mappable units, retaining the name Ayer as a designation for at least one of the subdivisions.

Billings (1956), Sriramadas (1966), and Sundeen (1971), use the name Ayer for quartz monzonite and granodiorite in southeastern New Hampshire. The New Hampshire rocks are correlated with Ayer Granodiorite as defined at the type locality by Jahns (1952, p. 108-112).

Jahns (1952, p. 108, 112) characterized his Ayer Granodiorite at the type locality as a composite mass of gneiss, migmatite, and plutonic igneous rocks showing wide textural variations. He reported ranges in composition from quartz diorite to quartz monzonite.

The present investigation shows that Jahns's unit at Ayer and Harvard, Massachusetts, (Fig. 2) can be divided into two geographically and lithologically distinct portions. The most distinctive subdivision is the porphyritic quartz monzonite, the rock most thoroughly discussed in Emerson's (1917, p. 224-226) Ayer Granite description. This unit shows a strong modal uniformity despite its coarse texture and variations in development of foliation.

The second subdivision of Jahns's Ayer Granodiorite is the dominant rock type in the town of Ayer and consists of a complex assemblage of gneiss, migmatite,

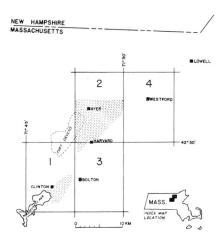

Figure 1. Location map. Hatched areas detailed in Figures 2 and 3. 1. Clinton quadrangle. 2. Ayer quadrangle. 3. Hudson quadrangle. 4. Westford quadrangle.

and possibly plutonic igneous rocks that show compositional and wide textural variations. Rocks ranging in composition from quartz diorite to quartz monzonite are present.

In light of these subdivisions, the Ayer Granodiorite of Jahns (1952) needs redefinition. The implications of this study suggest that a complete separation of the subdivisions on genetic grounds may be warranted. However, the presence within the gneiss complex of some rocks of possible comagmatic origin with the porphyritic quartz monzonite requires that any clearly defined genetic subdivision await further study. A partial redefinition of the term Ayer is offered here to accentuate differences found between rocks within Jahns's Ayer Granodiorite, to avoid the genetic and time-equivalency connotation of the older designation, and to avoid confusion with Hansen's (1956, p. 47-48) subdivisions of the Ayer Granodiorite (Granite) in the Hudson quadrangle, Massachusetts.

It is proposed that the Ayer Granodiorite be renamed the Ayer Crystalline Complex. This complex is divided into two units: the porphyritic quartz monzonite, which in this paper will be called the Clinton facies, and a gneiss assemblage here called the Devens–Long Pond facies.

METHODS

This paper reports observations and conclusions made primarily by standard field mapping methods augmented by petrographic and x-ray investigation. Portions of the 7-1/2' Clinton, Hudson, and Ayer, Massachusetts, quadrangle maps (Fig. 1) provided geographic control.

The modes of rock units were estimated by point counting on sodium cobaltinitrite stained sawed slabs varying in area from 39 to 195 cm^2. The porphyritic texture of the Clinton facies necessitated direct outcrop point counting to obtain a representative sampling of the K-feldspar megacryst population. To facilitate outcrop point counting, a 1-cm^2 grid was constructed on a 0.56-m^2 sheet of heavy flexible plastic.

Procedures used in making up K-feldspar samples for x-ray analysis were guided by an investigation (Gore, 1973, p. 150-157) of the variations in unit cell parameters between megacrysts within a single Clinton-facies outcrop. This work reveals that differences are generally small, but variations equivalent to 10 percent in the estimate of Or content (Wright and Stewart, 1968) may be present. These variations between megacrysts required that as many grains as feasible be used in making a representative sample for each location. The large grain size of the megacrysts of the Clinton facies resulted in a K-feldspar sample composed of from four to six megacrysts. The K-feldspar augen from the Devens–Long Pond facies were small enough to permit the preparation of a sample composed of more than 10 grains from a single slab. A drawback in using "averaged" samples is that the standard error associated with the cell-volume determinations (Tables 4, 6), particularly for the unhomogenized samples, is fairly large due to peak broadening in the diffraction patterns.

Smear mounts of the K-feldspar samples mixed with annealed CaF_2 were run in copper $K\alpha$ radiation on a Phillips vertical goniometer and diffractometer at a scanning speed of $1/2°$ 2θ/min. Three patterns were made for each sample. Least squares refined unit cell parameters were calculated using Appleman's modification of Evans and others' (1963) variable indexing, cell-parameter refinement computer program.

Homogenization of the perthite is required to determine its bulk composition from unit cell measurements. The microperthite examined for this study can be

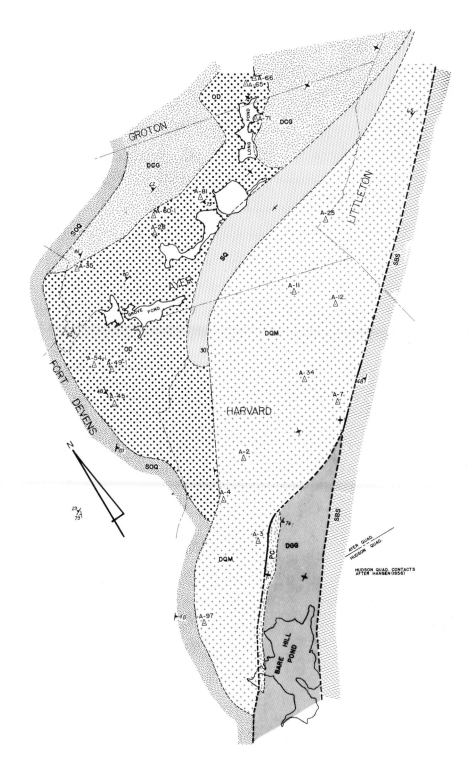

Figure 2. Geologic map of Harvard-Ayer area.

completely homogenized after dry heating for 6 hr at 975°C. This was verified by noting the disappearance of the albite reflections in powder patterns and *b*-axis oscillation photos.

CLINTON FACIES

Location

The Clinton facies occurs principally in one large body in the Harvard-Littleton-Ayer region (Fig. 2) and extends from just west of Bare Hill Pond in the northwest corner of the Hudson quadrangle to a pinch out just southwest of Forge Pond, Westford. This is the Harvard body of this paper.

The best exposures of this facies are found in the Clinton, Massachusetts, area. The rocks of these exposures were described by Crosby (1899, p. 74-75) and Emerson (1917, p. 225-226). In the Clinton area (Fig. 3), the facies occurs in two bodies.

Description

Megascopically, the Clinton facies is a porphyritic, slightly to moderately foliated, coarse-grained rock. The color of the rock varies from very light to medium gray, green-gray, to dark gray. Color is strongly dependent on the degree of facturing and its concomitant alteration. The darker color is primarily due to the development of chlorite and epidote group minerals.

The most distinctive aspect of the Clinton facies is the presence of megacrysts of microcline microperthite. Although most of these crystals appear to be magmatic in origin, a small portion may have crystallized after consolidation. This possible dual origin requires the use of the term "megacryst," which is genetically neutral. The crystals are large (averaging 5 cm in length), euhedral, and generally exhibit Carlsbad twins. Megacrysts up to 10 cm long are not uncommon. These megacrysts may contain unaltered, small (average length between 0.2 and 0.5 mm), euhedral, poikilitic plagioclase. This poikilitic plagioclase (An_{19}?) has no textural counterpart in the groundmass plagioclase. The K-feldspar megacrysts generally appear to predate the development of the major Clinton-facies foliation.

The other major minerals are saussuritized plagioclase (albite to sodic oligoclase), quartz, and biotite. The accessory minerals are apatite, zircon, and an unidentified opaque. Muscovite may be present in small amounts as well developed flakes, whereas sericite is universally present. Epidote, allanite, clinozoisite(?), chlorite, sphene, and a carbonate mineral occur in highly variable amounts. Gore (1973) provided detailed descriptions of each mineral.

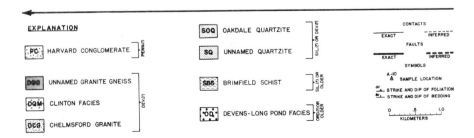

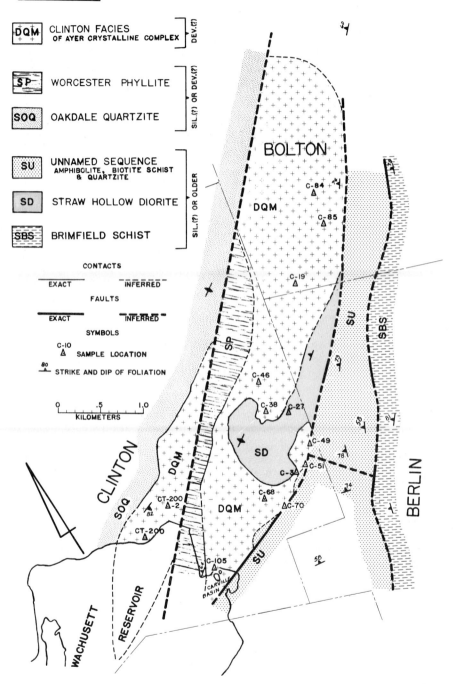

Figure 3. Geologic map of Clinton area.

A nonporphyritic variety frequently occurs in contact zones. Volumetrically, these rocks constitute only a small portion of the rock of the Clinton facies. Hansen (1956) mapped the southwestern end of the Harvard body (Fig. 2) of the Clinton facies and subdivided it into porphyritic and nonporphyritic facies. A re-examination of some outcrops, shown on Hansen's map as his nonporphyritic facies, revealed that these rocks had undergone granulation and recrystallization in a major shear zone. It is uncertain if any portion of these rocks is correlative with the well-exposed contact facies seen at Clinton.

Mode

The megacryst population was estimated by outcrop point counting at six locations. These locations (C-19, C-27, A-2, A-4, A-7, A-25) extended from Clinton to Littleton, Massachusetts (Figs. 2, 3). The K-feldspar megacryst modes ranged from 16.6 to 20.4 percent, averaging about 20 percent.

Slab point-count data from 13 locations in the Clinton bodies and 10 locations in the Harvard body were averaged and recalculated (Table 1), using 20 percent as the estimated volume of the K-feldspar megacryst population. A comparison of these data indicates that no significant difference exists between the modes of the Clinton and Harvard bodies. The rock of all three bodies of the Clinton facies is quartz monzonite.

Foliation

Two prominent foliations and a lineation exist in the Clinton facies. The lineation is produced by a preferred orientation, particularly in contact zones, of the xenoliths and K-feldspar megacrysts. This preferred orientation (Fig. 4) contributes to a coarse foliation produced by the alignment of mica, lens-shaped masses of mosaic quartz and the subparallel alignment of plagioclase.

Page (1968, p. 372, 377) included the Clinton facies of this study within the New Hampshire plutonic series and attributed their major foliation and lineation to crystallization under syntectonic conditions. This study supports, for the Clinton facies, Page's (1968, p. 377) view that the K-feldspar within these rocks is generally of magmatic origin but disagrees with his conclusion that crystallization and foliation development occurred simultaneously. It is my opinion that the original fabric of the Clinton facies was basically isotropic, possibly exhibiting a preferred orientation of the megacrysts near contacts. The major foliation may have been imposed soon after consolidation once the pore fluid had disappeared, which permitted the development of large intergrain stresses; however, a separate tectonic pulse other than that which emplaced the magma better explains the observations of this study.

Evidence for a secondary development of the major foliation can be seen best in the textures of the quartz and biotite. The quartz generally appears as lenticular

TABLE 1. AVERAGE MODES OF THE CLINTON FACIES

	Clinton bodies	Harvard body	Average
K-feldspar (phenocrysts)	20	20	20
K-feldspar (groundmass)	7.8	9.1	8.5
Plagioclase	38.2	34.1	36.1
Quartz	25	29.4	27.2
Mafic mineral	9	7.4	8.2

Figure 4. Outcrop shows preferred orientation of megacrysts and xenoliths in Clinton facies near Worcester Phyllite contact. Pen shows scale.

Figure 5. Slab shows Clinton-facies coarse foliation produced by lens-shaped quartz and elongate patches of biotite.

masses (Fig. 5), often with a biotite or chlorite clot or patch at the center of the mass. During this event, the K-feldspar megacrysts experienced rotation and, in some instances, elongation by slippage along fractures parallel to their (001) cleavage.

The most intense development of major foliation near contacts has been used as a criterion to support the hypothesis of a protoclastic origin for the foliation. In the Clinton area, where contacts are well exposed, pronounced foliation is developed only at the quartz monzonite-Worcester phyllite (Emerson, 1917) contact. The contact of the quartz monzonite and Emerson's (1917) Oakdale Quartzite is knife-edge sharp (Fig. 6) and does not show this pronounced foliation or a preferred orientation of the megacrysts.

The most pronounced development of the major foliation along the phyllite contact is best explained in terms of the differences in competency of the two rocks. The mechanical stress produced by the drag of a plastically flowing phyllite under metamorphic conditions may explain the apparent concentration of shearing stress in the quartz monzonite along their common boundary.

The other prominent foliation is clearly a shear foliation concentrated in narrow zones. This foliation was produced by a cataclastic event. Granulation at some places has reduced the rock to a very finely (1 mm) laminated gneiss, whereas at other locations the rock is only partly reduced to an augen gneiss or blastomylonite. The shear foliation is generally parallel to the more pervasive foliation and regional strike (N30°-60°E), but other orientations occur. It is not clear if the shear foliation and the major foliation were produced by separate tectonic episodes or are products of different pulses within a single event.

Contact Relationships

The Clinton facies intrudes three rock units (Fig. 3), the Straw Hollow Diorite, Worcester Phyllite, and Oakdale Quartzite of Emerson (1917). The nature of the contact (Fig. 2) between the Clinton facies and the Devens-Long Pond facies is uncertain.

The relationship of the Clinton facies to the Harvard Conglomerate has been a subject of considerable interest. The importance of this relationship rests on the correlation of the Harvard Conglomerate on Pin Hill (type locality) with a similar conglomerate in Worcester, Massachusetts, considered by Grew and others (1970) as possibly Pennsylvanian in age. Emerson (1917) and Hansen (1956) maintained that the Clinton facies (Ayer Granite) was Pennsylvanian or post-Pennsylvanian on the basis of their belief that it intrudes the presumed Pennsylvanian Harvard Conglomerate. Grew (1973, p. 127), however, contended that the Harvard Conglomerate is unconformably deposited on the Clinton facies.

An examination of the exposures and samples from the west side of Pin Hill, including the outcrop discovered by P. Robinson (Grew, 1973, p. 127), has led the author to conclude that the Clinton facies and Harvard Conglomerate are in fault contact separated by a narrow mylonite zone.

The Clinton facies and Brimfield Schist (Tadmuck Brook Schist of Bell and Alvord, 1974) of Emerson (1917) are separated by a shear zone that marks the trace of the regional Clinton-Newbury fault. This shear zone is approximately 1,000 ft wide in the vicinity of Route 2 and is marked by a series of phyllonite units having considerable variation in quartz content.

Three fault zones (Fig. 3) pass through the mapped portion of the Clinton area. Two of these faults cut the Clinton facies. The existence of these sheared zones was first reported by Crosby (1899) from observations made in the Wachusett

Figure 6. Outcrop shows sharp contact between Clinton facies and Oakdale Quartzite. Pen shows scale.

Figure 7. Outcrop shows nonporphyroblastic gneiss with coarse banding. Scale shown by lens cap.

Aqueduct. He characterized only the southernmost zone (Rattlesnake Hill fault zone, Skehan, 1968, p. 287) as a fault. Skehan (1968), working in the Wachusett-Marlborough Tunnel recognized the importance of faulting along all the shear zones and extended into the Clinton area the regional fault that can be traced from Newburyport.

Skehan (1968) originally considered the main trace of the Clinton-Newbury fault zone to pass through the Nashua River Gorge at Clinton. Later (1974, written commun.), Skehan reinterpreted this shear zone in the gorge as a branch of the major fault in the upper plate and located the main part of the trace of the Clinton-Newbury fault along the fault through the Carville basin of the Wachusett Reservoir (Fig. 3). I agree that the Nashua River Gorge fault is not the extension of the major regional fault because all significant masses of Ayer-like rocks occur north or northwest of this fault from Newburyport to Harvard, Massachusetts. If the Nashua River Gorge fault was the main branch of the Clinton-Newbury fault, the Clinton-facies body southeast of that fault would be the only rock of its kind, except for very small lens-shaped bodies (D. Alvord, 1975, personal commun.), to occupy that tectonic position.

All three faults are closely related and probably define an imbricate structural belt at Clinton. However, the Rattlesnake Hill fault zone displays relationships that rank it above the other faults in importance and suggest that it marks the main trace of the Clinton-Newbury fault. The structural importance of the Rattlesnake Hill fault zone east of Clinton was reported by Crosby (1899, p. 93) when he noted that the strike on the foliation abruptly changed from nearly due east on the western side of the fault zone to northeast on the east side. The same structural relationships were verified by my surface mapping.

The only lithologic unit continuously present in the vicinity of the Clinton-Newbury fault zone from the Lowell area southwest through the Clinton area is the Brimfield Schist. This unit marks the top of the eastern side of the fault zone in the Lowell and Harvard areas and is the unit southeast of the Rattlesnake Hill fault zone. The Rattlesnake Hill fault may also be a boundary separating rocks of highly contrasting metamorphic grade, but pronounced retrograde metamorphism, particularly on the west side of the fault, has obscured relationships.

DEVENS–LONG POND FACIES

Location

This facies constitutes almost all the outcrops of the Ayer Crystalline Complex in the town of Ayer (Fig. 2). The name "Devens–Long Pond" was chosen for the excellent exposures of porphyroblastic gneisses about Shepleys Hill on the Fort Devens Army Reservation and for the exposures, particularly those of the equigranular gneiss, west of Long Pond in Ayer, Massachusetts.

Description

The gneiss of this belt has considerable textural and modal variations. The feldspathic gneiss of this belt can be subdivided into equigranular (Fig. 7) and porphyroblastic (Fig. 8) varieties. Intermediate textures do occur but are subordinate. Both major textural types occur intermixed, but the porphyroblastic type is more abundant in the western portion of the belt. The porphyroblasts are composed of microcline and, to a much lesser extent, plagioclase. The microcline porphyroblasts

Figure 8. Outcrop of porphyroblastic gneiss with interlayered (upper left and lower right) nonporphyroblastic gneiss. Scale shown by 6-in. white rule.

are dominantly lenticular, imparting an ophthalmitic structure to the rock. The porphyroblastic rocks have xenoblastic microtextures that appear to have been produced by intense granulation. The porphyroblastic gneiss is light to medium gray in color. The average length of the quartz and plagioclase ranges between 3 and 8 mm. The microcline porphyroblasts average between 1 to 2 cm in length. Most of the porphyroblastic gneiss is mineralogically and modally similar (Table 2). All samples contain albite, microcline perthite, quartz, and biotite. Other minerals usually present are chlorite, sericite, sphene, and an opaque. The following minerals are commonly present: zircon, apatite, muscovite, a carbonate mineral, and limonite(?). Allanite and tourmaline occur more rarely. The porphyroblastic gneiss has the composition of a quartz monzonite.

The nonporphyroblastic gneiss is usually medium-gray, medium-grained rock. Color may vary from light gray to almost black, with greenish hues not uncommon.

TABLE 2. MODAL ANALYSES AND AVERAGE GRAIN SIZES OF SELECTED SAMPLES OF PORPHYROBLASTIC GNEISS OF THE DEVENS—LONG POND FACIES

	Samples*			
	A-45	A-49-1	A-54-1	Average
Quartz	29.9 (5 mm)	21.0	21.0 (2-5 mm)	26.0
Plagioclase	36.7 (15 mm)^L	40.3 (4-7 mm)	43.0 (3-8 mm)	40.2
K-feldspar	29.9 (10-20 mm)	27.45 (10 mm) (20 mm)^L	30.6 (4-8 mm) (17 mm)^L	29.4
Mafic mineral†	3.5(B)	4.0(B,C)	5.4(B)	4.3
Other		1.4		

Note: Superscript L refers to largest grains.
* Locations plotted on Figure 2.
† (B) = biotite; (C) = chlorite.

Most of these rocks are biotite gneiss, but amphibole-bearing biotite gneiss also occurs. The texture of the nonporphyroblastic gneiss is foliated crystalloblastic with slight to moderate cataclasis. The average grain size of the major felsic minerals ranges from 5 to 8 mm. The mineralogy of most of these rocks is similar, but the modal percentage (Table 3) is highly variable. The dominant minerals are quartz, albite, and biotite. Microcline, sphene, and allanite are common. Blue-green pleochroic amphibole (actinolite?), muscovite (commonly sericite), epidote, zircon, apatite, garnet, tourmaline(?), and chlorite are frequently present.

The nonporphyroblastic gneiss shows distinct compositional variations over short distances. A coarse banding (sometimes a regular schlieric structure) can often be observed within an outcrop. Some of the gneiss is fairly homogeneous and may be plutonic igneous rocks. The nonporphyroblastic rocks range in composition from quartz monzonite to quartz diorite (Table 3).

Contact Relationships

The Devens-Long Pond facies belt (Fig. 2) is bounded on the west by the Oakdale Quartzite of Emerson (1917). The nature of the contact is uncertain, but the possibility exists that the Oakdale is unconformably deposited on parts of the Devens-Long Pond rocks. Observations supportive of this interpretation are that pegmatite and aplite dikes are common in the Devens-Long Pond facies but are lacking in the Oakdale. Primary features are preserved in the Oakdale, a condition that contrasts sharply to the dominantly crystalline character of the Devens-Long Pond. The succession of lithologies around Ayer, Massachusetts, are somewhat analogous to the geologic framework within the belt of nappes and gneiss domes on the west side of the Merrimack synclinorium, where an unconformity exists between the Ammonoosuc Volcanics and Clough Quartzite.

The Chelmsford granite, (Currier, 1937), which actually has the composition of a quartz monzonite, intrudes the Devens-Long Pond facies. The contact is transitional over a zone varying from 30 to more than 100 m wide. The central and eastern portions of the Devens-Long Pond belt are cut by numerous pegmatite dikes that seem to be directly related to the Chelmsford granite. Prominent myrmekite shells around K-feldspar grains have been found at many of the Devens-Long Pond locations that are within 0.8 km of the Chelmsford granite. The close relationship of these myrmekite shells to the Chelmsford granite contact suggests that they are part of an aureole associated with the formation of the granite.

POTASSIUM FELDSPAR ANALYSIS

The similarity of mineralogy and mode between the Clinton facies and the porphyroblastic rocks of the Devens-Long Pond facies has traditionally suggested a comagmatic origin. The extreme differences in texture could be explained by intrusion of magma into areas having somewhat different local stress environments.

The prominence of K-feldspar in these rocks and the dependence of its crystallography on composition and thermal history make it excellent for studying genetic similarities. Cell-volume data were exclusively used in making comparisons between the Clinton facies and the porphyroblastic gneiss of the Devens-Long Pond facies. The use of cell volumes was dictated by the presence of some samples with anomalous (defined by Wright and Stewart, 1968, p. 40, as cells were a [observed] $- a$ [estimated from b and c] exceeds 0.02 A) cell dimensions.

The large standard error associated with the cell-volume determination for some

TABLE 3. MODAL ANALYSES AND AVERAGE GRAIN SIZES OF SELECTED NONPORPHYROBLASTIC GNEISS OF THE DEVENS—LONG POND FACIES

	Samples*				
	A-60-1	A-66	A-65	A-71	A-81
Quartz	27.1 (10 mm)^L	20.9 (1 mm)	25.5 (2-5 mm)	16.7 (5 mm)^L	29.5 (2 mm)
Plagioclase	59.4 (2-5 mm)§	53.0 (3-4 mm)	44.1 (3-6 mm)	44.6 (3-6 mm)	53.0 (2-4 mm)
K-feldspar	8.9	18.0 (1-2 mm)	21.4 (10 mm)	23.8 (10 mm)	4.7 (5 mm)
Muscovite	0.6				0.7
Mafic mineral$^+$	3.9(B)	6.9(B) (1 mm)	9.0(B) (2-8 mm)	14.9(B,A) (2-3 mm)	12.1(B)

Note: Superscript L refers to largest grains.
* Locations plotted on Figure 2.
+ (B) = biotite; (A) = amphibole.
§ Refers to size range.

samples required that any inferences drawn from this data be statistically supported. Two test models were used in the analysis of the data. Model 1 is a standard F test on the null hypothesis that observed variations in unit cell volumes between samples (V_B) are no greater than those that would be expected by randomly sampling a common population having a population variance similar to the error variance (V_W). Rejection of the Model 1 null hypothesis means that a significant difference can be inferred to exist in the data. Model 2 is a standard F test that compares the cell-volume variance within defined groups to the cell-volume variance between these groups. The discovery of a significant difference in this application would mean that the cell-volume character of the defined groups differed more than that expected if the groups were composed of individual samples randomly drawn from a common population.

K-rich Phase of Unhomogenized Perthite

The K-rich phase of the megacryst perthite samples from the Clinton bodies of the Clinton facies ranged in composition from Or 86.5 percent to 100 percent. A Model 1 test (test 1, Table 5) failed to show any significant difference in the data (Table 4). An estimate of mean composition of the population based on the mean cell volume of 11 samples is 94.5 percent Or.

A Model 1 test (test 2, Table 5) was conducted on the cell-volume data (Table

TABLE 4. UNIT CELL VOLUMES OF POTASSIUM-RICH PHASE OF PERTHITE

Samples*	Mean cell volume	Variance$^{1/2}$ (standard deviation)	Samples$^+$	Mean cell volume	Variance$^{1/2}$ (standard deviation)
		Clinton facies			
C-46	719.815	1.032	A-2	721.061	0.694
C-49	718.929	0.305	A-3	721.508	0.739
C-51	718.968	1.747	A-4	720.100	2.362
C-68	722.689	0.769	A-11	719.058	0.522
C-70	716.813	1.748	A-12	718.369	0.878
C-84	719.848	1.697	A-25	721.504	0.956
C-85	722.360	0.993	A-32-2	720.255	0.439
C-105	720.376	1.427	A-34	721.939	1.441
CT-200	720.063	0.450	A-80	721.833	0.859
CT-200-2	720.797	0.608	A-97	722.628	0.784
			Devens—Long Pond facies		
			A-28	721.952	0.228
			A-35	720.010	0.563
			A-49-1	719.059	0.424

* Locations plotted on Figure 3.
+ Locations plotted on Figure 2.

4) of 10 Harvard body samples. No significant difference was found to exist in the data. The composition of the K-rich phase of these megacryst perthite samples ranged from Or 90.5 percent to 100 percent. The mean cell volume of the group is equivalent to an Or composition of 96 percent.

All the data for the unhomogenized samples from the Clinton and Harvard bodies of the Clinton facies were combined and given a Model 1 test (test 4, Table 5). Again, no significant difference was found to exist.

K-feldspar samples from porphyroblastic gneiss of the Devens–Long Pond facies were collected at three widely spaced locations (Fig. 2). The x-ray diffraction patterns of the three samples were of higher visual and statistical quality (criteria of Wright in Orville, 1967) than those obtained from the Clinton facies. This can be seen in the smaller standard errors (Table 4) associated with the cell-volume determinations.

The composition of the K-rich phase of the Devens–Long Pond perthite ranges from Or 92 to 98 percent, and they have a mean value of Or 94.5 percent. This mean is equal to that observed for the Clinton facies samples from the Clinton bodies and only Or 1.5 percent different from the observed mean of the Clinton facies samples from the Harvard body.

The structural states of the K-rich phase of the perthite were estimated using the Δbc (see Fig. 1, Crosby, 1971) parameter. No significant difference was found to exist in the structural states of the samples. The observed mean of this Δbc population is 0.987, a value indicative of the maximum Al/Si ordering of maximum microcline.

A Model 1 test (test 3, Table 5) was made on the cell-volume data of the K-rich phase of the perthite samples from the Devens–Long Pond facies. A significant difference at the 95 percent level of confidence was found to exist within this data. However, based on the similarity of this group mean composition to that of the Clinton-facies groups and the similar structural states of all samples, it seems probable that in this application, the associated error variance (0.366 A^6) is too low an estimate of the population variance and not the result of the samples coming from different populations.

A final Model 1 test (test 5, Table 5) was conducted on the cell-volume data of all the unhomogenized samples, and no significant difference was found. Field evidence, coupled with the similarity in the structural states and compositions of the K-rich phase of the perthite, indicates that the Devens–Long Pond facies and the Clinton facies experienced a similar final thermal history. The close relationship of K-feldspar structural state to cooling history has found support in work on regional metamorphic rocks in Maine (Guidotti and others, 1973).

Homogenized Perthite

Eleven samples from the Clinton bodies of the Clinton facies were individually homogenized to determine the total Or content of each. The estimated compositions ranged from Or 72 to 84 percent. The mean of the cell volumes (Table 6) is equivalent to a group mean composition of Or 78.5 percent.

A Model 1 test (test 1, Table 7) on these cell volumes showed a significant difference at the 95 percent confidence level. Tukey's test (Edwards, 1962, p. 332) for a "straggler" separated samples CT-200 and C-68 from the others, whereas sample C-84-1 was a borderline case. No obvious geological relationship exists to account for the compositional differences implied by the smaller cell volumes of these three samples.

Eight homogenized perthite samples from the Harvard body of the Clinton facies

TABLE 5. ANALYSIS OF VARIANCE ON DATA FROM UNHOMOGENIZED POTASSIUM FELDSPAR

Analysis of variance (F test)	Group mean	V_B/V_W*	D.F.[+]	F value[§]	Significance[#]
Test 1 on cell volumes of the K-rich phase of microperthite megacrysts from the Clinton bodies of the Clinton facies	720.063 (Or% = 94.5)	1.95	10,11	2.86	n.s.
Test 2 on cell volumes of the K-rich phase of microperthite megacrysts from the Harvard body of the Clinton facies	720.826 (Or% = 96)	1.505	9,10	3.02	n.s.
Test 3 on cell volumes of the K-rich phase of microperthite megacrysts from the Devens–Long Pond facies	720.340 (Or% = 94.5)	11.88	2,3	9.28	s.
Test 4 on combined Clinton and Harvard bodies data of the Clinton facies	720.426 (Or% = 95)	1.782	20,21	2.09	n.s.
Test 5 on combined data from the Clinton facies and the Devens–Long Pond facies	720.415	1.90	23,24	1.99	n.s.

* V_B = variance between samples; V_W = variance within samples.
[+] D.F. = degrees of freedom. First figure is the D.F. associated with V_B. The second figure is the D.F. associated with V_W.
[§] F is at the 5% level. F values from Edwards' (1962) Table 8.
[#] n.s. = not significant, s. = significant (5% level).

ranged in estimated composition from Or 77 to 85 percent. The mean composition of this group is Or 79.5 percent, which is only 1 percent different from the mean composition estimated for the samples from the Clinton bodies. A Model 1 analysis of variance (test 2, Table 7) performed on the Harvard body samples showed that no significant difference exists in the data.

A Model 1 test (test 4, Table 7) was run on the combined Harvard and Clinton body samples, excluding the stragglers, and no additional differences of significance were found.

To test the possiblity that the three stragglers separated from the samples of Clinton facies from the Clinton bodies may simply indicate that the population variance of megacryst cell volumes is larger than the error variance associated with their determination—and not a signal that two distinct populations may be present—the grouped Clinton bodies data were compared to the grouped Harvard body data. A test to check for homogeneity between the two group variances found no significant difference between the groups. A Model 2 test (test 5, Table 7) was performed comparing the Harvard body samples to the Clinton body samples, including stragglers. This test showed that no significant difference exists between these groups. The estimate of mean composition based on all the Clinton facies samples is Or 78.75 percent.

TABLE 6. UNIT CELL VOLUMES OF HOMOGENIZED PERTHITE

Samples*	Mean cell volume	Variance½ (standard deviation)	Samples[+]	Mean cell volume	Variance½ (standard deviation)
		Clinton facies			
C-3	714.098	0.371	A-3	716.650	0.861
C-19	714.537	0.415	A-4	714.381	0.919
C-38	715.591	0.529	A-7	714.174	0.813
C-46	715.068	0.442	A-11	714.127	0.468
C-49	716.391	0.783	A-12	715.181	0.781
C-51	715.647	0.494	A-25	714.077	0.418
C-68	712.390	1.420	A-34	715.267	0.450
C-70	714.127	0.483	A-97	714.280	0.849
C-84	712.874	0.388			
C-105	714.627	0.401			
CT-200	711.684	0.494			
CT-200-2	716.046	1.410			
			Devens–Long Pond facies		
			A-28	719.501	0.956
			A-35	718.487	0.419
			A-49-1	718.424	0.241

* Locations plotted on Figure 3.
[+] Locations plotted on Figure 2.

TABLE 7. ANALYSIS OF VARIANCE ON DATA FROM HOMOGENIZED POTASSIUM FELDSPAR

Analysis of variance (F test)	Group mean	V_B/V_W*	D.F.[†]	F value[§]	Significance[#]
Test 1 on cell volumes of the homogenized megacrysts from the Clinton body of the Clinton facies	714.423 (Or% = 78.5)	4.1	11,12	2.72(5%)	s.
Test 2 on cell volumes of the homogenized megacrysts from the Harvard body of the Clinton facies	714.768 (Or% = 79.5)	1.53	7,8	3.50(5%)	n.s.
Test 3 on cell volumes of the homogenized megacrysts from the Devens—Long Pond facies	718.804 (Or% = 91.5)	0.956	2,3	9.28(5%)	n.s.
Test 4 on combined Clinton and Harvard bodies homogenized samples, excluding "stragglers"	714.957 (Or% = 79.75)	1.52	16,17	2.29(5%)	n.s.
Test 5 to compare homogenized data of grouped Clinton samples (including "stragglers") to grouped Harvard samples	714.595 (Or% = 78.75)	0.358	1,18	3.60(5%)	n.s.
Test 6 to combine Clinton, Harvard, and Devens homogenized data, excluding Clinton "stragglers"	715.534	5.35	19,20	2.97(1%)	v.s.
Test 7 to compare homogenized data of grouped Clinton body samples, including "stragglers," to grouped Harvard body samples to grouped Devens—Long Pond samples	715.114	20.0	2,20	5.85(1%)	v.s.

* V_B = variance between samples; V_W = variance within samples.
[†] D.F. = degrees of freedom. First figure is the D.F. associated with V_B. The second figure is the D.F. associated with V_W.
[§] Confidence level in parentheses after F value. F values from Edwards (1962).
[#] n.s. = not significant, s. = significant (5% level), v.s. = very significant (1% level).

Three homogenized K-feldspar samples from the porphyroblastic gneiss of the Devens-Long Pond facies have cell volumes (Table 6) equivalent to compositions ranging from Or 90.5 to 93 percent. A Model 1 test (test 3, Table 7) failed to reveal any significant variation between samples. The three samples had a mean composition of Or 91.5 percent.

A Model 1 test (test 6, Table 7) on the combined data from the Devens-Long Pond facies and the Clinton facies, excluding the Clinton body stragglers, showed that a significant difference at the 99 percent confidence level exists within the group. However, this difference, as noted with the stragglers from the Clinton facies, does not preclude the possibility that the samples come from a common population having a population variance larger than the error variance. A Model 2 test (test 7, Table 7) to check this possibility also showed that a significant difference at the 99 percent confidence level existed in the data. Because it has already been shown that no excessive variability exists between the Clinton bodies data and the Harvard body data, it can be concluded without further testing that there is a better than 99 percent probability that the Devens-Long Pond samples represent a different population of homogenized unit cell volumes than do the Clinton facies samples.

The Or 12.75 percent difference between the mean bulk compositions of the Clinton facies and Devens-Long Pond facies perthite suggests that final between-grain migration of alkalis took place at lower temperatures in Devens-Long Pond facies. Barth's (1962, p. 250-252) two-feldspar geothermometer places the temperature of final intergrain alkali migration 130°C lower in the Devens-Long Pond facies. Depending on the assumed pressure and chemical environment, the binodal curves of Crosby (1971, p. 1804) place this temperature 100° to 200°C lower.

ORIGIN

The Clinton facies has locally discordant contacts and apophyses, sharply bounded xenoliths, and a fine-grained contact facies. This plus the development of contact metamorphic effects in the country rock, homogeneity of the Clinton facies over wide areas, and a chemical composition (Gore, 1973, p. 254-257) coincident with the low temperature trough in the system $Ab\text{-}Or\text{-}SiO_2\text{-}H_2O$ (Tuttle and Bowen, 1958) all combine to identify the Clinton facies as plutonic in origin.

The presence of euhedral microcline megacrysts in syntectonic rocks has often been interpreted (Marmo, 1971, p. 42-48) to be generally related to hydrothermal granitization and not magmatic origin. Although some outcrops contained a few crystals indicating a late magmatic or postconsolidation origin, the majority appear to have crystalized during the early and main period of magmatic consolidation. The apparent contradiction may indicate conditions described by Mehnert (1968, p. 289-295), who cited observations that indicate that K-feldspar formation may begin as an early magmatic phase and continue crystallization beyond consolidation under metasomatic conditions. The bulk composition of the megacrysts suggests temperatures of final crystallization at or just below the lowest temperatures under which a granitic melt may exist. Barth's (1962, p. 250-252) two-feldspar geothermometer indicates a temperature of 490°C. Depending on the environmental assumptions, Crosby's (1971, p. 1804) binodal curves indicate temperatures between 600° and 650°C.

The origin of all the rocks within the Devens–Long Pond facies is not satisfactorily explained as the product of a single intrusive igneous event. The variations in texture and composition of the Devens–Long Pond rocks are better explained as the metamorphic product of a layered sequence. Evidence strongly suggestive of a specific protolith sequence is not now available. However, if isolated quartzite outcrops (not shown in Fig. 2) within the Devens–Long Pond belt are interlayered with the feldspathic gneiss, this would suggest a protolith sequence of surficial, primarily subaqueous, origin.

The bulk compositions of the K-feldspar porphyroblasts indicate that final crystallization took place in the porphyroblastic gneiss considerably below magmatic temperatures. The fabric elements defined by the dimensional orientation of each of the major mineral phases strongly suggests a blastic origin, with the K-feldspar blastesis producing the most distinctive change. It is uncertain whether some of the apparently concordant gneissic granitoid rocks within the gneiss belt formed in situ by metasomatism or anatexis or were intruded.

GEOLOGIC IMPLICATIONS

The Devens–Long Pond rocks are the oldest in the area of study and are probably of Ordovician age or older. This age assignment finds some support in the relationship of the Devens–Long Pond facies to the presumable Silurian or older (Grew, 1973, p. 126) Oakdale Quartzite that overlies it, possibly unconformably.

In the Devonian Period, the Devens–Long Pond facies underwent a regional metamorphism (Acadian) to produce its blastic character. Conditions compatible to the higher P-T portion of the greenschist facies of the Barrovian type facies series were present. This P-T estimate is indicated by the bulk composition of the K-feldspar porphyroblasts and the mineral assemblages common to the rocks of the lower portion of the crustal block immediately west of the Clinton-Newbury fault.

The regionally concordant character of both the Clinton facies and Chelmsford granite suggests that they were intruded into the axial regions of developing folds. The Silurian or Devonian age of the youngest country rocks suggests that the Chelmsford granite and Clinton facies magmas were generated and emplaced under the tectonic conditions of the Acadian orogeny.

Intrusion was followed by a dominantly cataclastic metamorphic event producing the foliation seen in both the Chelmsford granite and Clinton facies. Myrmekite is common and well developed in the Chelmsford granite and rare in the Clinton facies. The presence of the myrmekite shells around the K-feldspar porphyroblasts at locations close to the Chelmsford granite strongly indicates that the porphyroblasts were present when the Chelmsford intruded. The postconsolidational development of foliation in the two main igneous units, combined with observation of the myrmekite shell, lend support to the contention (Lyons and Faul, 1968) that the Acadian orogeny was characterized by at least two distinctive pulses.

The first pulse was the most intense, being characterized by blastesis in the Devens-Long Pond facies and the anatectic development of large quantities of magma. These magmas were probably emplaced into somewhat higher crustal levels and crystallized during the waning stages of the initial pulse. The second pulse was a lower temperature, predominantly cataclastic event. During this event, homogenization of both the composition of the K-rich phase and structural state of the perthite occurred. Most of the chemical activity during the pulse appears to have been restricted mainly to intragrain adjustments or very short range intergrain migrations.

The relationship of the Clinton facies to the Chelmsford granite is still not entirely clear, but calculated chemical compositions of the two rocks are almost identical (Gore, 1973, p. 255) if the chemical contribution of the modal biotite and chlorite is removed before calculation. A model explaining the known facts would begin with the anatectic development of large quantities of magma. The present crustal level exposed on the surface and marked by the Devens-Long Pond—facies outcrop belt was probably slightly above the level of widespread anatexis. The Chelmsford may represent the upper, more hydrous and felsic portion of this magma development. The Clinton facies probably represents a slightly deeper, higher temperature magma containing a greater quantity of dissolved biotite. The parent rocks undergoing anatexis could have been extensions of the Devens-Long Pond facies. The compositional similarities of the Clinton facies and the Devens-Long Pond facies could be explained by postulating that the Clinton-facies magma was generated by the almost complete melting of the more felsic portions of the Devens-Long Pond belt.

The more hydrous character of the Chelmsford granite generally creates transitional country rock contacts of varying thickness. The country rock as well as the granite itself is cut by numerous pegmatite dikes. In contrast, the Clinton facies nearest to the Chelmsford granite lacks pegmatite and shows no signs of alteration. This suggests that the Clinton facies may be younger than the Chelmsford granite, having intruded after consolidation of the Chelmsford.

The last major tectonic event is a Permian disturbance (Zartman and others, 1970). The crust exhibited brittle behavior because this disturbance is characterized by the fracture tectonics (Skehan, 1968) that produced the major regional faults. This event produced intense cataclastic modifications within narrow zones and, locally (Grew, 1970), some prograde metamorphism.

CONCLUSIONS

The Ayer Granodiorite at the type location has been divided into two lithologically and geographically distinct units. The youngest of these facies, the Clinton facies, is a homogeneous, clearly plutonic igneous rock. This facies is composed of a coarse porphyritic rock of quartz monzonite composition containing large microcline perthite megacrysts. The older facies, the Devens-Long Pond facies, is composed of a complex belt of felsic gneiss of metamorphic and metasomatic origin. Some of the rocks within this belt may be of plutonic igneous origin, but the extensive development of coarse schlieren suggests that if a true melt phase was present, it was generated in place. An analysis of the K-feldspar megacrysts of the Clinton facies and the porphyroblasts of the Devens-Long Pond facies revealed that a significant difference exists in the bulk composition between the two sample groups. These data indicate that final between-grain migration of alkalis occurred at higher temperatures in the Clinton facies. The similar composition of the K-rich phase of all tested perthite samples and their similar maximum microcline structural states suggest that both facies underwent a similar final thermal history.

The age of the Devens-Long Pond facies is probably pre-Silurian because it is overlain, possibly unconformably, by the presumably Silurian or Devonian Oakdale Quartzite. The Devens-Long Pond facies is intruded by the Devonian(?) Chelmsford granite and may be intruded by the Clinton facies.

The Clinton facies is probably Devonian in age, being a product of anatexis during the main pulse of the Acadian orogeny. The Clinton facies intrudes the Oakdale Quartzite, Worcester Phyllite, and Straw Hollow Diorite.

The interpretation that the foliation of the Clinton facies and the Chelmsford granite was produced postconsolidationally suggests that two distinct deformational pulses may have occurred within the Acadian orogeny. The initial pulse was the higher temperature, more intense event, whereas the later one was dominated by cataclastic processes in which chemical adjustment was primarily restricted to intragrain migrations.

ACKNOWLEDGMENTS

I am grateful to James W. Skehan, S. J., of Boston College for originally introducing me to the area of this study; to Arthur H. Brownlow and C. Wroe Wolfe of Boston University for their suggestions and aid during the course of this study; and to John H. Peck, Paul C. Lyons, James W. Skehan, S. J., Arthur H. Brownlow, and Nigel Harris for suggestions regarding the improvement of this manuscript.

I thank the Boston University Department of Geology for funds, facilities, and aid while completing my Ph.D. dissertation, part of which constitutes the data reported in this paper.

This study was partly supported by a Penrose Grant from the Geological Society of America and a Science Faculty Fellowship from the National Science Foundation.

REFERENCES CITED

Barth, T.F.W., 1962, Theoretical petrology (2nd ed.): New York, John Wiley & Sons, Inc., p. 250-252.

Bell, K., and Alvord, D., 1974, U.S. Geological Survey Open-File Map 74-356.

Billings, M. P., 1956, The geology of New Hampshire, Part II—Bedrock geology: New Hampshire State Planning and Devel. Comm., 203 p.

Crosby, P., 1971, Composition and structural state of alkali feldspars from charnockitic rocks on Whiteface Mountain, New York: Am. Mineralogist, v. 56, p. 1788-1811.

Crosby, W. O., 1899, Geology of the Wachusett dam and Wachusetts aqueduct tunnel of the Metropolitian Water Works in the vicinity of Clinton, Massachusetts: Tech. Quarterly, v. 12, p. 68-96.

Currier, L. W., 1937, The problem of the Chelmsford, Massachusetts granite: Am. Geophys. Union Trans., 18th Ann. Mtg., pt. 1, p. 260-261.

——1952, Geology of the "Chelmsford Granite" area: Massachusetts field trip no. 3, in Geol. Soc. America guidebook for field trips in New England: p. 103-108.

Edwards, A. L., 1962, Statistical methods for the behavioral sciences: New York, Holt, Rinehart & Winston, Inc., 542 p.

Emerson, B. K., 1917, Geology of Massachusetts and Rhode Island: U.S. Geol. Survey Bull. 597, 289 p.

Evans, H. T., Jr., Appleman, E. E., and Handwecker, D. S., 1963, The least squares refinement of crystal unit cells with powder diffraction data by an automatic computer indexing method [abs.]: Cambridge, Mass., Am. Crystallog. Assoc., Ann. Mtg. Program, p. 42-43.

Gore, R. Z., 1973, Geology of the porphyritic Ayer Quartz Monzonite and associated rocks in portions of the Clinton and Ayer quadrangles, Massachusetts [Ph.D. thesis]: Boston, Boston Univ., 299 p.

Grew, E. S., 1970, Geology of the Pennsylvanian and pre-Pennsylvanian rocks of the Worcester area, Massachusetts [Ph.D. thesis]: Cambridge, Mass., Harvard Univ., 263 p.

——1973, Stratigraphy of the Pennsylvanian and pre-Pennsylvanian rocks of the Worcester area, Massachusetts: Am. Jour. Sci., v. 273, p. 113-129.

Grew, E. S., Mamay, S. H., and Barghoorn, E. S., 1970, Age of plant fossils from the Worcester coal mine, Worcester, Massachusetts: Am. Jour. Sci., v. 268, p. 113-126.

Guidotti, C. V., Herd, H. H., Tuttle, C. L., 1973, Composition and structural state of K-feldspars from K-feldspar + sillimanite grade rocks in northwestern Maine: Am. Mineralogist, v. 58, p. 705-716.

Hansen, W. R., 1956, Geology and mineral resources of the Hudson and Maynard quadrangles, Massachusetts: U.S. Geol. Survey Bull. 1038, 104 p.

Jahns, R. H., 1952, Stratigraphy and sequence of igneous rocks in the Lowell-Ayer region, Massachusetts, in Geol. Soc. America guidebook for field trips in New England, 1952 Ann. Mtg.: p. 108-112.

Lyons, J. B., and Faul, H., 1968, Isotope geochronology of the northern Appalachians, in Zen, E-an, White, W. S., Hadley, J. B., and Thompson, J. B., Jr., eds., Studies in Appalachian geology: Northern and Maritime: New York, Interscience Publs., p. 305-318.

Marmo, V., 1971, Granite petrology and the granite problem: New York, Elsevier Pub. Co., p. 42-48.

Mehnert, K. R., 1968, Migmatites and the origin of granitic rocks: New York, Elsevier Pub. Co., p. 289-295.

Orville, P. M., 1967, Unit cell parameters of the microcline-low albite and the sanidine-high albite solid solution series: Am. Mineralogist, v. 52, p. 55-86.

Page, L. R., 1968, Devonian plutonic rocks of New England, in Zen, E-an, White, W. S., Hadley, J. B., and Thompson, J. B., Jr., eds., Studies in Appalachian geology: Northern and Maritime: New York, Interscience Publs., p. 371-383.

Skehan, J. W., 1968, Fracture tectonics of southeastern New England as illustrated by the Wachusett-Marlborough Tunnel, east-central Massachusetts, in Zen, E-an, White, W. S., Hadley, J. B., and Thompson, J. B., Jr., eds., Studies in Appalachian geology: Northern and Maritime: New York, Interscience Publs., p. 281-290.

Sriramadas, A., 1966, The geology of the Manchester quadrangle: New Hampshire Dept. Resources and Econ. Devel., 78 p.

Sundeen, D. A., 1971, The bedrock geology of the Haverhill 15' quadrangle New Hampshire: New Hampshire Dept. Resources and Econ. Devel. Bull. 5, 125 p.

Tuttle, O. F., and Bowen, N. L., 1958, Origin of granite in the light of experimental studies in the system $NaAlSi_3O_8$-$KAlSi_3O_8$-SiO_2-H_2O: Geol. Soc. America Mem. 74, 153 p.

Wright, T. L., and Stewart, D. B., 1968, X-ray and optical study of alkali feldspar: 1. Determination of composition and structural state from refined unit-cell parameters and 2V.: Am. Mineralogist, v. 53, p. 38-87.

Zartman, R. E., Hurley, P. J., Krueger, H. W., and Giletti, B. J., 1970, A Permian disturbance of K-Ar radiometric ages in New England: Its occurrence and cause: Geol. Soc. America Bull., v. 81, p. 3359-3374.

MANUSCRIPT RECEIVED BY THE SOCIETY JULY 1, 1974
REVISED MANUSCRIPT RECEIVED FEBRUARY 3, 1975
MANUSCRIPT ACCEPTED FEBRUARY 20, 1975

Printed in U.S.A.

Geological Society of America
Memoir 146
© 1976

Stratigraphy and Petrography of the Volcanic Flows of the Blue Hills Area, Massachusetts

ULDIS KAKTINS*
*Department of Geology, Boston University
Boston, Massachusetts 02215*

ABSTRACT

The volcanic rocks of the Blue Hills area of Massachusetts, referred to in the earlier literature as "aporhyolites," have been divided informally into six units on the basis of textural and compositional features and stratigraphic position. The Great Dome Trail flow is the oldest, followed by the Wampatuck Hill flow, Chickatawbut Road flow, and the pyritic volcanic rocks. The stratigraphic positions of the Pine Hill and Hemenway Hill units are still uncertain.

Glass shards, eutaxitic structure, severely fractured crystals, and vertical zonal features found in the Great Dome Trail, Wampatuck Hill, and Chickatawbut Road flows, in the pyritic volcanic rocks, and in one of the units on Pine Hill, indicate an ash-flow origin. Lack of gradational bedding and the presence of axiolitic structures and substantial amounts of pumice also show the tuffs to be of ash-flow rather than ash-fall origin. Pine Hill contains a possible vent area and the only known exposure in the area of a true rhyolite flow.

Chemical analyses are reported for samples from 15 locations. All the rocks are low-Ca rhyolites and have some characteristics of peralkaline rhyolites. However, only parts of one flow are actually peralkaline in terms of $Al_2O_3/Na_2O + K_2O$.

GEOLOGIC SETTING

The Blue Hills igneous complex of eastern Massachusetts includes the Quincy Granite, the Blue Hill Granite Porphyry, and associated volcanic rocks (Figs. 1, 2). The Quincy Granite is an aegirine-riebeckite granite which is the principal rock of the eastern part of the complex. Along the south side of the complex, the Blue Hill Granite Porphyry, a chemically and mineralogically related rock, forms the highest hills of the Blue Hills area. The volcanic rocks occupy the south-central

---

*Present address: Earth and Planetary Sciences Department, University of Pittsburgh, Johnstown, Pennsylvania 15904.

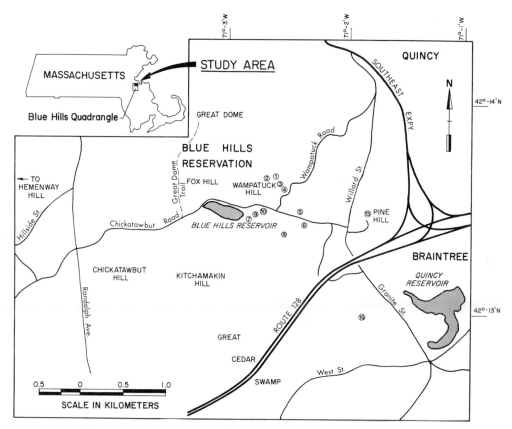

Figure 1. Index map showing study area, in eastern half of Blue Hills Reservation, Massachusetts. Circled numbers indicate sample locations for analyses in Tables 1 and 2.

part of the complex and are closely associated with the granitic rocks in several places. These volcanic rocks have been correlated with the nearby Mattapan Volcanic Complex of the Boston Basin by LaForge (1932) and Chute (1964, 1969), but Billings (1976) questions this. Similar but as yet uncorrelated volcanic rocks extend southward from the Blue Hills toward Rhode Island.

The area has been affected by a late Paleozoic (Permian disturbance) regional low-grade metamorphism of the quartz-albite-muscovite-chlorite subfacies of the greenschist facies (Chute, 1966). The granitic rocks of the Blue Hills complex contain higher weight percent values of silica, the alkalis, and iron than Nockolds' (1954) average chemical composition of similar igneous rocks. Calcium and magnesium are low, and there is slightly more potash than soda. The closest fit in Nockolds' classification places these rocks with the alkalic granites.

PREVIOUS WORK

The first geologic survey of the area was made by Crosby (1900), and later detailed petrographic and chemical work was done by Warren (1913). Since then, Emerson (1917), Billings (1929), LaForge (1932), and Chute (1940, 1964, 1969) have described the area in varying detail.

LaForge (1932) assigned a pre-Pennsylvanian (probably Mississippian) age to

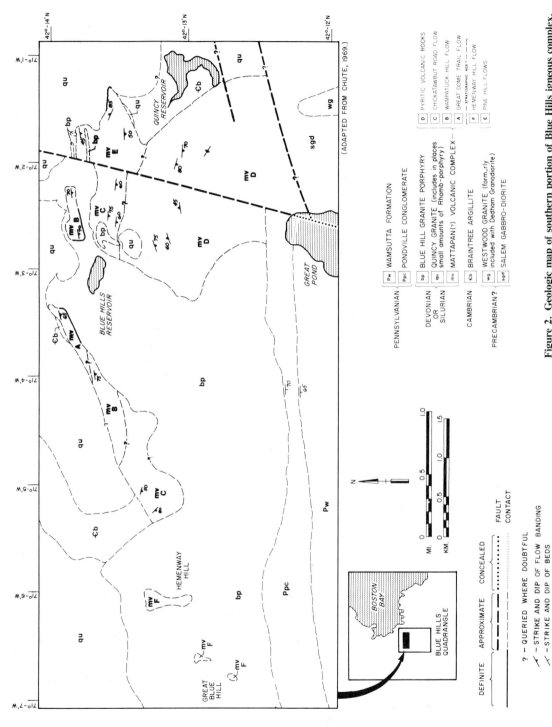

Figure 2. Geologic map of southern portion of Blue Hills igneous complex.

the Quincy Granite on the basis of boulders of the associated granite porphyry that were found in the overlying Pondville Conglomerate. Attempts to more closely define the age of the Quincy Granite by radiometric dating have yielded conflicting results (Bottino and others, 1970; Zartman and Marvin, 1971). The most recent work (Lyons and Kreuger, 1976) suggests an age of 400 to 450 m.y. for the Quincy Granite.

Crosby (1900) noted "dikes" of Quincy Granite cutting the Blue Hill Granite Porphyry and therefore concluded that the slower crystallization of the granite enabled it to intrude into the granite porphyry, a supposed chilled margin of the granite. However, observations by Chute (1964) have indicated that the "dikes" are xenoliths in the granite. Rb-Sr dating of the granite porphyry is inconclusive; the substantial scatter of data points indicates that the porphyry acted as an open system with respect to Rb and Sr (Bottino and others, 1970). Whatever the exact age relation, the granite and the porphyry seem to belong to the same magmatic suite.

Crosby (1900) and Bascom (1900) classified the volcanic rocks as "aporhyolites," the prefix "apo" meaning altered. Crosby thought that the volcanic rocks were the youngest of the alkalic rocks, on the basis of what appeared to be tongues of volcanic rocks in the granite porphyry. However, Warren (1913) rightly concluded that these are not tongues but xenoliths. The presence of these numerous small volcanic xenoliths in the porphyry is the best evidence for the younger age of the porphyry. Rb-Sr dates by Bottino and others (1970) show a scatter of points so wide that an accurate isochron cannot be drawn for the volcanic rocks. However, some of this scatter may be caused by incorrect assumptions about the homogeneity of the volcanic rocks. There seems to be a direct relation between apparent ages and the characteristics of zones within the flows. That is, the greater the induration and degree of welding within a zone, the older the apparent age. The oldest age (420 m.y.) was calculated by Bottino and others (1970) for the Pine Hill rhyolite, and, if it can be assumed that the rhyolite acted more like a closed system than the associated ash-flow tuffs, then 420 m.y. should be considered the best approximation for the actual age of the volcanic rocks.

FIELD RELATIONS AND PETROGRAPHY

Stratigraphy and Structure

The volcanic rocks of the Blue Hills area have been divided into six units on the basis of textural variations, compositional variations, and stratigraphic position (Fig. 2). The Great Dome Trail ash flow is the oldest, followed by the Wampatuck Hill and Chickatawbut Road ash flows and the pyritic volcanic rocks. The stratigraphic positions of the Pine Hill and Hemenway Hill flows are still uncertain.

The Great Dome Trail, Wampatuck Hill, and Chickatawbut Road ash-flow tuffs all show consistent east-trending strikes and nearly vertical southerly dips. There is no evidence in these flows of overturning; on the contrary, there is a discernible upward gradation of cooling zones (especially in the Wampatuck Hill and Chickatawbut Road flows) similar to that described by Ratté and Steven (1964): that is, the ash-flow tuffs grade from a phenocryst-poor, clast-rich basal zone to phenocryst-rich, clast-poor, densely welded upper zones. Therefore, the tuffs must have been tilted southward as a block (see also Billings, 1976).

Nowhere in the Blue Hills area is there a visible contact between adjacent ash-flow tuffs; therefore, a considerable amount of interpretation is required. Assuming the simplest case possible, it can be concluded that the Great Dome Trail flow is relatively older than the Wampatuck Hill flow because it is farther north and should be stratigraphically below the Wampatuck Hill flow. Since early flows would be greatly controlled by the existing relief and topography, the somewhat limited exposure of the Great Dome Trail flow is not surprising.

The Chickatawbut Road flow occurs south of the Wampatuck Hill flow in the western part of the mapped area, and the pyritic volcanic rocks are found south of the Chickatawbut Road flow in the eastern part of the area (Fig. 2). Thus they are interpreted to be younger than both the Great Dome Trail and the Wampatuck Hill flows.

The Pine Hill flows are found only in the eastern part of the mapped area. Their relationship to the other units is unknown. The Hemenway Hill flow occurs as isolated masses in the western part, and these masses are interpreted to be roof pendants in the Blue Hill Granite Porphyry. It is not clear whether these flow remnants represent a lava or a pyroclastic eruption. If the unit is a tuff, the high phenocryst content (Table 1) and scarcity of pumice suggest it was once an upper zone of an ash-flow tuff. It is not possible to indicate the relative stratigraphic position of this flow.

A detailed description of the volcanic units is given in the following sections. A summary of the characteristics of these units can be found in Table 1.

Great Dome Trail Ash Flow

This densely welded ash-flow tuff is exposed along the Great Dome Trail, west and northwest of Fox Hill. Most of the outcrops occur along the contact with the Blue Hill Granite Porphyry, where they are heavily sericitized.

The basal zone of the Great Dome Trail flow is not exposed, and the grayish-blue lower part of the exposed tuff represents the densely welded middle zone of the flow. Megascopic feldspar phenocrysts and a few minute spherulites characterize

TABLE 1. CHARACTERISTICS OF BLUE HILLS VOLCANICS

Rock unit	Phenocrysts (%)	Pumice (%)	Lithic clasts (%)	Color Fresh	Weathered	Thickness
Pyritic volcanic rocks						
Type D-1	12	1	17	greenish gray	brownish gray	
Type D-2	6	8	..	greenish gray	reddish brown	350+ m
Chickatawbut Road ash flow						
Upper zone	9	1	..	bluish black	reddish gray	
Eutaxitic zone	6	19	<1	bluish black	bluish gray	240 m
Densely welded zone	3	<1	..	bluish gray	light gray	
Basal zone	4	14	4	dark bluish gray	light bluish gray	
Wampatuck Hill ash flow						
Upper zone	13	3	..	bluish black	red	
Eutaxitic zone	7	18	2	bluish black	pinkish gray	210 m
Densely welded zone	3	<1	..	bluish gray	light gray	
Basal zone	3	12	6	dark bluish gray	light bluish gray	
Great Dome Trail ash flow						
Upper zone	7	<1	<2	bluish gray	pinkish gray	
Lower zone	4	<1	<1	grayish blue	light gray	170 m
Pine Hill volcanic rocks*						
Rhyolite	13	..	..	reddish black	reddish brown	45+ m
Breccia	..	..	10	greenish gray	light brown	15+ m
Ash-flow tuff						
upper zone	<5	..	..	bluish gray	gray	60+ m
eutaxitic zone	23	10	..	gray blue	pinkish gray	
Hemenway Hill flow*	22	..	2	purplish gray	red	100+ m

*Stratigraphic position of these units is uncertain.

this otherwise massive rock. The phenocryst content is evenly divided between quartz and microperthite. Both types of phenocryst are subhedral and average slightly less than 1 mm in greatest diameter. The microperthite consists of albite and sanidine; micropoikilitic structures rim many of the crystals. Groundmass makes up over 90 percent of the rock, and consists of a cryptocrystalline intergrowth of sanidine, quartz, sericite, magnetite, hematite, and traces of illite. Rarely, distorted and well-compacted pumice fragments can be distinguished in the matrix. They are similar in composition to the groundmass but have a granophyric texture. Heavily sheared porphyritic felsite inclusions, of a much coarser texture than the host rock, average 2 cm in diameter.

A 15-cm-wide gradational contact, with well developed eutaxitic structure and some spherulites, separates the middle zone from the upper one. The upper zone has a slightly greater concentration of lithic clasts, which average 2 cm in diameter and are irregularly dispersed throughout the zone. Quartz and microperthite phenocrysts are also more abundant and show only slight rounding (Table 1). Although chemical analyses are not available for this unit, the equal amounts of quartz and feldspar phenocrysts suggest an especially silicic composition (Enlows, 1955).

Wampatuck Hill Ash Flow

This ash-flow unit extends from Wampatuck Hill eastward past the Blue Hills Reservoir aquaduct; to the west, the flow is interrupted by exposures of the Blue Hill Granite Porphyry, and it is not evident again until north of Chickatawbut Hill (Fig. 2). Four definite zones can be identified within the tuff. Starting at the base, the tuff consists of a clast-rich zone which has increasingly better developed eutaxitic structure toward the top of the zone. This basal zone grades into a completely homogenized, densely welded zone characterized by low phenocryst content and lack of spherulitic structure. The tuff then grades upward into a eutaxitic zone with well-flattened pumice, and finally into a phenocryst-rich upper zone with progressively less, and less flattened, pumice.

Basal Zone. The lower part of the basal zone can be recognized by partly collapsed pumice, some spherulites, and fragments of previous flow material that range in diameter from 2 to 4 cm. Partial devitrification structures, such as spherulites, and the partly collapsed pumice indicate rapid cooling at the base of the tuff. This partially welded part of the basal zone grades upward into more densely welded material with 1- to 11-mm-thick eutaxitic banding that curves around phenocrysts and is contorted into drag folds. Such structure is typical of the basal part of a highly welded section where excessive compaction and heat retention cause pumice to deform plastically (Cook, 1962). Both the matrix and the pumice have recrystallized to a microcrystalline aggregate of sanidine and quartz. Rounded and embayed quartz and fractured microperthite phenocrysts average less than 1 mm in diameter and compose 2 to 5 percent of the total volume. The microperthite consists of albite and sanidine. Magnetite, some of it with hematite rims, composes about 3 percent of the rock.

Densely Welded Zone. The basal zone is overlain by 60 to 90 m of bluish-gray, massively welded tuff. Partly resorbed quartz phenocrysts make up about 1 percent of the rock. Microperthite phenocrysts, often with clear borders, total about 2 percent. Glass shards are scarce, and their margins have been almost totally obliterated by devitrification. The cryptocrystalline matrix is a mosaic of feebly undulating grains that are mainly sanidine. Euhedral magnetite is ubiquitous. Small spherulites and disc-shaped pumice compose an insignificant part of the total volume.

Eutaxitic Zone. The densely welded zone grades upward into a zone containing

30 to 35 percent discoid pumice fragments. The dark, almost obsidianlike pumice indicates strong welding (Ross and Smith, 1961). Toward the top of the zone, the pumice content decreases somewhat and the fragments become progressively less flattened. Average width of the fragments is about 1 to 2 cm, but the long dimension in some cases exceeds 10 cm. Pumice near the top of the eutaxitic zone is almost completely dominated by spherulites (Fig. 3). As a result, the cellular structure of the pumice has been destroyed, and only in a few cases is a highly contorted tubular structure visible. Deformation of this type is the only positive criterion of welding in recrystallized tuffs (Smith, 1960b). The pumice is often wrapped around primary phenocrysts and previous flow fragments (Fig. 4); these fragments range in diameter from 1 to 12 cm. Within the pumice, quartz phenocrysts are more common than microperthite, and both are smaller in size than their equivalents in the cryptocrystalline quartz and sanidine matrix where microperthite phenocrysts predominate. In places, small pumice pieces and completely flattened shard fragments show microscopic flow structure. Magnetite (1 to 3 percent) occurs as wispy streaks, octahedra, and as prismatic pseudomorphs.

Upper Zone. A spherulite-rich layer marks the gradational contact between the eutaxitic zone and the upper zone. This layer varies in thickness from 0.15 to 5 m. The spherulites are seldom greater than 6 mm in diameter and decrease slightly in size toward the top of the layer. Often, quartz or microperthite crystals serve as nuclei for the spherulites. The radial structure of large spherulites extends without interruption across older fragments. This relatively coarse texture may be due to vapor-phase crystallization, which reaches maximum development in pore spaces within the upper zones of ash-flow tuffs (Smith, 1960a). Minute spherulites found above the coarse spherulite layer are probably the products of incipient devitrification that was halted by rapid cooling of the top layers. Recognizable pumice fragments are only partly collapsed.

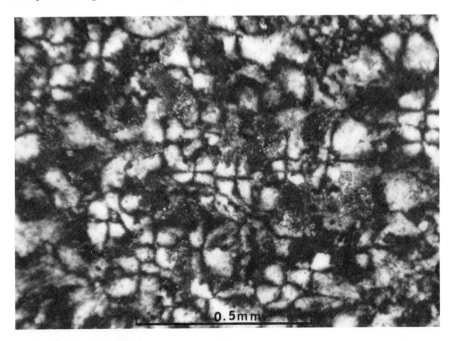

Figure 3. Small spherulites completely obscuring original structure of pumice fragment; eutaxitic zone, Wampatuck Hill flow. Polarized light.

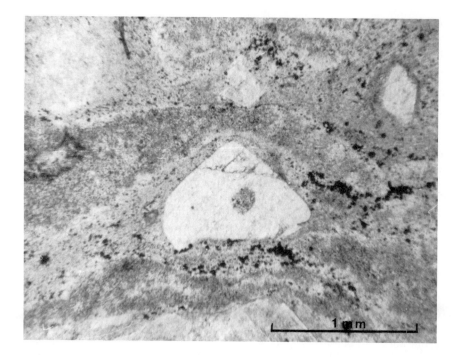

Figure 4. Pumice fragments bent around phenocrysts, with magnetite grains disseminated throughout the matrix; eutaxitic zone, Wampatuck Hill flow. Plain light.

The spherulite layer grades into a reddish tuff that exhibits little or no eutaxitic structure. This lack of eutaxitic structure might be explained by deposition of relatively cool, already solidified material near the top of the flow (Enlows, 1955). Microperthite, both as semiequant and rounded grains, ranges in diameter from 1 mm to 7 mm, and averages a little over 5 percent by volume. Quartz phenocrysts vary from 1 to 4 mm in diameter. The cryptocrystalline matrix of quartz and sanidine has a few discernible tricuspid shards. Also present are some octahedral pseudomorphs of hematite after magnetite.

The presumably nonwelded top has not been preserved and the next exposure, 10 m downdip, contains the contact between this unit and the Blue Hill Granite Porphyry. Near the contact, the volcanic rocks are heavily sericitized and occur as stringerlike xenoliths (as much as 0.6 m in length) in the granite porphyry, which has a very fine grained marginal zone.

Chickatawbut Road Ash Flow

The type locality of this ash-flow tuff is south of the intersection of Chickatawbut and Wampatuck roads (Fig. 1). Westward, the continuity of the flow is interrupted for about 2 km by the Blue Hill Granite Porphyry. The porphyry at the contact is fine grained and has abundant partially absorbed elongated xenoliths. Some of the included tuff material still has recognizable pumice fragments. A few heavily altered granitic (Quincy Granite?) fragments are also present. The tuff is again exposed in an area extending from the west flank of Chickatawbut Hill to just west of Hillside Street (Fig. 1). The Chickatawbut Road ash-flow tuff has been subdivided into four zones on the basis of the degree of welding and crystallization.

Basal Zone. The estimated thickness of this zone is 50 m. It grades upward from an incompletely welded clast-rich tuff, to a phenocryst-poor, pumice-rich tuff with fluidal eutaxitic structure (Fig. 5). Felsitic fragments range in diameter from 1 to 9 cm. The pumice fragments near the base of the zone are poorly compacted and average about 6 percent by volume; however, pumice content increases to as much as 19 percent toward the top of the zone. The fragments show distinct flattening and are often wrapped around phenocrysts. Minute spherulites and barely visible relict glass shards are confined to the lower portions of the zone. Microperthite phenocrysts dominate over quartz; both types are rounded and embayed. Magnetite is much more abundant than hematite.

Densely Welded Zone. This massive zone has an average thickness of 80 m. Pumice and other clastic fragments are scarce (Table 1); pumice flattening is great. The fragments show the usual recrystallization intergrowth of quartz and sanidine. Semiequant and fractured microperthite phenocrysts constitute 2 percent by volume, and partly resorbed quartz accounts for 1 percent. Magnetite is relatively more abundant than hematite. Small discontinuous lines within the sanidine-quartz groundmass probably represent compressed shards.

Eutaxitic Zone. A gradational contact separates the densely welded zone from the eutaxitic zone above. The latter averages 70 m in thickness. A lenticular rubble zone 5 m thick is exposed at one outcrop near the lower part of the eutaxitic zone, but few rock fragments are present in the eutaxitic zone itself. Well-flattened pumice fragments impart a well-defined eutaxitic structure to the tuff. The pumice is often bent around phenocrysts, indicating that the phenocrysts are primary. Total phenocryst content is 5 percent, with low sanidine and high albite microperthite predominating over quartz. The matrix is composed of very finely crystallized quartz and sanidine. Opaque minerals consist of hematite after magnetite octahedra

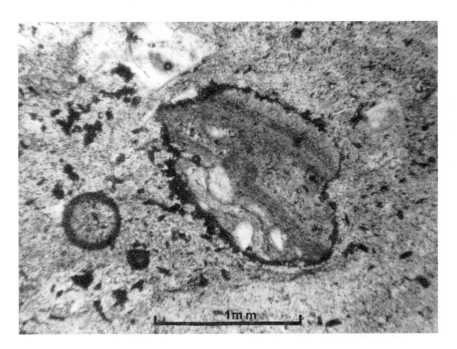

Figure 5. Inclusion of previous flow material in basal zone, Chickatawbut Road flow. Plain light.

and hematized siderite rhombs. Relict shards, and spherulites outlined by finely grained magnetite, are present.

Toward the upper part of the zone, megascopic spherulites appear (Fig. 6). They are concentrated for the most part in pumice fragments and mark the gradational contact between the eutaxitic and the upper zone. The spherulite layer shows a greater concentration of phenocrysts than does the zone as a whole—as much as 5 percent quartz and 6 percent semiequant microperthite. This layer is 10 m thick on the average.

Upper Zone. The average thickness of this zone is 40 m. Rounded and embayed quartz averages about 3 percent, and microperthite averages about 5 percent. Magnetite and hematite grains pepper the cryptocrystalline quartz and sanidine matrix. As a rule, spherulites, glass shards, and rock fragments are missing. Recognizable pumice composes only about 1 percent of the tuff. Toward the top of this zone, there is a band very rich in volcanic bombs as much as 1 m in length, dark felsite clasts, and potassium feldspar phenocrysts; the band is 4 to 5 m thick.

Pyritic Volcanic Rocks

The largest and most continuous expanse of volcanic rocks is situated on either side of the Willard Street fault zone. On the western side of the fault they extend into Great Cedar Swamp; the eastern boundary is along the west flank of Pine Hill. The pyritic volcanic rocks have been tentatively subdivided into type D-1 and type D-2 flows based on mineralogic composition and texture. The nomenclature is approximate because the extensive alteration, staining, and bleaching made it impossible to distinguish consistently between flows and flow zones or to trace them with any degree of accuracy.

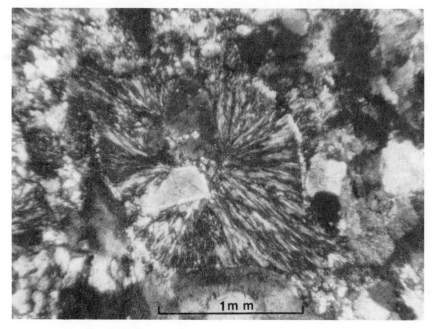

Figure 6. Phenocrysts included within the radiating structure of a spherulite, upper zone, Chickatawbut Road flow. Polarized light.

Type D-1. This flow type is most readily recognized by green stringers of finely divided sericite, as much as 31 cm in length. The degree of flattening of pumice varies from minimal to severe. In places, the unit has the characteristics of a lithic ash-flow tuff, with a felsitic fragment content of 15 to 20 percent. Rounded and embayed quartz phenocrysts, up to 2 mm in diameter, make up as much as 10 percent of the unit. The microperthite fragments often have cores of albite (An_8). The matrix consists of a microcrystalline aggregate of sanidine and quartz. Infrequently, flattened glass shard ghosts are visible. Siderite rhombs and hematite pseudomorphs after siderite compose as much as 6 percent of the rock; magnetite, hematite, and pyrite compose 10 percent. Originally the magnetite was probably intergrown with ilmenite, which has now been replaced by leucoxene.

Type D-2. The majority of these welded tuffs are distinguished by a well-developed eutaxitic structure, with the pumice fragments usually distorted into flamelike shapes (Fig. 7). Microperthite fragments, 1 mm in average size, compose 3 to 5 percent of the rock. Rounded and embayed quartz phenocrysts are usually present in smaller amounts than is microperthite, although in certain sections quartz ranges from 5 to 7 percent. Sericite and pyrite, which is usually concentrated in pumice, is minor; siderite is present in trace amounts. The groundmass, which consists of a dense intergrowth of quartz and sanidine, is more coarsely recrystallized than in type D-1. The absence of plagioclase and leucoxene also serves to distinguish between the two types. Spherulites are confined within pumice fragments. In certain sections, radiating fibers take on an axiolitic configuration within elongated shard outlines.

Pine Hill Flows

Rhyolite. Pine Hill has the only known exposure of a lava flow in the Blue Hills area. Brownish-red to black flow banding is continuous over several feet, and in places the bands are contorted into drag folds 60 to 90 cm in diameter. More intricate banding, varying in thickness from 0.1 to 1.0 cm can be seen on the fresh surface. Such intense foliation is common near a vent (Christiansen and Lipman, 1966), and the possibility exists that a vent or fissure was in the vicinity.

The most striking aspect of the matrix is the large number of spherulites, which completely obliterate all microscopic evidence of flowage and crystallite orientation. These spherulites, unlike most found in nearby ash-flow tuffs, do not have crystal centers. The maximum size of microperthite crystals is 4 mm; quartz phenocrysts are somewhat smaller. Phenocrysts do not show the degree of fracturing common in the tuffs. Partly resorbed quartz blebs account for 5 percent of the flow; microperthite phenocrysts total 8 percent. Magnetite pseudomorphs in the form of crystallites are intergrown with radiating aggregates of alkali feldspar and quartz. Magnetite is also present as octahedral and hypidiomorphic grains. Sanidine, quartz, and traces of hematite, siderite, and sericite make up the matrix. No evidence of pumice fragments or glass shards was observed.

Breccia. Down the southwestern flank of Pine Hill, the rhyolite interfingers with and overlies a breccia. Most of the breccia fragments are tuffaceous, but there is also some rhyolite and some fragments with granitoid texture. The angular inclusions range in size from 2 to 30 cm. According to Fisher's (1960) classification, the rock is an agglomerate. Since breccias with large angular fragments are common in the upper parts of volcanic necks (Williams, 1936), the possibility is increased that Pine Hill could have been the source for the nearby volcanic rocks.

Ash-flow Tuff. A completely devitrified unit of ash-flow tuff is exposed on the

Figure 7. Eutaxitic structure, pyritic volcanics. Note stretching and bending of pumice fragments around lithic clasts.

southern slope of Pine Hill. There is no visible contact of the tuff with the rhyolite or the breccia, but approximately 60 m east of the hill, the tuff has a north-trending contact with a fine-grained phase of the Quincy Granite. Correlation with units to the west is impossible because the amount of displacement along the Willard Street fault (Chute, 1969) is unknown (Figs. 1, 2).

The upper zone of the flow consists of a fine-grained rock containing no visible pumice or other fragments. This zone does not exceed 5 to 9 m in thickness before it grades downward into a coarsely recrystallized zone. Fairly good but rather faint eutaxitic structure is visible in the densely welded tuff. Lithoidal pumice content increases toward the base of the exposure, and microperthite phenocrysts account for 15 to 18 percent of the rock. Quartz composes another 7 percent. The groundmass is coarsely recrystallized and all vestiges of vitroclastic texture have been destroyed.

Hemenway Hill Flow

This flow type is exposed on Hemenway Hill and as two small, heavily altered patches on the east side of Great Blue Hill. A minimum thickness of 100 m can be estimated from exposures on Hemenway Hill. Fairly continuous reddish-purple bands alternate with a lighter red variety and show good flowage structures around altered granitoid and felsitic inclusions.

Quartz phenocrysts compose 10 percent of the unit, and microperthite makes up another 12 percent. Both phenocryst types average 1 mm in greatest diameter and exhibit poikilitic texture along the borders. Also present, in amounts as much as 9 percent, are tabular crystals of magnetite after an amphibole, now mostly altered to hematite. The groundmass consists of a granophyric intergrowth of small quartz and sanidine crystals and is unlike that found in any of the known ash-flow tuffs of this area.

Since glass shards or pumice fragments cannot be identified in this flow, evidence of pyroclastic origin is lacking. Although the Hemenway Hill unit has certain similarities with the Pine Hill lava flow, the overall evidence is not conclusive. Without satisfactory proof of origin, the unit can only be classified according to its composition, which is rhyolitic (see Table 3).

DISCUSSION OF CHEMICAL RESULTS

Rock samples from 13 locations in the Blue Hill igneous complex were analyzed by flame photometry (Na_2O), atomic absorption (MgO), wet chemistry (SiO_2), and x-ray fluorescence (all other oxides). The results are given in Tables 2 and 3. Sample locations (except for sample 14 from Hemenway Hill) are shown in Figure 1. Analyses 11 and 12 of Table 3 are averages of analyses for locations 2 and 3 (Wampatuck Hill flow) and locations 6 and 7 (Chickatawbut Road flow), respectively. For both the Wampatuck Hill and Chickatawbut Road flows, only analyses from the more indurated middle part of the tuffs were averaged, since both the upper and basal zones are more altered and would not be as representative of the original composition. Analyses of different samples from any given location indicated that the precision of the results is ±3 percent.

All the Blue Hills volcanic rocks are rhyolitic in composition. With the exception

TABLE 2. MAJOR-ELEMENT ANALYSES OF COOLING ZONES WITHIN WAMPATUCK HILL AND CHICKATAWBUT ROAD ASH FLOWS

	1	2	3	4	5	6	7	8	9	10
SiO_2	74.92	75.49	74.67	75.63	75.58	74.32	74.93	75.94	74.80	75.65
Al_2O_3	13.50	12.90	12.60	12.03	13.60	12.10	12.80	12.20	12.70	12.30
FeO*	3.50	3.11	4.50	3.79	2.87	3.44	3.28	3.20	4.49	3.34
MgO	<0.10	<0.10	<0.10	<0.10	<0.10	<0.10	<0.10	<0.10	<0.10	<0.10
CaO	0.02	0.08	0.09	0.06	0.19	0.07	0.04	0.10	0.08	0.05
Na_2O	2.37	4.11	3.85	3.44	2.54	5.21	3.83	6.35	3.41	3.69
K_2O	5.45	4.36	4.40	4.24	4.92	4.49	4.99	2.30	3.98	4.18
TiO_2	0.26	0.25	0.26	0.25	0.25	0.22	0.25	0.23	0.28	0.26
MnO	0.07	0.08	0.08	0.11	0.09	0.09	0.08	0.10	0.10	0.08
Total	100.09	100.38	100.45	99.55	100.04	99.94	100.20	100.42	99.84	99.55

Note: Na_2O determined by flame photometry, MgO by atomic absorption, SiO_2 by wet chemistry (Japan Analytical Chemistry Consultants Co., Ltd.), all other oxides by x-ray fluorescence.

*All iron was calculated as FeO; Fe^{+2} and Fe^{+3} values were not determined.

Column
1. Wampatuck Hill ash flow, basal zone, average of 2 samples.
2. Wampatuck Hill ash flow, densely welded zone, average of 4 samples.
3. Wampatuck Hill ash flow, eutaxitic zone, average of 6 samples.
4. Wampatuck Hill ash flow, upper zone, average of 6 samples.
5. Chickatawbut Road ash flow, basal zone.
6. Chickatawbut Road ash flow, densely welded zone, average of 2 analyses.
7. Chickatawbut Road ash flow, eutaxitic zone, average of 2 analyses.
8. Chickatawbut Road ash flow, upper zone.
9. Chickatawbut Road ash flow at contact with Blue Hill Granite Porphyry.
10. Chickatawbut Road ash flow, tuff inclusion in basal zone.

of the flow on Pine Hill, the rocks are ignimbrites. Ross and Smith (1961) have found that for any single average ignimbrite eruption the chemical composition is nearly constant, although the upper layers of thick flows tend to be relatively more mafic. The Wampatuck Hill flow shows little vertical variation in bulk chemical composition, except for Na_2O, which decreases upward from the densely welded zone (Table 2, analyses 1 through 4). This variation probably does not reflect chemical differences in the magma itself. Lipman (1965) and Noble (1965, 1967) have shown that almost all secondarily hydrated glasses have lost Na. Ignimbrites that have undergone deuteric alteration (reactions of aqueous and magmatic phases during cooling) show chemical variations between different zones of the same flow (Scott, 1966). The more densely welded zones contain somewhat greater amounts of Na_2O and lesser amounts of K_2O than the more poorly welded zones. Table 2 indicates a similar variation in the Wampatuck Hill flow. Another explanation for Na loss during or shortly after crystallization and cooling has been demonstrated by Noble (1970): if the parent melt was peralkaline, then the available Al would not be able to accommodate all the Na in alkali feldspar. Therefore, excess Na may be lost during devitrification. However, calculation of $Al_2O_3/K_2O + Na_2O$ ratios for the Wampatuck Hill flow shows it is not peralkaline. Parts of the Chickatawbut Road flow are peralkaline (Table 2, analyses 6, 8).

The Chickatawbut and Wampatuck Hill flows are very similar overall, both physically and chemically (Tables 2, 3). Weathering effects may have masked any minor variations in the original composition. The most striking difference between the two flows is the presence of volcanic bombs in the upper zone of the Chickatawbut Road flow. Rittman (1962) has interpreted bombs as evidence of a largely devolatilized magma in its final stages of eruption.

TABLE 3. COMPARISON OF CHEMICAL ANALYSES OF BLUE HILLS IGNEOUS COMPLEX

	11	12	13	14	15	16	17	18
SiO_2	75.08	74.62	74.52	73.63	71.50	76.37	75.08	72.88
Al_2O_3	12.75	12.35	13.20	10.90	15.50	12.15	11.57	12.30
Fe_2O_3						1.65	2.25	1.67
FeO	3.80	3.36	2.94	8.04	3.65	1.06	0.93	2.10
MgO	<0.10	<0.10	<0.10	<0.10	<0.10	0.10	0.03	0.09
CaO	0.08	0.06	0.11	0.05	0.24	0.17	0.44	0.87
Na_2O	3.98	4.52	2.13	3.14	3.58	3.64	4.21	4.43
K_2O	4.38	4.78	5.88	4.62	5.67	4.68	4.62	4.90
H_2O^+	..	..	..	..	..	0.13	0.19	0.31
H_2O^-	..	..	..	..	..	0.08	0.04	0.15
TiO_2	0.25	0.24	0.27	0.27	0.35	0.18	0.20	0.35
P_2O_5	..	..	..	..	..	..	tr.	tr.
MnO	0.08	0.08	0.04	0.06	0.16	0.07	tr.	0.10
ZrO_2	..	..	..	..	..	..	0.20	0.10
CO_2	..	..	..	..	..	..	..	0.30
Total	100.40	100.01	99.09	100.71	100.65	100.28	99.76	100.55

Note: For analyses 11 through 15, Na_2O determined by flame photometry, MgO by atomic absorption, SiO_2 by wet chemistry (Japan Analytical Chemistry Consultants Co., Ltd.), all other oxides by x-ray fluorescence. All iron was calculated as FeO; Fe^{+2} and Fe^{+3} values were not determined.

Column
11. Wampatuck Hill ash flow, average of columns 2 and 3 of Table 2.
12. Chickatawbut Road ash flow, average of columns 6 and 7 of Table 2.
13. Pine Hill rhyolite, average of 2 samples.
14. Hemenway Hill flow, average of 2 samples.
15. Pyritic volcanic rocks, type D-2.
16. "Aporhyolite," Wampatuck Hill, analyst Warren (1913).
17. Medium-gray Quincy Granite, Hitchcock quarry, North Common Hill, Quincy, Mass., analyst Warren (1913).
18. Blue Hill Granite Porphyry, east side of Rattlesnake Hill, analyst Warren (1913).

A relatively high Or:Ab ratio for the pyritic volcanic rocks indicates that these rocks are similar to the previously described tuffs in that there has been a probable loss of Na. However, it is apparent that the pyritic volcanic rocks are relatively more mafic than the flows described earlier (Table 3, analysis 15). Since ignimbrite eruptions often become more mafic toward the end of an eruptive sequence, the relative basicity of these flows is consistent with their apparent stratigraphic position above the previously described flows.

The presence of the Pine Hill rhyolite would suggest degassing of the magma to the point where extremely violent eruptions are followed by lava outpouring. A number of workers, among them Kennedy (1955) and Sparks and others (1973), have noted that volatile content decreases toward the end of an ignimbrite eruptive sequence. Ratté and Steven (1964) have observed that a typical ignimbrite sequence begins with predominantly ash-flow eruptions, which give way to alternating ash-flow and lava-flow eruptions and finally to predominantly lava-flow eruptions. Of course, the presence of one lava flow does not by itself prove any sequential position, but it does increase the possibility that the nature of the Blue Hills eruptions varied with time. Analyses of the Pine Hill rhyolite and of the Hemenway Hill flow are given in Table 3 (analyses 13, 14).

Also included in Table 3, for comparison with the new analyses of this paper, are analyses by Warren (1913) of "aporhyolite" from Wampatuck Hill (analysis 16), Quincy Granite (analysis 17), and Blue Hill Granite Porphyry (analysis 18). The present analyses of the Wampatuck Hill ash-flow sheet are consistent with Warren's "aporhyolite," and fall within the compositional range of low-Ca rhyolites. The analyses in Table 3 show that all the Blue Hills volcanic rocks have low CaO and MgO values and relatively high TiO_2 and total Fe values. Such composition, along with high Na_2O values, is characteristic of comendites (peralkaline rhyolites) (Noble, 1968; Noble and others, 1969). Assuming a loss of Na during devitrification, all the ash flows, with the exception of the pyritic volcanic rocks, may originally have been comendites.

ACKNOWLEDGMENTS

This study was undertaken at the suggestion of N. E. Chute, whose assistance throughout the course of the work is greatly appreciated. The field work was done during 1966 and 1967 and was part of a M.S. thesis at Syracuse University. The writer is indebted to members of the Boston University Geology Department, particularly to A. H. Brownlow and W. C. Park, who contributed generously of their time and experience. I also thank M. P. Billings, P. C. Lyons, R. S. Naylor, and D. C. Noble for constructive criticisms and suggestions.

REFERENCES CITED

Bascom F., 1900, Volcanics of Neponset Valley, Mass.: Geol. Soc. America Bull., v. 11, p. 115-126.

Billings, M. P., 1929, Structural geology of the eastern part of the Boston basin: Am. Jour. Sci., v. 18, p. 97-137.

———1976, Geology of the Boston basin, in Lyons, P. C., and Brownlow, A. H., eds., Studies in New England geology: Geol. Soc. America Mem. 146, p. 5-30 (this volume).

Bottino, M. L., Fullagar, P. D., Fairbairn, H. W., Pinson, W. H., and Hurley, P. M., 1970, The Blue Hills igneous complex, Massachusetts: Whole rock Rb-Sr open systems: Geol. Soc. America Bull., v. 81, p. 3739-3746.

Christiansen, R. L., and Lipman, P. W., 1966, Emplacement and thermal history of a rhyolite lava flow near Fortymile canyon, southern Nevada: Geol. Soc. America Bull., v. 77, p. 671-684.

Chute, N. E., 1940, Preliminary report on geology of the Blue Hills: Massachusetts Dept. Public Works and U.S. Geol. Survey Coop. Geol. Proj. Bull., no. 1, 50 p.

———1964, Geology of the Norfolk Basin Carboniferous sedimentary rocks and the various igneous rocks of the Norwood and Blue Hills quadrangles: New England Intercoll. Geol. Conf. Guidebook, 56th Ann. Mt., p. 91-114.

———1966, Geology of the Norwood quadrangle, Norfolk and Suffolk Counties, Massachusetts: U.S. Geol. Survey Bull. 1163-B, 78 p.

———1969, Bedrock geologic map of the Blue Hills quadrangle, Norfolk, Suffolk, and Plymouth counties, Mass.: U.S. Geol. Survey Geol. Quad. Map GQ-796.

Cook, E. F., 1962, Ignimbrite bibliography and review: Idaho Bur. Mines and Geology, Inf. Cir., no. 13, 64 p.

Crosby, W. O., 1900, Geology of the Boston basin: Occasional Papers, Boston Soc. Nat. History, v. 1, no. 4, pt.3, p. 289-563.

Emerson, B. K., 1917, Geology of Massachusetts and Rhode Island: U.S. Geol. Survey Bull. 597, 289 p.

Enlows, H. E., 1955, Welded tuffs of Chiricahua National Monument, Arizona: Geol. Soc. America Bull., v. 66, p. 1215-1246.

Fisher, R. V., 1960, Classification of volcanic breccias: Geol. Soc. America Bull., v. 71, p. 973-982.

Kennedy, G. C., 1955, Some aspects of the role of water in rock melts: Geol. Soc. America Spec. Paper 62, p. 489-503.

LaForge, L., 1932, Geology of the Boston area, Massachusetts: U.S. Geol. Survey Bull. 83a, 105 p.

Lipman, P. W., 1965, Chemical comparison of glassy and crystalline volcanic rocks: U.S. Geol. Survey Bull. 1201-D, p. 1-24.

Lyons, P. C., and Kreuger, H. W., 1976, Petrology, chemistry, and age of the Rattlesnake pluton and implications for other alkalic granite plutons of southern New England, in Lyons, P. C., and Brownlow, A. H., eds., Studies in New England geology: Geol. Soc. America Mem. 146, p. 71-102 (this volume).

Noble, D. C., 1965, Gold Flat Member of the Thirsty Canyon tuff—A pantellerite ash-flow sheet in southern Nevada: U.S. Geol. Survey Prof. Paper 525-B, p. 885-890.

———1967, Sodium, potassium, and ferrous iron contents of some secondarily hydrated natural silicic glasses: Am. Mineralogist, v. 52, p. 280-286.

———1968, Systematic variation of major elements in comendite and pantellerite glasses: Earth and Planetary Sci. Letters, v. 4, p. 167-172.

———1970, Loss of sodium from crystallized comendite welded tuffs of the Miocene Grouse Canyon member of the Belted Range tuff, Nevada: Geol. Soc. America Bull., v. 81, p. 2677-2688.

Noble, D. C., Haffty, J., and Hedge, C. E., 1969, Strontium and magnesium contents of some natural peralkaline silicic glasses and their petrogenic significance: Am. Jour. Sci., v. 267, p. 598-608.

Nockolds, S. R., 1954, Average chemical composition of some igneous rocks: Geol. Soc. America Bull., v. 65, p. 1017-1032.

Ratté, J. C., and Steven, T. A., 1964, Magmatic differentiation in a volcanic sequence related to the Creede caldera, Colorado: U.S. Geol. Survey Prof. Paper 475-D, p. D49-D53.

Rittman, A., 1962, Volcanoes and their activity: New York, John Wiley & Sons, Inc., 305 p.

Ross, C. S., and Smith, R. L., 1961, Ash-flow tuffs: Their origin, geologic relations and identification: U.S. Geol. Survey Prof. Paper 366, 81 p.

Scott, R., 1966, Origin of chemical variations within ignimbrite cooling units: Am. Jour. Sci., v. 264, p. 273-288.

Smith, R. L., 1960a, Ash-flows: Geol. Soc. America Bull., v. 71, p. 795-842.

——1960b, Zones and zonal variations in welded ash-flows: U.S. Geol. Survey Prof. Paper 354-F, 10 p.

Sparks, R.S.J., Self, S., and Walker, G. P., 1973, Products of ignimbrite eruptions: Geology, v. 1, p. 115-118.

Warren, C. H., 1913, Petrology of the alkali-granites and porphyries of Quincy and the Blue Hills, Massachusetts: Am. Acad. Arts and Sci. Proc., v. 49, p. 201-333.

Williams, H., 1936, Pliocene volcanoes of the Navajo-Hopi country: Geol. Soc. America Bull., v. 47, p. 111-171.

Zartman, R. E., and Marvin, R. F., 1971, Radiometric age (Late Ordovician) of the Quincy, Cape Ann, and Peabody granites from eastern Massachusetts: Geol. Soc. America Bull., v. 82, p. 937-958.

Manuscript Received by the Society June 7, 1974
Revised Manuscript Received November 20, 1974
Manuscript Accepted December 5, 1974

Printed in U.S.A.

Geological Society of America
Memoir 146
© 1976

Fossil Plants of Pennsylvanian Age from Northwestern Narragansett Basin

JOHN OLEKSYSHYN
Professor of Geology, Emeritus
Boston University
Boston, Massachusetts 02215

ABSTRACT

Fossil plants collected in the lower beds of the Rhode Island Formation in Plainville, Massachusetts, consist of two species of lycopods, ten species of sphenopsids, three species of ferns, twelve species of gymnosperms, and one type of rootlike structure, probably arthrophyte. Two species of gymnosperms, *Palmatopteris narragansettensis* and *Palmatopteris plainvillensis*, are considered new. Known stratigraphic occurrences suggest that the Plainville assemblage is Early Alleghenyan (corresponding to Westphalian C) and is equivalent in age to flora of the early strata (Minto Formation) of the Pictou Group in New Brunswick and Nova Scotia, Canada. The upper beds of the Rhode Island Formation and the Dighton and Purgatory conglomerates are probably Late Alleghenyan (corresponding to Westphalian D) and are contemporaneous with upper beds of the Pictou Group in the Maritime Provinces of Canada.

INTRODUCTION AND PURPOSE

Fossil plants in the sedimentary strata of the Narragansett basin were first reported by Jackson (1840), Hitchcock (1841), Teschemacher (1846), and Lesquereux (1884, 1889). Based on his identification of more than 50 plant species, Lesquereux assigned a Pennsylvanian age to these strata. Later collections of fossil plants were reported by Round (1924, 1927), Knox (1944), and Lyons (1969, 1971). Correlation with similar fossil plants of the Maritime Provinces of Canada and the Appalachian and central regions of the United States indicated a Pennsylvanian (upper Carboniferous) age for the strata of the Narragansett and Norfolk basins. The poor preservation, usually fragmental fossil remains commonly without diagnostic features, and absence of detailed descriptions of previously studied fossil plants did not permit precise determinations of the age of the strata in this region.

The purpose of this paper is to establish more precisely the age of the Rhode Island Formation based on a new collection of fossil plants from Plainville, Massachusetts.

OUTLINE OF GEOLOGY OF THE NARRAGANSETT BASIN

The Narragansett basin is elongated northeast-southwest and is approximately 100 km long and 35 km wide (Fig. 1). The Pennsylvanian sediments in this basin were deposited unconformably upon older metamorphic and plutonic rocks.

The oldest rocks exposed along the western margin of the Narragansett basin belong to the Blackstone Series, which consists of quartz-mica schist, quartzite, greenstone, and marble of Precambrian or early Paleozoic age. The rocks of this series were later intruded by magma that cooled into plutonic rocks of Devonian or earlier age. The Quincy Granite, together with related intrusive and volcanic rocks, is exposed along the western margin of the Narragansett basin and along the northern side of the Norfolk basin.

The oldest rocks cropping out along the northern, northeastern, and eastern margins of the Narragansett basin are presumably Precambrian rocks, consisting of Dedham quartz monzonite (Dedham Granodiorite of Emerson, 1917), gabbro-diorite, and metamorphic rocks. On the islands in the southwestern part of the basin, the Blackstone Series, and granitic and metamorphic rocks unconformably underlie Pennsylvanian strata. In the vicinity of Hoppin Hill (Fig. 1). Precambrian(?) Dedham quartz monzonite (Hoppin Hill granite) is overlain by Lower Cambrian slates with preserved fossil remains. Near Bristol, Rhode Island, Pennsylvanian strata overlie the Metacom granite gneiss and quartz-mica schist (Quinn and Springer, 1954).

The sedimentary rocks in the Narragansett basin include conglomerate, sandstone, gray and black shale, and meta-antracite. These deposits are of continental origin. On the basis of fossil plants preserved in these rocks, they are known to be Pennsylvanian in age (Fig. 2).

The oldest formation in the Narragansett basin is the Pondville Conglomerate, which rests unconformably on the pre-Pennsylvanian plutonic and metamorphic rocks. This poorly sorted basal conglomerate, consisting primarily of quartzite and granitic clasts, is exposed at many places along the margin of the basin (Towe, 1959).

Overlying the Pondville Conglomerate in the northern area of the basin, and partly equivalent to it, is the Wamsutta Formation, which consists of poorly sorted red conglomerate, sandstone, siltstone, and shale. The red coloring of the particles is due to an iron oxide stain.

Toward the south, the red beds of the Wamsutta Formation pass into equivalent strata of the lower part of the Rhode Island Formation. The Wamsutta Formation around Hoppin Hill, North Attleboro (Fig. 1), rests unconformably on Cambrian strata, and in some areas of the Brockton quadrangle, it lies directly on pre-Pennsylvanian rocks. The Wamsutta Formation west of Attleboro contains intercalated layers of rhyolite and basalt flows (Quinn and Oliver, 1962).

The most extensive and the thickest stratigraphic unit in the Narragansett basin is the Rhode Island Formation, which in many areas overlies the Precambrian basement rocks. In other areas it overlies the Pondville Conglomerate, and in the northwestern sector of the basin, it interfingers with or overlies the Wamsutta Formation.

The Rhode Island Formation consists of thick sequences of poorly sorted

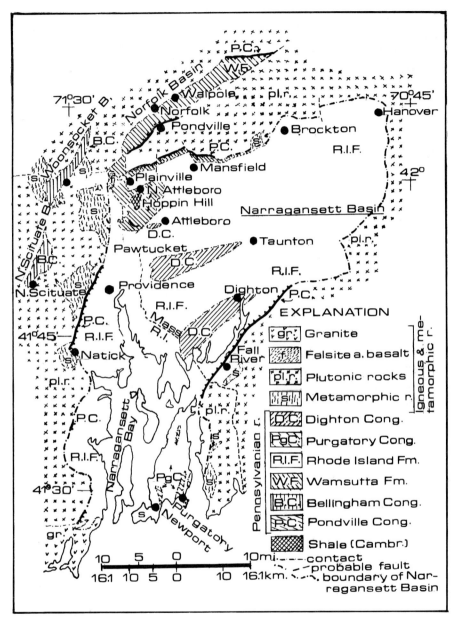

Figure 1. Geologic map of the Narragansett basin and associated basins in southeastern Massachusetts and eastern Rhode Island.

conglomerate, graywacke, lithic graywacke, subgraywacke, subarkose types, and black and gray shale. In some places, for example, at Plainville in the northwestern section of the basin, it contains fossil plants. In places, lenses of meta-anthracite occur between the black and gray shales. Such meta-anthracite was mined in the basin during the nineteenth century, but its high ash content and the low content of volatile matter, together with competition with a better grade of coal from the Pennsylvania coalfields, caused the interruption of meta-anthracite mining in this region during the early part of the twentieth century.

Geol. Time Unit:	ROCK STRATIGRAPHIC UNITS: Appalachian region	New England region	Maritime Prov. of Canada	TIME STRATIGPHIC UNITS: 'Canada'		Appalachian region
PENNSYLVANIAN (UPPER CARBONIFEROUS)	Monongahela Group	Hiatus	Hiatus	Stephanian	C	Monongahelan
				Stephanian	B	
	Conemaugh Group			Stephanian	A	Conemaughian
	Allegheny Group	Dighton Congl. / Rhode Island Fm.	Pictou Group	Westphalian	D	Late Alleghenyan
				Westphalian	C	Early Alleghenyan
	Pottsville Supgr.: Upper Pottsville Group	Wamsutta Fm. / Pondville Congl.	Hiatus	Westphalian	B²	Late Pottsvillian
	Pottsville Supgr.: Middle Pottsville Group	Hiatus	Cumberland Group	Westphalian	B¹	Medial Pottsvillian
	Pottsville Supgr.: Lower Pottsville Group		Riversdale Group	Westphalian	A	Early Pottsvillian
	Pocahontas Group		Canso Group	Namurian	C	Pocahontasan
				Namurian	B	

Figure 2. Correlation chart of Pennsylvanian strata in the Appalachian region, New England, Maritime Provinces of Canada and Western Europe.

The youngest Pennsylvanian deposit in the Narragansett basin is the Dighton Conglomerate, cropping out in three synclinal areas—the Attleboro, the Taunton, and the Dighton synclines (Fig. 1). The Dighton Conglomerate consists mainly of poorly sorted pebbles of quartzite and granite and scattered rounded fragments of metamorphic and volcanic rocks. The conglomerate in many places interfingers with lenses of lithic graywacke.

A small area in the southern section of the basin in the vicinity of Newport is occupied by very coarse Purgatory Conglomerate that is gradational into the underlying Rhode Island Formation. About 95 percent of the pebbles and boulders of the Purgatory Conglomerate consists of quartzite and about 5 percent consists of granite (Perkins, 1920). Cross-stratification is locally evident in thin lenses of sandstone found in the Conglomerate (Towe, 1959).

The principal source area of the sediments deposited in the Narragansett basin, as indicated by lithologic character and cross-stratification in sandstone, was probably to the northeast in the region of Massachusetts Bay. Here some highlands probably existed, presumably a consequence of the Acadian orogeny. This uplifted region was composed mainly of Cambrian and Ordovician quartzite and Dedham quartz monzonite. Only in the northwestern section of the Narragansett basin, where the rocks of the Wamsutta Formation are reddish, were the rocky fragments probably derived from some highlands located to the north and northwest.

Orogenic movements during early and middle Paleozoic time, especially during the Acadian orogeny, created a mountain relief in the regions of New England and the Maritime Provinces of Canada. These earth-crust movements usually were accompanied in many places with epeirogenic movements, which resulted in many small intermontane basins.

The Narragansett basin was probably one such intermontane basin, located in southeastern Massachusetts and eastern Rhode Island. The basin was associated with smaller nearby or connected basins, such as the North Scituate basin, the Woonsocket basin, and the Norfolk basin (Fig. 1). In all these basins, Pennsylvanian (upper Carboniferous) sediments were deposited, which later during the Appalachian orogeny were downfolded or downfaulted into the older pre-Pennsylvanian rocks and were intruded by magma in the southern part of the Narragansett basin (Quinn and Oliver, 1962).

The Narragansett basin was a deeply subsiding epieugeosynclinal trough (Kay, 1951) with limited volcanism, associated with narrow uplift and overlying a deformed and magmatically intruded eugeosyncline. This trough was characterized by rapid subsidence contemporaneous with rapid deposition of sediments. The thick sequences of lithic graywacke, arkose, and graywacke conglomerate indicate a fluvial origin for these sediments. The coarseness of the conglomerate indicates that the source area was of considerable relief. The black shale indicates stagnant conditions in swamps most likely associated with flood plains. The presence of fossil land plants and coal deposits indicates a terrestrial origin for these sediments. The red color of the Wamsutta Formation in the northern section of the basin indicates that this area was higher, better drained, and exposed to oxidizing conditions.

The most fossiliferous Pennsylvanian rocks in the Narragansett basin are the black shales of the Rhode Island Formation, which are exposed in the northwestern sector of the basin. One place where such shale is exposed is in the Herman G. Protze, Masslite, Inc. quarry, located about 2 km northwest of Plainville Center,

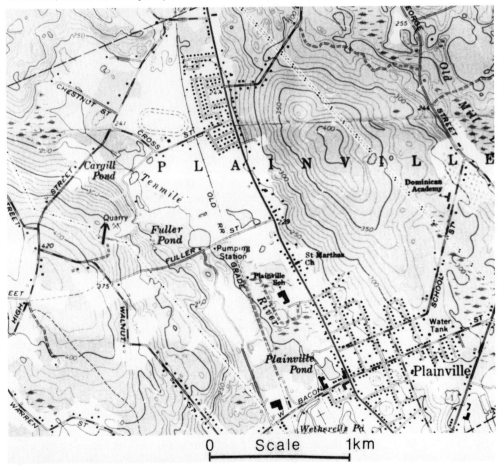

Figure 3. Map of the Plainville area, Massachusetts.

Massachusetts. Here sandstone and coal of the Rhode Island Formation are quarried (Fig. 3). The plant fossils described in this paper were found in gray and black shales of this formation on the north side of the quarry.

SPECIMEN NUMBERS

Specimens numbered are reposited in the permanent collections of the Paleobotanical Collections of the Botanical Museum of Harvard University (PC BM HU).

DESCRIPTION OF FOSSIL PLANTS
KINGDOM PLANTAE
Phylum TRACHEOPHYTA
Subphylum LYCOPODOPHYTA (= LYCOPSIDA)
Class LYCOPODINAE
Order LEPIDODENDRALES
Family LEPIDODENDRACEAE
Genus *Lepidodendron* Sternberg, 1820
Lepidodendron cf. *pictoense* Dawson
(Fig. 4A)

Lepidodendron lycopodioides Sternberg, 1820, Bd. 1, p. 26, Pl. 16, Figs. 1, 2, 4; Bd. 4, Pl. 68, fig. 1.
Lepidodendron pictoense Dawson, 1863, p. 449.
Lepidodendron pictoense Dawson, 1868, p. 454, 487, Figs. 169A–A7.
Lepidodendron lanceolatum Lesquereux, 1880, p. 369; Atlas (1879), Pl. 63, figs. 3, 4.
Lepidodendron lycopodioides Sternberg. Zeiller, 1888, p. 464–466; Atlas (1886), Pl. 69, figs. 2, 3; Pl. 70, fig. 1.
Lepidodendron lanceolatum Lesquereux. White, 1899; p. 192, Pl. 53, fig. 2.
Lepidodendron lycopodioides Sternberg. Bell, 1938, p. 93, Pl. 96, figs. 2, 3; Pl. 97, figs. 1-3; Pl. 98, fig. 13.
Lepidodendron lanceolatum Lesquereux. Bell, 1940, p. 122.
Lepidodendron lanceolatum Lesquereux. Bell, 1944, p. 88, Pl. 48, fig. 3.
Lepidodendron pictoense Dawson. Bell, 1962a, p. 52-53, Pl. 46, fig. 1; Pl. 49, figs. 1, 3; Pl. 50, figs. 1-3.
Lepidodendron pictoense Dawson. Bell, 1962b, p. 41.

Description. Species is represented by several fragments of leafy shoots, branching at acute angles. The lanceolate uninerved leaves, about 15 mm long and 2 mm broad near the base depending on the size of the stem, are crossed by short transverse wrinkles. The straight or slightly curved ends of the leaves are acute.

Remarks. Species is very similar to *L. lanceolatum* Lesquereux, *L. lycopodioides* Sternberg, and other related species (Bell, 1962).

Distribution. Species is also known from the Clifton Formation (Pictou Group), which in the opinion of Bell (1962) is of the same age or a little younger than the Minto Formation. Both the Minto and Clifton Formations belong to the lower part of the *Linopteris obliqua* Zone, which corresponds to Westphalian C.

Specimen. PC BM HU hypotype 51345 A-C.

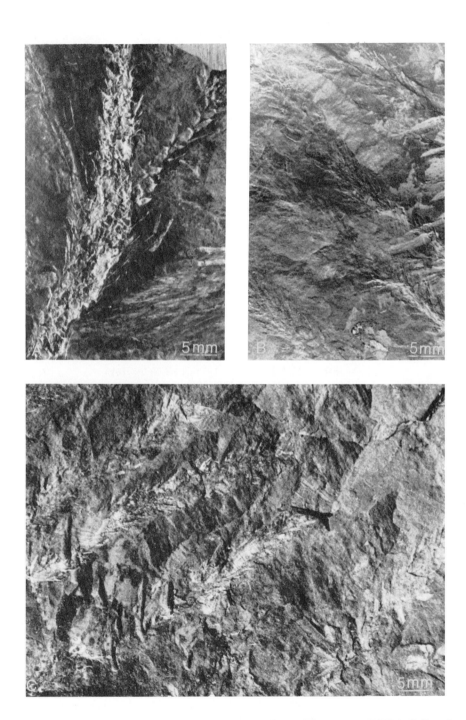

Figure 4. A, *Lepidodendron* cf. *pictoense* Dawson, PC BM HU hypotype 51345 A-C; B and C, *Lepidodendron* cf. *lanceolatum* Lesquereux, PC BM HU hypotype 51342, nontype 51344 A-C.

Lepidodendron cf. lanceolatum Lesquereux
(Figs. 4B, 4C)

Lepidodendron dichotomum Sternberg, 1820, p. 177, Pls. 1-3.
Lepidodendron lanceolatum Lesquereux, 1880, p. 369; Atlas (1879), Pl. 63, figs. 3, 4.
Lepidodendron lanceolatum Lesquereux. White, 1899, p. 192, Pl. 53, fig. 2.
Lepidodendron lycopodioides Sternberg. Bell, 1938, p. 93, Pl. 90, figs. 2, 4; Pl. 97, figs. 1-3.
Lepidodendron lanceolatum Lesquereux. Bell, 1940, p. 122.
Lepidodendron lanceolatum Lesquereux. Bell, 1944, p. 88, Pl. 48, fig. 3.
Lepidodendron lanceolatum Lesquereux. Bell, 1962b, p. 40.

Description. Species is represented by several dichotomously dividing branches with lanceolate, commonly more or less falcate leaves, as much as 20 mm long by 2 to 3 mm wide.

Remarks. The shape of this species is very similar to *L. lycopodioides* Sternberg, *L. simile* Kidston, *L. acutum* Presl, and *L. pictoense* Dawson (Bell, 1962); thus, there is often confusion in the identification of this lycopod.

Distribution. Species is also known from the Riversdale, Cumberland, and Pictou groups in the Maritime Provinces of Canada (Westphalian A through D).

Specimen. PC BM HU hypotype 51342; nontype 51344 A-C.

Subphylum ARTHROPHYTA (= SPHENOPSIDA)
Class EQUISETINAE
Order EQUISETALES
Family CALAMITACEAE
Genus *Calmites* Suckow, 1784
Calamites undulatus Sternberg
(Fig. 5A)

Calamites undulatus Sternberg, 1825, Bd. 1, Teil 4, p. 26, Pl. 1, fig. 2.
Calamites undulatus Brongniart, 1828, v. 1, p. 127, Pl. 17, figs. 1-4.
Calamites undulatus Dawson, 1873, p. 30, Pl. 8, figs. 66-69.
Calamites undulatus Brongniart. Bell, 1944, p. 98, Pl. 50, fig. 2; Pl. 53, figs. 1, 5, 6.
Calamites undulatus Sternberg. Gothan and Remy, 1957, p. 46-48, text-fig. 36, Chart 5.
Calamites undulatus Brogniart. Bell, 1966, p. 12, Pl. 5, fig. 16.
Calamites undulatus Sternberg. Abbott, 1968, p. 21, Pl. 14, figs. 1, 2.
Calamites undulatus Sternberg. Crookall, 1969, v. 4, pt. 5, p. 665-673, Pl. 121, figs, 2, 3; Pl. 122, figs. 1-3; Pl. 124, figs. 1, 2; text-fig. 195.

Description. Species is represented by a pith cast with low flat longitudinal ribs, about 3 mm wide, separated by shallow furrows, about 0.5 mm broad. Pointed ends of undulating ribs form a zigzag pattern at nodes.

Remarks. Species is distinguished from *C. suckowii* by the zigzag junction of the ribs at the nodes and by the common undulation of the ribs.

Distribution. *Calamites undulatus* is also known from the Riversdale and lower Cumberland groups in Nova Scotia, Canada (Westphalian A and Early B). In Western Europe this species ranges from Late Namurian C to Stephanian C. According

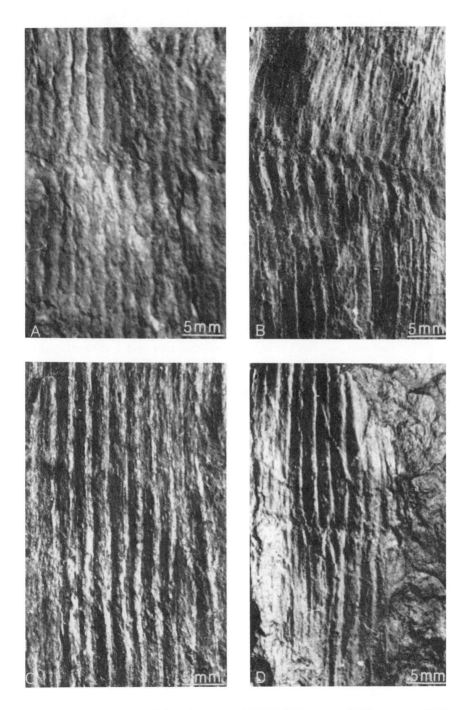

Figure 5. A. *Calamites undulatus* Sternberg, PC BM HU hypotype 51331, nontype 51332; B, *Calamites carinatus* Sternberg, PC BM HU hypotype 51327, nontype 51328 A-C; C and D, *Calamites cistii* Brongniart, PC BM HU hypotype 51329, nontype 51330.

to Darrah (1970) this species is more frequent in the United States than published records indicate.
Specimen. PC BM HU hypotype 51331; nontype 51332.

Calamites carinatus Sternberg
(Fig. 5B)

Calamites carinatus Sternberg, 1823, Bd. 1, Teil 3, p. 36, 39, Pl. 32, fig. 1.
Calamites ramosus Artis, 1825, p. 18, Pl. 2, fig. 6.
Calamites ramosus Artis. Lesquereux, 1880, v. 1, p. 22-23; Atlas (1879), Pl. 1, fig. 2.
Calamites ramosus Artis. Bell, 1938, p. 84, Pl. 86, fig. 1.
Calamites carinatus Sternberg. Gothan and Weyland, 1954, p. 194, text-fig. 175.
Calamites carinatus Sternberg. Gothan and Remy 1957, p. 48-49, text-fig. 37, Chart 5.
Calamites ramosus Artis. Bell, 1962, p. 49, Pl. 44, fig. 5; Pl. 45, fig. 6.
Calamites carinatus Sternberg. Abbott, 1968, p. 21, Pl. 15, figs. 1, 2.
Calamites carinatus Sternberg. Darrah, 1970, p. 170, Pl. 35, fig. 2.

Description. Species is represented by a few fragments of pith casts with long internodes covered by flattened longitudinal ribs, about 1 mm broad, separated by slightly narrower, delicately striated furrows that are rounded or bluntly pointed at the nodes. The ribs near the nodes are slightly bent.

Remarks. Species, prior to the publication of the Monograph of the *Calamites* of Western Europe by Kidston and Jongmans (1915-1917), was often designated by paleobotanists as *C. ramosus*.

Distribution. Species was found rarely in the Riversdale Group but was found more commonly in the Cumberland Group and Minto Formation (the base of Pictou Group) (Westphalian A through C). In Western Europe this species has a similar range.

Specimen. PC BM HU hypotype 51327; nontype 51328 A-C.

Calamites cistii Brongniart
(Figs. 5C, 5D)

Calamites cistii Brongniart, 1828, p. 129-130, Pl. 20, figs. 1-5.
Calamites cistii Brongniart. Geinitz, 1855, p. 7, Pl. 12, figs. 4, 5; Pl. 13, fig. 7.
Calamites cistii Dawson, 1873, p. 29, Pl. 8, fig. 65.
Calamites cistii Brongniart. Bell, 1944, p. 97, Pl. 55, fig. 3; Pl. 64, fig. 1.
Calamites cistii Brongniart. Gothan and Remy, 1957, p. 46-47, text-fig. 35, Chart 5.
Calamites cistii Brongniart. Abbott, 1968, p. 21, Pl. 11, fig. 5.
Calamites cistii Brongniart. Crookall, 1969, pt. 5, p. 639-643, Pl. 123, fig. 1; text-fig. 183.

Description. Species is represented by a few fragments of pith casts covered by fine, flattened longitudinally striated ribs, about 1 mm broad, separated by straight furrows with a prominent line on the sides of each furrow. The ends of the alternately arranged ribs on opposite sides of the nodes taper to subacute points. Internodal distance could not be established, and there is no evidence of tubercles.

Remarks. This species differs from *C. suckowii* Brongniart by the narrower ribs and subacute ends of the ribs. Lack of undulations of narrower ribs separate this species from *C. undulatus* Sternberg.

Distribution. Species is also known from the Riversdale, Cumberland, and Pictou groups in Nova Scotia, Canada (Westphalian A, B, and C). It is also found in the Appalachian region and in central regions of the United States and in Western Europe, where it ranges from Namurian B to Permian.

Specimen. PC BM HU hypotype 51329; nontype 51330.

Calamites sp.
(Fig. 6A)

Description. This plant is represented by a pith cast 26 cm long and 3 cm broad. The flat longitudinal ribs, about 1.5 mm wide, are separated by narrower furrows. Absence of nodal structures makes specific identification impossible.

Remarks. The specimen is similar to *Asterocalamites scrobiculatus* (Schlotheim) Zeiller, known from the Namurian age Canso Group in New Brunswick, Canada.

Specimen. PC BM HU hypotype 51333.

FOLIAGE OF ARTHROPHYTA

The leaves of arthrophytes may be divided into two main groups: one represented by genera *Asterophyllites* and *Annularia* and the other group represented by genus *Sphenophyllum*.

Asterophyllites and *Annularia* are identified by foliage of probable calamitean affinity and represent articulated delicate stems bearing whorls of leaves at each articulation or node (Fig. 6B, after Abbott, 1958).

The uninerved leaves of *Asterophyllites* are falcate, linear, or linear-lanceolate with nearly parallel margins, more or less united at the base, usually cupped at the stem, and partially overlap the upper internode. The area of leaf attachment is elliptical.

The uninerved leaves of *Annularia* are usually lanceolate, linear-lanceolate, oblanceolate, or spatulate with convex margins radiating from the nodes on alternate sides of the stem, more or less overlapping both the upper and lower internodes. The leaves, equal in some species and unequal in others, are united at the base. The area of leaf attachment is circular.

The determination of species within these two genera is based on details of the width/length ratio, the number of leaves in a whorl, and the shape, size, and position of the widest portion of the leaf.

Characteristic of the second group, *Sphenophyllum*, are cuneate to broadly ovate leaves that occur in whorls usually six in number, not alternating with those of the whorl below. Usually a single vein enters the base of the leaf and dichotomizes several times so that each of the teeth or lobes at the distal margin of the leaves is provided with a single terminal veinlet (Abbott, 1958).

Genus *Annularia* Sternberg, 1821
Annularia mucronata Schenk
(Figs. 6C, 6D, 6E)

Annularia spinulosa Sternberg, 1821, Bd. 1, Teil 2, p. 28, 32, Pl. 19, fig. 4.
Annularia mucronata Schenk, 1883, p. 226, Pl. 30, fig. 10; text-fig. 10.
Annularia sphenophylloides var. *intermedia* Lesquereux, 1884, v. 3, p. 724.
Annularia stellata Zeiller (in part), 1888, p. 398; Atlas (1886), Pl. 61, figs. 3-6.
Annularia stellata forma *mucronata* Bell (in part), 1938, p. 85, Pl. 91, fig. 1.
Annularia mucronata Schenk. Abbott, 1958, p. 315-317, Pl. 35, fig. 5; Pl. 36, figs. 10, 11, 13, 14, 16-18; Pl. 41, fig. 57; Pl. 42, figs. 59, 61; Charts 2, 5.

Description. The specimens consist of verticils with 10 to 15 spatulate leaves, 4 to 8 mm long and 2.5 mm wide near the apex. The leaves in verticils are of different lengths; those lying more or less at right angles to the axis are almost twice as long as those lying nearly parallel to the axis. Such arrangement of leaves gives the verticils an elliptical shape. The margins of the strongly divergent leaves are straight. The width of the leaves increases from the base to the broad, rounded, and strongly mucronate tip. The prominent midvein extends from the base of the leaf to the mucronate tip.

Remarks. The leaves of a single whorl of *A. mucronata* are more widely spaced and the margins seldom touch one another, while the margins of adjacent leaves of *A. sphenophylloides* are typically close and parallel. *A. mucronata* was described by Lesquereux (1884) as *A. sphenophylloides* var. *intermedia* and by Bell (1938) as *A. stellata* forma *mucronata*. The leaves of *A. mucronata* are widest near the apex, while those of *A. stellata* are widest near the middle.

Distribution. This species is also known from the *Linopteris obliqua* Zone of the Pictou Group in the Sydney coalfield, Nova Scotia, Canada, which corresponds to Early Alleghenyan (Westphalian C), from the Appalachian and central regions of the United States, Europe, and China (Schenk, 1883). This widely distributed species ranges throughout Alleghenyan to Early Permian (Abbott, 1958).

Specimen. PC BM HU hypotype 51323.

Annularia sphenophylloides (Zenker) Gutbier
(Fig. 7A)

Galium sphenophylloides Zenker, 1833, p. 398-400, Pl. 5, figs. 6-9.
Annularia sphenophylloides (Zenker) Gutbier, 1837, Isis v. Oken, p. 436.
Annularia sphenophylloides (Zenker) Geimitz, 1855, p. 11, Pl. 18, fig. 10.
Annularia sphenophylloides Zenker. Lesquereux, 1879, v. 1, p. 48-49; Atlas (1880), Pl. 2, figs. 8, 9.
Annularia sphenophylloides (Zenker) Gutbier. White, 1899, p. 163-165.
Annularia sphenophylloides (Zenker) Gutbier. Bell, 1938, p. 84, Pl. 85, fig. 3; Pl. 87, fig. 1.
Annularia sphenophylloides (Zenker) Gutbier. Gothan and Remy, 1957, p. 53; text-fig. 42, Chart 5.
Annularia sphenophylloides (Zenker) Gutbier. Abbott, 1958, p. 319-321, Pl. 37, fig. 24; Chart 2.
Annularia sphenophylloides (Zenker) Gutbier. Lyons, 1969, Pl. I, fig. B; Pl. II, fig. A.

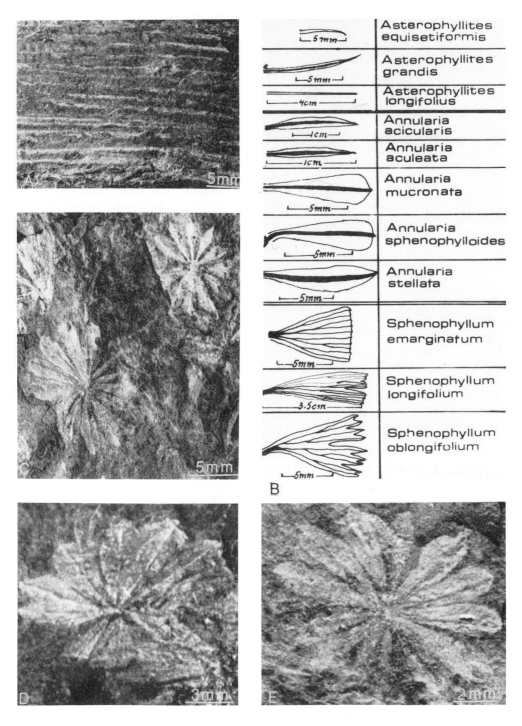

Figure 6. A, *Calamites* sp., PC BM HU hypotype 51333; B, Foliage of *Arthrophyta*; C, D, and E, *Annularia mucronata* Schenk, PC BM HU hypotype 51323.

Annularia sphenophylloides (Zenker) Gutbier. Darrah, 1970, p. 172, Pl. 23, figs. 1, 2; Pl. 43, fig. 7.

Description. This plant is represented by a few verticils of leaves, about 14 in number, borne on thin stems. The verticils are close to each other. The leaves in verticils are of different sizes, and those directed away from the stem are slightly longer than those nearly parallel to it. The length of the leaves ranges from 5 to 10 mm and the width from 0.5 to 1.5 mm in different verticils. The leaves are uninerved, spatulate, and straight margins are prolonged and have a broadly rounded or mucronate tip.

Distribution. Species is also found in the Rhode Island Formation at Foxboro, Massachusetts (Lyons, 1969). It is also known from the Dunkard Group (Early Permian) of central United States (Abbott, 1958; Darrah, 1970) and from the Pictou (Morien) Group of New Brunswick and Nova Scotia (Westphalian C and D) (Bell, 1938). In Europe this species is known from Westphalian C and D, Stephanian A, B, and C, and rearely from Early Permian strata (Gothan and Remy, 1957, Chart 5.).

Specimen. PC BM HU hypotype 51324 A, B, 51342; nontype 51325 A-P.

Annularia aculeata Bell
(Fig. 7B)

Annularia aculeata Bell, 1944, p. 101, Pl. 60, figs. 3, 4; Pl. 62, fig. 2; Pl. 63, fig. 4; Pl. 65, figs. 1, 4; Pl. 66, figs. 1, 3; Pl. 68, figs. 1-4; Pl. 69, figs. 1-3, 6; Pl. 74, figs. 4, 7.
Annularia aculeata Bell. Abbott, 1958, p. 308-310; Charts 2, 5.

Description. Species is represented by seven linear-lanceolate leaves, acutely pointed at the apex and not constricted at the base. The leaves are of equal length, about 14 mm long and 1 mm wide near the middle, and have a thick midvein.

Distribution. Species is also known from the Riversdale and Cumberland groups of Nova Scotia, Canada (Westphalian A and B).

Specimen. PC BM HU hypotype 51322.

Annularia acicularis (Dawson) D. White
(Figs. 7C, 7D)

Asterophyllites acicularis Dawson, 1871, p. 28, Pl. 5, fig. 54.
Annularia acicularis (Dawson) White, 1900, p. 898.
Annularia acicularis (Dawson) White. Bell, 1944, p. 101, Pl. 58, figs. 2, 5; Pl. 60, fig. 6; Pl. 63, fig. 3; Pl. 64, fig. 5; Pl. 65, fig. 2; Pl. 69, fig. 5.
Annularia acicularis (Dawson) White. Abbott, 1958, p. 306-308, Charts 2, 5.
Annularia acicularis (Dawson) White. Lyons, 1969, Pl. I, fig. A.

Description. This species is represented by verticils of lanceolate, acutely pointed leaves with a strong midvein, 15 in number, about 15 mm long and 0.7 mm wide. The leaves are widest at middle, taper toward their apex and are less constricted at the base. The lateral leaves are slightly longer than those more or less parallel to the axis.

Figure 7. A, *Annularia sphenophylloides* (Zenker) Gutbier, PC BM HU hypotype 51324 A, B, 51342; nontype 51325 A–P; B, *Annularia aculeata* Bell, PC BM HU hypotype 51322; C and D, *Annularia acicularis* (Dawson) D. White, PC BM HU hypotype 51320 A–B, nontype 51321 A–D.

Distribution. This species is also found in the Rhode Island Formation at Foxboro, Massachusetts. It is also known from the Middle and Upper Pottsville and Lower Allegheny groups of the Appalachian region and comparable strata of the central regions of the United States. It is also known from the Riversdale and Cumberland groups in the northern region of Nova Scotia, Canada (Westphalian A and B).

Specimen. PC BM HU hypotype 51320 A-B; nontype 51321 A-D.

Order SPHENOPHYLLALES
Family SPHENOPHYLLACEAE
Genus *Sphenophyllum* Brongniart, 1822
Sphenophyllum emarginatum Brongniart
(Fig. 8A)

Sphenophyllum emarginatum Brongniart, 1822, p. 209, 234, Pl. 2, figs. 8a, 8b.
Sphenophyllum emarginatum Brongniart. Koenig, 1825, Pl. 12, fig. 149 (copy of Brongniart's fig., 1822).
Sphenophyllum emarginatum Brongniart. Lesquereux, 1880, p. 53, Pl. 2, fig. 6.
Sphenophyllum emarginatum Brongniart. White, 1899, p. 177-180, Pl. 59, fig. 1d.
Sphenophyllum emarginatum Brongniart. Bell, 1938, p. 89, Pl. 93, figs. 1-3.
Sphenophyllum emarginatum Brongniart. Bell, 1940, p. 130.
Sphenophyllum emarginatum Brongniart. Gothan and Remy, 1957, p. 59-60. text-fig. 50; Chart 5.
Sphenophyllum emarginatum Brongniart. Abbott, 1958, p. 339-342, Pl. 38, figs. 29, 34; Pl. 44, fig. 66; Pl. 45, fig. 72; Charts 3, 5.
Sphenophyllum emarginatum Brongniart. Lyons, 1969, Pl. IV, fig. A.
Sphenophyllum emarginatum Brongniart. Crookall, 1969, pt. 5, p. 586-591, Pl. 107, fig. 6; text-figs. 164, 171 B.
Sphenophyllum emarginatum Brongniart. Darrah, 1970, p. 177-178, Pl. 55, fig. 1.

Description. Species is represented by a whorl of six leaves, each about 10 mm long and 2.5 mm wide. The wedge-shaped leaves have nearly straight sides, are gently arched and finely dentate with obtusely rounded teeth on the distal margin. A single vein enters the base of each leaf and after three or four asymmetrical dichotomous divisions, one veinlet enters each tooth, where it becomes enlarged at the margin.

Distribution. Species is also found in the Rhode Island Formation at Foxboro, Massachusetts (Lyons, 1969). It is also known from the Allegheny and Conemaugh groups and rarely from the early part of the Monongahela Group in the Appalachian region and from the Pictou Group in Nova Scotia, Canada (Westphalian C and D). In Europe this species is common in the Westphalian C and D and in the earlier part of Stephanian A (Gothan and Remy, 1957).

Specimen. PC BM HU hypotype 51335.

Sphenophyllum cf. *longifolium* (Germar) Geinitz
(Figs. 8B, 8C, 8D)

Sphenophyllites longifolius Germar, 1844, p. 426, Pl. 2, figs. 2a, 2b.
Sphenophyllum longifolium (Germar) Geinitz, 1855, p. 12, Pl. 20, figs. 15, 16.
Sphenophyllum longifolium Germar. Lesquereux, 1880, v. 1, p. 53; Atlas (1879), Pl. 91, fig. 6.

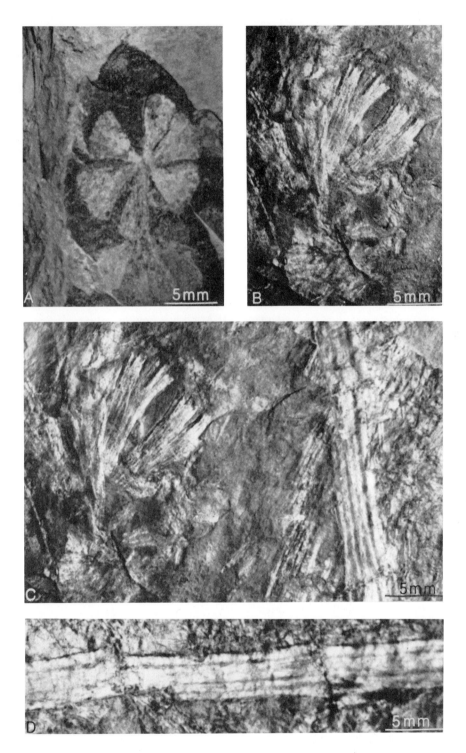

Figure 8. A, *Sphenophyllum emarginatum* Brongniart, PC BM HU hypotype 51355; B, C, and D, *Sphenophyllum* cf. *longifolium* (Germar) Geinitz, PC BM HU hypotype 51356.

Sphenophyllum longifolium (Germar) Geinitz. Abbott, 1958, p. 347-349, Pl. 45, figs. 70, 73; Chart 3.
Sphenophyllum longifolium (Germar) Geinitz. Darrah, 1970, p. 178-179.

Description. Specimens of this species consist only of some leaves, 2.5 cm long and 0.6 cm wide, extending from the node. The lateral margins of the leaves are nearly straight; their distal margins are bluntly toothed to laciniate in the upper one-fifth to one-third of the leaf. One vein enters the base of the leaf and after some asymmetric dichotomous divisions, a veinlet branch enters each tooth on the distal margin of the leaf.

Distribution. Species is also known from the Allegheny, Conemaugh, and Monongahela groups in western Pennsylvania, northwestern West Virginia, northwestern Maryland, and eastern Ohio (Westphalian C and D and Stephanian A, B, and C; Abbott, 1958). In Europe this species is known from Westphalian D and Stephanian A, B, and C (Gothan and Remy, 1957).

Specimen. PC BM HU hypotype 51356.

Subphylum PTEROPHYTA (= PTEROPSIDA)
Class FILICINEAE
Order FILICALES
Family PECOPTERIDACEAE
Genus *Pecopteris* (Brongniart) Sternberg, 1825
Pecopteris clarkii Lesquereux
(Fig. 9A)

Filicites (Pecopteris) pennaeformis Brongniart, 1822, v. 8, p. 236, Pl. 2, fig. 3.
Pecopteris pennaeformis Brongniart, 1828, v. 1, p. 345, Pl. 118, figs. 3, 4.
Pecopteris pennaeformis (Brongniart) Sternberg, 1825, Bd. 1, Teil 4, p. 152.
Pecopteris clarkii Lesquereux, 1880, v. 1, p. 261; Atlas (1879), Pl. 41, figs. 10, 10a.
Pecopteris clarkii Lesquereux. Bell, 1938, p. 82, Pl. 83, fig. 3; Pl. 84, fig. 1.
Pecopteris clarkii Lesquereux. Bell, 1962, p. 51.

Description. This true fern is represented by a pinna with wide flattened rachis to which five pairs of confluent pinnules are alternately attached. The apical lobe of the pinna is not evident in this specimen. The pinnules have slightly undulating margins deeply impressed in the stone. The flexuous midvein of a pinnule and the dichotomously branching lateral veins are very prominent. This type of preservation indicates that the venation of the pinnules was very thin and for the most part escaped carbonization.

Distribution. This species is also known from the Morien (Pictou) Group of the Sydney coalfield in Nova Scotia, Canada (Westphalian C) (Bell, 1962).

Specimen. PC BM HU hypotype 51350.

Pecopteris miltoni Artis
(Figs. 9B, 9C)

Filicites miltoni Artis, 1825, p. 14, Pl. 14.
Pecopteris miltoni Brongniart (pars), 1828, p. 333-335, Pl. 114, figs. 1-8.
Pecopteris (Asterotheca) miltoni Zeiller, 1888, no. 21, p. 32.

Figure 9. A, *Pecopteris clarkii* Lesquereux, PC BM HU hypotype 51350; B and C, *Pecopteris miltoni* Artis, PC BM HU hypotype 51349; D and E, *Pecopteris hemitelioides* Brongniart, PC BM HU hypotype 51351.

Asterotheca miltoni Artis. Kidston, 1924, v. 2, pt. 5, p. 501-508, Pl. 120, figs. 1-5; Pl. 121; Pl. 122, fig. 1; text-figs. 48, 54-57.
Pecopteris miltoni Artis. Gothan and Remy, 1957, p. 114, text-figs. 107, 201; Chart 5.
Pecopteris (Asterotheca) miltoni Artis. Bell, 1962, p. 30-32, Pl. 20, fig. 3; Pl. 21, figs. 1, 2; Pl. 22, figs. 1-3; Pl. 23; Pl. 24, fig. 2.
Percopteris (Asterotheca) miltoni Artis. Bell, 1962b, p. 52.
Pecopteris miltoni Artis. Darrah, 1970, p. 135-136, Pl. 64, fig. 1.

Description. Species is represented by a frond consisting of straight pinnae alternately and almost perpendicularly attached to a rachis. Each pinna consists of about 13 pairs of pinnules alternately and slightly obliquely attached to a rachis. The suboblong pinnules, 5 to 6 mm long and 3 mm wide, are a little confluent at their base and broadly rounded at their apex. The pinnules located near the apex of a pinna possess the typical crenulated margins.

The midvein of pinnules arising from the rachis is very thick, straight, and extending almost to the apex and terminates in a dichotomy. The lateral veins arise at an open angle, usually divide once near the base, slightly curve toward the margins, and sometimes bifurcate again.

Distribution. Species is also known from the Minto and Clifton Formations (Pictou Group) in New Brunswick, Canada (Westphalian C). In the Appalachian region of the United States this species is known from the uppermost Pottsville and lowest Allegheny strata (Late Westphalian B and C). In Europe this species is known from Westphalian A, B, and C and rarely from Westphalian D.

Specimen. PC BM HU hypotype 51349.

Pecopteris hemitelioides Brongniart
(Figs. 9D, 9E)

Pecopteris hemitelioides Brongniart, 1828, p. 314-316, Pl. 108, figs. 1, 2.
Pecopteris (Asterotheca) hemitelioides Zeiller, 1888, livr. 2, p. 133, Pl. 11, figs. 6, 7.
Pecopteris hemitelioides Potonié, 1893, p. 51, Pl. 5, fig. 7.
Asterotheca hemitelioides Brongniart. Kidston, 1924, pt. 5, p. 519-522, Pl. 117, figs. 1, 2, 2a, 4, 5; text-fig. 60.
Asterotheca robbi Bell, 1938, p. 74-75, Pl. 72, figs. 3-6; Pl. 73, figs, 1, 2; Pl. 74, fig. 1; Pl. 76, fig. 1.
Pecopteris (Asterotheca) hemitelioides Brongniart. Bell, 1962, p. 34, Pl. 21, fig. 3.
Pecopteris (Asterotheca) hemitelioides Brongniart. Bell, 1962b, p. 52.
Pecopteris hemitelioides Brongniart. Darrah, 1970, p. 132-133, Pl. 19, fig. 2.

Description. Specimen is represented by a fragment of a sterile pinna with closely spaced lanceolate pinnules alternately and almost perpendicularly attached to a thick, straight rachis. The pinnules, about 6 mm long and 2 mm wide, slightly contracted at the base, with parallel sides for about two-thirds of their length, then become gradually convergent to a broadly rounded apex. The thick, straight median vein extends nearly to the apex of the pinnule; the simple lateral veinlets arise from the midvein at a slight angle. Because sporangia are not preserved, the taxon *Pecopteris* is preferable to *Asterotheca*.

Remarks. Similar fossil plants from the Sydney coalfield were previously described by Bell (1938, 1940) as *Asterotheca robbi*. Later description of venation and epidermal characters of *Pecopteris hemitelioides* from the Saar coal basin by P. Corsin (1951) indicates that the specimens of *Asterotheca robbi* must be considered a synonym of *P. hemitelioides*.

Distribution. Species is known from the Pictou Group of New Brunswick, Canada, corresponding to Westphalian C. (Bell, 1962). In the Appalachian region of the United States this species is known from the Lower Allegheny, Conemaugh, Monongahela, and Lower Permian strata (Darrah, 1970). In Europe this species ranges from Westphalian D to Permian (Gothan and Remy, 1957).

Specimen. PC BM HM hypotype 51351.

Class GYMNOSPERMAE
Order PTERIDOSPERMALES
Family ALETHOPTERIDACEAE
Genus *Alethopteris* Sternberg, 1825
Alethopteris serlii (Brongniart) Goeppert
(Fig. 10A)

Pecopteris serlii Brongniart, 1828, v. 1, p. 292, Pl. 85, figs. 1-8.
Alethopteris serlii Goeppert, 1836, p. 301, Pl. 21, figs. 6, 7.
Alethopteris serlii (Brongniart) Goeppert. Zeiller, 1878, v. 4, p. 75 (1879); Atlas (1878), Pl. 163, figs. 1, 2.
Alethopteris serlii Brongniart. Lesquereux, 1880, p. 176 (1880); Atlas (1879), Pl. 29, figs. 1-5.
Alethopteris serlii var. *missouriensis* D. White, 1899, p. 118-120, Pl. 37, figs. 1, 2; Pl. 42, fig. 5.
Alethopteris serlii Noé pars, 1925, p. 17, Pl. 38, figs. 1-4; Pl. 39, figs. 2, 3.
Alethopteris serlii (Brongniart). Bell, 1938, p. 67-68, Pl. 61, figs. 6, 7; Pl. 62, fig. 1.
Alethopteris serlii Brongniart. Gothan and Remy, 1957, p. 120, text-fig. 112; Chart 5.
Alethopteris serlii (Brongniart). Bell, 1962, p. 36-37, Pl. 39, fig. 1; Pl. 41, fig. 4; Pl. 42, fig. 1.
Alethopteris serlii Brongniart. Crookall, 1965, pt. 1, p. 17-22, Pl. 7, figs. 1, 2, 2a, 3; text-figs. 6, 16B, 17C.
Alethopteris serlii (Brongniart) Goeppert. Lyons, 1969, Pl. VI, fig. B.
Alethopteris serlii (Brongniart) Goeppert. Darrah, 1970, p. 113, Pl. 4, fig. 2; Pl. 31, fig. 3; Pl. 50, fig. 1.

Description. Species is represented by two pinnae with oblong-lanceolate pinnules that are confluent at their base. A thick midvein is deeply sulcate and ascends to the apex; numerous lateral veins arise nearly at right angles from the midvein and are simple or bifurcate once near the middle, and extend at right angles to the margins.

Distribution. Species is also known from the Upper Pottsville and Allegheny groups of the Appalachian region, from contemporaneous strata of Illinois (Darrah, 1970), and from the Pictou Group of Nova Scotia, Canada (Westphalian C and D) (Bell, 1962). In Europe this species is known from Westphalian C and D (Gothan and Remy, 1957).

Specimen. PC BM HU hypotype 51326.

Family NEUROPTERIDACEAE
Genus *Neuropteris* (Brongniart) Sternberg 1825
Neuropteris heterophylla Brongniart
(Fig. 10B)

Filicites (Sect. *Nevropteris*) *heterophyllus* Brongniart, 1822, p. 209, 239, Pl. 2, figs. 6a, 6b.
Nevropteris heterophylla Brongniart, 1828, v. 1, p. 243, Pls. 71, 72, fig. 2.
Neuropteris heterophylla Brongniart. Sternberg, 1826, Bd. 1, Teil 4, p. 17; Bd. 2, p. 72-73.
Neuropteris heterophylla Brongniart. Bell, 1938, p. 56, Pl. 50, fig. 2.
Imparipteris (Neuropteris) heterophylla Brongniart. Gothan, 1941, p. 427, figs. 1-3.
Imparipteris (Neuropteris) heterophylla Brongniart. Gothan and Weyland, 1954, p. 147-150, text-fig. 132.
Imparipteris (Neuropteris) heterophylla Brongniart. Gothan and Remy, 1957, p. 125-128, text-figs. 118, 119, Chart 5.
Neuropteris heterophylla Brongniart. Crookall, 1959, p. 96-104, Pl. 25, figs. 1-3; Pl. 26; Pl. 27, figs. 1-5; Pl. 28, figs. 1-5; Pl. 29, fig. 5; Pl. 32, figs. 1, 2; Pl. 49, fig. 7; Pl. 54, fig. 2; Pl. 55, figs. 1, 2; text-figs. 33-35, 63e, 63h, 63j.
Neuropteris heterophylla Brongniart. Lyons, 1969, Pl. VIII, fig. B.
Neuropteris heterophylla Brongniart. Darrah, 1970, p. 97-99.

Description. Species is represented by many pinnae alternately attached to a thick rachis. Subcordate or ovate pinnules are alternately attached to the rachis; pinnules are wider near their base and greatly contracted in their upper part. The midvein of pinnules if flexuous in its upper half and is divided close to the apex. The lateral veins divide dichotomously two or three times, somewhat flexuous, arched to the borders.

Distribution. Species is also known from the Rhode Island Formation of Foxboro, Massachusetts, from the Appalachian and Illinois coal basins, and from the Pictou Group (*Linopteris obliqua* Zone) of New Brunswick and (the *Ptychocarpus unitus* Zone) of the Sydney coalfield in Nova Scotia, Canada (Westphalian C and D). In Europe this species is known from Westphalian A through D.

Specimen. PC BM HU hypotype 51348.

Family SPHENOPTERIDACEAE
Genus *Sphenopteris* (Brongniart) Sternberg, 1826
Sphenopteris cf. *rhomboidea* Ettingshausen
(Fig. 11A)

Filicites elegans Brongniart, 1822, p. 209-233, Pl. 2, figs. 2a, 2b.
Sphenopteris elegans (Brongniart) Sternberg, 1826, Teil 4, p. 15, p. 56, Pl. 23, figs. 2a, 2b.
Sphenopteris elegans Brongniart, 1828, v. 1, p. 172-173, Pl. 53, figs. 1, 2.
Cyclopteris rhomboidea Ettingshausen, 1855, p. 12, Pl. 2, fig. 5.
Sphenopteris rhomboidea Ettingshausen. Bell, 1944, p. 59, Pl. 3, figs. 8-10, Pl. 4, fig. 6.
Sphenopteris rhomboidea (Ettingshausen). Bell, 1962b, p. 61.

Description. Species is represented by only a few pinnules. The terminal, deltoidal

Figure 10. A, *Alethopteris serlii* (Brongniart) Goeppert, PC BM HU hypotype 51326; B, *Neuropteris heterophylla* Brongniart, PC BM HU hypotype 51348.

pinnule is larger than those immediately preceding it; lateral pinnules are inequilateral, rhomboidal with basal lobe on one side. The midvein runs to about half the length of a pinnule and gives off numerous, strongly ascending, nearly straight, dichotomously divided secondary veins.

Distribution. Species is also known from the Riversdale and Cumberland groups in Nova Scotia, Canada (Westphalian A and B) (Bell, 1944, 1962).

Specimen. PC BM HU hypotype 51357.

Sphenopteris cf. *hirticula* Bell
(Fig. 11B)

Sphenopteris hirticula Bell, 1962, p. 22-23, Pl. 9, figs. 1, 2, 4.

Description. Species is represented by a pinna with some pinnules alternately attached to a thin, slightly flexuous rachis. Pinnules have two or three pairs of alternate, oblique, flatly rounded lobes with slightly emarginate margins. Terminal pinnule is small, flatly rounded or slightly asymmetrically bilobate, or obscurely emarginate. Nervation is not evident.

Distribution. Species is also found in the Minto Formation of the Pictou Group in New Brunswick, Canada (Westphalian C).

Specimen. PC BM HU hypotype 51358.

Genus *Mariopteris* Zeiller, 1879
Mariopteris cf. *paddocki* D. White
(Fig. 11C)

Pecopteris nervosa Brongniart, 1828, p. 297-298, Pl. 94; Pl. 95, figs. 1, 2.
Mariopteris nervosa (Brongniart) Zeiller, 1879, p. 97, Pl. 5, figs. 1, 2.
Mariopteris nervosa (Brongniart) Zeiller, 1880, p. 69, Pl. 167, figs. 1, 2.
Mariopteris paddocki D. White, 1943, p. 86-87, Pl. 14, figs. 1-6.

Description. Species is represented only by a few pinnae with obovate pinnules alternately and obliquely attached to a strong, slightly flexuous, obscurely lineate rachis. The obovate, trifid lobate pinnules taper to a slender, acute apex and are constricted at the base. Nervation is not evident.

Distribution. Species is also known from the Upper Pottsville and Allegheny groups in the Appalachian region and comparable strata of the central regions of the United States (Westphalian A, B, and C).

Specimen. PC BM HU hypotype 51346 A, B.

Mariopteris sp.
(Fig. 11D)

Description. Species is represented only by impressions (cast and mold) of a fossil plant. The evident rachis of a pinna composed of oval-pyriform pinnules that are tapered to the apex and slightly contracted at the base. The pinnules are alternately and obliquely attached to the rachis. Venation is not evident.

Remarks. These impressions are similar to *Mariopteris pottsvillea* D. White (1943).

Specimen. PC BM HU hypotype 51347 A, B.

Figure 11. A, *Sphenopteris* cf. *rhomboidea* Ettingshausen, PC BM HU hypotype 51357; B, *Sphenopteris* cf. *hirticula* Bell, PC BM HU hypotype 51358; C, *Mariopteris* cf. *paddocki* D. White, PC BM HU hypotype 51346 A, B; D, *Mariopteris* sp., PC BM HU hypotype 51347 A, B.

Genus *Eremopteris* Schimper, 1869
Eremopteris lincolniana D. White
(Fig. 12A)

Sphenopteris artemisiaefolia Sternberg, 1826, Bd. 1, Teil 4, p. 15, Pl. 54, fig. 1; v. 2, Teil 5-6, p. 58.
Sphenopteris artemisiaefolia Brongniart, 1828, p. 176-177, Pl. 46; Pl. 47, figs. 1, 2.
Eremopteris artemisiaefolia (Sternberg) Schimper, 1869, p. 416, Pl. 30, fig. 5.
Eremopteris artemisiaefolia Brongniart. Lesquereux, 1879, p. 293-294, Pl. 53, figs. 5, 6.
Eremopteris lincolniana D. White, 1900, pt. 2, p. 869, Pl. 192, figs. 1, 1a.
Eremopteris lincolniana D. White, 1943, p. 91, Pl. 24, figs. 1-8.

Description. Species is represented by a few distant pinnae alternately attached to a longitudinally striated rachis. Pinnules alternate in the lower portion of the pinna and are broadly triangular, approaching a palmate form. The uppermost pinnules are narrow and deeply dissected into compound lobes. Such subdivided pinnules have a slightly trifoliate, cuneate, or obovate-cuneate shape. The distal sides of dissected pinnules usually are denticulo-truncate at the apex or are deeply cut into two or three unequal, short, obtuse teeth. Primary coarse venation is derived from the depressed axis of the rachis, forking at a very open angle at the base of each lobe, and later forked at the base of each compound lobe to supply a vein to each ultimate lobe or tooth.

Remarks. This genus in the opinion of Darrah (1970) is of uncertain relationship. The generic name *Eremopteris* was proposed by Schimper (1869) for sphenopteroid fronds with open pinnae irregularly deeply dissected into laciniate pinnules of obovate, wedged, or narrowly elongated shape. The most important characteristic of this genus is the laciniate lobation and evident fine venation.

Distribution. A closely related species, *E. artemisiaefolia*, was found in the *Linopteris obliqua* Zone (Pictou Group), corresponding to Westphalian C of the Sydney and Pictou coalfields in Nova Scotia, Canada.

Specimen. PC BM HU hypotype 51340, nontype 51341.

Diplothmema cf. *cheathami* (Lesquereux) D. White
(Fig. 12B)

Eremopteris cheathami Lesquereux, 1884, p. 770, Pl. 104, figs. 2-4.
Diplothmema cheathami (Lesquereux) D. White, 1943, p. 95-96, Pl. 28, figs. 1-3, figs. 5-10.

Description. Species is represented by a small upper part of a pinna with cuneiform pinnules obliquely attached to the rachis. Pinnules alternate and are decurrent at the base. They are deeply dissected, usually into three lobes. A vein extends from the rachis dichotomously and forks a few times in the lower portion of the pinnule; and one veinlet passes into the distal teeth.

Distribution. Species is also known from coal basins in Tennessee and Alabama (White, 1943).

Specimen. PC BM HU hypotype 51337.

Figure 12. A, *Eremopteris lincolniana* D. White, PC BM HU hypotype 51340, nontype 51341; B, *Diplothmema* cf. *cheathami* (Lesquereux) D. White, PC BM HU hypotype 51337; C, *Diplothmema geniculatum erectum* Bell, PC BM HU hypotype 51338 A, B, nontype 51339 A-C; D, *Palmatopteris narragansettensis* sp. nov., PC BM HU holotype 51352.

Genus *Diplothmema* Stur, 1877
Diplothmema geniculatum erectum Bell
(Fig. 12C)

Sphenopteris geniculata Germar and Kaulfuss, 1831, p. 224-230, Pl. 65, fig. 2.
Diplothmema subgeniculatum Stur, 1877, p. 135-137, Pl. 12, figs. 8-10.
Diplotmema geniculatum Germar and Kaulfuss. Kidston, 1928, p. 224, Pl. 65, fig. 2.
Diplotmema geniculatum var. *erectum* Bell, 1938, p. 33-34, Pl. 7, fig. 2.

Description. Species is represented by a few secondary pinnae, obliquely attached to a strong, slightly geniculate rachis with central cordlike swelling. The pinnae bear many stalked pinnules of nearly equal diameter that dichotomously branch into many ultimate, linear, fingerlike, bluntly pointed lobes, each containing a single vein.

Remarks. Stur (1877) first proposed the genus *Diplothmema* for fronds that dichotomously divide twice (Andrews, 1955). Later investigators found that similar branching occurs in different species of *Mariopteris*, *Sphenopteris*, *Eremopteris*, and *Palmatopteris*. Recent investigations on more complete and better preserved fosil remains of the above mentioned 'genera' should clarify the taxonomy of these fossils.

Distribution. Species is also known from the earliest strata of *Ptychocarpus unitus* Zone (Pictou Group) of the Sydney coalfield, Nova Scotia, Canada (Bell, 1938, 1962) (Westphalian D.)

Specimen. PC BM HU hypotype 51338 A, B; nontype 51339 A-C.

Genus *Palmatopteris* H. Potonié, 1893

Type Species

Sphenopteris furcata Brongniart, 1828, p. 50.
Sphenopteris furcata Brongniart, 1829, p. 179, Pl. 49, figs. 4, 5.
Palmatopteris furcata (Brongniart) H. Potonié, 1893, p. 1-21, Pl. 1, text-figs. 1, 5.

In the Narragansett Basin there are two rare forms that show a resemblance to *Palmatopteris furcata*; they differ from each other in the shapes of the pinnule segments. One of these forms is known only from a single specimen. Possibly a larger number of specimens would show gradation between these two forms, but without such information they should be considered distinct.

The generic name *Palmatopteris* first was proposed by Potonié (1893) for sphenopterids with pinnules dichotomously dividing once or twice, forming narrow linear forks. This genus in its major architectural feature and habit of growth is similar to such genera as *Rhodea*, *Eremopteris*, and *Diplothmema*. They differ mainly by the shape of pinnae and their leaves. All of them, as indicated by their thin flexuous aerial stems, probably were climbers or lianas.

Palmatopteris narragansettensis sp. nov.
(Figs. 12D, 13A, 13C, 13D)

Diagnosis. Sphenopterid with frond of unknown size and small ultimate pinnae with fan-shaped pinnules bifurcating one to three times. Tips of pinnules are sharply pointed but without spine. Midrib is evident in lower portion of pinnules but is not visible in the upper portion. Pinnules arise alternately and obliquely from a slender rachis, about 0.7 mm in diameter, with evident longitudinal cordlike swelling.

Holotype (Fig. 13A). PC BM HU 51352; paratype 51353 A-C. Massachusetts, Plainville, northwestern area of Narrangansett Basin.

Geological Age. Pennsylvanian (Westphalian C).

Remarks. *Palmatopteris narragansettensis* is closely allied with *P. furcata*, the type species of this genus. *P. furcata* has more elongated pinnae with shorter stalks than *P. narragansettensis*. The pinnae of this last species are fan-shaped with slightly longer stalks. The reniform pinnules of *P. furcata* are narrower and have more sharply pointed tips, while the forklike pinnules of *P. narragansettensis* are of equal width and are pointed only near the tips. Kidston (1923) described *P. furcata* as having close but scarcely touching pinnules, whereas *P. narragansettensis* has an open aspect. He also mentions that *P. furcata* has prominent veins in each segment. The pinnules of *P. narragansettensis* are longer than of *P. sturi* Gothan.

Palmatopteris plainvillensis sp. nov.
(Fig. 13B)

Diagnosis. This sphenopterid is represented by a single specimen showing a few pinnae of triangular shape, alternately and obliquely attached to a thin slightly flexuous rachis, about 0.7 mm in diameter, with evident longitudinal cordlike swelling. The wedge-shaped pinnules divide dichotomously once or twice to forklike structures, widening to the blunt tips. A thin middle longitudinal cordlike swelling, which marks the main vein in the rachis, enters each pinna and later each wedge-shaped pinnule.

Holotype (Fig. 13B). PC BM HU 51354. Massachusetts, Plainville, northwestern area of Narragansett basin.

Geological Age. Pennsylvanian (Westphalian C).

Remarks. Species differs from *P. narragansettensis* because it has narrower triangular pinnae, which are more obliquely attached to a thin flexuous rachis. The wedge-shaped pinnules of *P. plainvillensis* end more bluntly at the tips, while those of *P. narragansettensis* have a more equal diameter and taper to the tips.

Order CORDAITALES
Family CORDAITACEAE
Genus *Cordaites* Unger, 1850
Cordaites principalis Germar
(Fig. 14 A)

Flabellaria borassifolia Sternberg, 1822, Bd. 1, Teil 2, p. 27, 32, Pls. 18, 41.
Flabellaria principalis Germar, 1848, Teil 5, p. 55-56, Pl. 23.
Cordaites borassifolia (Sternberg) Unger, 1850a, p. 277.
Cordaites principalis Geinitz pars, 1855, p. 41, Pl. 21, figs. 1, 2, 2a, 2b.

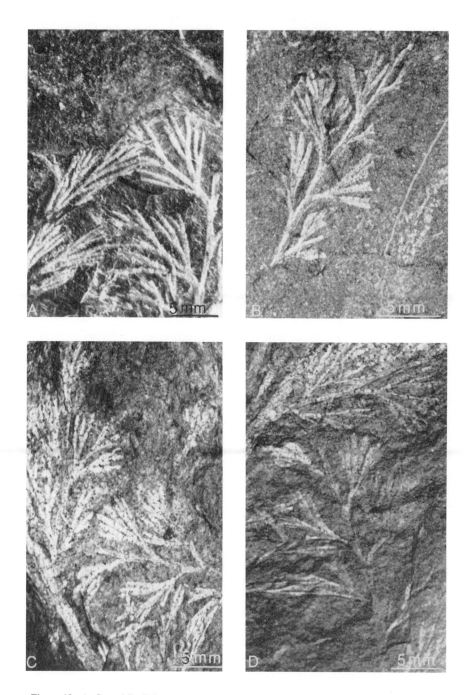

Figure 13. A, C, and D, *Palmatopteris narragansettensis* sp. nov., PC BM HU holotype 51352, paratype 51353 A–C; B, *Palmatopteris plainvillensis* sp. nov., PC BM HU holotype 51354.

Cordaites principalis Germar. Stopes, 1914, p. 84, Pl. 20, fig. 51; Pl. 21, fig. 53.
Cordaites principalis (Germar). Bell, 1938, p. 103, Pl. 105, fig. 1; Pl. 106, fig. 1.
Cordaites principalis (Germar). Bell, 1940, p. 130-131, Pl. 9, figs. 1, 2; Pl. 10, figs. 5, 6.
Cordaites principalis (Germar). Bell, 1944, p. 106.
Cordaites principalis Germar. Gothan and Remy, 1957, p. 152-153, text-fig. 153; Chart 5.
Cordaites principalis (Germar). Bell, 1962, p. 59, Pl. 55, fig. 2.
Cordaites principalis (Germar). Bell, 1962b, p. 36.
Cordaites principalis (Germar) Geinitz. Bell, 1966, p. 44, Pl. 21, fig. 1.
Cordaites principalis Germar. Lyons, 1969, Pl. 17.
Cordaites principalis (Germar). Darrah, 1970, p. 185.

Description. Species is represented by a few fragments of leaves with parallel venation, consisting of coarse longitudinal primary veins separated by furrows with usually two or three fine longitudinal secondary veins.

Remarks. The main difference between this species and *C. borassifolia* is the venation of leaves. The venation of *C. principalis* consists of coarse longitudinal primary veins, separated by two or three, rarely four or five, fine secondary veins. The venation of *C. borassifolia* consists of distant, alternately changing, thick and fine longitudinal veins.

Distribution. *Cordaites principalis* is also known from the Rhode Island Formation of Foxboro, Massachusetts (Lyons, 1969), from the Allegheny, Conemaugh, and Monongahela groups in the Appalachian region (Darrah, 1970), and from the Minto Formation of the Pictou Group of Nova Scotia, Canada (Westphalian C). In Europe this species ranges from Namurian B to Permian (Gothan and Remy, 1957).

Specimen. PC BM HU hypotype 51335; nontype 51336 A-C, 51334.

Composita (Incertae sedis)
"Radicites" Potonie, 1893
(Figs. 14B, C.)

Radicites Potonie, 1893b, p. 260.

Description. The specimens are rootlike structures, probably of some arthrophyte.
Specimen. PC BM HU hypotype 51359 A-C; nontype 51360.

CONCLUSIONS

Correlation with similar fossils in the Appalachian region, in the Maritime Provinces of Canada and in Western Europe indicates that the Rhode Island Formation at Plainville, Massachusetts, is Early Alleghenyan and is correlative with early strata of the Pictou Group in Nova Scotia, Canada (Westphalian C; see Fig. 15).

The absence of *Neuropteris scheuchzeri, Sphenophyllum cuneifolium,* and *Lepidostrobophyllum majus,* which Lyons found in the Rhode Island Formation at Foxboro, Massachusetts, and which he concluded (1969, 1971) are of Early Alleghenyan age (Lower Westphalian C), suggest that the Plainville beds may be a little younger than the strata of Foxboro, but still of Early Alleghenyan age (Westphalian C).

The later sedimentary rocks of the Rhode Island Formation in the Narragansett basin are probably of Early Alleghenyan age (Westphalian C) and partially of Late Alleghenyan age (Westphalian D). The Dighton and Purgatory conglomerates are probably of Late Alleghenyan age (Westphalian D).

The Pondville Conglomerate, which forms the base of Pennsylvanian deposits around the Narragansett basin, is of Late Pottsvillian age (late Westphalian B; see Lyons and others, 1976). The coarse Bellingham Conglomerate found in the Woonsocket and the North Scituate basins, about 9.5 km west of the Narragansett

Figure 14. A, *Cordaites principalis* Germar, PC BM HU hypotype 51335, nontype 51336 A-C, 51334; B and C, *"Radicites"* Potonie, (rootlike structure)/Incertae sedis/. PC BM HU hypotype 51359 A-C, nontype 51360.

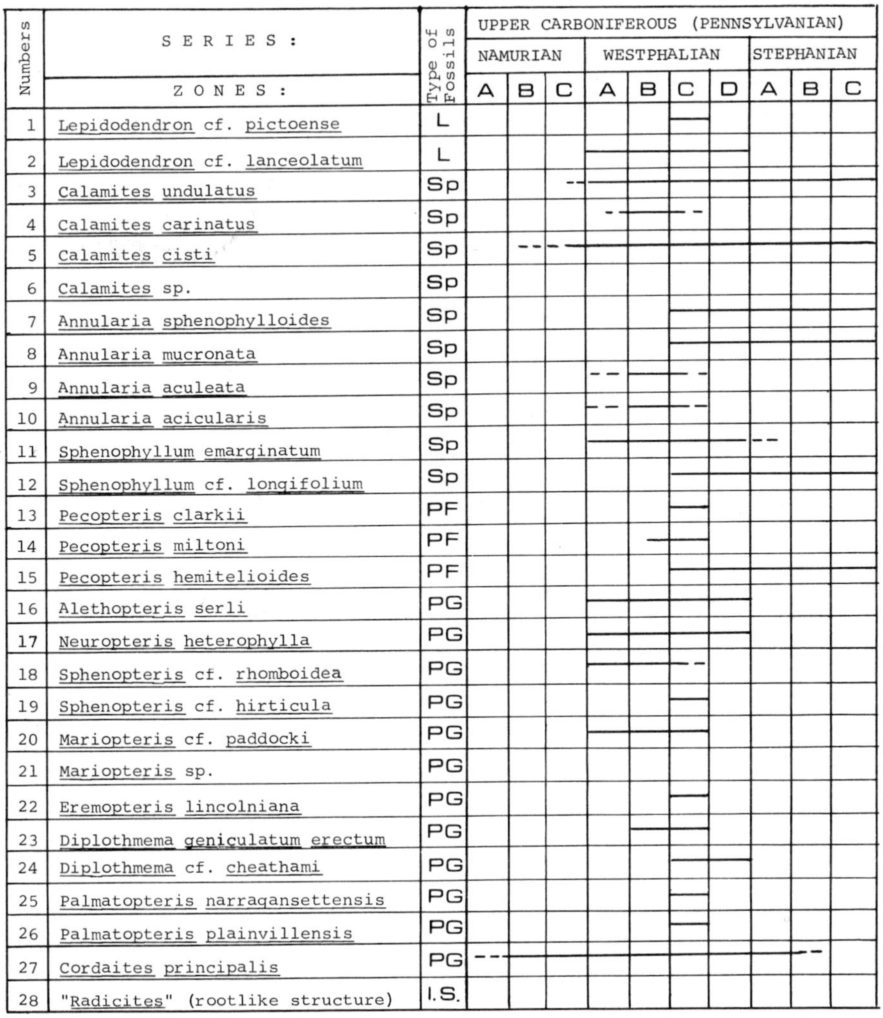

Numbers	SERIES: ZONES:	Type of Fossils	UPPER CARBONIFEROUS (PENNSYLVANIAN)									
			NAMURIAN			WESTPHALIAN				STEPHANIAN		
			A	B	C	A	B	C	D	A	B	C
1	Lepidodendron cf. pictoense	L					—	—				
2	Lepidodendron cf. lanceolatum	L						—	—			
3	Calamites undulatus	Sp			--	--	—	—				
4	Calamites carinatus	Sp					--	—	--	--		
5	Calamites cisti	Sp		--	--	—	—					
6	Calamites sp.	Sp					—	—				
7	Annularia sphenophylloides	Sp					—	—				
8	Annularia mucronata	Sp					—	—				
9	Annularia aculeata	Sp				--	--	—	—	--		
10	Annularia acicularis	Sp				--	--	—	—			
11	Sphenophyllum emarginatum	Sp					—	—		--	--	
12	Sphenophyllum cf. longifolium	Sp					—	—				
13	Pecopteris clarkii	PF					—	—				
14	Pecopteris miltoni	PF					—					
15	Pecopteris hemitelioides	PF					—	—				
16	Alethopteris serli	PG					—	—				
17	Neuropteris heterophylla	PG					—	—				
18	Sphenopteris cf. rhomboidea	PG					—	—	--	--		
19	Sphenopteris cf. hirticula	PG					—	—				
20	Mariopteris cf. paddocki	PG					—	—				
21	Mariopteris sp.	PG					—	—				
22	Eremopteris lincolniana	PG					—					
23	Diplothmema geniculatum erectum	PG					—	—				
24	Diplothmema cf. cheathami	PG					—	—				
25	Palmatopteris narragansettensis	PG					—	—				
26	Palmatopteris plainvillensis	PG					—	—				
27	Cordaites principalis	PG	--	--	—	—	—	—	—	—	--	--
28	"Radicites" (rootlike structure)	I.S.										

L Lycopods, Sp Sphenopsids, PF Ferns, PG Gymnosperms, I.S. Incertae sedis

Figure 15. Geologic range of Pennsylvanian fossil plants in the Rhode Island Formation, Plainville, Massachusetts.

basin, may be of Late Pottsvillian age. The Wamsutta Formation, which interfingers with the lowermost part of the Rhode Island Formation, is partly of Late Pottsvillian and partly of Early Alleghenyan age (late Westphalian B and lower part of C; Fig. 15).

ACKNOWLEDGMENTS

I thank C. Wroe Wolfe, Professor of Geology Emeritus, Boston University, for showing me the outcrop of the fossiliferous strata in Plainville, Massachusetts. Fossil plants described in this report were collected by me at this site.

I am indebted to Sergius Mamay, U.S. Geological Survey, Washington, D.C., for checking my identifications of the Plainville fossil plants and for valuable suggestions. I also thank Arthur Watt, U.S. Geological Survey, Washington, D.C., for help with classification of the fossil plants by comparing them with those in systematic collections in the Museum of the U.S. Geological Survey.

I am grateful to William C. Darrah, Gettysburg College, Gettysburg, Pennsylvania, who examined my collection and clarified the identification of *Pecopteris miltoni* and *Pecopteris hemitelioides*; thanks are given to James M. Schopf, U.S. Geological Survey, Columbus, Ohio; Paul C. Lyons, Boston University; and James W. Skehan, S.J., Boston College, for critical review of my manuscript and for several helpful comments and suggestions during the preparation of this paper.

I also thank Elso S. Barghoorn, Fisher Professor of Natural History and Curator of Paleobotanical Collections of Harvard University, for examining the fossil plants described in this paper. The specimens have been accepted into the permanent collections and have been given the numbers of Paleobotanical Collection of the Botanical Museum of Harvard University for holotypes, paratypes, hypotypes, and nontype specimens.

REFERENCES CITED

Abbott, M. L., 1958, The American species of *Asterophyllites, Annularia,* and *Sphenophyllum*: Am. Paleontology, Bull. v. 38, no. 174, p. 289-390, Pls. 35-49.

―――1968, Lycopsid stems and roots and sphenopsid fructifications and stems from the upper Freeport coal of southeastern Ohio: Am. Palaeontography, v. 6, no. 38, p. 1-49, Pls. 1-18.

Andrews, H. N., Jr., 1955, Index of generic names of fossil plants, 1820-1950: U.S. Geol. Survey Bull. 1013, p. 1-262.

Artis, E. T., 1825, Antediluvian phytology, illustrated by a collection of fossil remains of plants, peculiar to the coal formations of Great Britain: London (1838), p. I-XIII, p. 1-24, Pls. 1-24.

Bell, W. A., 1938, Fossil flora of Sydney coalfield, Nova Scotia: Canada Geol. Survey Mem. 215, 334 p., Pls. 1-107, 2 charts.

―――1940, The Pictou coalfield, Nova Scotia: Canada Geol. Survey Mem. 225, 161 p., Pls. 1-10, map 619 A, 1 diagram.

―――1944, Carboniferous rocks and fossil floras of northern Nova Scotia: Canada Geol. Survey Mem. 238, 276 p. Pls. 1-79.

―――1962a, Flora of Pennsylvanian Pictou Group of New Brunswick: Canada Geol. Survey Bull. 87, 71 p., Pls. 1-56.

―――1962b, Catalogue of types and figured specimens of fossil plants in the Geological Survey of Canada collections: Canada Geol. Survey, 154 p.

―――1966, Illustrations of Canadian fossils. Carboniferous plants of eastern Canada: Canada Geol. Survey Paper 66-11, 76 p., Pls. 1-36.

Brongniart, A., 1822, Sur la classification et la distribution des végétaux fossiles en général, et sur cent des terrains dé sediment supérier en particulier: Mus. Natl. Histoire Nat. Mém., v. 8, p. 203-348.

―――1828, Prodrome d'une Histoire des végétaux fossiles: Paris, Dictionnaire des Sciences Naturelles, v. 57, 223 p.

―――1828, Histoire des végétaux fossiles ou recherches botaniques et géologiques sur les végétaux renfermes dans les diverses couches du globe: v. 1, 488 p., Pls. 1-166; v. 2, 72 p., Pls. 1-29 (Text and Atlas reprinted 1965, Amsterdam, H. Asher and Co.).

Corsin, P., 1951, Bassin houiller de la Sarre et de la Lorraine: Flore Fossile, 4e fasc. Pecopterides Loos-Nord, v. 1.

Crookall, R., 1955-1970, Fossil plants of the Carboniferous rocks of Great Britain: Great Britain Geol. Survey, Mem. Palaeontology (2nd Sect.), v. 4, pts. 1-6, 839 p., Pls. 1-159.

Darrah, W. C., 1970, A critical review of Upper Pennsylvanian floras of eastern United States with notes on the Mazon Creek Flora of Illinois: Gettysburg, Pennsylvania, 221 p., Pls. 1-80.

Dawson, J. W., 1863, Synopsis of the flora of the Carboniferous Period in Nova Scotia: Canadian Naturalist, v. 8, Art. xxx, p. 431-457.

―――1868, Acadian geology, 2nd ed.: London, MacMillan and Co., Chapters X-XX, p. 128-496.

―――1871, The fossil plants of the Devonian and Upper Silurian formations: Canada Geol. Survey, 92 p., Pls. 1-20.

―――1873, Report on the fossil plants of the Lower Carboniferous and Millstone Grit Formations of Canada: Montreal, Canada Geol. Survey, p. 1-47, Pls. 1-10.

Emerson, B. K., 1917, Geology of Massachusetts and Rhode Island: U.S. Geol. Survey Bull. 597, 289 p.

Ettingshausen, C., 1852b, Die Steinkohlenflora von Radnitz in Böhmen: Abh. K.-k. Geol. Reichsanst., Bd. 2, Heft 3, p. 1-74, Pls. 1-29.

Ettingshausen, C., 1852, Die Steinkohlenflora von Stradonitz in Böhmen: Abh. K-k. Geol. Reichsanst., Bd. 1, Abt. 3, no. 4, 18 p., 6 pls.

Geinitz, H. B., 1855, Die Versteinerungen der Steinkohlenformation in Sachses: Leipzig, Gaea von Sachses, p. 1-61, Pls. 1-36.

Germar, E. F., 1844-1853, Die Versteinerungen der Steinkohlengebirges von Wettin und Löbejün im Saalkreise: Halle, Isis von Oken, Teils 1-8, p. 1-116, Pls. 1-40.

Germar, E. F., and Kaulfuss, F., 1828, Über einige merkwürdige Pflanzenabdrücke aus der Steikohlenformation: Nova Acta Leopoldina, Bd. 15, Heft 2, p. 224-230, Pl. 65, Fig. 2.

Goeppert, H. R., 1836, Die fossilen Farrenkräuter (Systema filicum fossilium): Breslau, Nova Acta Leopoldina, Bd. 17, 487 p., Pls. 1-44.

Gothan, W., 1941, Paläobotanische Mitteilung der karbonischen Neuropteriden: Palaeont. Zeitschr., Bd. 22, no. 3/4, p. 421-428, text-figs. 1-3a.

Gothan, W., and Remy, W., 1957. Steinkohlenpflanzen: W. Th. Webels, Verl. Glückauf GMBH—Essen, 248 p., 221 text-figs., Pls. 1-6, 6 Tafel.

Gothan, W., and Weyland, H., 1954, Lehrbuch der Paläobotanik: Akad. Verl. Berlin, 535 p.

Grew, E. S., Mamay, S. H., and Barghoorn, E. S., 1970, Age of plant fossils from the Worcester coal mine, Worcester, Massachusetts: Am. Jour. Sci., v. 268, p. 113-126.

Gutbier, A., 1837, Abdrücke und Versteinerungen des Zwickauer Schwartzkohlengebirges und seinen Umgebung: Zwickau., 80 p., Pls. 1-11.

Hitchcock, Edward, 1841, Final report on geology of Massachusetts: Amherst and Northhampton, Mass., 2 v., 831 p.

Jackson, C. T., 1840, Report on the geological and agricultural survey of the State of Rhode Island: Providence, Rhode Island, p. 1-312.

Kay, Marshall, 1951, North American geosynclines: Geol. Soc. America Mem. 48, 143 p., 16 Pls.

Kidston, R., 1923-1925, Fossil plants of the Carboniferous rocks of Great Britain: Great Britain Geol. Survey, Mem. Palaeontology, v. 2, pts. 1-6, 681 p., Pls. 1-153.

Kidston, R., and Jongmans, W. J., 1917, A monograph of the Calamites of western Europe: Gravenhage, Meded. Rijksoporing van Delstoffen., v. 1, no. 7, 207 p.; Atlas, Pls. 1-158.

Knox, A. S., 1944, A Carboniferous flora from the Wamsutta Formation of southeastern Massachusetts: Am. Jour. Sci., v. 242, p. 130-138.

Koenig, Ch., 1825, Icones fossilium sectiles: London, p. 1-4, Pls. 1-19.

Lesquereux, L., 1879, 1880-1884, Description of the coal flora of the Carboniferous formation in Pennsylvania and throughout the United States: Pennsylvania 2nd Geol. Survey, Prog. Rept., v. 1-3, 977 p.; Atlas (1879), Pls. 1-85.

———1884, The Carboniferous flora of Rhode Island: Am. Naturalist, v. 18, p. 921-923.

———1889, Fossil plants of the coal-measures of Rhode Island: Am. Jour. Sci., v. 37, no. 219, p. 229-230.

Lyons, P. C., 1969, Bedrock geology of the Mansfield quadrangle, Massachusetts [Ph.D. dissert.]: Boston, Mass., Boston Univ., 283 p., Pls. 1-18.

———1971, Correlation of the Pennsylvanian of New England and the Carboniferous of New Brunswick and Nova Scotia: Geol. Soc. America Abs. with Programs (Northeastern Sec.), v. 3, no. 1, p. 43-44.

Lyons, P. C., Tiffney B., and Cameron, B., 1976, Early Pennsylvanian age of the Norfolk basin, southeastern Massachusetts, based on plant megafossils, in Lyons, P. C., and Brownlow, A. H., eds., Studies in New England geology: Geol. Soc. America Mem. 146, p. 181-198 (this volume).

Noé, A. C., 1925, Pennsylvanian flora of northern Illinois: Illinois State Geol. Survey Bull. 52, 18 p., Pl. 38, figs. 1-4; Pl. 39, fig. 2-4.

Moore, R. C., chm., 1944, Correlation of Pennsylvanian formations of North America: Geol. Soc. America Bull., v. 55, p. 657-706, 1 Pl.

Mutch, T. A., 1968, Pennsylvanian nonmarine sediments of the Narragansett Basin, Massachusetts-Rhode Island in Klein, G. deVries, Late Paleozoic and Mesozoic continental sedimentation, northeastern North America: Geol. Soc. America. Spec. Paper 106, p. 177-209.

Perkins, E. H., 1920, The origin of the Dighton Conglomerate of the Narragansett Basin of Massachusetts and Rhode Island: Am. Jour. Sci., 4th ser. v. 64, p. 61-75.

Potonié, H., 1893, Über einige Carbonfarne: K.-Preuss. geol. Landesanst. u. Bergakademie Abh. Teil 3, Bd. 12, p. 1-36, Pls. 1-4.

———1893b, Flora des Rothligenden von Thúringen: Abh. d. K. Preuss. geol. Landesanstult, K. F., Bd. 9, Teil 2, 298 p., Pls. 1-24.

Quinn, A. W., and Springer, G. H., 1954, Bedrock geology of Bristol quadrangle and vicinity, Rhode Island-Massachusetts: U.S. Geol. Survey Geol. Quad. Map GQ-42, scale 1:24,000.

Quinn, A. W., 1971, Bedrock geology of Rhode Island: U.S. Geol. Survey Bull. 1295, p. 1-65; map, scale 1:24,000.

Quinn, A. W., and Oliver, W. A., Jr., 1962, Pennsylvanian rocks of New England: Reprinted from Penn. Syst. in the U.S., Am. Assoc. Petroleum Geologists, p. 60-73.

Round, E. M., 1924, Correlation of fossil floras of Rhode Island and New Brunswick: Bot. Gaz., v. 78, no. 1, p. 116-118.

―――1927, Correlation of coal floras in Henry County, Missouri and the Narragansett Basin: Bot. Gaz., v. 83, p. 61-69.

Schenk, A., 1883, Pflanzen aus der Steinkohlenformation, in Richthofen, F. F., China: Berlin, Palaeontologischer Theil, Bd. 4, Abt. 2, Abh. 9, p. 211-269, Pls. 30-45.

Schimper, W. P., 1869-1874, Traité de paléontologie végétale ou la flore du monde primitif: Paris, tome 1, p. 1-740, Pls. 1-56 (1869); tome 2, p. 1-522, Pls. 57-84 (1870); p. 523-968, Pls. 85-94, (1872); tome 3, p. 1-896, Pls. 95-110, (1874).

Schlotheim, F. V., 1804, Beschreibung merkwürdiger Kräuter-Abdrücke und Pflanzen-Versteinerungen: Gotha, Ein Beitrag zur Flora der Vorwelt, 68 p., Pls. 1-14.

―――1820, Die Petrefactenkunde auf ihrem jetzigen Standpunkte durch die Beschreibung seiner Sammlung versteinerten und fossiler Überreste des Thier- und Pflanzenreichs der Vorwelt erläuter: Gotha, 437 p., Pls. 15-29.

―――1822-1823, Nachträger zur Petrefactenkunde: Gotha, Abh. 1, p. 1-100, Pls. 1-21 (1822); Abh. 2, p. 1-114, Pls. 22-37 (1823).

―――1832, Merkwürdige Versteinerungen aus der Petrefactensammlung: Gotha, p. 1-40, Pls. 1-37.

Schopf, J. M., Wilson, L. R., and Bentall, R., 1944, An annotated synopsis of Paleozoic fossil spores and the definition of generic groups: Illinois State Geol. Survey Rept. Inv. 91, p. 1-66, Pls. 1-3.

Shaler, N. S., Woodworth, J. B., and Foerste, A. E., 1899, Geology of the Narragansett basin: U.S. Geol. Survey Mon. 33, 402 p.

Sternberg, G. K., 1820-1826, Essai d'un Exposé geognostico-botannique de la Essai flore monde primitif: Leipzig and Prague, Bd. 1, 1820, 26 p; Ratisbonne, Bd. 2, 1825, 37 p.; Ratisbonne, Bd. 3, 1824, 45 p.; Ratisbonne, Bd. 4, 1826, 53 p., + tent. 36 p., Pl. 1-69, A-E.

Sternberg, G. K., 1820-1838, Versuch einer geognostisch- botanischen Darstellung der Flora der Vorwelt: Leipzig und Prague, Bd. 1, Teil 1, p. 1-24 (1820); Teil 2, p. 1-33 (1822); Teil 3, p. 1-39 (1823); Teil 4, p. 1-46 (1825); Bd. 2, Teile 5, 6, p. 1-80 (1833); Teile 7, 8, p. 81-220 (1838); Tentamen, p. I-VIII, 1825.

Stanley, D. J., 1968, Graded bedding-sole marking-graywacke assemblage and related sedimentary structures in some Carboniferous flood deposits, eastern Massachusetts in Klein, G. deVries, Late Paleozoic and Mesozoic continental sedimentation, northeastern North America: Geol. Soc. America Spec. Paper 106, p. 211-239.

Stopes, M. C., 1914, The "Fern Ledges" Carboniferous Flora of St. John. New Brunswick: Canada Geol. Survey Mem. 41, p. 1-142, Pls. 1-25.

Stur, D. R. J., 1875, Beitrage zur Kenntniss der Flora der Vorwelt—Die Culm-Flora, Teil 1, Die Culm-Flora des märisch-schlesischen Dachschiefers: Vienna, Abh. K.-k. Geol. Reichsanst., Bd. 8, 106 p. Pls. 1-17.

―――1877, Beitrage zur Kenntniss der Flora der Vorwelt—Die Culm-Flora, Teil 2, Die Culm-Flora der Ostrauer und Waldenburger Schichten: Vienna, Abh. K.-k. Geol. Reichsanst., Bd. 8, p. 107-472, Pls. A-C, 18-44.

―――1883, Zur Morphologie und Systematic der Culm und Carbonfarne: Vienna, Sitzungsber. k. Akad. Wiss., Bd. 88, p. 633-846.

―――1885, Beitrage zur Kenntniss der Flora der Vorwelt—Die Carbon-Flora der Schatzlarer Schichten: Wien, Abh. K.-k. Geol. Reichsanst., Bd. 2, I. Abt., p. 283-410, Pls. 1-65.

Teschemacher, J. E., 1846, On the fossil vegetation of America: Boston Soc. Natl. Hist. Jour., v. 5, p. 370-385 (Abs. B.S.N.H. Proc., v. 2, p. 146-147).

Towe, K. M., 1959, Petrology and source of sediments in the Narragansett basin of Rhode Island and Massachusetts: Jour. Sed. Petrology, v. 29, no. 4, p. 503-512.

Unger, F., 1850a, Genera et species plantarum fossilium: Vienna, 627 p.
——1850b, Blätterabdrücke aus dem Schwefelflötze von Swoszowice in Galizien: Vienna, Haidinger W., Natur. Abh., Bd. 3, Teil 1, p. 121-128, Pls. 13, 14.
White, D., 1899, Fossil flora of the Lower Coal Measures of Missouri: U.S. Geol. Survey Mon. 37, 467 p., Pls. 1-73.
——1900, The stratigraphic succession of the fossil floras of the Pottsville Formation in the southern Anthracite Coal Field, Pennsylvania: U.S. Geol. Survey, 20th Ann. Rept., pt. 2, p. 749-930, Pls. 180-193.
——(posth. edited by C. B. Read), 1943, Lower Pennsylvanian species of *Mariopteris*, *Eremopteris*, *Diplothmema* and *Aneimites* from the Appalachian region: U.S. Geol. Survey Prof. Paper 197-C, p. 85-140, Pls. 8-39.
Woodworth, J. B., 1894, Carboniferous fossils in the Norfolk County Basin: Am. Jour. Sci., v. 148, p. 145-148.
Zeiller, R., 1879, Explication de la carte géologique de la France: Second Partie, Végétaux fossiles du terrain houiller de France, Paris, v. 4, Atlas p. 1-185, Pls. 159-176.
——1888, Études des gites minéraux de la France—Basin houiller de Valenciennes, description de la flore fossile: Etudes, Gites Min. France, p. 1-731 (1888); Atlas (1886), Pls. 1-96.
——1888, Études sur le terrain houiller de Commentry: Livre II, Flore Fossile, pt. 1; Indus. Min. Soc. Bull., ser. 3, v. II, p. 1-366, Pls. 1-42.
Zenker, F. C., 1833, Beschreibung von *Galium sphenophylloides* Zenk.: Neues Jahrb. Mineralogie Geognosie, Geologie und Petrefactenkunde, p. 398-400, Pl. 5, figs, 6-9.

MANUSCRIPT RECEIVED BY THE SOCIETY MAY 14, 1974
REVISED MANUSCRIPT RECEIVED DECEMBER 4, 1974
MANUSCRIPT ACCEPTED JANUARY 3, 1975

Printed in U.S.A.

Geological Society of America
Memoir 146
© 1976

Early Pennsylvanian Age of the Norfolk Basin, Southeastern Massachusetts, Based on Plant Megafossils

PAUL C. LYONS*
Division of Science, College of Basic Studies
Boston University
Boston, Massachusetts 02215

BRUCE TIFFNEY
Department of Biology, Paleobotanical Laboratory
Harvard University
Cambridge, Massachusetts 02138

AND

BARRY CAMERON
Department of Geology
Boston University
Boston, Massachusetts 02215

ABSTRACT

Plant megafossils were collected from the Pondville Conglomerate of the Norfolk basin in southeastern Massachusetts, at Woodworth's (1894) Canton Junction locality. Before this only conflictingly identified, poorly described, and unfigured plant fossils had been reported from the Norfolk basin. The fossil association suggests a late Pottsvillian age, presumably equivalent to the late Westphalian B of Maritime Canada and Europe. This indicates that the Pondville Conglomerate of the Norfolk basin was deposited during a time interval that is represented by a hiatus in Maritime Canada. *Neuropteris obliqua*, *Neuropteris* cf. *scheuchzeri*, *Cordaites principalis*, *Calamites cisti*, *Cordaicarpus* cf. *cordai*, a ?*Samaropsis* species, a ?decorticated *Sigillaria*, and a probable *Lonchopteris* species have now been identified.

---

*Present address: 18 Northwood Drive, Walpole, Massachusetts 02081.

INTRODUCTION

The age of the sedimentary rocks of the Norfolk basin of southeastern Massachusetts (Fig. 1) has not been convincingly demonstrated, although the adjoining Narragansett basin has been clearly established as Pennsylvanian in age on the basis of nearly 200 plant species. Several different ages have been proposed for the rocks of the Norfolk basin (Crosby and Barton, 1880; Shaler and others, 1899; Mutch, 1968). However, only a few poorly preserved and inadequately identified plant fossils have been reported from the basin (Crosby and Barton, 1880; Woodworth, 1894). There is general agreement that these plant fossils are Pennsylvanian in age, but detailed evidence supporting this claim is lacking (Chute, 1966, p. 32) because of the poor preservation and absence of diagnostic features of the plant fossils, and the lack of any detailed study on them.

This paper attempts to remedy these paleobotanical uncertainties and to evaluate their implications for southern New England's late Paleozoic history. We describe newly discovered, poorly preserved to well-preserved fragmented plant remains, collected by us and a group of Boston University students in 1971 at the little-known Canton Junction locality. Over 100 plant specimens were collected, primarily at the horizon (locality A of this report) where Lyons (1969) discovered *Cordaites principalis*, and also at Woodworth's 1894 exposure (locality B of this report). Additional collections were made at locality A by Lyons and Tiffney.

GEOLOGIC SETTING

The Norfolk basin is one of a number of late or presumably late Paleozoic basins located in eastern or southeastern Massachusetts (Fig. 1). These basins

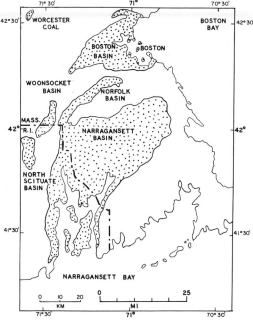

Figure 1. Late Paleozoic sedimentary basins of southeastern Massachusetts and Rhode Island. All but Boston Basin are Pennsylvanian or presumably Pennsylvanian.

are characterized by predominant graywacke suites with arkose, plutonic pebbles in the coarser sedimentary rocks, and few orthoquartzites or volcanic rocks (Quinn and Oliver, 1962). Volcanic rocks are almost totally lacking except locally in the Wamsutta Formation of the Narragansett basin. Most of the rocks have undergone low-grade metamorphism. Some of the stratigraphic units contain poorly preserved to well-preserved plant fossils.

The Narragansett basin contains a rich Pennsylvanian flora, primarily associated with coal seams along its northern and western margins. Some of these seams were mined during the 19th and 20th centuries. The degree of metamorphism in this basin increases from chlorite grade in the north to garnet and staurolite grade in the south (Quinn, 1971).

The small Woonsocket and North Scituate basins are located about 10 km to the west of the Narragansett basin (Fig. 1). These basins contain the Bellingham Conglomerate, which is believed to correlate with the Pondville Conglomerate of the Narragansett basin (Fig. 2), but they have yet to yield plant fossils that prove their inferred Pennsylvanian age (Quinn and Oliver, 1962).

The Norfolk basin lies to the north of the Narragansett basin (Fig. 1) and is connected to it at its southwestern end. The two basins have been assumed to be of similar Pennsylvanian age on the basis of their physical connection, similar lithologies and stratigraphies, and similar plant fossils.

The Boston basin, to the north of the Norfolk basin (Fig. 1), is probably either Mississippian or Pennsylvanian in age, though it may range as far back as the Devonian. Burr and Burke (1900) discovered in the Roxbury Conglomerate of the Boston basin casts of *Artisia*, a form genus often allied to *Cordaites* and characterized by cyclindrical casts of transverse diaphragms. One specimen collected by Burr and Burke (Harvard University specimen 2775) was examined by Lyons, and found to be correctly identified. According to Barghoorn (oral commun. to Tiffney, 1974), its large size indicates an affinity to the Devonian genus *Callixylon*. As several different Paleozoic genera—and even some Mesozoic and modern plants—show this transversely septate character (Seward, 1917, p. 246-247), age determinations based on this evidence are inconclusive. Pollard (1965) reported an ill-defined spiriferoid brachiopod and poorly preserved cortical and seed impressions in the Mattapan Volcanic Complex, which directly underlies the Roxbury Conglomerate (Fig. 2). From this evidence he concluded that the assemblage "is almost indisputably . . . Mississippian." Although some of the figured structures (Pollard, 1965) seem to be plant fossils, others could be pseudofossils produced by inorganic processes. This collection should be critically re-examined to clarify identifications and, thus, the age of the Boston basin.

The Worcester coal bed to the west of the Boston basin (Fig. 1) was dated as Early to Middle Pennsylvanian on the basis of new collections of plant fossils in phyllite from the old Worcester coal mine (Grew and others, 1970). Most of these fossils are poorly preserved. Some of Grew's specimens were examined by Lyons, who observed that the better preserved fossils indicate an Alleghenian flora comparable to that of the Rhode Island Formation in the northern part of the Narragansett basin (Lyons, 1969, 1971).

STRATIGRAPHY OF THE NORFOLK BASIN

The Norfolk basin contains two nonmarine formations which are also present in the Narragansett basin to the south. In ascending order, these are the Pondville Conglomerate and the Wamsutta Formation.

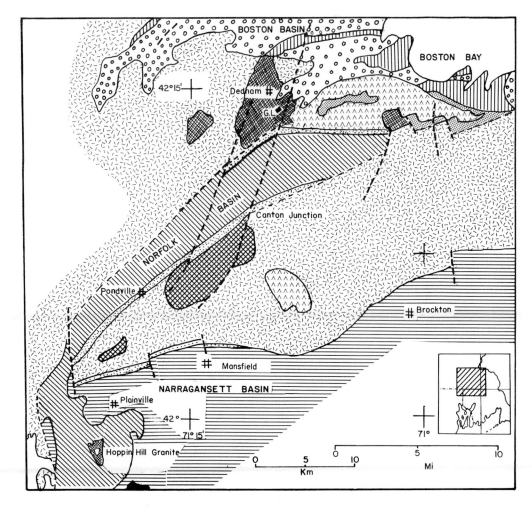
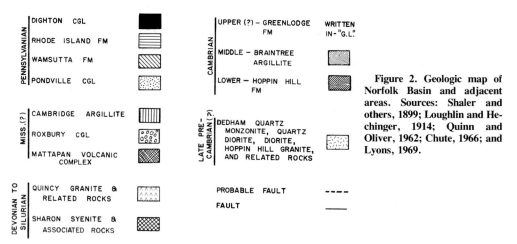

Figure 2. Geologic map of Norfolk Basin and adjacent areas. Sources: Shaler and others, 1899; Loughlin and Hechinger, 1914; Quinn and Oliver, 1962; Chute, 1966; and Lyons, 1969.

Pondville Conglomerate

The Pondville was originally termed the Pondville Group by Shaler and others (1899) but Emerson (1917) and later workers applied the term Pondville Conglomerate. The type section is at Pondville Station in the town of Norfolk, where Shaler and others (1899) described about 80 to 90 m of red and green slate, arkose, and quartz pebble conglomerate. Crosby and Barton (1880) noted plant fossils here. Chute (1966) subdivided the Pondville into two members, of which the type section belongs to the upper member.

At the northern end of the Norfolk basin, exposures of the Pondville consist of about 125 m of unbedded boulder conglomerate unconformably overlying older igneous and metamorphic rocks (Stanley, 1968). This conglomerate corresponds to Chute's (1966) lower member of the Pondville, which reaches a thickness of 300 to 510 m. This member grades into the upper member over an 8- to 30-m interval.

The upper member of the Pondville, which contains the plant fossils described in this paper, consists of mudstone, shale, gray sandstone and granule to pebble conglomerate. The sedimentary structures present include scour and fill, graded bedding and cross-bedding. The thickness of this member varies from about 180 m near the Canton Junction locality to 300 to 450 m farther west in the basin.

The rocks bearing the fossiliferous horizons are exposed at a cut on the Pennsylvania Central Railroad's main line, approximately 731 m north of the Canton Junction railroad station. The railroad cut follows a probable strike-slip fault parallel to the nearby Stony Brook fault (Chute, 1966; Billings, 1976). Locality A, which contains the fossiliferous horizon where we collected most of our specimens, is found on the west side of the tracks, and locality B occurs on the east side (Fig. 3). Both outcrops have a black coating of railroad soot which covers a limonitic weathering surface.

Locality A has a southernmost shale and graywacke zone (8.5 m thick), a central conglomeratic zone (9.5 m thick), and a northern graywacke zone (55 m thick) containing the fossiliferous horizon. The beds dip an average of 65°S. The southernmost zone consists of dark gray green sheared shale and graywacke with coarse grains of quartz and feldspar (roughly 10 percent). The central zone consists of gray green pebble to cobble conglomerate with lenses of quartz sandstone and a quartz sand matrix. The northern zone consists of coarse quartz sandstone and graywacke with thin beds and laminae of sheared shale ("slate") and a few lenses of granule conglomerate. The fossiliferous horizon itself, located 34 m from the south end of the cut, is a shaly laminated graywacke (0.3 m thick) with the best fossils in the shaly laminae.

A modal analysis (700 points) of the graywacke at this horizon indicates 74.2 percent clay minerals, 11.0 percent quartz, 7.4 percent feldspar, 5.0 percent muscovite, 1.1 percent rock fragments, 0.7 percent feldspar (plagioclase?), and 0.6 percent undetermined.

Wamsutta Formation

The Wamsutta Formation, originally termed the Wamsutta Group by Shaler and others (1899), was proposed (Emerson, 1917) for the reddish sediments exposed in North Attleboro in the northwestern part of the Narragansett basin. It is characterized by its red color, especially in the fine-grained components; the coarse-grained components are often gray in color. The contact between the Pondville and Wamsutta in the Norfolk basin is gradational and is characterized by alternating gray and red beds (Chute, 1966). Shaler and others (1899) estimated a thickness for the Wamsutta of about 300 m in the Narragansett basin, but this cannot be measured directly because of rapid facies changes (Mutch, 1968) and the extensive glacial cover of the area. However, Chute (1966) estimated a thickness of at least 900 m in the Norfolk basin (Norwood quadrangle). The original thickness before erosion is unknown. In the Norfolk basin the Wamsutta is composed of many lithologies: polymictic conglomerate containing rip-up clasts; granule conglomerate; poorly sorted to well-sorted, coarse- to fine-grained sandstone; siltstone; mudstone; and shale (Chute, 1966; Stanley, 1968). Sedimentary structures include horizontally laminated and rippled sandstone, graded bedding, channels, scour and fill, mud cracks, raindrop prints and cyclic sedimentation (Stanley, 1968). Sandstone dikes and intraformational breccia were observed in the northern part of the Norfolk basin.

In the Narragansett basin, the Pondville Conglomerate is also the basal unit. It is succeeded by the Wamsutta which is in turn succeeded by and interfingers with the extensive, gray, coal-bearing, fossiliferous sandstone, siltstone and shale of the Rhode Island Formation. Lyons (1969) found no evidence of the Wamsutta between the Pondville Conglomerate and the Rhode Island Formation in the Mansfield quadrangle, although Shaler and others (1899) mention red beds in this area. It may be that there are several so-called "Pondville Conglomerates," or that the reddish character of the Wamsutta is a geographical rather than a time indicator. The Rhode Island Formation is believed to be overlain by the gray conglomerate, lithic graywacke and quartz sandstone of the Dighton Conglomerate, presumably the youngest formation in the Narragansett basin (Fig. 2).

PREVIOUS WORK ON BIOSTRATIGRAPHY

The similar stratigraphy and lithology and apparent physical connection between the Norfolk and Narragansett basins have indicated a Pennsylvanian age for the Norfolk basin (Crosby and Barton, 1880, p. 416). This inferred age was inadequately supported by the discovery of such wide-ranging Mississippian to Permian genera as "*Sigillaria*" and possibly "*Calamites*" or "*Lepidodendron*" in a fine conglomerate at Rockdale, near Pondville Station (Fig. 2) in the southeastern part of the town of Norfolk. Poor descriptions of these fossils and the absence of illustrations make definite determination of critical genera and species impossible. The plant fossils were apparently stem axes; the authors (Crosby and Barton, 1880, p. 419) claimed that some of the specimens were "unquestionably *Sigillaria*." Woodworth visited the Pondville Station locality with Barton and described one of the specimens as a ". . . solid cast of longitudinally, coarsely striated stem, probably but not certainly identifiable with a species of *Sigillaria* or *Calamites*" (Woodworth, 1894, p. 145). He further commented (p. 145), "In none of these specimens, however, are afforded characters which enable one to discriminate them from plant remains which occur elsewhere in rocks of Devonian age, the period to which the red beds [Wamsutta Formation] of this basin were referred by Prof. Edward Hitchcock."

A second fossil plant locality was discovered by Dodge (1875) in a railroad cut about 731 m north of Canton Junction (Fig. 3). The plants were later described by Woodworth (1894, p. 146) as "Compressed stems of *Calamites* . . . about one inch wide and finely striated. They are ill preserved, but one form suggests *C. Cistii* Brgt. . . . *Sigillaria* is also found in the same bed."

Round (1927) reported *Calamites cisti* from the Narragansett basin; Hitchcock (1841) described and illustrated a somewhat similar *Calamites* species from Wren-

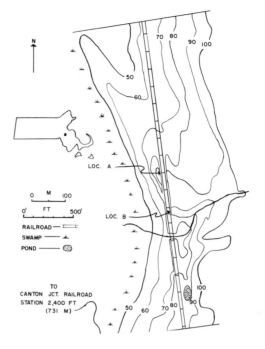

Figure 3. Location of Canton Junction fossil localities A and B.

tham, Massachusetts. Woodworth (1894, p. 148) also mentioned occurrences of *Calamites* and plant stems in two other beds, and in pebbles and boulders of glacial drift, within or near the margins of the Norfolk Basin.

There is some doubt about the nature of these fossils due to the lack of detailed descriptions and illustrations. Nowhere is there mention of nodes in these stem axes that would definitely indicate *Calamites*. The presence of lineation in itself is not diagnostic of *Calamites*, because imperfectly preserved *Cordaites* or inorganic structures such as slickensides can resemble the ribbing of *Calamites*. Those specimens identified as *Sigillaria* also lack definitive characteristics. Crosby (1900, p. 469) also mentioned the Canton Junction locality, referring to a C. M. Endicott who found "other forms of vegetation," but no further details were given about these fossils.

There has been no apparent disagreement with Woodworth's conclusion on the Carboniferous (that is, Pennsylvanian) age of the Norfolk basin. No further details about the plant fossils were provided by subsequent investigators (Crosby, 1900; Emerson, 1917; Quinn and Oliver, 1962; and Chute, 1966), except for a brief statement by Lyons (1969), who reported *Cordaites principalis* from the Canton Junction locality.

In contrast to the Norfolk basin, the Narragansett basin was clearly established as Pennsylvanian on the basis of about 200 plant species (Hitchcock, 1841; Teschemacher, 1846; Lesquereux, 1884, 1889; Round, 1924, 1927; Lyons, 1969, 1971). The flora from the northern part of the Narragansett basin (some 50 species) is closely related to the Westphalian C floras reported by Bell (1962) from the Sydney coalfield and the Pictou Group of New Brunswick (Lyons, 1969, 1971). Oleksyshyn (1976) has documented a Westphalian C flora in the Rhode Island Formation of Plainville, Massachusetts (see Fig. 2). Recent work (P. C. Lyons, unpub. data) indicates that even younger floras (corresponding to Westphalian D and Stephanian A floras) are also present in the Rhode Island Formation.

No such detailed age determinations were made for the Norfolk basin. Thus, although the evidence tentatively supported a Pennsylvanian age for the Norfolk basin, biostratigraphic evidence was conflicting and inconclusive.

SYSTEMATIC PALEOBOTANY

Neuropteris obliqua Brongniart
(Figs. 4A-4H, Figs. 5A-5C)

N. (Mixoneura) obliqua, Bell, 1940, Pl. III, figs. 1, 2.
N. obliqua, Bell, 1944, Pl. XXXIV, fig. 1; Pl. XXXV, figs. 2, 3, 5; Pl. XXXVI, figs. 2, 4; Pl. XXXVII, fig. 5; Pl. LXVII, figs. 1, 2.
N. obliqua Brgt, Crookall, 1959, Pl. XLII, fig. 2; Pl. XLV, figs. 1, 2; Pl. XLVI, figs. 1-4; Pl. XLVII, figs. 4, 5.
N. obliqua Brgt, Laveine, 1967, Pl. L, fig. 3a; Pl. LIV, figs. 2, 2a, 3a.

The specimens are highly variable, consisting of detached pinnules and fragmented pinnae. The most definitive characteristics are (1) decurrent venation, with three to six veins derived directly from the rachis (Figs. 4A, 5C); (2) bifurcation of the primary vein into two equal secondary veins about halfway to two-thirds of the way toward the apex (Figs. 4B, 4H); (3) dichotomy of the lateral venation two to three times (rarely four) prior to the margin (Fig. 4A); (4) extreme polymorphism of the pinnules; (5) longitudinally striated rachis (Fig. 5C); and (6) pinnules broadly attached to the rachis (Figs. 4A, 5C).

The lateral pinnules are subopposite to alternate, usually contiguous, variable in shape, commonly lanceolate to ovate, and rounded at the tip; the lanceolate and ovate forms are usually on opposite sides of the rachis (Fig. 4A). The length varies from 5 to 13 mm in the lanceolate forms, 4 to 7 mm in the ovate forms, and is 10 mm in the cordate form (Fig. 4B). The venation meets the margin at right angles at the apical end, grading to small angle intercepts in the central portion of the pinnule; the veins are 24 to 26 per cm at the margin in the lanceolate forms, 24 to 38 per cm in the ovate forms, and 36 to 42 per cm in the cordate form; the base of the pinnules is usually slightly extended, but tapers in some forms (Fig. 4A). The odontopterid form has no medial veins and is short and broad; the venation is thin and close (52 per cm; Fig. 4C). The midvein extends about halfway to the apex in the lanceolate and ovate forms, and nearly to the apex in the cordate form.

The cyclopterid form (Figs. 4D-4F) is incomplete, with the basal section missing. The apical section is broadly triangular and 28 mm wide; the apex is obtuse. The venation is asymmetrical, close and fine, bifurcating five times along the midline toward the apex; the lateral venation meets the margin at right angles; and the veins are spaced 30 per cm at the margin (Fig. 4E).

The detached terminal(?) pinnule (Figs. 4G, 4H) is acuminate. It is 14 mm long, with the veins spaced 30 to 35 per cm at the margin. Another terminal(?) pinnule has a broadly rounded apex and is 8 mm long, with an extremely thick, trunklike medial vein extending nearly to the apex.

The rachis is thin, with two to three striae (Fig. 5C; often absent, presumably because it has not been preserved), and it is sometimes seen continuous with the basal veins of the pinnules.

Some peculiar properties of Canton Junction *N. obliqua* are the shape of the terminal(?) pinnules; the tapering toward the base of some lateral pinnules; parallelism

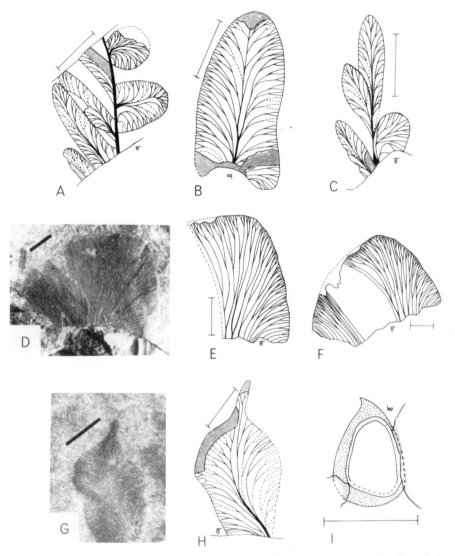

Figure 4. A to H: *Neuropteris obliqua;* C, G, and H: terminal pinnules; D, E, and F: "cyclopterid" pinnule; I: *Cordaicarpus* cf. *cordai*. Line scale equal to 0.5 cm.

of the secondary and tertiary venation to the midvein (Fig. 4B); and the presence of ovate and lanceolate forms on different sides of the pinnae.

Extensive synonymies of *Neuropteris obliqua* can be found in Crookall (1959) and Laveine (1967).

Illustrations and a specimen were sent to W. C. Darrah, who confirmed the identification of this species (Darrah, 1973, written commun.).

Neuropteris cf. *scheuchzeri* Hoffmann
(Fig. 5D)

N. scheuchzeri, Crookall, 1959, Pl. XLI, figs. 1, 2; Pl. XLII, figs. 4–7; Pl. LVII, figs. 1, 2, 5.

N. scheuchzeri Hoffmann, forma *angustifolia*, Bell, 1962, Pl. XXXVII, figs. 1, 2, 5.
N. scheuchzeri Hoffmann, Lyons, 1969, Pl. X, figs. A, B.

The specimen (Fig. 5D) is a fragmented pinnule, large (30 mm long and 18 mm wide at the base) and gradually tapering asymmetrically toward the apex. There is a distinct medial vein about 0.5 mm wide and extended to one side at the truncate(?) base; the lateral venation is obscure. The size, base, and general character are very similar to *N. scheuchzeri* from Foxboro, Massachusetts. (Lyons, 1969, Pl. X, fig. A).

Calamites cisti Brongniart
(Fig. 5E)

C. cisti, Bell, 1944, Pl. LV, fig. 3, Pl. LXIV, fig. 1.
C. cisti, Crookall, 1969, Pl. CXXIII, fig. 1.

These are fragmented casts as much as 10 cm long with fine ribs about 1 mm broad, spaced about 1.5 mm apart; the ribs are alternating or almost straight at the nodes, and are rounded and swollen at the nodes. The furrows are longitudinally striated. The internodal distance could not be determined but is at least 5.2 cm and as much as 4 cm broad; there is no evidence of tubercles. The articulation is at about 75° to the ribs, a porperty long ago noted by Hitchcock (1841) in forms from Wrentham, Massachusetts, and apparently a consistent property found in specimens from the Narragansett and Norfolk basins. Our description of *C. cisti* almost fits the description of *Calamites cistiiformis* Stur (Crookall, 1970, p. 624), except for the inflated ribs at the node and the oblique articulation observed in Figure 5E.

Cordaites principalis Germar
(Figs. 5F–5H)

C. principalis, Bell, 1962, Pl. LV, fig. 2.
C. principalis, Lyons, 1969, Pl. XVII.
C. principalis, Crookall, 1970, Pl. CLIII, fig. 6.

These fragmented specimens are as much as 9.8 cm long and 4.1 cm broad and taper gradually. The primary venation is parallel, fine, and spaced about 1/4 to 1/3 mm apart. The secondary venation (very rarely preserved) consists of two extremely fine veins (Figs. 5G, 5H) between and paralleling each pair of primary veins; they are visible only under low-power magnification. The Canton Junction impressions of *C. principalis* are dwarfed compared to those from Foxboro, Massachusetts (Lyons, 1969), which are as much as 46 cm long and 9 cm wide.

Cordaicarpus cf. *cordai* Geinitz
(Fig. 4I)

Cardiocarpus minor Newb., Lesquereux, 1879, Pl. LXXXV, fig. 38.
Cordaicarpus cordai Geinitz, Arber, 1914, Pl. VI, fig. 29.

These seeds are small (3 by 5 mm and 4 by 6 mm) and ovate, with a rounded base and acuminate apex, and are surrounded by a 0.5 to 1 mm border (?sclerotesta).

The rusty nucellus (3 by 4 mm in the larger seed) is ovate with a well-rounded base, tapering to an acuminate, curved apex.

Lonchopteris? species
(Figs. 5I, 5J)

This species is represented by a single fragmented pinna with two pinnules (Fig. 5I). The pinnules are 22 to 24 mm long and 5 mm wide at the middle. They are lanceolate and contiguous, with a decurrent base. The medial vein is deep and pronounced, extending about two-thirds of the way toward the apex; the venation is netted, consisting of roughly diamond-shaped meshes with about 4 to 5 meshes between the medial vein and the margin. The rachis is 1 mm wide. The netted

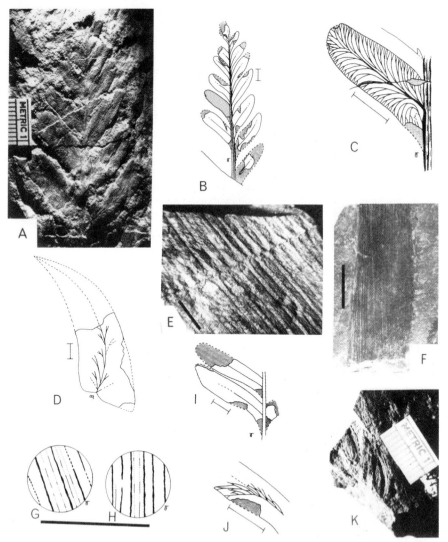

Figure 5. A, B, and C: *Neuropteris obliqua*; C: enlargement of diagonally shaded pinnule in B; D: *Neuropteris* cf. *scheuchzeri*; E: *Calamites cisti*; F, G, and H: *Cordaites principalis*; I and J: ?*Lonchopteris* species; K: ?*Samaropsis* species. Line scale equal to 0.5 cm.

venation and the alethopteroid pinnules almost certainly indicate a *Lonchopteris* species, but because of the fragmentary nature of the specimen and the lack of good preservation, it is assigned to this genus with a question. The specimen was examined carefully by W. C. Darrah, who agrees with this provisional assignment and notes the absence of this genus in the Appalachian beds (Darrah, 1973, written commun.). *Lonchopteris* is a Westphalian species common to the upper Westphalian A and Westphalian B in Europe.

?*Sigillaria* sp. (Brongniart genus)

This is a poorly preserved, fragmented specimen, 16 mm by 52 mm. It has straight parallel margins between which are two prominent longitudinal ribs (1.5 to 2 mm broad) spaced 5 mm apart and connected by transverse, undulating striations spaced about 2/3 mm apart. This specimen is most likely a decortiated *Sigillaria*, but it may be a *Calamites* or *Artesia* species.

?*Samaropsis* sp. (Goeppert genus)
(Fig. 5K)

These two elongate seed impressions (the external margins are dashed in the photograph) are 13 mm by 3.5 mm and 8 mm by 2.5 mm. The base of the larger is broken, and the outer end tapers asymmetrically to an acuminate tip. A central (?nucellar) body is present within an enclosing (?sclerotestal) layer; this consists of at least two indistinct sublayers. The smaller seed is entire, having a rounded base and a sub-acuminate tip; a central (?nucellar) body and enclosing (?sclerotestal) sheath are again visible, but no obvious subdivisions exist in the latter.

AGE OF THE CANTON JUNCTION FLORA

As a group, the Canton Junction species are found in the Upper Carboniferous of Europe and Canada and in the Pennsylvanian of the United States. *Cordaites principalis* and *Calamites cisti* are too long ranging to be of any assistance in precise age determination. *Neuropteris obliqua*, on the other hand, is a relatively short-ranging species which occurs in the Westphalian A and B in Europe and is suggestive of the Medial Pottsvillian in the central Appalachians, which corresponds to the Westphalian B in Europe and Canada. *Neuropteris scheuchzeri* first occurs in the upper Westphalian B and is common in the Westphalian C of Europe; it is common in the Alleghenian and Conemaughian of the Central Appalachians, roughly corresponding to the Westphalian C and D in Europe. Thus, the association of *Neuropteris obliqua* and *Neuropteris scheuchzeri* indicates that the Canton Junction flora is transitional between the Westphalian B and Westphalian C and is most likely a late Pottsvillian flora; it could be as old as medial Pottsvillian or as young as early Alleghenian.

The Pondville Conglomerate is younger than the Riversdale and Cumberland Groups of Maritime Canada and older than the Pictou Group, marking a hiatus in Maritime Canada (Bell, 1944). This break presumably coincides with the beginning of New England Pennsylvanian sedimentation, which began in the Norfolk and Narragansett basins, and with part of the upper Pottsville. Thus it appears that the Canton Junction flora is the oldest Pennsylvanian flora yet discovered in New England.

A chart summarizing the stratigraphic, age, and geographic distribution of the Canton Junction flora in eastern North American is shown in Figure 6.

PALEOENVIRONMENT AND PALEOGEOGRAPHY

The similar ages, lithologies, stratigraphy, structure and geographic proximity of most of the late Paleozoic basins of southeastern New England indicate that their origins are related. It has been suggested that they were united at one time (Quinn and Oliver, 1962), and it is conceivable that a single basin may have extended as far as Worcester, Massachusetts (Fig. 1). However, faulting and large-scale erosion make it impossible to define the original boundaries of such a hypothetical basin. The Boston basin was most likely independent from the other basins of southeastern Massachusetts and Rhode Island (Fig. 1), as indicated by its different lithologies and stratigraphy, and probably was an isolated intermontane basin where abundant volcanic ashes and flows were deposited. The smaller quantity of volcanic rocks and different stratigraphy of the Norfolk and Narragansett basins (Quinn and Oliver, 1962; Chute, 1966; Mutch, 1968), along with scanty floral and faunal evidence from the Boston basin (Burr and Burke, 1900; Pollard, 1965), indicate that these basins were not synchronous. On the other hand, floral evidence does support the view that the Worcester coal and the Norfolk and Narragansett basin sediments were relatively contemporaneous Pennsylvanian deposits.

A large number of sedimentary features have been noted in the Narragansett and Norfolk basins, including graded bedding, scour and fill, mud cracks, raindrop impressions, cross-bedding (Chute, 1940; Quinn and Oliver, 1962; Stanley, 1968), sandstone dikes (Lyons, 1969) and intraformational breccia. As a group, these structures, along with the paleontological evidence to be discussed later, indicate subaerial exposure, fluvial conditions, and a nonmarine environment.

Quinn and Oliver (1962) and Mutch (1968) suggested that the depositional environment of the Pondville conglomerate in the southern Narragansett basin

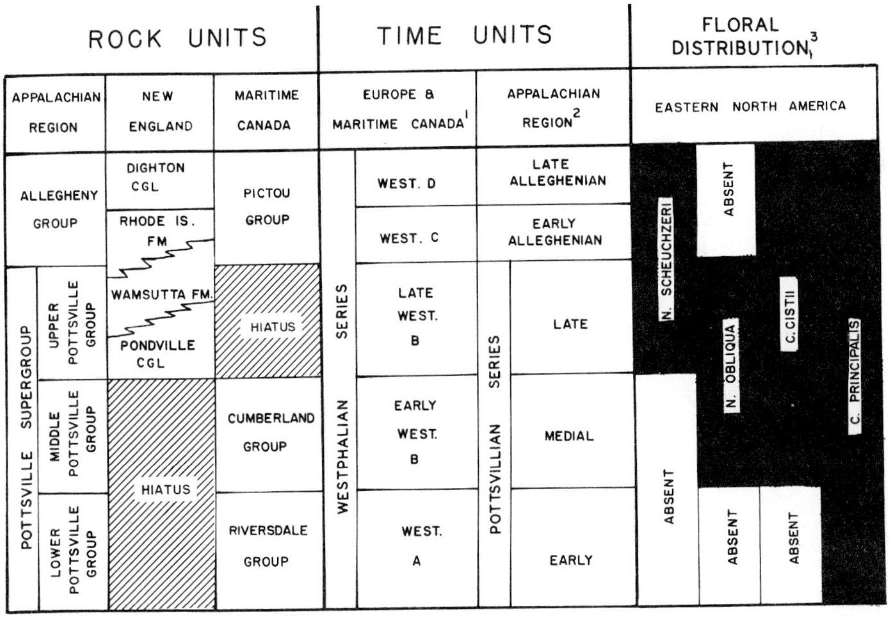

Figure 6. Stratigraphic chart showing time distribution of important Canton Junction plant species in eastern North America. Sources: [1]Bell, 1944, 1962, slightly modified; [2]Darrah, 1970, slightly modified; [3]Darrah, 1970; Bell, 1944, 1962.

was a swampy lowland because of the lack of clear-cut channels, the carbonaceous character of the fine-grained sediments, and the occurrence of plant fossils. This interpretation is also applicable to the fine-grained, plant-bearing rocks of the upper Pondville of the Norfolk basin. However, the conglomerate and sandstone of the lower Pondville indicate fluvial conditions and the presence of nearby source areas of high relief.

The Wamsutta Formation was interpreted as alluvial by Shaler and others (1899), and as fluvial with Flood-plain overbank deposits by Quinn and Oliver (1962). Repeated subaerial exposure is indicated by the mud cracks and raindrop impressions (Stanley, 1968). The red iron oxide pigment may also indicate deposition on a well-drained upland or piedmont flood plain where most plant debris was oxidized (Towe, 1959; Quinn and Oliver, 1962; Mutch, 1968), ot it may be of diagenetic origin (Walker, 1967; Van Houten, 1972). The presence of amphibian footprints in the Wamsutta of the Narragansett basin also indicates subaerial exposure (Willard and Cleaves, 1930).

All the reported plant fossils of the Norfolk and Narragansett basins are terrestrial and include some in situ specimens in the Wamsutta, indicating sudden burial, possibly in floodplain deposits (Knox, 1944). The different nature of the terrestrial floral assemblages in the Norfolk and Narragansett basins indicates somewhat different sedimentary environments. The *Calamites-Cordaites*-Pteridosperm association at Canton Junction indicates a flood-plain–swamp margin to upland habitat for the Norfolk basin, as opposed to the Lycopod-Pteridosperm association of the Narragansett basin which indicates a swampy lowland (Cridland and Morris, 1963; Peppers and Pfefferkorn, 1970). The scarcity of plant fossils in the Norfolk basin also seems to indicate a lack of extensive lowlands, in contrast to the Narragansett basin where a rich Pennsylvanian flora developed.

Of the two localities at Canton Junction, locality B, with its coarse-grained graywacke, may have been a stream margin, considering the dominance of *Calamites*; locality A, with its fine-grained graywacke, may have been a swamp margin, considering the dominance of seed ferns and *Cordaites*. Both were probably deposited in at least slightly upland habitats, in contrast to the Rhode Island Formation of the Narragansett basin.

Mutch (1968) suggested that the source areas of the sediments in the Narragansett basin were to the north, northwest and west of the present boundaries of the basin. However, paleocurrent analysis based on cross-bedding, ripple marks and alignment of plant fossils in the Narragansett basin indicates a northeast to southwest drainage pattern (Towe, 1959). One of the distinguishing lithologic aspects of the Pennsylvanian basins of southeastern New England that may be useful in this analysis is the extensive development of conglomerate (e.g., Pondville, Dighton, Purgatory, and Bellingham Conglomerates; Quinn and Oliver, 1962), particularly at the margins of the basins. These conglomerates are frequently dominated by quartzite pebbles and cobbles comprising 80 to 95 percent of the clasts (Perkins, 1920; Lyons, 1969). In the Narragansett basin the percentage of these clasts increases southward, whereas that of other compositions, particularly granite, decreases in abundance southward (Perkins, 1920). The pale gray to white quartzite pebble and cobble clasts resemble the Westboro Quartzite found to the west of the Narragansett basin, and the basal Early Cambrian quartzite found at Hoppin Hill (Fig. 2), an inlier of basement rocks within the Narragansett basin (Dowse, 1952). These quartzite clasts, therefore, indicate westerly and (or) southwesterly source areas. Rocks similar to those of the Quincy group (Fig. 2) are absent in these conglomerates, except along the northern margin of the Norfolk basin where boulders up to 60 cm in diameter from the underlying Blue Hill Granite Porphyry are found in a basal boulder conglomerate of the lower member of the Pondville Conglomerate (Chute, 1940, 1966), indicating a northern marginal source area of considerable relief. Rocks similar to those of the Dedham quartz monzonite (Fig. 2) have been found in small quantities (a few percent) in several places in both the Norfolk and Narragansett basins (Chute, 1940, 1966; Lyons, 1969; Quinn, 1971), most likely indicating a northerly source area. The conglomerates of the Norfolk basin are polymictic, commonly having less quartzite and more volcanic and granitic clasts than those of the Narragansett basin (Chute, 1940, 1966). In general, however, distant source areas rather than local ones are indicated by the nature of the clasts in these conglomerates.

In summary, the reconstruction of the source areas and paleodrainage patterns has presented a particularly difficult problem. Source areas for the clasts of the conglomerates have been suggested to lie on the northern, western, southern and possibly even to the eastern, or at least northeastern, margins of the basin(s?). The superposition of this problem of multiple source areas upon a presumably southerly to southwesterly drainage pattern suggests the strong possibility that the drainage was centripetal, with a southerly outlet. This is not an uncommon pattern in present-day intermontane basins, as C. Wroe Wolfe of Boston University has suggested (1974, oral commun.).

CONCLUSIONS

The new collections of plant fossils from near Canton Junction confirm the previously assumed Pennsylvanian age of the Norfolk basin. The Pondville Conglomerate of the Norfolk basin is probably equivalent to part of the Upper Pottsville of the central Appalachians and coincides with a hiatus in the Upper Carboniferous of Maritime Canada.

There is no solid evidence to indicate that the so-called "Pondville Conglomerate" of the Narragansett basin is equivalent in age to the Pondville of the Norfolk basin. It may be that the Pondville Conglomerate of the northern section of the Narragansett basin is somewhat younger, as it appears to be continuous into the lower beds (early Alleghenian) of the Rhode Island Formation (Lyons, 1969). The stratigraphic evidence indicates that the Wamsutta Formation is partly equivalent in age to the Pondville Conglomerate and the Rhode Island Formation. This is shown graphically in Figure 6.

ACKNOWLEDGMENTS

We are indebted to several former undergraduate and graduate students in the Geology Department at Boston University who helped find and collect the plant fossils described in this report: Stephen Mangion, Robert Titus, Alan Weiss, Dorothy Richter, Margaret O'Neill, and Mike Kelly. Stephen Mangion, now associated with the Department of Geological Sciences, Brown University, made a thin section of the graywacke at the fossilferous horizon and helped in the preparation of many of the specimens. Alan Weiss photographed some of the specimens. William C. Darrah of Gettysburg College, Pennsylvania, examined drawings and a specimen of both *Neuropteris obliqua* and our presumably *Lonchopteris* species; Elso Barhoorn of Harvard University supplied us with a specimen of *Artisia*. Thanks are given to Barghoorn, A. W. Quinn, and H. W. Pfefferkorn for critically reviewing the paper and making helpful suggestions for its improvement.

REFERENCES CITED

Arber, E. A. N., 1914, A revision of the seed impressions of the British Coal Measures: Annals Botany, v. 28, p. 81-108, Pls. 6-8.
Bell, W. A., 1940, The Pictou coalfield, Nova Scotia: Canada Geol. Survey, Mem. 225, p. 1-160, Pls. 1-10.
——1944, Carboniferous rocks and fossil floras of northern Nova Scotia: Canada Geol. Surv., Mem. 238, p. 1-276, Pls. 1-79.
——1962, Flora of Pennsylvanian Pictou Group of New Brunswick: Canada Geol. Surv., Bull. 87, p. 1-71, Pls. 1-56.
Billings, M. P., 1976, Geology of the Boston basin, *in* Lyons, P. C., and Brownlow, A. H., eds., Studies in New England Geology: Geol. Soc. America Mem. 146, p. 5-30 (this volume).
Burr, H. T., and Burke, R. E., 1900, The occurrence of fossils in the Roxbury Conglomerate: Boston Soc. Natl. Hist. Proc., v. 29, p. 179-184.
Chute, N. E., 1940, Preliminary report of the geology of the Blue Hills Quadrangle, Massachusetts: Mass. Dept. Public Works, U.S. Geol. Survey Cooperative Geologic Project, Bull., no. 1, Boston, 52 p.
——1966, Geology of the Norwood quadrangle, Norfolk and Suffolk Counties, Massachusetts: U.S. Geol. Survey Bull. 1163-B, 78 p.
Cridland, A. A., and Morris, J. E., 1963, Taeniopteris, Walchia and Dichophyllum in the Pennsylvanian System of Kansas: Univ. Kans. Sci. Bull., v. 44, no. 4, p. 71-85.
Crookall, Robert, 1959, Fossil plants of the Carboniferous rocks of Great Britain: Great Britain Geol. Surv., Mem. Palaeontology, v. 4, pt. 2, p. 85-216, Pls. 25-58.
——1969, Fossil plants of the Carboniferous rocks of Great Britain: Great Britain Geol. Survey, Mem. Palaeontology, v. 4, pt. 5, p. 573-791, Pls. 107-150.
——1970, Fossil plants of the Carboniferous rocks of Great Britain: Great Britain Geol. Survey, Mem. Palaeontology, v. 4, pt. 6, p. 793-840, Pls. 151-159.
Crosby, W. O., 1900, Geology of the Boston Basin: The Blue Hills Complex: Boston Soc. Natl. Hist., Occ. Pap. IV, v. 1, pt. 3, p. 289-563.
Crosby, W. O., and Barton, G. N., 1880, Extension of the Carboniferous Formation in Massachusetts: Am. Jour. Sci., v. 120, p. 416-420.
Darrah, W. C., 1970, A critical review of the Upper Pennsylvanian floras of Eastern United States with notes on the Mazon Creek flora of Illinois: Gettysburg, Pa. [privately printed] 221 p., 80 Pls.
Dodge, W. W., 1875, Note on the geology of eastern Massachusetts: Boston Soc. Natl. Hist., v. 17, p. 414.
Dowse, A. M., 1952, New evidence on the Cambrian contact at Hoppin Hill, North Attleboro, Massachusetts: Am. Jour. Sci., v. 248, no. 2, p. 95-99.
Emerson, B. K., 1917, Geology of Massachusetts and Rhode Island: U.S. Geol. Survey Bull. 597, 289 p.
Grew, E. S., Mamay, S. N., and Barghoorn, E. S., 1970, Age of plant fossils from the Worcester coal mine, Worcester, Massachusetts: Am. Jour. Sci., v. 268, p. 113-126.
Hitchcock, Edward, 1841, Final report on the geology of Massachusetts: Amherst, Mass., v. II, p. 301-714.
Knox, A. S., 1944, A Carboniferous flora from the Wamsutta Formation of southeastern Massachusetts: Am. Jour. Sci., v. 242, p. 130-138.
Laveine, Jean-Pierre, 1967, Les Neuroptéridées du Nord de la France [Ph.D. thesis]: Faculté des Sciences de l'Université de Lille, 344 p., Pls. 1-83.
Lesquereux, Leo, 1879, Coal flora of Pennsylvania: Pennsylvania Geol. Survey Atlas, 2d, Pls. 1-85.
——1884, The Carboniferous flora of Rhode Island: Am. Naturalist, v. 18, p. 921-923.
——1889, Fossil plants of the Coal-measures of Rhode Island: Am. Jour. Sci., v. 137, p. 229-230.
Loughlin, G. F., and Hechinger, L. A., 1914, An unconformity in the Narragansett Basin of Rhode Island and Massachusetts: Am. Jour. Sci., v. 38, p. 45-64.

Lyons, P. C., 1969, Bedrock geology of the Mansfield Quadrangle, Massachusetts [Ph.D. dissert]: Boston University, 283 p., 18 Pls.

——1971, Correlation of the Pennsylvanian of New England and the Carboniferous of New Brunswick and Nova Scotia: Geol. Soc. America, Abs. with Programs (Northeastern Sec.), v. 3, no. 1, p. 43-44.

Mutch, T. A., 1968, Pennsylvanian nonmarine sediments of the Narragansett Basin, Massachusetts-Rhode Island: Geol. Soc. America Spec. Paper 106, p. 177-209.

Oleksyshyn, John, 1976, Fossil plants of Pennsylvanian age from northwestern Narragansett basin, *in* Lyons, P. C., and Brownlow, A. H., eds., Studies in New England geology: Geol. Soc. America Mem. 146, p. 143-180 (this volume).

Peppers, R. A., and Pfefferkorn, H. W., 1970, A comparison of the floras of the Colchester (No. 2) Coal and the Francis Creek Shale: Illinois State Geol. Survey Guidebook Ser. no. 8, p. 61-74.

Perkins, E. H., 1920, The origin of the Dighton conglomerate of the Narragansett basin of Massachusetts and Rhode Island: Am. Jour. Sci., 4th ser., v. 49, no. 1, p. 61-75.

Pollard, Melvin, 1965, Age, origin, and structure of the post-Cambrian Boston strata, Massachusetts: Geol. Soc. America Bull., v. 76, p. 1065-1068.

Quinn, A. W., 1971, Bedrock geology of Rhode Island: U.S. Geol. Survey Bull. 1295, 65 p. and map.

Quinn, A. W., and Oliver, W. A., 1962, Pennsylvanian rocks of New England, *in* Pennsylvanian System in the United States, C. C. Branson, ed.: Am. Assoc. Petroleum Geol., p. 60-73.

Round, E. M., 1924, Correlation of fossil floras of Rhode Island and New Brunswick: Bot. Gaz., v. 78, p. 116-118.

——1927, Correlation of coal floras in Henry County, Missouri, and the Narragansett Basin: Bot. Gaz., v. 83, p. 61-69.

Seward, A. C., 1917 (reprinted 1969), Fossil plants: Cambridge, England, Cambridge Univ. Press, v. III, 656 p.

Shaler, N. S., Woodworth, J. B., and Foerste, A. F., 1899, Geology of the Narragansett Basin: U.S. Geol. Survey Mon. 33, 402 p.

Stanley, D. J., 1968, Graded bedding-sole marking-graywacke assemblage and related sedimentary structures in some Carboniferous flood deposits, Eastern Massachusetts: Geol. Soc. America Spec. Paper 106, p. 211-239.

Teschemacher, J. E., 1846, Fossil vegetation of America: Boston Soc. Nat. Hist. Jour., p. 370-385.

Towe, K. M., 1959, Petrology and source of sediments in the Narragansett Basin of Rhode Island and Massachusetts: Jour. Sed. Petrology, v. 29, p. 503-512.

Van Houten, F. B., 1972, Iron and clay in tropical savannah alluvium, northern Columbia: A contribution to the origin of red beds: Geol. Soc. America Bull., v. 83, p. 2761-2771.

Walker, T. R., 1967, Color of Recent sediments in tropical Mexico—A contribution to the origin of red beds: Geol. Soc. America Bull., v. 78, p. 917-919.

Willard, B., and Cleaves, A. B., 1930, Amphibian footprints from the Pennsylvanian of the Narragansett Basin: Geol. Soc. America Bull., v. 41, p. 321-327.

Woodworth, J. B., 1894, Carboniferous fossils in the Norfolk County Basin: Am. Jour. Sci., v. 148, p. 145-148.

Manuscript Received by the Society March 12, 1974
Revised Manuscript Received July 18, 1974
Manuscript Accepted August 28, 1974

Printed in U.S.A.

PART II
GEOLOGY OF NORTHERN NEW ENGLAND
Arthur M. Hussey II
Co-ordinator

Introduction

The geology of northern New England is diverse and complex. The six papers in this section deal with regional stratigraphy (Pankiwskyj and others), regional tectonics and high-grade metamorphism (Moench and Zartman), emplacement of plutonic rocks (Nielson and others; Chapman), low-grade metamorphism (Richter and Roy), and sulfide mineralization (Rainville and Park). The first four papers represent major syntheses of regional data, and the other two are significant because they represent detailed mineralogical studies of unique areas in New England.

The paper by Pankiwskyj and others summarizes extensive mapping and paleontological studies of the southeast limb of the Merrimack synclinorium. The report covers a large portion of central Maine and is a major contribution to our knowledge of the stratigraphy of New England. The stratigraphy of the northwest limb of the synclinorium is known as a result of earlier work by Moench and other workers. In this volume, Moench and Zartman integrate structural, metamorphic, and igneous features of the northwestern part of the Merrimack synclinorium. This area has been subject to multiple deformation, extensive plutonism, and high-grade metamorphism. Moench and Zartman summarize the probable timing of the major events and relate them to the Acadian orogeny.

Rocks of the Devonian New Hampshire Plutonic Series occur in the Merrimack synclinorium in both Maine (as discussed by Moench and Zartman) and in New Hampshire. Many of these plutons are apparently thin concordant sheets rather than large bottomless masses. Nielson and others use gravity data in conjunction with field mapping to place constraints on the shapes and structural relations of several major plutons of the New Hampshire Plutonic Series. Chapman discusses another major group of plutonic rocks, the White Mountain Magma Series of Jurassic to Cretaceous age. These rocks occur mostly in central and northern New Hampshire and form stocks, ring dikes, and volcanic masses. Chapman presents a detailed petrogenetic model to explain the variety of structural relations, shapes, compositions, and relative ages of these alkalic-magmatic rocks.

It has long been known that the grade of metamorphism decreases to the north in New England. However, prehnite-pumpellyite-facies metamorphism in New England was not known until 1970, when an area in northern Maine was shown to have this type of subgreenschist metamorphism. The paper by Richter and Roy is a detailed petrographic study that greatly amplifies and extends our knowledge of this region of prehnite-pumpellyite metamorphism. The authors have delineated three zones corresponding to increasing grade within the facies and assign the metamorphism to the Acadian orogeny. The paper by Rainville and Park is also a detailed mineralogical study, in this case of nickeliferous pyrrhotite deposits of southeastern Maine. These deposits, which occur in peridotite and gabbro-diorite, differ significantly in their mineralogic composition from other nickel deposits. Particularly interesting is the presence of chromium ulvospinel and of large amounts of graphite. The overall mineralogic relations of the deposit favor an early magmatic origin under conditions of low sulfur and oxygen fugacities.

PAUL C. LYONS
ARTHUR H. BROWNLOW

Geological Society of America
Memoir 146
© 1976

Chronology and Styles of Multiple Deformation, Plutonism, and Polymetamorphism in the Merrimack Synclinorium of Western Maine

ROBERT H. MOENCH

AND

ROBERT E. ZARTMAN
U.S. Geological Survey
Federal Center
Denver, Colorado 80225

ABSTRACT

This report attempts to synthesize the results of recent geologic mapping and petrologic studies in western Maine, an area that lies on the transition zone between the sillimanitic portion of the Merrimack synclinorium in New Hampshire and the greenschist-facies portion in central Maine. Results of two fundamentally different deformations are recognized in the transition zone. The first or early deformation produced northeast-trending tight, upright passive-flow folds; longitudinal premetamorphic faults of large displacement; and slaty cleavage formed also by the first metamorphism (M-1). These features were produced throughout the whole geosyncline that was ancestral to the Merrimack synclinorium; they characterize the structure of the greenschist terrane. But southwestward across the transition zone, they are increasingly blurred by younger superposed structural and metamorphic features. Inferred origins are geosynclinally controlled slump deformation and tectonic dewatering culminating in low-grade metamorphism.

The younger features collectively define the late deformation and at least two late metamorphic events (M-2 and M-3). The late deformation produced a complex pattern of slip cleavages; schistosity derived from slip cleavage, small recumbent folds, large flexural folds, and domes of varied shapes and trends; and conspicuous northwest-trending flexural cross folds. In the northeast part of the transition zone, M-2 and M-3 metamorphic events produced overlapping metamorphic zones approximately concentric to granitic plutons. To the southwest, where metamorphic grade is regionally in the sillimanite or potassium feldspar plus sillimanite zones, M-2 and M-3 zones coalesce and become indistinguishable. Many structural features are directly magma-generated, but the northwest-trending cross folds and related cleavages probably express regional strain that is only temporally magma-related.

Nine samples of granite and aplite have yielded an isotopic age of 379 ± 6 m.y. by the Rb-Sr whole-rock method. This age dates the youngest intrusive rocks of the New Hampshire Plutonic Series in the report area, M-3 metamorphism, and the late deformations that accompanied M-3. Because all the plutons intrude previously deformed Devonian and older strata, the isotopic age narrowly restricts the duration of the early and late deformations, plutonism, and M-1 to M-3 metamorphisms.

We speculate that the Mooselookmeguntic pluton and the larger Sebago pluton farther south are irregular but broadly arched subhorizontal sheets only a few kilometres thick that were emplaced at depths of 11 to 15 km. The transition zone is where these and possibly other sheetlike bodies dive monoclinelike to the northeast. Northeast of the transition zone, the tops of the sheets are now a few kilometres below the surface. Associated with them are many of the late structural and metamorphic features that characterize the transition zone and the sillimanite terrane. Above them is a partly eroded thick slate layer. Numerous plutons now exposed northeast of the transition zone are cupolas and possibly isolated bodies that rose above the sheets. In the sillimanite terrane southwest of the transition zone, a subhorizontal granitic sheet complex is exposed. The thick slate layer that is now exposed only in the northeast once extended over the sillimanite terrane.

INTRODUCTION

The Merrimack synclinorium is a major northeast-trending tectonic feature of the northern Appalachians in New England that extends at least 400 km from Connecticut into central Maine (Billings, 1956; Osberg and others, 1968). The synclinorium coincides with a geosynclinal ancestor defined stratigraphically by an enormous thickness of clastic metasedimentary rocks having some calcareous units but only sparse metavolcanic rocks; these deposits accumulated continuously at least from Ordovician to Early Devonian (?) time. This ancestral geosynclinal tract may in fact extend from Connecticut to the Gaspé Pensinula (Rodgers, 1970, p. 131). The area of this report is on a major structural, metamorphic, plutonic, and geophysical transition zone—a belt 100 km or so wide astride line G-H (Fig. 1)—on the trend of the Merrimack synclinorium in western Maine. The area within the sillimanite isograd is a broad terrane of sillimanite-bearing schist, gneiss, and migmatite characterized by polyphase deformation, varied trends, and a strong tendency toward recumbency of bedding and foliation in large areas. Metamorphic mineral assemblages (Thompson and Norton, 1968) represent an intermediate-pressure-facies series. In contrast, the northeastern part of the synclinorium is a broad area of slate characterized by greenschist-facies metamorphism and in places by a remarkably straight longitudinal beltlike pattern of tightly folded subvertical formations (Osberg, 1968). Minor kink banding and slip cleaving are ubiquitous there, but the results of one major deformation are predominant. Unlike the commonly elongate concordant plutons in New Hampshire (Fig. 1), granitic plutons in Maine are sharply discordant and have diverse trends. In the area northeast of the sillimanite isograd, the plutons have contact-metamorphic aureoles of the low-pressure Buchan type (Boone, 1973; Osberg, 1971), indicating crystallization at shallower depths than plutons of about the same age in the sillimanite terrane.

In the transition zone, the large Mooselookmeguntic and Sebago plutons (Fig. 1) trend northwest across the synclinorium, parallel to a belt of cross folds described below. Geophysically, the transition zone is expressed by a gradational but major change in the character of the regional gravity field that takes place across line G-H (Fig. 1; Kane and Bromery, 1968; Kane and others, 1972).

This paper synthesizes the results of recent geologic mapping, isotopic and petrologic studies, and available gravity data in the transition zone. This zone is probably the best place to seek answers to important tectonic problems concerning

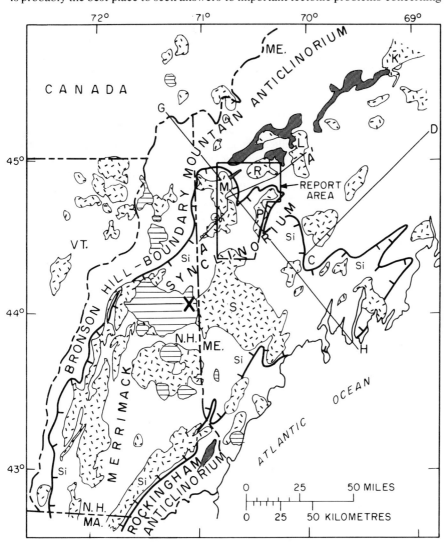

Figure 1. Index map showing reported area, plutons (modified from White, 1968), and sillimanite (Si) isograd (modified from Thompson and Norton, 1968). Random pattern is Devonian silicic and intermediate rocks and may include Ordovician plutons near coast. Dark shading is Devonian gabbroic rocks. Lined pattern is Mesozoic alkalic rocks. X indicates body of slate of Littleton Formation (from Billings, 1928, 1956). Line A-A', section shown in Figure 9. Lines C-D and G-H from Kane and Bromery (1968) mark changes in regional gravity field. Pertinent plutons: M, Mooselookmeguntic; R, Redington; P, Phillips; L, Lexington; S, Sebago; K, Katahdin.

the Merrimack synclinorium. It is commonly held that the whole synclinorium is an expression of a predominantly single-phase compressional Acadian orogeny whose structural and metamorphic results are more complex in the southwest because deeper levels in the orogen are exposed there. The orogen, according to this view, behaved as a "giant vise" (Rodgers, 1970, p. 224), and the deeper levels were cooked and squeezed the most. In contrast, Hamilton and Myers (1967, p. C13–C15; 1974, p. 375) have proposed that the sillimanite terrane outlined in Figure 1 formed beneath a large surficial batholith and consists of rocks that flowed around downward and beneath many ascending and spreading plutons; metamorphic and structural complexity are largely magma-generated. The northeast limit of the sillimanite terrane (the area of this report), according to Hamilton and Myers (1967, p. C15), "represents the limit of initially nearly continuous granitic plutons in the overlying batholithic complex that has since been largely eroded away." We offer alternatives to both views, but we recognize that no satisfactory answers have been proved. These questions are fundamental and must be solved before the role of the synclinorium can be assessed properly in terms of Plate Tectonic theory.

This paper is based largely on mapping by Moench (1971) and Moench and Hildreth (1974, 1976) in the Phillips, Rangeley, and Rumford quadrangles, supported by isotopic dating by R. E. Zartman. Incorporated also are the pertinent results of Guidotti's petrologic studies (Guidotti, 1968, 1970a, 1970b, 1974) and Moench's interpretations of mapping by Milton (1961), Fisher (1962), Guidotti (1965), Pankiwskyj (1964), and Guidotti (Green and Guidotti, 1968) in adjacent quadrangles. Geophysical data are incomplete in the report area, but we have integrated available gravity maps and studies (Kane and Bromery, 1966, 1968; Kane and others, 1972) with the geological relationships.

STRATIGRAPHY OF METASEDIMENTARY ROCKS

At least 13 km of predominantly clastic metasedimentary rocks are exposed in the area of Figure 2 and nearby. For convenience these rocks are divided into a lower sequence of Ordovician and Ordovician (?) age (about 7 km thick), a middle sequence of Silurian and Silurian (?) age (about 4 km thick), and an upper sequence of Devonian (?) age (about 2 km thick). These data on thicknesses are from Harwood (1973, p. 8), Osberg and others (1968, Table 18-1), and Moench (1970, Table 1). The middle and upper sequences may be considerably thicker, however, because the thicknesses were measured normal to bedding on the limbs of tight folds, where bedding has been thinned by factors as large as four. Bedding is well preserved everywhere except in the areas of migmatitic gneiss with disrupted bedding (Figs. 2, 3).

Figure 2 incorporates some remapping and stratigraphic reassignment by Moench in the Old Speck Mountain, Bethel, and Bryant Pond quadrangles. These areas originally were mapped before stratigraphy was worked out in detail in the simpler, lower grade terrane farther northeast. Thus, rocks assigned to the middle sequence in Figure 2 in the Old Speck Mountain and Bethel quadrangles were previously considered members of the Littleton Formation (Milton, 1961; Fisher, 1962). The metaconglomerate and associated pelitic schist of Milton (1961) are now (Moench, 1971) identified as parts B and C of the Rangeley Formation and as the Perry Mountain Formation. Rocks considered approximately correlative to the Littleton Formation are restricted in the Old Speck Mountain and Bethel quadrangles to the areas underlain by the upper sequence, as shown in Figure 2. In the Bryant Pond quadrangle, the rusty-weathering Billings Hill and Thompson Mountain

Formations of Guidotti (1965) are correlated with the Smalls Falls Formation. Guidotti's (1965) Shagg Pond Formation and Concord Pond Member of the Littleton are correlated with one another on lithologic and structural grounds; both belong to the upper sequence. The Wilbur Mountain Member and the nonsulfidic part of the Howard Pond Member of the Littleton (Guidotti, 1965) are identified, respectively, as the Smalls Falls and Madrid Formations, here included in the middle sequence. The sulfidic part of the Howard Pond belongs to the Smalls Falls.

Lower Sequence

The northwest corner of the area of this report is underlain by a broad belt of pre-Silurian rocks that reappear to the south in the core of the Brimstone Mountain anticline; they are exposed also in the northeast corner of the Old Speck Mountain quadrangle (Fig. 2). Pre-Silurian rocks shown in the Old Speck Mountain quadrangle are part of an assemblage of metapelite, metabasalt, and euxinic black metashale that Milton (1961) assigned to the Albee (Lower and Middle Ordovician), Ammonoosuc (Middle Ordovician), and Partridge (Middle Ordovician) Formations. They define the northeast end of the Bronson Hill anticlinorium (Fig. 1), and are approximately equivalent to the pre-Silurian rocks exposed northwest of Rangeley Lake on the southeast flank of the Boundary Mountain anticlinorium (Fig. 1). There the pelitic Albee and Aziscohos (Lower Ordovician) Formations are overlain by metabasaltic greenstone and weakly metamorphosed Middle Ordovician black shale dated by graptolites found a few kilometres to the north of the map boundary (Harwood and Berry, 1967; Harwood, 1973). Harwood (1973) assigned the dated shale to the Dixville Formation, but for an alternative interpretation, see Boone and others (1970). The dated black shale is overlain by about 900 m of Upper Ordovician (?) metagraywacke, felsic metavolcanic rocks, and black or dark-gray metashale (Quimby Formation), which is overlain by about 90 to 180 m of laminated calcareous metasiltstone (Greenvale Cove Formation). These younger pre-Silurian rocks are exposed mainly east and south of Rangeley Lake and in the core of the Brimstone Mountain anticline. Where unfaulted, the boundary between the lower and middle sequences is conformable. The Taconic unconformity, located between rocks equivalent to the lower and middle sequences, is expressed northwest of the area of this report area along the Bronson Hill-Boundary Mountain anticlinorium.

Middle Sequence

About 4,400 m of Silurian and Silurian(?) rocks are exposed in a broad central belt and in anticlines southeast of the Blueberry Mountain fault (Fig. 2). The lower two-thirds of this mass is the Rangeley Formation of Early Silurian age, dated by a shelly fauna discovered a few kilometres north of the area of this report (Moench and Boudette, 1970; Harwood, 1973). The Rangeley is composed of interbedded and intertonguing metashale, feldspathic metasandstone, polymictic metaconglomerate, and quartz metaconglomerate. The formation becomes more pelitic toward the southeast. The Rangeley is overlain by rather potassic metashale with cyclic interbeds of quartz-rich metasandstone (Perry Mountain Formation). The Perry Mountain Formation is overlain by black sulfidic metashale with thin interbeds of quartzose metasandstone and sulfidic calc-silicate rocks (Smalls Falls Formation). The Smalls Falls Formation is in turn overlain by thin-bedded calc-silicate rocks and thick-bedded calcareous metasandstone and metasiltstone (Madrid Formation). The Perry Mountain and Smalls Falls Formations are considered Silurian(?)

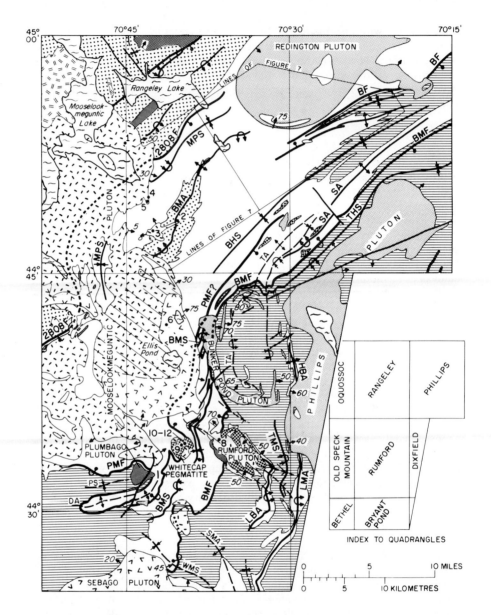

Figure 2. Geologic map showing principal features of early and late deformations and sample sites for isotopic age determinations. Names of structural features: Early faults: BF, Barnjum; BMF, Blueberry Mountain; 2808F, Hill 2808; PMF, Plumbago Mountain. Early folds: BHS, Bear Hill syncline; BMS, Black Mountain syncline; BMA, Brimstone Mountain anticline; HBA, Hutchinson Brook anticline; LMA, Lovejoy Mountain anticline; LBA, Logan Brook anticline; MPS, Mountain Pond syncline; SA, Salem anticline; TMS, Thompson Mountain syncline; THS, Tory Hill syncline. Late folds: DA, Dimmock antiform, PS, Plumbago synform; SMA, Spruce Mountain antiform; TA, Tumbledown antiform; WMS, Wilbur Mountain synform. Geology modified from Green and Guidotti (1968, Fig. 19-2), Milton (1961), Fisher (1962), Guidotti (1965), Pankiwskyj (1964), Moench (1971), and Moench and Hildreth (1974).

in age, on the basis of correlation with fossiliferous rocks exposed along the southeast limb of the Merrimack synclinorium (Osberg and others, 1968; Ludman and Griffin, 1974; Pankiwskyj and others, 1976). The Madrid Formation could be earliest Devonian in age, but it is considered Silurian(?) by Moench (1970, 1971).

Upper Sequence

Overlying the Madrid Formation are dominantly pelitic Lower Devonian(?) rocks that are exposed in most of the southeast part of the area of this report along the approximate axial trace of the Merrimack synclinorium. Terminology of these rocks conforms to that of Moench and Hildreth (1974, 1976). Three principal units (listed in ascending order) are recognized: (1) unit A, rather massive metashale,

EXPLANATION

NEW HAMPSHIRE PLUTONIC SERIES (DEVONIAN)

Pegmatite

Two-mica granite--Lined where inclusions of sphene-bearing granodiorite are abundant

Two-mica granodiorite; small body 3.5 km south of Plumbago pluton is trondhjemite

Sphene-bearing tonalite--Body in northern Old Speck Mountain quadrangle possibly Ordovician

Gabbroic and sparse ultramafic rocks

HIGHLANDCROFT PLUTONIC SERIES (ORDOVICIAN)

Granodiorite and granite

METASEDIMENTARY ROCKS

Upper sequence (Devonian?)

Middle sequence (Silurian? and Silurian)

Lower sequence (Ordovician? and Ordovician)

Contact--Showing dip or dip direction. Dashed where indefinite

Postmetamorphic fault

Premetamorphic early normal fault--Bar and ball on downthrown side; upright and overturned. Dotted where projected across plutons

Crestline of upright and overturned late antiform

Troughline of upright and overturned late synform

Trace of axial plane of late cross fold--Showing trend of bedding on limbs

Crestline of upright and overturned early anticline

Troughline of upright and overturned early syncline

Site and number of sample analyzed for isotopic age determination commonly thin bedded in the upper part; (2) Hildreths Formation, thin but extensive feldspathic, probably volcanisclastic metagraywacke, thin-bedded calc-silicate rocks, and local impure to pure marble; and (3) unit B,, cyclically interbedded metashale and metasandstone. The whole upper sequence is correlated approximately with the Littleton Formation of New Hampshire, which is of Early Devonian age, and the Seboomook Formation of northwestern Maine, which is of Early Devonian age. In detail, however, unit A is correlated with the massive metapelite and thinly bedded schist members of the Carrabassett Formation of Boone (1973, Fig. 8), the Hildreths is correlated with the upper member of the Carrabassett, and unit B is correlated with Boone's restricted definition of the Seboomook.

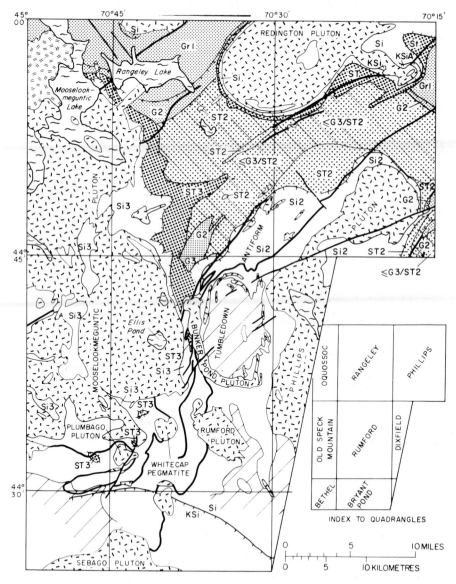

Figure 3. Map showing metamorphic zones. Metamorphic zones from Guidotti (1970b, Figs. 2, 3), Moench (1971), and Moench and Hildreth (1974), with modifications. Premetamorphic and postmetamorphic faults as in Figure 2.

PLUTONIC ROCKS

Most plutons of the area of this report belong to the New Hampshire Plutonic Series of Devonian age (Billings, 1956). On the west is the Mooselookmeguntic pluton (after Mooselookmeguntic Lake; Fig. 2), which extends west to the New Hampshire border (Fig. 1). In the northeast is the Redington pluton (after Redington Pond), and in the east is the Phillips pluton (after the town of Phillips). The Sebago pluton (Hussey, 1968, Fig. 22-2), whose northern tip extends into the reported

EXPLANATION

New Hampshire Plutonic Series (Devonian)

Highlandcroft Plutonic Series (Ordovician)

Metasedimentary rocks (Devonian? to Ordovician)—Shown by patterns for structural features and metamorphic zones

Contact

Postmetamorphic fault

Premetamorphic fault

Area of migmatitic gneiss with disrupted bedding

Metamorphic zones and isograds--Hachures point to areas of higher metamorphic grade

First metamorphism (M-1)

Gr1--Greenschist facies

Second metamorphism (M-2)

G2--Garnet

ST2--Staurolite

Si2--Sillimanite

Third metamorphism (M-3)

G3--Garnet

≤G3/ST2--Garnet or lower grade superposed on M-2 staurolite

Si3--Sillimanite

Undivided

ST--Staurolite plus andalusite of Redington pluton

Si--Sillimanite

KSi--Potassium feldspar plus sillimanite

KSiA--Potassium feldspar plus sillimanite or andalusite

area, is by far the largest intrusive body in western Maine (Fig. 1). Names for smaller bodies of the New Hampshire Plutonic Series (Fig. 2) are applied as needed in this report. Igneous rock names used in this report conform to the classification and nomenclature recommended by the IUGS subcommission on the systematics of igneous rocks (Streckeisen and others, 1973).

Most of the plutons are composite, but they show no conspicuous concentric zoning. The distribution of the principal rock types is shown in Figure 2. Almost invariably the order of emplacement was from mafic to silicic. As described later the sequence of emplacement appears to be tied to the sequence and styles of the varied expressions of the late deformations and metamorphic events. All the granitic rocks are mesozonal two-feldspar types.

Not emphasized in this report are granitic bodies of the Ordovician Highlandcroft Plutonic Series that intrude pre-Silurian rocks in the northwest (Fig. 2; Green and Guidotti, 1968). In addition, many thin Mesozoic hypabyssal dikes are exposed.

Gabbroic and associated dioritic and ultramafic rocks are exposed in several small bodies that appear to be related to the northeast-trending belt of gabbroic rocks in Maine (Fig. 1). Many small bodies of gabbro crop out in a narrow northwest-trending belt along the trend of the Rumford pluton; some are inclusions in granitic rock, others are small intrusive bodies in metasedimentary rocks. All have been variably altered, uralitized, metamorphosed, and deformed. The Plumbago pluton (Fig. 2) is a deformed body of faintly to conspicuously layered metagabbro composed largely of tremolite, calcic plagioclase, and magnesian chlorite (Moody, 1974). The Plumbago pluton and a dike of gabbro to the south are closely associated with the Plumbago Mountain early fault (Fig. 2).

A small body of trondhjemite (light-colored tonalite) is exposed immediately south of the gabbroic Plumbago pluton. A close genetic relation between the trondhjemite and the gabbro is suggested by the occurrence of both on the trace of the early Plumbago Mountain fault.

Sphene-bearing biotite granodiorite, tonalite, and quartz diorite—the Songo Granodiorite of Guidotti (1965)—underlie the northern end of the Sebago pluton (Fisher, 1962; Guidotti, 1965). Reconnaissance by Guidotti (1965, p. 50, 55) indicates that these rocks extend about 30 km south of the northern tip of the Sebago; by far the greater area of the pluton to the south is underlain by lighter colored, more silicic rocks. The sphene-bearing rocks crop out in the southeastern part of the Mooselookmeguntic pluton and in much of the Rumford pluton. The large body shown in the northern part of the Old Speck Mountain quadrangle is the Umbagog Granodiorite of Green and Guidotti (1968, p. 262) and of Milton (1961), tentatively correlated by them with the Highlandcroft Plutonic Series on the basis of lead-alpha age determinations. The Umbagog is here tentatively correlated with the New Hampshire Plutonic Series because it apparently intrudes the Rangeley Formation and is similar to intermediate rocks elsewhere in the Mooselookmeguntic and Sebago plutons; absent in the Umbagog is the cataclastic texture that characterizes granite of the Highlandcroft Series exposed near Mooselookmeguntic Lake. Fine-grained, well-foliated quartz diorite near the edges of the Rumford pluton and along the northwest-trending cross folds is deformed and has a metamorphic schistosity; commonly, thin irregular veinlets and gash fractures in the rock are filled with aplite. In the Mooselookmeguntic pluton northwest of Ellis Pond (Fig. 2), medium-grained granodiorite and tonalite occur as large disoriented blocks in a matrix of two-mica granite.

The dominant rock of the Bunker Pond, Phillips, and Redington plutons, and of smaller bodies nearby is two-mica granodiorite, but the rocks range from two-mica granite to two-mica tonalite. Both porphyritic and equigranular types are present.

These rocks tend to be slightly sheared and have not yielded suitable material for isotopic dating.

The most abundant rock of the Mooselookmeguntic pluton is fine-grained two-mica granite of rather uniform compositon. It also is present in the Rumford pluton and in numerous small bodies in the western part of the area. The rock is fine grained, equigranular, and rarely sheared; commonly, it has a faint flow structure defined by subparallel tabular crystals of microcline.

The youngest rocks that are certainly related to the New Hampshire Plutonic Series are innumerable small bodies of pegmatite and pegmatitic granite. The Whitecap pegmatite (Fig. 2) is a subhorizontal sheetlike body of pegmatitic granite. This body is approximately central to a northeast-trending belt of rare-mineral pegmatite dikes that extends across the Plumbago pluton and the northern end of the Rumford pluton. The rare-mineral pegmatites of Newry mines are extensive thin flat-lying sheets that cut the gabbroic Plumbago pluton.

SHAPES OF PLUTONS

Field mapping is rarely adequate to define the three-dimensional shapes of plutons. In this section a qualitative attempt is made to relate the published gravity maps and interpretations of Kane and Bromery (1966, 1968) and Kane and others (1972) to evidence obtained from geologic mapping on the dips of contacts of the principal plutons. The available gravity maps are not terrane-corrected. Because topographic relief of as much as 1,100 m exists in the reported area, and relief of as much as 380 m exists vertically between gravity stations that are critical to our study, no quantitative analysis of the gravity data was attempted. Thus the subhorizontal granitic sheet model proposed for the Mooselookmeguntic and Sebago plutons is necessarily speculative. The tentative conclusions drawn in this section are no more than questions for future geophysical studies.

Mooselookmeguntic Pluton

The outer contacts of the Mooselookmeguntic pluton apparently dip outward in almost all directions. A probable exception is the east-trending length of contact northwest of the Whitecap pegmatite (Fig. 2). In the Rangeley quadrangle and in the northernmost part of the Rumford quadrangle, primary foliation, measured dips of contacts, topographic relationships, and drilling cited by Guidotti (1970b, p. B-2, B-25) all indicate a gentle eastward dip of the eastern contact (Moench, 1971; Moench and Hildreth, 1974). The small tadpole-shaped granitic body (site of samples 2 and 3) east of the main mass is a window of the Mooselookmeguntic pluton exposed beneath a cover of metamorphic rocks; the eastward dip here is very shallow (Fig. 2). In the Rumford quadrangle farther south, the eastward dip steepens, but the southernmost contact (near the Plumbago pluton) appears to dip fairly gently south. In the Old Speck Mountain quadrangle west of the area of Figure 2, topography is strongly controlled by the irregular, but rather gently dipping, contacts of the Mooselookmeguntic pluton (Milton, 1961, geologic map). There, with some exceptions, granitic rocks are exposed in the valleys, and more resistant metamorphic rocks form the mountains. The central part of the Mooselookmeguntic pluton in that area appears to be flat-topped; the southern contact dips south, and the northwestern contact dips northwest. Farther north, the western contact of the Mooselookmeguntic apparently dips west (Green, 1964, Pl. 5, secs. A-A' to C-C').

The gently eastward dipping eastern contact of the Mooselookmeguntic pluton may join the upper contact of a subhorizontal granitic sheet inferred by Kane and Bromery (1968, p. 422) to extend outward at depth from the area of the Lexington pluton, a large body exposed a few kilometres east of the area of this report (Fig. 1). These relationships are illustrated in Figures 7 and 9. According to Kane and Bromery (1968, p. 422), the Lexington pluton consists of two parts: (1) an exposed cupolalike mass wider than it is thick and (2) a subhorizontal sheetlike mass whose upper contact is at an average depth from about 1.6 to 3.2 km. The calculated average thickness of the sheet is about 1.6 km, based on an assumed density contrast of 0.2 g/cm^3. A density contrast of 0.17 g/cm^3, based on density determinations in the reported area, is probably more appropriate and would yield a thickness of about 2 km. The horizontal dimensions, defined by the gradient around the primary low, would be nearly 50 × 80 km, elongate northeast and centering in the Lexington pluton. Mapping by Boone (1973) suggests that the east contact of the Lexington pluton dips steeply west—not east, as Kane and Bromery (1968) inferred solely from gravity-field relationships.

A northwest-trending ridge of low amplitude in the gravity field extends through the Phillips and Rangeley quadrangles approximately where shown in Figure 7, and it must be explained. Although defined by only three widely spaced gravity stations (Kane and Bromery, 1966), at least part of the ridge would survive terrane correction and the establishment of more gravity stations. Qualitatively, the available data permit the interpretation shown in Figures 7 and 9 that the metamorphic rocks thicken over relatively less dense granitic rocks. Two alternatives are shown.

Sebago Pluton

Mapping by Guidotti (1965) and Fisher (1962) indicates that the northern part of the Sebago pluton dips west, north, and east (Fig. 2). Both authors interpreted this part of the Sebago pluton as a broadly arched feature that dips beneath a cover of metamorphic rocks.

Although the Sebago pluton is by far the largest granitic body in southwestern Maine (Fig. 1), it has no obvious expression in the gravity field. Southwest of line G-H in Maine (Fig. 1), the gravity field dips more or less uniformly westward across the Sebago pluton and into a large gravity low in north-central New Hampshire (Kane and Bromery, 1968). Kane and others (1972, p. 2) suggested that this ramplike character of the field can be explained by one or more of the following: (1) a lower density contrast between granitic and metamorphic rocks; (2) granitic rock present at shallow depth beneath a cover of metamorphic rocks; and (3) a deep-seated source for the regional field—the lack of local anomalies would indicate that the exposed granitic plutons are thin. Because at least the northern part of the Sebago dips outward beneath a cover of metamorphic rocks, we prefer the second alternative. We view the Sebago pluton as a thin subhorizontal sheet that extends eastward at shallow depth at least to the sillimanite isograd. Such a sheet might extend farther east at greater depth and might help to explain the gradient along line C-D (Fig. 1). According to Kane and Bromery (1968, p. 419), the maximum depth to the top of the source of this gradient is about 5 km, and the minimum vertical extent is about 2.6 km. The cause of the gradient is unknown, but the likeliest source at that shallow depth would be the southeast margin of a granitic sheet. Such a sheet—possibly wedge-shaped to explain the absence of a gradient on the northwest—could extend to the northwest beneath the level of Kane and Bromery's (1968) inferred sheetlike portion of the Lexington pluton.

A low northeast-trending ridge in the gravity field exists between the gradient along line C-D and the Lexington pluton (Fig. 1; Kane and Bromery, 1968). This ridge, interestingly, exactly coincides with a sharp, southwest-pointed V of the sillimanite isograd (Kane and others, 1972). If, as proposed in this paper, widespread sillimanite-zone metamorphism is related to subhorizontal granitic sheets, the isogradic V and the gravity ridge could express a thickening of metamorphic cover approximately along the line of overlap of the two postulated sheets.

Density contrasts in the area of the Sebago pluton are subdued by the presence of abundant pegmatite and migmatite in the metamorphic country rocks and by abundant tonalite and related rocks (approximate $d = 2.73 \text{g}/\text{cm}^3$) in the northern part of the pluton. According to Guidotti (1965, p. 55), however, the main southern portion of the Sebago is composed largely of silicic rocks, which are similar to the granite of the reported area (approximate $d = 2.64 \text{g}/\text{cm}^3$). A density contrast of at least $0.1 \text{g}/\text{cm}^3$ between the main portion of the Sebago and the country rocks is reasonable.

Smaller Plutons

The many small plutons exhibit a great variety of shapes, apparently controlled by the local structural environments in which they were emplaced. The Rumford pluton is a complex northwest-trending body that approximately conforms to the late cross folds. This body has nearly all the recognized varieties of rocks of the New Hampshire Plutonic Series; the rocks were emplaced in a definite sequence apparently tied to the history of the cross folds. The Bunker Pond pluton is the largest body of a semiconcentric, steeply dipping sheet complex apparently related to the most intensely deformed part of the Tumbledown antiform. The Redington pluton intrudes a lower grade metamorphic area. It is funnel shaped in the southwest, but it has an irregular northeast dip in the northeast. Like parts of the Lexington pluton to the east (Boone, 1973), but unlike other plutons of the reported area, the Redington pluton is strongly porphyritic and probably represents the roof phase of a cupola in a relatively high-level environment. The Phillips pluton is a roughly conformable body; its blunt northeast end apparently dips steeply northeast. Both the Redington and Phillips may be cupolas above the margins of the inferred subhorizontal sheet related to the Lexington pluton.

The common northeast dips of the contacts of plutons may express a control by the widespread pattern of late slip cleavage that dips northeast and by the axial planes of the late cross folds, described in the following sections.

EARLY DEFORMATION

The fundamental structure of the Merrimack synclinorium in western Maine is defined by northeast-trending tight upright folds with axial-plane foliation and by extensive longitudinal premetamorphic faults. The premetamorphic faults were first mapped in detail and described by Moench (1970, 1973), but Wolfe (1956) was first to recognize them as important features. The folds, faults, and foliation characterize the early deformation in the synclinorium in areas of staurolite and lower grade metamorphism and are longitudinal to the synclinorium as a whole. Although the early fault-fold pattern is extremely redeformed southwest of the principal sillimanite isograd (Figs. 2, 3), it has been mapped more than 20 km into the sillimanite terrane. The principal early faults and folds are named in Figure 2. Most are described and illustrated elsewhere and have been mapped in detail

(Moench, 1970, 1971, 1973; Moench and Hildreth, 1974, 1976); new information is summarized here.

In previous papers by Moench (1966, 1970), bedding was labeled S_1, early cleavage S_2, and all later cleavages S_3, contrary to current practice in New England. These features might be relabeled S_0, S_1, and S_2, respectively, to correlate with bedding, early folds (F_1), and late folds (F_2). For simplicity in this paper, however, features of early deformation are called early foliation (slaty cleavage and schistosity), early folds, and early faults. All features of subsequent deformation are called late foliation (kink bands, slip cleavage, fracture cleavage, and schistosity) or late folds, even though two or even more generations of late folds and foliations can be identified in a few outcrops. Because different generations of late features cannot be correlated from one setting to another, it would be misleading to label them S_2, S_3, and so on.

The conformable contact between the lower and middle sequences exposed at the southeast corner of Rangeley Lake (Fig. 2) is truncated farther southwest by the Hill 2808 fault, which has been interpreted as a major normal fault downthrown on the southeast (Moench, 1970). The fault is truncated by the Mooselookmeguntic pluton, but it is interpreted as reappearing in the Old Speck Mountain quadrangle along the boundary between the lower and middle sequences. That boundary was interpreted as an unconformity by Milton (1961), but it is here interpreted as a major premetamorphic fault on the basis of missing stratigraphic units and outcrop characteristics. Missing along the boundary are the Quimby and Greenvale Cove Formations. The only known outcrop of the boundary was found by Milton (1961, p. 68) a short distance southwest of the map area and was visited recently by Moench and Thomas Vehrs. Here, polymictic Rangeley conglomerate (unit A) of the middle sequence on the southeast is in contact with thinly bedded dark schist of the Partridge Formation on the northwest. The actual contact could easily be called an unconformity, except that graded beds found in the conglomerate within 1 m of the contact become finer grained northwest toward the older rocks. Most clasts in the conglomerate are flattened and extremely stretched vertically in the foliation plane, which crosses the contact at a low angle.

The Plumbago Mountain fault (Moench, 1973) marks the boundary between the middle and upper sequences in the southwest and is defined by locally abrupt truncations of formations on both sides of the fault. The fault is clearly identified only south of the Mooselookmeguntic pluton. It probably extends, however, from the Rangeley quadrangle southwest at least to the New Hampshire border. In the Old Speck Mountain quadrangle, the fault is inferred to separate a conglomerate-bearing sequence on the north (Rangeley Formation of the middle sequence) from a more pelitic sequence (upper sequence) having some calcareous rocks (Madrid Formation) on the south. The extremely sinuous map patterns of the Plumbago Mountain fault and the other early faults express younger deformation: measured dips of the faults are steep, and the map patterns show little or no relation to topographic features.

When originally defined (Moench, 1970), the Blueberry Mountain fault was mapped only as far south as the Bunker Pond pluton (Fig. 2). Since then, it has been mapped in detail to the southern border of the Rumford quadrangle (Moench and Hildreth, 1974) and extended by reconnaissance to the northern tip of the Sebago pluton.

The Black Mountain syncline is the principal early synclinal fold between the Plumbago Mountain and Blueberry Mountain early faults. The syncline is truncated by the Mooselookmeguntic pluton and is slightly en echelon to the Bear Hill syncline of the Rangeley and Phillips quadrangles (Fig. 2).

The Logan Brook anticline southeast of the Blueberry Mountain fault is cored by rusty-weathering migmatitic gneiss of the Smalls Falls Formation of the middle sequence and is flanked by pelitic gneiss of the upper sequence. Calcareous metasandstone and calc-silicate rocks of the Madrid Formation, which normally lie between the Smalls Falls and the upper sequence, are well exposed only on the northern nose and on a smaller anticline on the west limb (Moench and Hildreth, 1974). Elsewhere along the contact, only local fragments of the Madrid are preserved. They are set in a matrix of migmatitic pelitic gneiss, and much of the contact is only a boundary between rusty and nonrusty pelitic gneiss. In migmatitic gneiss throughout the Rumford quadrangle, calcareous rocks of both the Madrid and Hildreths Formations are characteristically disrupted and appear to have been more brittle than the enclosing pelitic rocks during migmatization and late deformation. The southern part of the Logan Brook anticline is the Little Zircon Mountain synform of Guidotti (1965, p. 3). Guidotti recognized, however, that the Little Zircon could be an anticline whose southern end is inverted and plunges north. This shape, which Guidotti (1965, p. 69) did not favor, is accepted here because of the stratigraphic identification of the core and flanking rocks of the whole feature. Overturning of the axis probably was produced by late cross deformation.

The Lovejoy Mountain anticline is named here for Lovejoy Mountain in the southwest corner of the Dixfield quadrangle. The core of the anticline is defined by a belt of rusty-weathering pelitic schist and gneiss, mapped by Pankiwskyj (1964) and Guidotti (1965) as the Billings Hill Formation, but here called the Smalls Falls Formation of the middle sequence. This belt is flanked by rocks of the upper sequence. As on the limbs of the Logan Brook anticline, only fragments of the Madrid Formation are preserved on the limbs of the Lovejoy Mountain anticline. This anticline appears equivalent to the Hutchinson Brook anticline farther north, however, where the Madrid is well exposed (Fig. 2).

LATE DEFORMATION

Major features of late deformation are expressed in Figure 2 by obvious distortions of the early fault-fold pattern. In outcrops, early foliation, bedding, and small early folds are deformed by late slip cleavage (microfaults and axial planes of microfolds), crenulation cleavage, or schistosity derived from slip cleavage (White, 1949). With some exceptions, the late foliations are subparallel to axial planes of the major late folds. The trends and styles of some late folds are diverse and appear to be related to individual plutons; others trend rather uniformly northwest. All the late folds, however, differ widely from the general northeast trend and character of the early faults and folds and from the trend of the Merrimack synclinorium as a whole.

The named late folds are called antiforms or synforms because their stratigraphic expressions are not necessarily consistent with their forms. Thus, the Tumbledown antiform has an exposed core of upper sequence rocks southeast of the Blueberry Mountain early fault and a northern nose and western limb of older middle-sequence rocks northwest of the fault.

The northwest-trending cross folds are defined by abrupt deflections of older structural trends. The axes of the cross folds commonly plunge steeply northeast, parallel to intersections between northeast-dipping axial planes of the cross folds and the northeast-trending subvertical early folds, faults, foliation, and bedding. Cross sections of the cross folds are best seen in plan view.

Also mapped but not described here are several younger postmetamorphic faults that are characterized by silicified shear zones and breccia and by slip cleavages in the wall rocks. Some of these faults probably began to form during the final stages of Devonian plutonism, as shown by the sheared granite that is exposed locally along them.

Dimmock Antiform and Plumbago Synform

This pair of folds is shown by the S-shaped fault boundary between the middle and upper sequences south of the Mooselookmeguntic pluton (Fig. 2). Early foliation and bedding are subparallel here; they are warped by the antiform-synform pair and show intense slip cleavage. The Plumbago Mountain fault is deformed by flexural folds that probably plunge west. The fault separates an upper plate of young rocks on the south and west from a lower plate of older rocks on the north and east. Several formations are truncated on both sides of the fault, and there is no correlation of early structure across the fault. In detail, the "heel" of the Plumbago synform (at the gabbroic Plumbago pluton) has a subvertical axis, whereas the axis of the eastern part of the Dimmock antiform is subhorizontal. The overall shape of the superposed antiform-synform pair, however, is necessarily predetermined by the original attitude of the Plumbago Mountain fault, which is not known. The largest early faults are interpreted as normal faults by Moench (1970, 1973), and for reasons discussed by Moench (1970, p. 1490), these faults are thought to have flattened in depth prior to late deformation. The map pattern of the deformed Plumbago Mountain fault is consistent with the interpretation that it dipped south prior to late folding, flattened in depth, and then was folded by the antiform-synform pair. If the original dip was fairly gentle to the south, the map pattern could represent a rather open west-plunging flexural fold that is complex only in detail.

Whatever its shape in detail, the Dimmock antiform–Plumbago synform pair probably formed by crowding between the Mooselookmeguntic and Sebago plutons, the two largest plutons in western Maine (Fig. 1). The style of this synform-antiform pair is unique in this area and is best explained by the unique position of the pair between these plutons. The Black Mountain syncline (Fig. 2) is subparallel to the Dimmock antiform but is an early feature and unrelated to the formation of the antiform.

Tumbledown Antiform

The Tumbledown antiform is a large complex feature about 22 km long and 7 km wide, named after Tumbledown Mountain at the north edge of the Rumford quadrangle. In the Rangeley quadrangle, the antiform and its complementary synform on the southeast plunge northeast and cross the northwest limb of the early Salem anticline at a low angle. Here, subparallel bedding and early schistosity are flexurally warped over the crest of the antiform and are creased by the synform. The sillimanite isograd extends around the nose of the antiform (Figs. 2, 3). Exposed in this area are several bodies of pegmatite and granitic rock, including a small body of two-mica granodiorite (not shown in Fig. 2) exposed in a canyon bottom near the crestline of the antiform (Moench, 1971). These features suggest that a granitic pluton responsible for the arching and metamorphism is present at shallow depth along at least the northern part of the antiform.

To the south near the Rumford-Rangeley quadrangle boundary, the Tumbledown antiform crosses the crestline of the Salem anticline and deforms the Blueberry Mountain fault and the Tory Hill syncline (Fig. 2). In this area rocks grade southward from schist to coarsely crystalline gneiss. Bedding is well preserved in the gneiss near the quadrangle boundary, but early foliation is commonly obliterated. Bedding and remnants of early foliation are broadly arched by the antiform. Extensive pavement crops out on Tumbledown Mountain and provides a spectacular display of intersecting early and late folds and foliations. Late boudinage is conspicuous and suggests that stretching occurred along the crest of the antiform.

About 0.5 to 2 km south of the Blueberry Mountain fault, the antiform passes into a broad area of coarse-grained migmatitic sillimanitic two-mica gneiss in which bedding is thoroughly disrupted (Figs. 2, 3). The structure here is a broad, asymmetrical domelike feature overturned to the west. Near the southern end, the antiform apparently intersects a tight early anticline cored by a narrow belt of middle-sequence rocks (Fig. 2). The margins of the asymmetric domelike feature approximately coincide with the margins of the area of migmatitic gneiss and with the conspicuously concentric pattern of granitic sheets (Fig. 3); the sheets truncate the margins of the migmatitic gneiss at low angles. The sheet complex may be visualized as a tube-shaped pattern of incipient ring dikes that plunges steeply northeast but is pinched on the north side. A narrow belt of cataclastically deformed rocks conforms to the northern part of the sheet complex. Some of the two-mica granodiorite and tonalite exposed there are cataclastically deformed.

As noted by Guidotti (1970b, trip B-2, p. 18), the migmatitic gneiss (Noisy Brook Gneiss of Guidotti, 1970b) may have been metamorphosed to the potassium feldspar plus sillimanite zone and later downgraded to or below the upper sillimanite zone in an environment of little or no stress. Downgrading affected a much larger area than the area of the migmatitic gneiss.

The Tumbledown antiform probably formed over a rising body of felsic magma. The following events are postulated: (1) regional early deformation forming the Salem anticline, the Tory Hill syncline, and the Blueberry Mountain fault, and pervasive steeply dipping, northeast-trending slaty cleavage; (2) heating, weakening, and doming over a rising body of magma; (3) stretching, migmatization, and sillimanite or potassium feldspar plus sillimanite–grade metamorphism over the magma, producing gneissic fabric, boudinage, and disruption of bedding; (4) penetration by magma near the fringes of the most intensely migmatized area, forming incipient ring dikes; (5) slight metamorphic downgrading; and (6) slight resurgence of doming, producing cataclasis in a narrow arcuate belt along the northern sheets.

Northwest-trending Cross Folds, Recumbent Folds, and Late Foliations

Guidotti (1965), Fig. 3) mapped several northwest-trending folds in a belt about 10 to 20 km wide immediately east of the Sebago pluton. The folds, some of which are shown in Figure 2, conform to the northeastern border of the pluton, and many are overturned to the southwest toward the pluton. These folds, according to Guidotti (1965, p. 82), are temporally related to emplacement of the Sebago pluton and are distinctly younger than the northeast-trending folds that he mapped farther east in his area. Evidence that supports Guidotti's interpretation (but not discussed by him) is that subparallel bedding and foliation (probably early) are deformed by the northwest-trending folds and help to define the folds (Guidotti, 1965, Pl. 3). In this report, Guidotti's northeast-trending folds are termed longitudinal early folds, whereas his northwest-trending folds are called late cross folds.

The style of flexural cross folding in the south-central part of the Rumford quadrangle is shown in Figure 4, which was traced from the geologic map of the Rumford quadrangle (Moench and Hildreth, 1974). Although the cross fold here is nearly 5 km across, it resembles kink bands seen in outcrops or microfolds

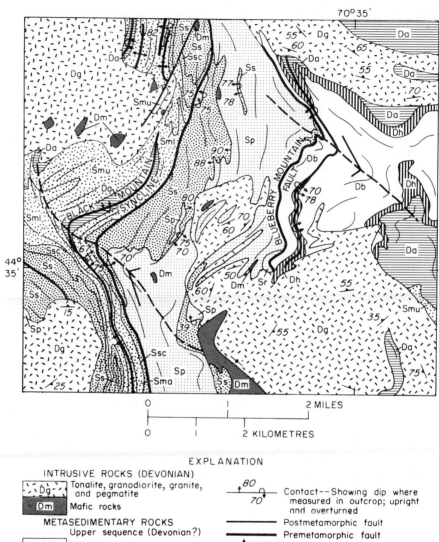
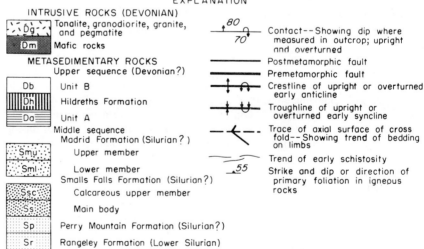

Figure 4. Map showing northwest-trending flexural cross folds between the Rumford and Mooselookmeguntic plutons. From Moench and Hildreth (1974).

seen in thin sections. Axes of the cross fold plunge steeply northeast. The fold sharply deforms the Black Mountain syncline, the Blueberry Mountain fault, early foliation, and several lesser features of the early deformation.

A complex pattern of late foliations is associated with the cross folds near the Rumford pluton (Fig. 5A). The two maximums in the southwest quadrant represent late foliations (dominantly slip cleavage) that strike northwest and dip northeast and east-northeast; they are slightly nonparallel to the axial planes of the mapped plan folds. These late foliations are probably conjugate sets related to flexure that produced the larger cross folds, although the few observed apparent offsets are not entirely consistent with this interpretation. A third set, shown in the northwest quadrant of Figure 5A, strikes northeast and is nearly vertical. This set is most conspicuous between the two principal axial traces shown on Figure 4; apparent offsets relate to flexural slip that accompanied formation of the mapped cross fold.

As shown in Figures 2 and 4, the Rumford pluton conforms to the northwest trend of mapped cross folds that are most conspicuous near the pluton. Sphene-bearing biotite quartz diorite near the fringes of the pluton is deformed and has a metamorphic fabric, whereas the younger two-mica granite and pegmatitic rocks in and around the pluton are not significantly deformed. Cross folding is thus interpreted as having overlapped emplacement of these rocks; except locally, folding did not occur after the most silicic magmas solidified. The distribution of many small bodies of first-emplaced gabbroic rocks—in a northwest-trending belt along the trend of the Rumford pluton (Moench and Hildreth, 1974)—indicates some control by an early expression of the cross fold.

A larger northwest-trending cross fold is shown by a left-lateral deflection of the regional longitudinal grain (Fig. 2). The Blueberry Mountain early fault, for example, is a convenient marker surface. In the northeast, the fault is straight and conforms to the trends of early foliation and folds. To the southwest, the fault is bent around the north end of the Tumbledown antiform; thence, it trends sinuously south at least to the Sebago pluton. Early foliation and recognized early folds conform to these deflections. Assuming that the Blueberry Mountain fault was nearly straight prior to cross folding, about 11 to 13 km of left-lateral offset can be inferred. The Hill 2808 fault and its inferred continuation to the southwest across the Mooselookmeguntic pluton (Fig. 2) show a similar deflection, although it is confined to a narrower northwest-trending belt. A probable southern extension of the Mountain Pond syncline, shown in a large roof pendant in the Oquossoc quadrangle (Fig. 2; C. V. Guidotti, 1964, unpub. mapping), conforms to the deflected projection of the Hill 2808 fault. Assuming that the trace of the Hill 2808 fault was straight originally, about 10 km of left-lateral offset can be inferred; this is in approximate accord with the inferred deflection of the Blueberry Mountain fault by the same cross fold.

In the sillimanite zone near the gently dipping eastern contact of the Mooselookmeguntic pluton are many small but tight and conspicuous recumbent folds having a rather coarse-grained late schistosity (Fig 6A). The axial planes of the folds strike northwest and dip 10° to 35° northeast. The axes necessarily plunge gently northeast parallel to the intersections of the axial planes with the northeast-trending subvertical early foliation. The late schistosity forms the principal maximum in Figure 5C; its center strikes N60°W and dips 25°NE. In the outcrop shown in Figure 6, early foliation is almost obliterated. It is preserved, however, within the lenticular porphyroblasts of staurolite (Fig. 6B). The conspicuous bands of inclusions that are inclined about 40° to the groundmass schistosity are the axial planes of late crinkles that deform early foliation. The history shown here is (1) formation of early foliation, (2) crinkling by late slip cleavage, (3) growth of staurolite, and (4) rotation of staurolite with included crinkles 40° out of alignment with late schistosity in the groundmass. Because the thin section was cut from an overturned limb of a recumbent fold just outside the area of Figure 6A (tops to left according to graded bedding), the counterclockwise rotation is consistent with shearing that produced the fold. A few other porphyroblasts enclose crinkles whose axial planes have been slightly folded into S-shaped patterns, indicating rotation during growth of the porphyroblasts. Although the groundmass schistosity wraps around the porphyroblasts, individual plates of muscovite that define the schistosity are straight and undeformed. This feature is an example of arched polygonal patterns described by Zwart (1962) that he interpreted as continued recrystallization after deformation.

Elsewhere in the outcrop (Fig. 6), early foliation is preserved as flattened clasts in conglomerate that dip steeply, strike northeast, and cross bedding at a wide angle. These brittle fragments and associated beds of quartzite are deformed by

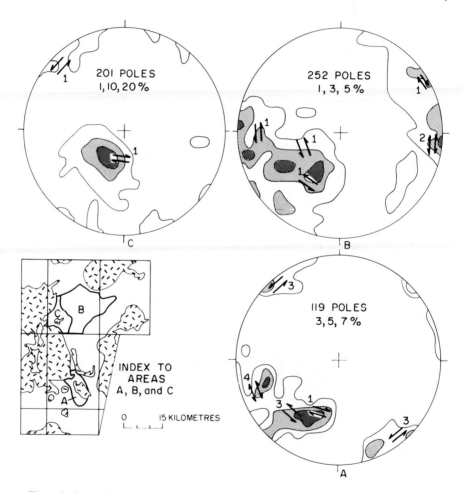

Figure 5. Lower hemisphere equal-area diagrams showing contours on poles of late foliations. Features included are slip cleavage, axial planes of crenulations and small folds, and schistosity derived from slip cleavage. Paired arrows indicate apparent offset parallel to strike—right-lateral, left-lateral, and both motions for two or more determinations. Numbers indicate number of observations. Patterned areas in index are plutons.

Figure 6. Late recumbent folds exposed in roadcut, Maine Rt. 17 about 2 km southeast of Mooselookmeguntic Lake. A. Outcrop, showing bedding (tops to left) and late schistosity. B. M-3 staurolite from overturned limb showing rotated axial planes of included crinkles.

subhorizontal fracture cleavage that is continuous with the late schistosity of Figure 6.

All the features shown in Figure 6 demonstrate a close tie between late recumbent folding; the formation of late fracture cleavage, slip cleavage, and schistosity; and M-3 metamorphism (Fig. 3).

The late schistosity of Figure 6 is part of the pattern that grades eastward to slip cleavage in a manner identical to that shown by White (1949) in Vermont. Most of this change takes place near the M-3 sillimanite isograd. Farther east, in the area of lower grade M-3 metamorphism (Fig. 3), slip cleavages define a complex pattern (Fig. 5B). The flattest maximum in Figure 5B has the attitude N58°W, 37°NE. The principal girdle in Figure 5B shows that eastward dips of slip cleavage steepen as strikes swing to the north.

The cross folds and associated foliations appear to be related to widespread regional strain of unknown origin that may have affected much of the northern Appalachians. In the sillimanite zone of western Maine, the cross folds are viscous-plastic in character; they formed during late high-grade metamorphism and plutonism. In the lower grade terrane farther east, the cross folds and associated slip cleavages are more brittle in character, and they conform to a pattern of northwest-trending faults. The slip cleavages formed during late lower grade metamorphism. A northwest-trending flexural synform has been mapped in low-grade rocks northwest of Rangeley Lake (Fig. 2); two northwest-trending faults are shown nearby, and another is shown northwest of the Phillips pluton. G. M. Boone (1974, written commun.) mapped several northwest-trending faults in the Pierce Pond quadrangle about 40 km northeast of the area of this report. In central Maine, Osberg (1968, p. 39–55) and Ludman and Griffin (1974, p. 166–167) described rather complex patterns of superposed small cross folds and cleavage bands. In northeastern Maine, Pavlides (1974, p. 71) defined the Houlton oroflex—a broad northwest-trending flexure in the northeast trend of folded rocks and slaty cleavage. According to the cited authors, probably all these features are of late Acadian age. They may be large and small expressions of major curvatures of the northern Appalachians.

The small recumbent folds are local expressions of the regional cross deformation, but their origin in detail may be unique. The restriction of these folds to the sillimanite zone above the gently dipping eastern contact of the Mooselookmeguntic pluton indicates a close genetic tie to the pluton. Moreover, the recumbent folds by definition have axial planes that are significantly flatter than those of other cross folds and late foliations. Tentatively, we suggest that the recumbent folds were produced by vertical gravitational flattening of rock that was heated and greatly weakened immediately above the pluton. Thus, in other regions where subhorizontal granitic sheets, recumbent folds, and schistosity are more widespread, as in New Hampshire (Nielson and others, 1976), the sheets may be indirectly responsible for the origin of the folds and schistosity, rather than vice versa. Nielson and others (1976) inferred that pre-existing foliation controlled emplacement of thin plutonic sheets in New Hampshire, but cause and effect relationships are difficult to prove.

METAMORPHISM RELATED TO DEFORMATION AND PLUTONISM

Three metamorphic events are distinguishable in the reported area. The first (M-1) was the regional slate-producing event that accompanied the early deformation. The second (M-2) produced a widespread higher grade zoning that was later modified retrogressively and progressively by metamorphism of the third event (M-3). These events are distinguished on the basis of the growth of porphyroblasts and their retrogressive and progressive recrystallization relative to the formation of the early and late foliations. Zwart's (1962) criteria for defining polymetamorphic sequences are applied to these events. With minor differences (cited where appropriate), the three events defined here agree with those outlined by Guidotti (1970b, p. B-2, B-13). The metamorphic zoning shown in Figure 3 combines mapping by Moench (1971) and Moench and Hildreth (1974) with petrologic work by Guidotti (1970a, 1970b, 1974) and Evans and Guidotti (1966). This zoning is an attempt to illustrate the complex polymetamorphic history on a map.

First Metamorphism (M-1)

The characteristic fabric of M-1 is the early foliation, or slaty cleavage, which is expressed by almost perfectly parallel fine-grained muscovite and chlorite, and lenticular grains of quartz, feldspar, and opaque minerals (Moench, 1966, Pl. 1; 1970, Fig. 4A). Where unaffected by M-2 or M-3, the characteristic mineralogy of M-1 is that of the lowest range of the greenschist facies. In Figure 3, only the areas labeled Gr1 represent M-1 where it is relatively unaffected by M-2 or M-3. Elsewhere, the M-1 fabric is typically coarser, variably deformed, and overprinted by M-2 or M-3 porphyroblasts.

Second Metamorphism (M-2)

M-2 is most clearly defined in a broad area of largely relict staurolite-zone metamorphism in the central parts of the Rangeley and Phillips quadrangles (Fig. 3). In most of the M-2 staurolite zone (Fig. 3), staurolite porphyroblasts are pseudomorphously replaced by chlorite and muscovite that are related to M-3 garnet or lower grade metamorphism. In the northeast and northwest corners of the area of this report and in the south-central part of the Rangeley quadrangle, the M-2 staurolite zone is bordered by M-2 garnet and lower grade zones, which encroach on the areas affected only by M-1. High-grade metamorphic zoning in the southeast around the Phillips pluton and Tumbledown antiform is tentatively included in M-2 as well. As shown in Figure 3, isograds of M-2 are truncated by the hornfels of the Redington pluton and by the M-3 staurolite and sillimanite isograds that conform to the Mooselookmeguntic pluton. Thus, only remnants of the M-2 zoning are preserved.

The presumed M-2 sillimanite zone wraps around the Phillips pluton and the Tumbledown antiform and may thus be related to the two-mica granodiorite and tonalite exposed there. If so, the wide M-2 staurolite zone may be related to a widespread body of similar rock at rather shallow depth. This speculation agrees with meager evidence, noted previously, that two-mica granodiorite and tonalite crystallized somewhat earlier than two-mica granite of the Mooselookmeguntic pluton, which here is considered responsible for the M-3 metamorphic overprinting.

Fresh euhedral M-2 staurolite is preserved in some areas of M-2 staurolite-zone metamorphism, in small areas in the Rangeley quadrangle, and near the east and northwest sides of the Phillips pluton (Fig. 3). In the Rangeley quadrangle, fresh M-2 staurolite is preserved where slip cleavage is sparse or absent. There, large euhedral porphyroblasts of staurolite and smaller crystals of garnet and biotite sharply cross and overprint the early foliation, which is planar and undeformed. Early foliation is preserved poikiloblastically in staurolite as parallel inclusions of quartz, opaque minerals, and dustlike particles of graphite (Moench, 1970, Fig. 4B).

Near margins of the small areas having fresh M-2 staurolite, slip cleavage is more conspicuous, and the staurolite is partly corroded by M-3 chlorite and muscovite. This habit is common in the lower M-3 staurolite zone of Guidotti (1970b, p. B-2, B-8), where the minerals may be in near equilibrium. In this paper, however, the corroded staurolite is considered a corroded relict of M-2 because the direction of reaction toward a lower grade assemblage is clear and because slip cleavage is more conspicuous than it is where M-2 staurolite is fresh.

Slip cleavage is most conspicuous in the large areas where M-2 staurolite is thoroughly replaced by coarse- or fine-grained muscovite and chlorite. The slip cleavage bends around the altered porphyroblasts; a few pseudomorphs are deformed. Undeformed tabular porphyroblasts of chlorite typically are aligned subparallel to slip cleavage, which is preserved poikiloblastically within the chlorite (Moench, 1966, Pl. 2, Fig. 1; 1970, Fig. 4C).

Thus, M-2 staurolite grew prior to most slip cleaving. During slip cleaving the staurolite was downgraded by low grade M-3, which outlasted the slip cleaving. Guidotti (1970b, p. B-2, B-13) inferred that most slip cleavage is older than M-2 staurolite, but textures and field relationships outlined above suggest otherwise. Minor slip cleavage formed prior to growth of M-2 staurolite, however, because euhedral pseudomorphs after staurolite locally are aligned subparallel to the cleavage. Near the Phillips pluton, sharply euhedral M-2 staurolite has grown over slip cleavage that is unusually conspicuous near the pluton and that is probably slightly older than the more widespread slip cleavage of the Rangeley quadrangle.

In the transition from the M-2 staurolite to M-2 sillimanite zones east and northwest of the Phillips pluton, staurolite porphyroblasts are pseudomorphously prograded to coarse-grained muscovite having fibrous sillimanite. Similar textures are characteristic of the M-3 transition zone and of lower sillimanite zones farther west (Guidotti, 1968). Along the northwest margin of the M-2 sillimanite zone (Fig. 3, near the southern boundary of the Rangeley quadrangle), M-2 staurolite is entirely retrograded; sillimanite-bearing rocks commonly have abundant chlorite, and the fibers of sillimanite show evidence of resorption. Evidence of disequilibrium continues southward in a belt that extends well into the Rumford quadrangle and covers most of the area of the Tumbledown antiform. Guidotti (1970b, p. B-2, B-17, B-18) listed several indications of disequilibrium in this area. He suggested that the migmatitic gneiss (the Noisy Brook Gneiss of Guidotti, 1970b) may have been K-feldspar plus sillimanite–grade gneiss, slightly downgraded to the sillimanite zone. Guidotti suggested that a fourth metamorphic event (M-4) caused these partial retrogressions. According to our interpretation, the retrogression was M-3, which in the Rumford quadrangle caused incomplete downgrading. Later kink bands that deform muscovite and biotite, noted by Guidotti (1970b, p. B-2, B-18), occur in the belt of cataclasis described in the section on the Tumbledown antiform.

Third Metamorphism (M-3)

High-grade metamorphic zoning related to M-3 is concentric with the eastern and southeastern margin of the Mooselookmeguntic pluton (Fig. 3). In the Rangeley quadrangle and in the northern part of the Rumford quadrangle, the M-3 sillimanite zone along the pluton is bordered on the east by the M-3 staurolite zone and farther east by a broad area where the M-2 staurolite zone has been almost pervasively downgraded by M-3 garnet or lower grade metamorphism. The isogradic surfaces in the Rangeley quadrangle have shallow dips, as noted by Guidotti (1974, p.

476), parallel to the east-dipping contact of the Mooselookmeguntic pluton. The M-3 sillimanite isograd is the lower grade boundary of Guidotti's transition zone, where chlorite persists in the assemblage sillimanite + staurolite + biotite + garnet for a short distance upgrade from the first appearance of sillimanite (Guidotti, 1974, Fig. 1). The M-3 staurolite isograd is east of the area of Guidotti's detailed sampling and is shown approximately in Figure 3 as being a short distance east of the boundary between the lower and upper staurolite zones of Guidotti (1970b, Fig. 3; 1974, Fig. 1). In the Rumford and Old Speck Mountain quadrangles, the M-3 staurolite zone is preserved in four small areas near the eastern and southern boundaries of the Mooselookmeguntic pluton. These areas may be erosional remnants of a once-continuous M-3 staurolite zone concentric with the pluton.

The isograd between the lower and upper sillimanite zones (not shown on Fig. 3) apparently is truncated locally by two-mica granite of the Mooselookmeguntic pluton. The most conspicuous truncation occurs in the large roof pendant on the southern border of the Oquossoc quadrangle (Guidotti, 1970a; 1970b, Figs. 2, 3). The higher grade rocks are on the south and are closer to that part of the pluton that contains abundant relatively calcic, sphene-bearing biotite tonalite and granodiorite (Fig. 2). The upper sillimanite zone may thus express higher grade metamorphism near a hotter part of the pluton. Apparent truncations of the same isograd are shown by Guidotti (1970b, Fig. 2) at the borders of the tonalite in the Rumford quadrangle, an area he did not study in detail. Mapping by Moench and Hildreth (1974) suggests that the upper sillimanite zone is in fact a continuous envelope around the southeast corner of the Mooselookmeguntic pluton.

Relationships between M-3 zoning, late foliations, recumbent folds, and the eastern (or upper) contact of the Mooselookmeguntic pluton are shown in Figure 7. The contact of the pluton dips moderately to gently east. Conformably above the pluton in the M-3 sillimanite zone are the recumbent folds with axial-plane late schistosity described previously (Fig. 6). Above the sillimanite zone is the M-3 staurolite zone, and above that is the M-3 garnet (or lower grade) zone, where remnants of M-2 staurolite are preserved as downgraded pseudomorphs.

The tie between retrogression at the top of the block (Fig. 7) and progression closer to the Mooselookmeguntic pluton is demonstrated by Guidotti's petrologic studies, together with the relations of porphyroblasts to slip cleavage and the change of slip cleavage to schistosity toward the pluton. As described under the second metamorphism and illustrated previously (Moench, 1970, Fig. 4C), the characteristic fabric of the downgraded M-2 staurolite zone has slip cleavage that is bent around porphyroblasts of M-2 staurolite, which are pseudomorphously replaced by M-3 chlorite and muscovite. Undeformed tabular crystals of chlorite are aligned subparallel to the slip cleavage, and they poikiloblastically include the slip cleavage. In the M-3 staurolite zone, porphyroblasts of new, euhedral, optically continuous staurolite have grown over slip cleavage, which is preserved poikiloblastically as crinkled trains of small quartz inclusions. Andalusite shows the same texture. Another common texture in the staurolite zone is shown by swarms of euhedral crystals of M-3 staurolite that have grown within larger pseudomorphs of chlorite and muscovite after M-2 staurolite (Guidotti, 1974, p. 476). In the M-3 sillimanite zone, M-3 staurolite is pseudomorphously replaced by coarse-grained muscovite with fibrous sillimanite (Guidotti, 1968).

The described textures show that M-3 recrystallization outlasted the principal events of slip cleaving, recumbent folding, and the formation of schistosity derived from slip cleavage. With exceptions, metamorphism took place in an environment of little or no stress. In the M-3 sillimanite zone, however, recumbent folding accompanied metamorphism, as shown previously (Fig. 6B). In addition, minor local shearing occurred after solidification of granite of the Mooselookmeguntic pluton but before the rocks had cooled significantly. Annealed cataclastic features have been found parallel to flow structures in the granite and in pegmatite dikes that cut late schistosity in the country rocks at a wide angle. The dikes are cut by annealed cataclastic foliation that is continued with late schistosity in the country rock, where the mica is polygonal.

LATE ALTERATION

Minor effects of undated late alteration are almost ubiquitous in the mapped area and, locally, late alteration is intense. Few specimens of granitic and metamorphic rocks are without at least slight chloritization of biotite. Garnet and staurolite may also be replaced by chlorite along rims or fractures. As noted by Guidotti (1970b, p. 17), the late chlorite may be richer in iron than associated chlorite of the equilibrium assemblage. Where biotite is intensely chloritized, the late chlorite is commonly interleaved with potassium feldspar, which may vein and embay plagioclase as well. Calcic plagioclase commonly is saussuritized, particularly in rocks with chloritized biotite.

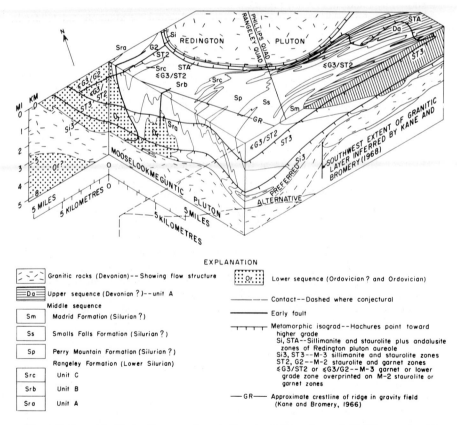

Figure 7. Isometric block diagram of central parts of Rangeley and Phillips quadrangles showing early folds and faults, preferred and alternative interpretations of plutonic structure, metamorphic zoning, and superposed late recumbent folds. Constructed from Moench (1971). Lines of section shown in Figure 2.

ISOTOPIC AGES

Western Maine lies in an area known to have K-Ar and Rb-Sr mineral ages that were affected by a subtle but pervasive late Paleozoic thermal event (Faul and others, 1963; Zartman and others, 1970). On the other hand, in central New Hampshire Rb-Sr whole-rock ages apparently were little affected by this disturbance (Naylor, 1969); thus, the Rb-Sr whole-rock method was adopted for the present study in an attempt to obtain valid primary crystallization ages for rocks of the Mooselookmeguntic pluton and related bodies.

Analyses were made on fresh samples weighing 5 to 10 kg; the samples were generally large compared to the grain size of the rock. Conventional methods of chemical processing and mass spectrometry were employed, and the resulting data are presented in Table 1. Element concentrations are considered accurate to ±1 percent, and the strontium isotopic composition routinely has a reproducibility of ±0.1 percent in the U.S. Geological Survey laboratory in Denver. Collection sites are shown in Figure 2. Samples 1 through 7 were obtained from the Mooselookmeguntic pluton. The granite samples are typical of large parts of the pluton. The aplite is from dikes a metre or less thick that cut the granite. Sample 8, from the Rumford pluton, is from a metre-thick dike that cuts sphene-bearing biotite tonalite and is cut by pegmatite. Sample 9 is a 0.6-m length of drill core of two-mica granite identical to that of most of the Mooselookmeguntic pluton. The sample is from the lower part of a hole that penetrates the Whitecap pegmatite, which overlies and intrudes the granite. Samples 10, 11, and 12 (Table 1) are drill cores from a nearby hole in pegmatite. Sample 10 is a composite from 177 m of core made by removing chips at 1.5-m intervals. Samples 11 and 12 are lengths of core, respectively 0.45 and 0.9 m long, from the indicated depths below the collar (Table 1).

TABLE 1. DATA FOR RUBIDIUM-STRONTIUM AGES OF WHOLE-ROCK SAMPLES FROM THE MOOSELOOKMEGUNTIC PLUTON AND RELATED BODIES

Sample No.	Rock type	Latitude	Longitude	Rb (ppm)	Sr$_n$ (ppm)	^{87}Rb/^{86}Sr	^{87}Sr/^{86}Sr	Age (m.y.)*
1[†]	2-mica granite	44°53'17"	70°45'07"	171.7	221.9	2.24	0.7172	
2[†]	2-mica granite	44°48'17"	70°40'16"	239.2	117.9	5.87	0.7385	
3[†]	Aplite	44°47'39"	70°42'35"	564.9	55.5	29.45	0.8615	
4[§]	2-mica granite	44°49'53"	70°43'02"	350.8	49.1	20.68	0.8134	
5[§]	Muscovite granite	44°49'40"	70°43'08"	285.5	23.1	35.74	0.8946	379 ± 6**
6[§]	Aplite	44°41'46"	70°41'25"	209.6	15.6	38.85	0.9112	
7[†]	2-mica granite	44°34'11"	70°43'10"	198.0	71.0	8.07	0.7500	
8[†]	Aplitic 2-mica granite	44°34'30"	70°36'25"	125.0	143.7	2.52	0.7207	
9 (110 m)[#]	2-mica granite	44°34'15"	70°39'57"	281.1	47.2	17.24	0.7974	
10 (4.6–181.4 m)[#]	Pegmatitic albite-microcline granite	44°34'20"	70°39'43"	253.0	9.4	77.55	1.093	358 ± 6[††]
11 (4.4 m)[#]	Aplite	44°34'20"	70°39'43"	233.7	5.5	123.7	1.333	364 ± 6[††]
12 (88 m)[#]	Pegmatitic albite-microcline granite	44°34'20"	70°39'43"	241.0	10.7	65.45	1.062	389 ± 6[††]

* Decay constant: $\lambda_\beta = 1.39 \times 10^{-11}$ yr^{-1}.

[†] Roadcut or blasted rubble near road.

[§] Natural outcrop.

[#] Drill core made available by Whitecap Mountain Syndicate. Numbers in parentheses indicate depths below collar. Sample 10 is from chips taken at 1.5-m intervals.

** Calculated from least-squares regression method of York (1966). See graphic representation of these data on Figure 8.

[††] Initial ^{87}Sr/^{86}Sr assumed to be 0.706.

The results from determinations made on samples 1 through 9 were pooled together into an isochron (Fig. 8) based on the least-squares regression method of York (1966). The excellent linear array (the slight scatter is within the stated analytical uncertainty) gives no evidence for geologic disturbance of these samples. Likewise, the initial $^{87}Sr/^{86}Sr$ ratio of 0.7065 ± 0.0005 calculated for these intrusive rocks is reasonable, suggesting neither an anomalous source material nor later alteration.

The three samples from the Whitecap pegmatite have very radiogenic strontium (Table 1); these data are used to calculate individual whole-rock ages by assuming an initial $^{87}Sr/^{86}Sr$ ratio of 0.706, as determined from the isochron. The pegmatite samples show significant scatter from the isochron age, but their average value of 370 m.y. is consistent with that obtained on the other intrusive rocks, and we do not doubt a close genetic relationship. Large pegmatites with high Rb/Sr ratios commonly are among the first rock types to reflect subsequent chemical migration from low-grade metamorphism or weathering. It is possible that these latter samples from a large sheetlike pegmatite did respond to a late Paleozoic thermal event, but a late Paleozoic origin for the pegmatite is not permitted by the data.

The isochron age of 379 + 6 m.y. for granite and aplite of the Mooselookmeguntic and two other nearby plutons also dates the M-3 metamorphism and some of the younger features of the late deformation. Using various estimates for the Devonian time scale (Friend and House, 1964; Bottino and Fullagar, 1966), this age is approximately late Early or Middle Devonian. The dated plutonic rocks intrude upper sequence strata, which are correlated approximately but conclusively with the Lower Devonian Littleton and Seboomook Formations. Prior to intrusion these strata were tightly folded, faulted, and cleaved by the early deformation and were weakly metamorphosed by the M-1 event. Thus, all the stratigraphic,

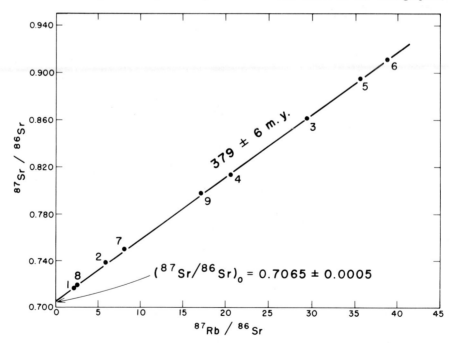

Figure 8. Rubidium-strontium isochron of whole-rock samples from the Mooselookmeguntic pluton and related bodies. Numbered points are samples listed in Table 1; collection sites are shown in Figure 2.

structural, and isotopic evidence restricts the early and late deformations and all three recognized metamorphisms to a rather brief interval of time. Arguments by Naylor (1971) and Boucot and others (1964) for limiting the most intense tectonic and metamorphic phase of the Acadian orogeny to a brief period of time apply equally here. A maximum of 30 m.y. (more probably, no more than 10 to 20 m.y.) intervened between deposition of the Devonian(?) upper sequence rocks and the emplacement of the late tectonic plutons in the area of this report. This restriction does not necessarily imply, however, the lack of younger tectonic and magmatic events that may be included justifiably in the Acadian orogeny elsewhere in the northern Appalachians.

SYNTHESIS

The transition zone at the northern end of the sillimanite terrane in New England corresponds with a tectonic and metamorphic superposition that increases southwestward. The early northeast-trending longitudinal fault-fold pattern is fundamental and continuous; it probably extends the full length of the Merrimack synclinorium but it is greatly deformed and blurred in the area of sillimanitic schist, gneiss, and migmatite. The chronology of major events in the reported area may be listed briefly as follows:

1. A vast thickness of dominantly pelitic deposits, having some coarse clastics and volcanic rocks, accumulated during Ordovician to Early Devonian time. Sedimentation occurred in a geosynclinal trough that approximately coincides with the present-day Merrimack synclinorium (Moench, 1970, p. 1469).

2. Early deformation and first metamorphism (M-1) took place in the lowest range of the greenschist facies, which was interpreted by Moench (1970, 1973) as geosynclinally controlled large-scale slump faulting, folding, and tectonic dewatering. No "giant vise" compression is required, but regardless of origin, early deformation and M-1 were largely complete before the onset of the following events.

3. Next, late deformation and at least two metamorphisms (M-2 and M-3) were accompanied by emplacement of successively younger gabbro and more abundant tonalite, granodiorite, granite, and pegmatite. The granite is dated by the 379 ± 6 m.y. Rb-Sr whole-rock isochron, and it also approximately dates the late structural and metamorphic events. Magma-generated superposed folds of diverse shape and trend were produced by crowding between or doming over rising plutons. Cross folds and associated foliations are temporally related to the plutons but express regional strain of unknown origin. Small recumbent folds in the sillimanite zone above the gently east dipping eastern margin of the Mooselookmeguntic pluton may have formed by vertical gravitational flattening of heated and weakened rock above the pluton.

4. Finally, there occurred a succession of relatively minor events not elaborated upon in this report: brittle postmetamorphic faulting, minor but widespread chloritization and saussuritization, and emplacement of many thin dikes in Mesozoic time. The chloritization and saussuritization may be the only visible effects of a late Paleozoic thermal event that disturbed K-Ar mica ages in a region overlapping the area (Faul and others, 1963; Zartman and others, 1970). Regardless of its cause (Zartman and others, 1970; Hamilton and Myers, 1974, p. 375), the late Paleozoic event did not significantly affect the Rb-Sr whole-rock systems.

What explains the northeast termination of the sillimanite terrane? Two alternatives are noted in the introduction to this paper. Rodgers' (1970) interpretation is supported by petrologic studies that show that deeper levels in the orogen are in fact exposed in the southern segment of the Merrimack synclinorium (Thompson and Norton, 1968). Evidence for superposition in the transition zone, however, indicates that more was involved in the origin of the sillimanite terrane than compression in a deep environment; at least two fundamentally different broad categories of processes are required—events 2 and 3 in the foregoing list. The alternative proposed by Hamilton and Myers (1967) deserves further study. However, tentative evidence against the former existence of a surficial batholith above the sillimanite terrane is exposed in east-central New Hampshire at the X in Figure 1. At that locality, which was visited by Moench, a large down-dropped block of low-grade slate, correlated by Billings (1928, 1956) with the Littleton Formation, is exposed well within the sillimanite terrane. The block of slate is in a cauldron filled with rocks of the Moat Volcanics, the extrusive phase of the White Mountain Plutonic Series (Billings, 1956). The slate is dark gray and slip-cleaved; it is cyclically interbedded with light-gray to white graded beds of impure quartzite. Slaty cleavage crosses bedding at a low angle. Although the slate could be related to Pennsylvanian meta-anthracite–bearing slate and phyllite exposed about 200 km farther south (Grew, 1973), bedding features are identical to those of unit B of the upper sequence in the reported area 50 to 100 km along strike to the northeast. If the downdropped block of slate is of Devonian age, as seems most likely, the central part of the sillimanite terrane was not overlain by a surficial batholith but rather by a layer of low-grade metamorphic rocks. Further evidence against the model by Hamilton and Myers is given by the fact that the inverted metamorphic zones and associated nappes cited by them (Hamilton and Myers, 1974) at the west edge of the sillimanite terrane in New Hampshire are well within the triple-point isobar, based on aluminum silicate minerals (Thompson and Norton, 1968, Fig. 24-1). During metamorphism the overburden in the area of the inverted zones was about 15 km or more. Thus, if the inverted zones were produced by contact metamorphism below a thin, westward-spreading batholith (Hamilton and Myers, 1974, p. 375), that "thin" marginal portion of the batholith was about 15 km thick. Hamilton and Myers (1967, 1974) preferred thicknesses of 5 km or 10 km at most for surficial batholiths, which presumably are much thinner at the margins.

Our interpretation is shown in Figure 9, a speculative longitudinal section that extends across the transition zone at the northeast end of the sillimanite terrane (Fig. 1). It needs testing by further detailed mapping and geophysical studies.

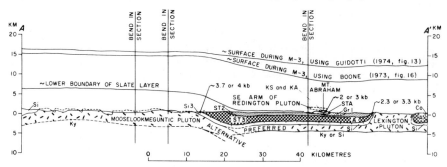

Figure 9. Speculative longitudinal section A-A' across the transition zone at the northeast end of the New England sillimanite terrane. Line of section is shown in Figure 1. Metamorphic symbols: Gr1, dominantly M-1 greenschist facies; ST2, M-2 staurolite in part downgraded by M-3; ST3, M-3 staurolite; Si3, M-3 sillimanite; Ky, probable kyanite; Si, sillimanite; KS and KA, K-feldspar plus sillimanite or andalusite; STA, staurolite and andalusite; A, andalusite; Co, cordierite. Approximate pressure in kilobars at sites indicated.

Qualitatively, Figure 9 is consistent with relationships described in this report and with the gravity and field studies of large intrusive bodies elsewhere in New England (Bothner, 1974; Nielson and others, 1973; Kane and Bromery, 1968).

A thick slate layer having structural characteristics of the greenschist terrane or "slate belt" of central Maine is inferred to have extended westward above the present erosion surface and over the higher grade zoning above the Mooselookmeguntic pluton (Fig. 9). Slate that was probably downdropped from this layer is exposed where shown in Figure 1.

If Figure 9 were continued to the northeast to include the Katahdin pluton (Fig. 1)—recently interpreted as a high-level subvolcanic body (Hon, 1974)—the Katahdin would be shown in the upper part of the slate layer. Although the Katahdin is one of the most extensive bodies of granite in Maine, the associated negative gravity anomaly is rather small (Kane and others, 1972). In contrast to the Lexington pluton, no deep central gravity low is present. From these relationships Kane and Bromery (1968, p. 423) suggested that the Katahdin pluton has a rather shallow depth and apparently does not widen in depth. Thus, the Katahdin may be a remnant of a surficial batholith somewhat analogous to the one postulated by Hamilton and Myers (1967, 1974) to have existed over the sillimanite terrane. However, the Katahdin pluton, and a probably thin low-pressure contact-metamorphic hornfels below it, would overlie a layer of slate perhaps 5 to 10 km thick.

Below the slate layer are higher grade rocks associated with at least one flat-lying granitic sheet, represented in Figure 9 by the broadly arched upper part of the Mooselookmeguntic pluton and by the subhorizontal sheet inferred by Kane and Bromery (1968) from gravity data. The single-sheet model is preferred for simplicity. The cover of metamorphic rocks is shown to thicken over the junction to accommodate qualitatively a northwest-trending ridge in the gravity field. The inferred sheet must be composite, composed of two or more rock types; at no time was the sheet entirely fluid. Extending above the sheet are many cupolas and possibly isolated bodies. High-grade metamorphism occurred around and over the cupolas and above and below the sheets. Probably several metamorphic events occurred—M-2 and M-3 of the area of this report—as the different injections crystallized. Metamorphic zones associated with the flat-lying sheets are thin but widespread because they are flat-lying.

From metamorphic petrologic studies, the inferred sheet was emplaced at depths to the upper contact of 11 to 15 km (Fig. 9), depending on the cited curves for pressure-dependent metamorphic reactions. Guidotti (1974, Fig. 13) and Boone (1973, Fig. 16) have constructed petrogenetic grids for the areas of the Mooselookmeguntic and Lexington plutons, respectively. Boone's diagram yields lower pressures and shallower depths than Guidotti's, as shown in Figure 9; an average overburden density at 2.7 g/cm^3 and no tectonic overpressure are assumed.

The pressure point shown at Mount Abraham (Fig. 9) is in hornfels at the end of the southeast arm of the Redington pluton; metamorphic zones here appear to be locally inverted beneath a lenticular projection of the pluton. Here, two assemblages close to one another are (1) quartz, sillimanite, K-feldspar, plagioclase, biotite with or without muscovite, cordierite, and andalusite and (2) quartz, andalusite, K-feldspar, biotite, cordierite, and about 5 percent muscovite. All the minerals are intergrown and texturally compatible, excepting andalusite in the sillimanite-bearing rocks. Assuming appropriate mineral compositions (not determined), these assemblages plot in the appropriate space for two feldspars, muscovite, and aluminum silicate of Evans and Guidotti's (1966) Figure 9C. The pertinent variable is pressure, which determines the aluminum silicate mineral that would appear during the breakdown of muscovite. Using curves plotted on Guidotti's and Boone's petrogenetic grids, about 3 kb and 2 kb, respectively, can be inferred for Mount Abraham.

In the southwest at A (Fig. 9), a depth of 14 or 15 km is shown for the present surface, in an area that is closer to the fossil isobaric surface defined by Thompson and Norton (1968) in southwestern New England. The northeastward slope of the inferred surfaces implies postmetamorphic tilting but probably not enough to produce visible strain features in the rocks.

Figure 9 may be viewed as a rudimentary longitudinal section through the upper crust immediately following emplacement of the New Hampshire Plutonic Series. The inferred granitic sheet may be one of several in New England (Bothner, 1974; Nielson and others, 1973, 1976) emplaced at depths of 15 km or so. The structural, metamorphic, and geophysical transition zone is here interpreted as occurring where the sheet complex dives monoclinelike to the northeast, carrying with it high-grade metamorphic zoning and complex superposed structural features. Obvious contact metamorphism is related to cupolas and isolated bodies at higher levels above the sheets, but regional staurolite- and sillimanite-zone metamorphism is related to the sheets; it is widespread because the sheets are flat lying. Obviously, a single sheet like the one shown can produce only a thin sillimanite zone (Fig. 9). In New Hampshire the sillimanite zone is at least as thick as the maximum local topographic relief (about 1.5 km). There, however, granitic rocks emplaced at slightly different levels and times are predominantly remnants of flat-lying sheets (Nielson and others, 1973, 1976). Inasmuch as temperatures of 350° to 400° C must have prevailed at a depth of 15 km prior to injection (assuming a normal gradient of 25° C/km), a complex of overlapping sheets injected at that depth or deeper could easily heat considerable thicknesses of rock to sillimanite-zone temperatures; metamorphism related to one granitic body or another would not be distinguishable. A sheet complex injected at that depth would have greater metamorphic expression than a similar complex injected at shallower depth and much greater expression than an extensive thin surficial batholith.

If the proposed sheets are shown to exist by future studies, several problems of origin must be addressed. Chief among them is how and why the magmas spread laterally at depths of 15 km or so below the surface. Assuming the intermediate to silicic magmas originated in the lower crust or mantle (pending petrologic studies), they probably rose largely because of their low density relative to the solid crust. Possibly, the approximate depth of 15 km was a transition zone between rather hot, weak, viscous-plastic lower crust and a higher layer of cooler, stronger, more brittle slate. Discrete bubblelike bodies of magma might have ascended through the lower crust, passing through weak rocks that flowed passively over, around, and under the rising plutons, as postulated by Hamilton and Myers (1967, 1974). But in the transition zone, the magma bodies encountered stronger rocks. Ascent to higher levels was inhibited and became increasingly restricted to slower brittle stoping mechanisms. The magmas thus tended to accumulate at that depth, further heating and weakening the crust there and permitting lateral spreading. In the reported area, we see the results of these processes. The northeast termination of the sillimanite terrane is the present approximate exposure of the former critical boundary between the weak lower crust and strong upper crust.

ACKNOWLEDGMENTS

During the mapping of three quadrangles since 1961 in western Maine, Moench has relied heavily on his colleagues in and around the area, particularly G. M.

Boone, E. L. Boudette, C. V. Guidotti, D. S. Harwood, C. T. Hildreth, and K. A. Pankiwskyj. Guidotti's classic work on metamorphic petrology has been invaluable to this report. This paper has been critically reviewed by J. R. Anderson, G. M. Boone, E. L. Boudette, H. R. Dixon, C. V. Guidotti, W. B. Hamilton, and Kenneth Segerstrom, whose penetrating comments are gratefully acknowledged by both authors. Paul C. Lyons offered many helpful suggestions at several stages of preparation of this manuscript. Gordon P. Eaton gave helpful advice on the interpretation of gravity data in the area. We are grateful also to Louis Moyd, who furnished a private report on drilling in the Whitecap pegmatite, and to Robert W. Bridgman, who made the drill cores available to us for research purposes.

Above all, Moench is grateful to C. Wroe Wolfe for introducing him to a geologic career, and to the joys and enigmas of Maine geology. Wolfe's phenomenal energy and his total devotion to his students and to the science of geology are a constant inspiration.

REFERENCES CITED

Billings, M. P., 1928, The petrology of the North Conway quadrangle in the White Mountains of New Hampshire: Am. Acad. Arts and Sci., v. 63, p. 80.
——1956, Bedrock geology, in The geology of New Hampshire, Pt. 2: New Hampshire Div. Econ. Devel. Mineral Resources Survey, 203 p.
Boone, G. M., 1973, Metamorphic stratigraphy, petrology, and structural geology of the Little Bigelow Mountain map area, western Maine: Maine Geol. Survey Bull. 24, 136 p.
Boone, G. M., Boudette, E. L., and Moench, R. H., 1970, Bedrock geology of the Rangeley Lakes-Dead River basin region, western Maine, in Boone, G. M., ed., The Rangeley Lakes-Dead River basin region, western Maine: New England Intercoll. Geol. Conf. Guidebook, p. 1-24.
Bothner, W. A., 1974, Gravity study of the Exeter pluton, southeastern New Hampshire: Geol. Soc. America Bull., v. 85, p. 51-56.
Bottino, M. L., and Fullagar, P. D., 1966, Whole-rock rubidium-strontium age of the Silurian-Devonian boundary in northeastern North America: Geol. Soc. America Bull., v. 77, p. 1167-1176.
Boucot, A. J., Field, M. T., Fletcher, R., Forbes, W. H., Naylor, R. S., and Pavlides, L., 1964, Reconnaissance bedrock geology of the Presque Isle quadrangle, Maine: Maine Geol. Survey Quad. Map Ser., no. 2, 123 p.
Evans, B. W., and Guidotti, C. V., 1966, The sillimanite-potash feldspar isograd in western Maine, U.S.A.: Contr. Minerology and Petrology, v. 12, p. 25-62.
Faul, H., Stern, T. W., Thomas, H. H., and Elmore, P.L.D., 1963, Ages of intrusion and metamorphism in the northern Appalachians: Am. Jour. Sci., v. 261, p. 1-19.
Fisher, I. S., 1962, Petrology and structure of the Bethel area, Maine: Geol. Soc. America Bull., v. 73, p. 1395-1420.
Friend, P. F., and House, M. R., 1964, The Devonian period, in Harland, W. B., ed., The Phanerozoic time-scale—A symposium dedicated to Professor Arthur Holmes: Geol. Soc. London Quart. Jour. Supp. 120, p. 233-236.
Green, J. C., 1964, Stratigraphy and structure of the Boundary Mountain anticlinorium in the Errol quadrangle, New Hampshire-Maine: Geol. Soc. America Spec. Paper 77, 71 p.
Green, J. C., and Guidotti, C. V., 1968, The Boundary Mountain anticlinorium in northern New Hampshire and northwestern Maine in Zen, E-an, White, W. S., Hadley, J. B., and Thompson, J. B., Jr., eds., Studies of Appalachian geology, northern and maritime: New York, Interscience Pubs., p. 225-266.
Grew, E. S., 1973, Stratigraphy of the Pennsylvanian and pre-Pennsylvanian rocks of the Worcester area, Massachusetts: Am. Jour. Sci., v. 273, p. 113-129.
Guidotti, C. V., 1965, Geology of the Bryant Pond quadrangle, Maine: Maine Geol. Survey Quad. Mapping Ser., no. 3, 116 p.
——1968, Prograde muscovite pseudomorphs after staurolite in the Rangeley-Oquossoc areas, Maine: Am. Mineralogist, v. 53, p. 1368-1376.
——1970a, The mineralogy and petrology of the transition from the lower to upper sillimanite zone in the Oquossoc area, Maine: Jour. Petrology, v. 11, pt. 2, p. 277-336.
——1970b, Metamorphic petrology, mineralogy and polymetamorphism in a portion of N.W. Maine, in Boone, G. M., ed., The Rangeley Lakes-Dead River basin region, western Maine: New England Intercoll. Geol. Conf. Guidebook, p. B-2, 1-29.
——1974, Transition from staurolite to sillimanite zone, Rangeley quadrangle, Maine: Geol. Soc. America Bull., v. 85, p. 475-490.
Hamilton, W. B., and Myers, W. B., 1967, The nature of batholiths: U.S. Geol. Survey Prof. Paper 554-C, 30 p.
——1974, Nature of the Boulder batholith of Montana: Geol. Soc. America Bull., v. 85, p. 365-378.
Harwood, D. S., 1973, Bedrock geology of the Cupsuptic and Arnold Pond quadrangles, west-central Maine: U.S. Geol. Survey Bull. 1346, 90 p.
Harwood, D. S., and Berry, W.B.N., 1967, Fossiliferous lower Paleozoic rocks in the Cupsuptic

quadrangle, west-central Maine, *in* Geological Survey research, 1967: U.S. Geol. Survey Prof. Paper 575-D, p. D16-D23.

Hon, Rudolph, 1974, New information on the geologic setting and petrogenesis of the Katahdin pluton: Geol. Soc. America Abs. with Programs, v. 5, p. 39.

Hussey, A. M., II, 1968, Stratigraphy and structure of southwestern Maine, *in* Zen, E-an, White, W. S., Hadley, J. B., and Thompson, J. B., Jr., eds., Studies of Appalachian geology, northern and maritime: New York, Interscience Pubs., p. 291-301.

Kane, M. F., and Bromery, R. W., 1966, Simple Bouguer gravity map of Maine: U.S. Geol. Survey Geophys. Inv. Map GP-580.

——1968, Gravity anomalies in Maine, *in* Zen, E-an, White, W. S., Hadley, J. B., and Thompson, J. B., Jr., eds., Studies of Appalachian geology, northern and maritime: New York, Interscience Pubs., p. 415-423.

Kane, M. F., Simmons, G. C., Diment, W. H., Fitzpatric, M. M., Joyner, W. B., and Bromery, R. W., 1972, Bouguer gravity and generalized geologic map of New England and adjoining areas: U.S. Geol. Survey Geophys. Inv. Map GP-839, 6 p. text.

Ludman, Allan, and Griffin, J. R., 1974, Stratigraphy and structure of central Maine, *in* Osberg, P. H., ed., Geology of east-central and north-central Maine: New England Intercoll. Geol. Conf. Guidebook, p. 154-179.

Milton, D. J., 1961, Geology of the Old Speck Mountain quadrangle, Maine: U.S. Geol. Survey Open-File Rept., 190 p.

Moench, R. H., 1966, Relation of S_2 schistosity to metamorphosed clastic dikes, Rangeley-Phillips area, Maine: Geol. Soc. America Bull., v. 77, p. 1449-1461.

——1970, Premetamorphic down-to-basin faulting, folding, and tectonic dewatering, Rangeley area, western Maine: Geol. Soc. America Bull., v. 81, p. 1463-1496.

——1971, Geologic map of the Rangeley and Phillips quadrangles, Franklin and Oxford Counties, Maine: U.S. Geol. Survey Misc. Geol. Inv. Map I-605.

——1973, Down-basin fault-fold tectonics in western Maine, with comparisons to the Taconic Klippe, *in* De Jong, K. A., and Scholten, Robert, eds., Gravity and tectonics: New York, Interscience Pubs., p. 327-342.

Moench, R. H., and Boudette, E. L., 1970, Stratigraphy of the northwest limb of the Merrimack synclinorium in the Kennebago Lake, Rangeley and Phillips quadrangles, western Maine, *in* Boone, G. M., ed., The Rangeley Lakes-Dead River basin region, western Maine: New England Intercoll. Geol. Conf. Guidebook, p. A-1, 1-25.

Moench, R. H., and Hildreth, C. T., 1974, Geologic map of the Rumford quadrangle and vicinity, Oxford and Franklin Counties, Maine: U.S. Geol. Survey Open-File Map 74-358.

——1976, Geologic map of the Rumford quadrangle, Oxford and Franklin Counties, Maine: U.S. Geol. Survey Geol. Quad. Map GQ-1272 (in press).

Moody, W. C., Jr., 1974, Origin of the Plumbago Mountain mafic-ultramafic pluton in the Rumford quadrangle, Maine, U.S.A. [M.S. thesis]: Madison, Wisconsin Univ., 90 p.

Naylor, R. S., 1969, Age and origin of the Oliverian domes, central-western New Hampshire: Geol. Soc. America Bull., v. 80, p. 405-428.

——1971, Acadian orogeny—An abrupt and brief event: Science, v. 172, p. 558-560.

Nielson, D. L., Lyons, J. B., and Clark, R. G., 1973, Gravity and structural interpretations of the mode of emplacement of the New Hampshire Plutonic Series: Geol. Soc. America Abs. with Programs, v. 5, p. 750.

Nielson, D. L., Clark, R. G., Lyons, J. B., Englund, E. J., and Borns, D. J., 1976, Gravity models and mode of emplacement of the New Hampshire Plutonic Series, *in* Lyons, P. C., and Brownlow, A. H., eds., Studies in New England geology: Geol. Soc. America Mem. 146, p. 301-318 (this volume).

Osberg, P. H., 1968, Stratigraphy, structural geology, and metamorphism in the Waterville-Vassalboro area, Maine: Maine Geol. Survey Bull. 20, 64 p.

——1971, An equilibrium model for Buchan-type metamorphic rocks, south-central Maine: Am. Mineralogist, v. 56, p. 569-576.

Osberg, P. H., Moench, R. H., and Warner, J., 1968, Stratigraphy of the Merrimack synclinorium in west-central Maine, *in* Zen, E-an, White, W. S., Hadley, J. B., and Thompson, J. B., Jr., eds., Studies of Appalachian geology, northern and maritime: New York, Interscience Pubs., p. 241-253.

Pankiwskyj, K. A., 1964, Geology of the Dixfield quadrangle, Maine [Ph.D. thesis]: Cambridge, Mass. Harvard Univ., 224 p.

Pankiwskyj, K. A., Ludman, Allan, Griffin, J. R., and Berry, W.B.N., 1976, Stratigraphic relationships on the southeast limb of the Merrimack synclinorium in central and west-central Maine, in Lyons, P. C., and Brownlow, A. H., eds., Studies in New England geology: Geol. Soc. America Mem. 146, p. 263-280 (this volume).

Pavlides, Louis, 1974, General bedrock geology of northeastern Maine, in Osberg, P. H., ed., Geology of east-central and north-central Maine: New England Intercoll. Geol. Conf. Guidebook, p. 61-85.

Rodgers, J., 1970, The tectonics of the Appalachians: New York, Interscience Pubs., 271 p.

Streckeisen, A. L., and others, 1973, Plutonic rocks—Classification and nomenclature recommended by the IUGS subcommission on the systematics of igneous rocks: Geotimes, v. 18, no. 10, p. 26-30.

Thompson, J. B., Jr., and Norton, S. A., 1968, Paleozoic regional metamorphism in New England and adjacent areas, in Zen, E-an, White, W. S., Hadley, J. B., and Thompson, J. B., Jr., eds., Studies of Appalachian geology, northern and maritime: New York, Interscience Pubs., p. 219-327.

White, W. S., 1949, Cleavage in east-central Vermont: Am. Geophys. Union Trans., v. 30, p. 587-594.

——1968, Geologic map of the northern and maritime Appalachian region, in Zen, E-an, White, W. S., Hadley, J. B., and Thompson, J. B., Jr., eds., Studies of Appalachian geology, northern and maritime: New York, Interscience Pubs., p. 453.

Wolfe, C. W., 1956, Pleated folding in northwestern Maine [abs.]: Geol. Soc. America Bull., v. 67, p. 1745.

York, D., 1966, Least-squares fitting of a straight line: Canadian Jour. Physics, v. 44, p. 1079-1086.

Zartman, R. E., Hurley, P. M., Krueger, H. W., and Giletti, B. J., 1970, A Permian disturbance of K-Ar radiometric ages in New England—Its occurrence and cause: Geol. Soc. America Bull., v. 81, p. 3359-3374.

Zwart, H. J., 1962, On the determination of polymetamorphic mineral associations, and its application to the Bosost area (central Pyrenees): Geol. Rundschau, v. 52, p. 38-65.

MANUSCRIPT RECEIVED BY THE SOCIETY MARCH 4, 1975
REVISED MANUSCRIPT RECEIVED JUNE 30, 1975
MANUSCRIPT ACCEPTED JULY 8, 1975

Prehnite-Pumpellyite Facies Metamorphism in Central Aroostook County, Maine

Dorothy A. Richter
Department of Earth and Planetary Sciences
Massachusetts Institute of Technology
Cambridge, Massachusetts 02139

AND

David C. Roy
Department of Geology and Geophysics
Boston College
Chestnut Hill, Massachusetts 02167

ABSTRACT

Lower Paleozoic eugeosynclinal sedimentary and volcanic rocks in central Aroostook County, Maine, have been metamorphosed to the prehnite-pumpellyite facies. Systematic variations in the distributions of the metamorphic mineral assemblages permit the mapping of three zones corresponding to increasing grade within the prehnite-pumpellyite facies described by Coombs. The lowest grade zone is characterized by prehnite, analcime, and pumpellyite; the intermediate zone contains similar assemblages but lacks analcime. In the highest grade zone, prehnite, pumpellyite, epidote, and actinolite are present, but actinolite is observed only in volcanic rocks containing pyroxene, whereas epidote is common in both volcanic rocks and graywacke. Chlorite, albite, sphene with or without quartz, white mica, and hematite are ubiquitous in all zones. Metamorphism increases in grade from northwest to southeast, and the zones appear to cut across formational contacts and trends of Acadian folds.

Local variations in μ_{CO_2}/μ_{H_2O} ratios are suggested to account for the neighboring of rocks of similar composition bearing calcium-aluminum silicates and those bearing an assemblage of calcite and chlorite. A regional increase in the μ_{CO_2}/μ_{H_2O} ratio to the east is inferred from the presence of only calcite-chlorite assemblages in the eastern part of the study area. All of the metamorphic assemblages seem to fit Seki's intermediate-pressure type of metamorphism. Although the possibility of Taconian metamorphism cannot be eliminated, most evidence suggests metamorphism during the Acadian orogeny.

INTRODUCTION

Regions of zeolite- and prehnite-pumpellyite-facies metamorphism have received increasing attention in the years since Coombs (1954, 1960) demonstrated that distinct low-grade metamorphic facies can develop in graywacke and intermediate to mafic volcanic rocks in a P-T regime intermediate between diagenesis and that of the greenschist facies. The prehnite-pumpellyite facies was defined by Coombs (1960) in the Triassic graywacke of the Taringatura Hills, New Zealand. Hashimoto (1966) subdivided the facies into a lower grade prehnite-pumpellyite facies and a higher grade pumpellyite-actinolite schist facies. The prehnite-pumpellyite facies has since been described in numerous regions, as summarized by Coombs and others (1970) and Nitsch (1971).

The first report of the prehnite-pumpellyite facies in lower Paleozoic rocks of the northern Appalachian Mountains is that of Coombs and others (1970), in which they described the facies in the Big Machias Lake region (Fig. 1). They found metamorphic mineral assemblages diagnostic of the facies in intermediate to mafic volcanic rocks and graywacke ranging in age from Middle Ordovician to Early Devonian.

Recently, subgreenschist-facies terranes have been described in other parts of the Appalachians. Zen (1974) reported prehnite-pumpellyite-facies mineral assemblages in lower Paleozoic mafic volcanic rocks from southeastern Pennsylvania, eastern New York, western Quebec, and western Newfoundland. Zen also observed the presence of analcime veins in pumpellyite-epidote-bearing rocks from the western Newfoundland locality.

Mossman and Bachinski (1972) described analcime-quartz assemblages in pyroclastic rocks of intermediate composition from the Devonian Dalhousie Group of New Brunswick. They do not report laumontite, prehnite, or other secondary calcium-aluminum silicates. Prehnite-pumpellyite-facies assemblages have also been found on the east coast of Newfoundland in lithic sandstone of the Precambrian Signal Hill Formation (Papezik, 1972).

The area included in this study (Fig. 1) is located in central Aroostook County, Maine. It is a terrain of low relief and generally poor bedrock exposure. The geology of central Aroostook County has been mapped by Boucot and others (1964), Horodyski (1968), Roy (1970), Boone (1970), and Roy and Mencher (1976). The main rock types are shale and slate, generally lithic graywacke and arenite, conglomerate (with abundant volcanic clasts), submarine mafic and intermediate volcanic rocks (including abundant pyroclastic rocks), and bedded chert. The rocks range in age from Middle Ordovician to early Middle Devonian.

Northern Maine has undergone two major periods of deformation: the Taconian and Acadian orogenies. Boucot and others (1964) recognized a third tectonic event in the Presque Isle region—the Salinic disturbance; this disturbance is marked by a disconformity in the Presque Isle and Fish River Lake areas, in which uppermost Silurian and lowermost Devonian strata are absent.

The Taconian orogeny occurred locally in New England at times ranging from Middle Ordovician to Early Silurian (Pavlides and others, 1968). Angular unconformities and distributions of Lower Silurian lithofacies are evidence that the Taconian deformation in central Aroostook County was characterized by locally

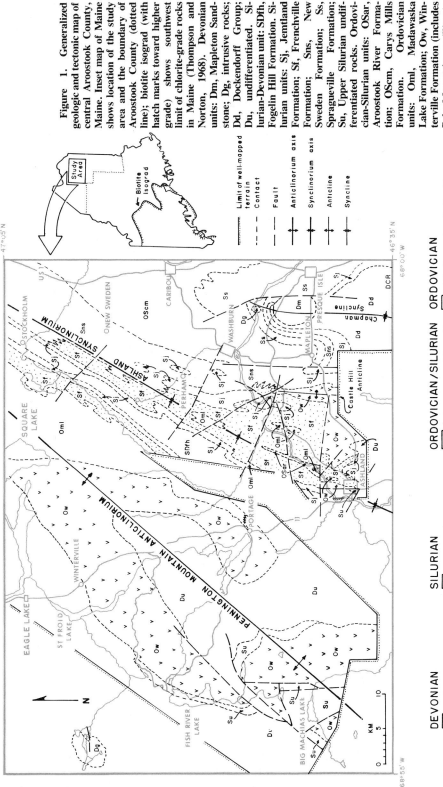

Figure 1. Generalized geologic and tectonic map of central Aroostook County, Maine. Inset map of Maine shows location of the study area and the boundary of Aroostook County (dotted line); biotite isograd (with hatch marks toward higher grade) shows southwest limit of chlorite-grade rocks in Maine (Thompson and Norton, 1968). Devonian units: Dm, Mapleton Sandstone; Dg, intrusive rocks; Dd, Dockendorff Group; Du, undifferentiated. Silurian-Devonian unit: SDfh, Fogelin Hill Formation. Silurian units: Sj, Jemtland Formation; Sf, Frenchville Formation; Sns, New Sweden Formation; Ss, Spragueville Formation; Su, Upper Silurian undifferentiated rocks. Ordovician-Silurian units: OSar, Aroostook River Formation; OScm, Carys Mills Formation. Ordovician units: Oml, Madawaska Lake Formation; Ow, Winterville Formation (includes Pyle Mountain argillite in Castle Hill anticline). Modified from Roy and Mencher (1976).

gentle folding and uplift during Ashgillian and early Llandoverian time (Fig. 2; Roy, 1970) in the western part of the region shown in Figure 1; to the east, however, no evidence of Taconian deformation is present (Pavlides and others, 1964; Roy, 1970). The anticlinoria and synclinorium, together with the smaller folds and slaty cleavage, are assigned to the Early Devonian Acadian orogeny (Pavlides and others, 1964; Boucot and others, 1964). An ancestral geanticline of Taconian origin, possibly an extension of the Somerset geanticline of Cady (1968, 1969), in west-central Maine was roughly coaxial with the present Pennington Mountain anticlinorium (Roy, 1970). This older geanticline probably influenced the later Acadian deformation in the western part of the area, as suggested by Pavlides and others (1964) and Cady (1968, 1969).

The Acadian orogeny is considered to be largely responsible for the patterns of regional metamorphism now observed in New England (Thompson and Norton, 1968). The question of earlier Taconian metamorphism is one of great interest in many parts of New England. In most areas the Acadian orogeny (including late plutonism) was either the only event to produce metamorphism or it was sufficient to overprint effects of previous metamorphism. For parts of Vermont, Albee (1968) and Rosenfeld (1968) suggested the possibility of pre-Silurian high-grade metamorphism (garnet- and kyanite-chloritoid zones) overprinted by Early Devonian (Acadian) metamorphism to the same or nearly the same grade.

Although the distribution of metamorphic grades is well established in more southerly parts of Maine, the scattered nature of detailed mapping in central and especially in northern Maine does not provide sufficient regional continuity to adequately define metamorphic patterns. Central and northern Maine have generally been regarded as a vast region of chlorite-greenschist-grade rocks with "subchlorite" rocks possibly present in the northernmost part of the state (Doyle, 1967). Until Coombs and others (1970) reported the presence of prehnite-pumpellyite-grade rocks in the Big Machias Lake area (Fig. 1), no firm petrographic basis for establishing subgreenschist metamorphism in Maine was available.

This paper presents the results of a petrographic study of rocks from a large region north and east of the area studied by Coombs and others (1970). Included in our study were thin sections examined by Coombs and others (1970); these samples allowed us to incorporate their assemblage information on the map shown in Figure 3. Our purpose was to determine the extent of the prehnite-pumpellyite facies in a region already extensively mapped by Roy and Mencher (1976) and shown to contain rocks of suitable composition to support mineral assemblages indicative of subgreenschist metamorphism. In addition, we wished to determine the distribution of subfacies within the prehnite-pumpellyite facies suggested but not mapped by Coombs and others (1970).

STRUCTURE

The largest tectonic features of central Aroostook County, the Pennington Mountain anticlinorium and the Ashland synclinorium (defined by Roy and Mencher, 1976), are northeast-trending structural elements with numerous smaller anticlines and synclines, such as the Castle Hill anticline and the Chapman syncline (Fig. 1). The Pennington Mountain anticlinorium is the northeast extension of the Munsungun anticlinorium of Hall (1970) and the Bronson Hill–Boundary Mountain anticlinorium farther southwest. The oldest rocks exposed in the core of the Pennington Mountain anticlinorium are of Ordovician age, but complex synclinal infolds of Silurian and Devonian rocks are present. Silurian and Devonian rocks

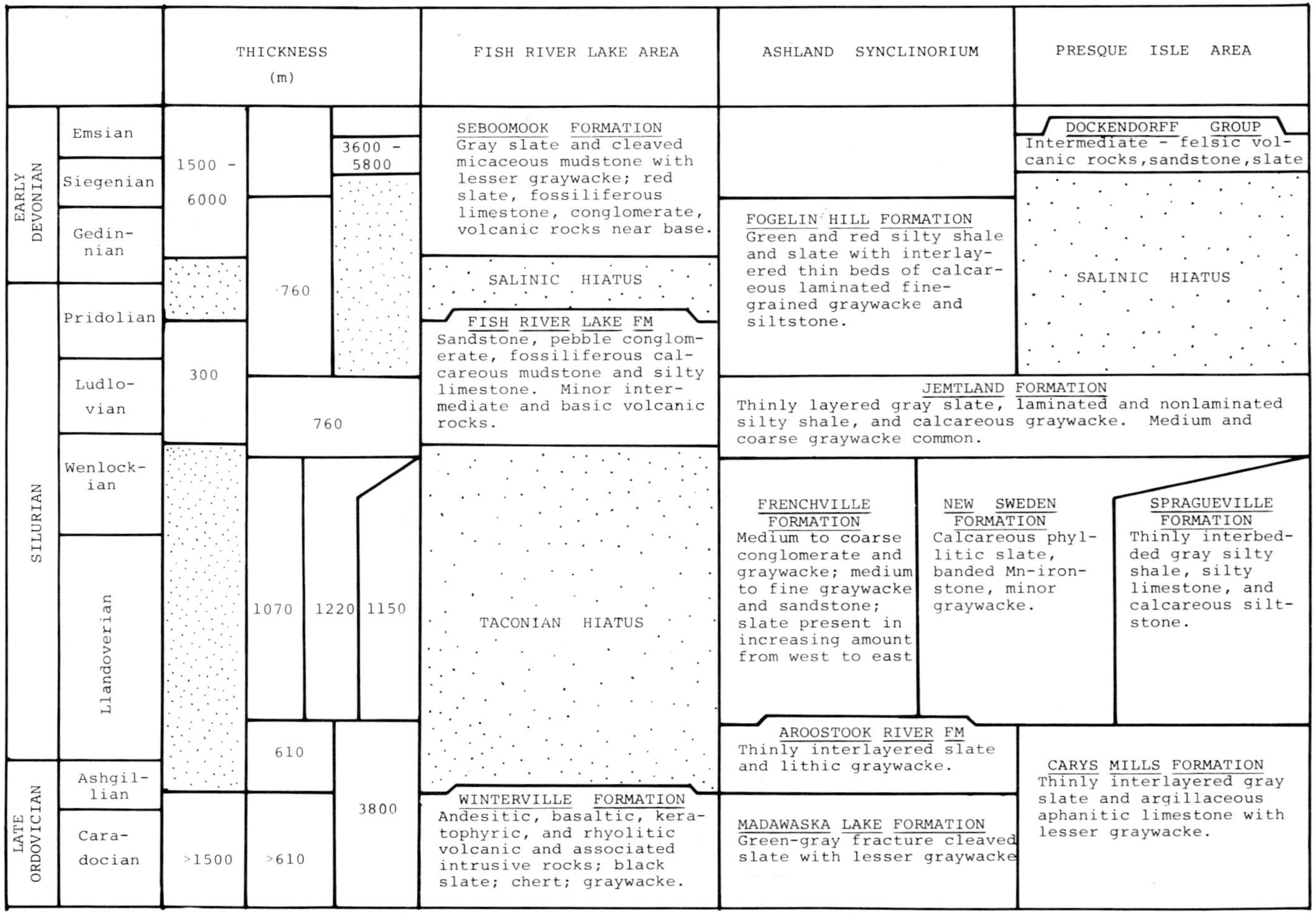

Figure 2. Composite stratigraphic column for central Aroostook County, Maine. Area is divided into three regions representing the western, central, and eastern parts. Stratigraphic nomenclature and lithologic data are from Roy and Mencher (1976) and Boone (1970). The Seboomook and Fish River Lake Formations are not differentiated in the generalized geologic map for the area (Fig. 1).

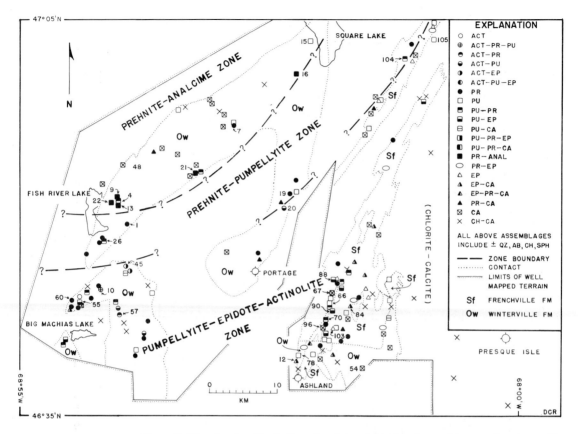

Figure 3. Map showing distribution of metamorphic mineral assemblages and metamorphic zones in central Aroostook County, Maine. Each symbol represents one assemblage. Symbols for more than one sample are combined where samples are crowded. Detailed descriptions of the location of samples are given in Richter (1973). Dotted lines outline Winterville and Frenchville Formations. Numbers refer to samples discussed in text and Tables 1 and 2. Abbreviations: Act = actinolite; Pr = prehnite; Pu = pumpellyite; Ep = epidote group; Ca = calcite; Sph = sphene; Ab = albite; Anal = analcime; Ch = chlorite.

predominate in the Ashland synclinorium, which lies to the east of the Pennington Mountain anticlinorium and to the west of the Aroostook-Matapedia anticlinorium of Pavlides and others (1964). Folds within the anticlinoria and synclinorium are generally tightly appressed, except in the Ashland and Presque Isle areas, where more open folds and more abundant faults are present.

In the region of this study, pelitic rocks of all ages are deformed by a single north to northeast cleavage. Nowhere have earlier or later cleavages been observed. The cleavage is subparallel to the axial planes of the principal folds. In noncalcareous slate, the foliation is a fracture cleavage, and the micaceous minerals largely retain the orientation of their primary bedding plane. Calcareous slate, abundant in the eastern part of the area, displays a foliation in which mica alignment is uniformly parallel to the cleavage; phyllitic appearance in outcrop is common in this slate. Volcanic rocks and sandstone of the Winterville and Frenchville formations show no cleavage in outcrop or thin section. Sandstone in the more slate-rich Aroostook River Formation has a widely spaced (greater than 3 cm) fracture cleavage, but in thin section there is no cleavage fabric.

STRATIGRAPHY

Figure 2 is a composite stratigraphic column for central Aroostook County, Maine. The following descriptions will be limited to those formations in which diagnostic metamorphic mineral assemblages have been found.

Ordovician

Winterville Formation. The Winterville Formation is exposed in the core of the Pennington Mountain anticlinorium and in the anticlines between Ashland and Presque Isle (Fig. 1). The internal stratigraphy of the Winterville Formation has not been determined because of poor exposure and scarcity of structural information. Our field and thin-section observations suggest that most of the exposed igneous rocks of the formation are spilitized pyroclastic rocks. The igneous rocks are generally fine grained and dark greenish gray to dark gray. Modal estimates of the composition of several spilitic rocks (with and without diagnostic metamorphic minerals) of the Winterville Formation are given in Table 1. Spilitic rocks comprise 50 to 70 percent of the exposures of this formation.

Horodyski (1968) and Roy and Mencher (1976) interpreted the volcanism represented by the Winterville Formation as largely submarine on the basis of a few exposures of pillowed lava and the interlaying of the volcanic rocks with apparently deep-water sedimentary rocks. The Winterville Formation is part of a belt of Ordovician volcanic-rich units, which includes the Ammonoosuc Formation in New Hampshire, as described by Naylor (1968). The Winterville rocks are the northernmost-exposed as well as least metamorphosed part of that belt.

Silurian

Aroostook River Formation. The Aroostook River Formation is a relatively thin (600 m) unit that is exposed along a 0.9-km stretch of the Aroostook River north of Ashland (Roy and Mencher, 1976). It is composed of calcareous slate interbedded with fine to coarse lithic graywacke. Many of the lithic clasts in the graywacke beds were derived from the Winterville Formation. The estimated composition of a prehnite-bearing graywacke (sample 70) of the Aroostook River Formation is included in Table 2. The compositions of the rock fragments as shown in Table 2 are as follows: mafic rock fragments—plagioclase, chlorite, ±pyroxene, ±sphene, ±carbonate; felsic rock fragments—quartz, plagioclase, ±chlorite, ±white mica; chert—quartz, ±chlorite, ±white mica; and pelite—chlorite, muscovite, quartz, ±plagioclase, ±calcite. Argillaceous matrix consists of finely divided (<20 μm) and recrystallized chlorite, white mica, quartz, feldspar, ±sphene, ±calcite.

Frenchville Formation. The Frenchville Formation is a sequence of approximately 1,100 m of sandstone, conglomerate, and graywacke that overlies the Winterville and Madawaska Lake Formations with angular unconformity or disconformity and conformably overlies the Aroostook River Formation (Boucot and others, 1964; Roy and Mencher, 1976). The formation has been subdivided into five provisional members (Roy, 1973): the graywacke member, conglomerate member, feldspathic sandstone member, sandstone-slate member, and the quartzose sandstone member. Along the west flank of the Ashland synclinorium the formation is composed entirely of the conglomerate member. Abundant pebbles of volcanic origin in the graywacke and conglomerate members are derived from the Winterville Formation; it is in those members that abundant metamorphic minerals characteristic of the prehnite-pumpellyite facies are found. Modal estimates of the composition of representative

TABLE 1. MODAL ESTIMATES OF THE COMPOSITION OF TYPICAL SAMPLES OF SPILITIC VOLCANIC ROCKS FROM THE WINTERVILLE FORMATION

	Prehnite-analcime zone				Prehnite-pumpellyite zone				Pumpellyite-epidote-actinolite zone			
	4	13	48	7	26	19	54	60	55	57	10
Plagioclase	43.7	35.4	47.4	35.6	32.6	45.8	48.2	45.2	39.6	39.6	47.8
Clinopyroxene	1.3	1.8	tr	11.3	19.0	3.2	16.2	14.6	16.0	13.6	3.8
Chlorite	17.8	11.0	18.6	31.0	12.6	20.6	22.0	16.6	23.4	19.6	5.3
Sphene*	18.7	7.4	13.6	8.4	12.2	11.0	7.0	8.8	14.4	3.8	2.8
Calcite	6.7	32.6	15.8	..	14.4	7.8	0.6	0.6	2.1	..	..
Opaque	tr	tr	0.4	3.1	tr	2.2	2.8	tr	tr	2.2	2.6
Quartz	3.8	2.2	4.2	2.2	5.8	6.0	3.2	2.0	1.0	2.2	1.4
White mica	..	tr	..	2.5	tr	tr	tr	tr	tr	tr	tr
Biotite	..	..	..	1.6	..	..	..	..	..	..	tr
Apatite	..	..	..	1.2	..	..	..	..	..	..	..
Prehnite	5.0	tr	..	3.1	3.4	3.4	..	12.2	..	11.0	1.0
Pumpellyite	3.0	..	..	..	tr	..	..	..	2.5	..	1.4
Analcime	..	9.6	..	..	..	..	..	..	..	..	..
Epidote	..	..	..	..	..	..	..	..	1.0	..	..
Actinolite	..	..	..	..	..	..	..	..	..	8.0	34.0
Total	100.0	100.0	100.0	100.0	100.0	100.0	100.0	100.0	100.0	100.0	100.0

Note: Estimates are based on 500 points. Numbers above columns of data refer to localities plotted in Figure 3: 4, recrystallized lithic tuff with felty matrix and vesicular clasts to 3 mm; 13, fine-grained mafic breccia, fragments average 2 to 3 mm; 48, recrystallized tuff, grain size 0.2 to 2.5 mm, calcite veins and amygdules; 7, ophitic diabase, grain size 1 to 2 mm; 26, subophitic basalt, grain size 0.5 mm; 19, recrystallized tuff, stringy albite xls, amygdules of quartz and chlorite; 54, recrystallized crystal tuff, plagioclase fragments to 2 mm, very fine matrix; 60, glomeroporphyritic andesite, phenocrysts to 4 mm, matrix 0.1 to 0.3 mm; 55, recrystallized tuff, felty matrix, clots of chlorite ± pumpellyite or epidote; 57, ophitic diabase, grain size 2 to 3 mm; 10, recrystallized ophitic diabase, actinolite pseudomorphs after pyroxene.
*Sphene is probably overestimated since it is usually finely mixed with leucoxene, hematite, and other alteration material.

TABLE 2. MODAL ESTIMATES OF THE COMPOSITION OF TYPICAL SANDSTONES FROM THE AROOSTOOK RIVER AND FRENCHVILLE FORMATIONS

	70*	104	96	88	103	67	66	90	78	84
Detrital minerals										
Quartz	24.3	18.6	18.9	34.8	18.5	12.7	22.1	10.0	26.4	16.3
Plagioclase	}6.4	}18.2	36.0	30.1	15.8	}20.7	18.5	15.9	11.2	31.0
K—feldspar			11.3	1.9	5.9		..	4.0	..	..
Mica	2.2	0.8	0.4	..	..	5.4	1.6	..	2.3	0.7
Pyroxene	..	5.0	..	1.2	2.2	..	..	1.0	..	0.4
Rock fragments										
Mafic volcanic	13.8	24.1	..	7.8	23.3	9.6	5.6	37.1	4.4	26.6
Felsic volcanic	6.8	2.0	0.4	2.6	10.1	10.8	9.8	..	7.6	3.6
Chert	4.3	1.0	0.6	8.8	1.8	2.8	6.6	6.1	9.0	2.8
Pelite	9.8	1.0	..	1.4	0.2	..	2.8	0.2	8.8	0.2
Detrital and (or) secondary minerals										
Prehnite	tr	11.5	..	..	..	..	..	..	..	..
Pumpellyite	tr	tr	..	tr	..	..	tr	tr	tr	..
Epidote	..	..	..	tr	tr	..	..	tr	..	tr
Carbonate	15.3	..	15.2	1.2	..	0.2	0.8	1.8	3.2	2.8
Sphene	0.6	5.5	1.2	..	..	tr	..	tr	tr	..
Opaque	1.6	tr	..	0.4	0.4	6.4	2.8	0.2	6.8	7.2
Argillaceous matrix	12.3	11.1	15.8	7.6	20.4	30.4	27.4	20.4	18.3	7.0
Unidentified	2.6	1.2	0.2	2.2	1.4	1.0	2.0	3.3	2.0	1.4
Total	100.0	100.0	100.0	100.0	100.0	100.0	100.0	100.0	100.0	100.0

Note: Estimates are based on 500 points. Numbers above columns of data refer to localities plotted in Figure 3: 70, 104, 103, 90, 84—lithic graywackes, clay-sized matrix, clasts 0.5 to 5.0 mm; 88, 66, 67—lithic sandstones, clasts 0.5 to 1.5 mm; 96, 78—sandstones, clasts 0.5 to 2.0 mm.

*Aroostook River Formation, all others are Frenchville Formation.

samples from the graywacke and conglomerate members of the Frenchville Formation are presented in Table 2.

Jemtland Formation. The Jemtland Formation is a unit of thinly bedded calcareous shale and calcareous graywacke. The formation conformably overlies the older Silurian units (Fig. 2). A marker horizon of devitrified aquagene tuff as thick as 20 m in the middle of the formation in the Stockholm area contains trace amounts of pumpellyite.

METHODS OF STUDY

Samples

Most of the thin sections used in this study were collected and lent by previous workers who had mapped in the area. Most of these thin sections, as well as rocks collected by us, are from exposures along lumbering roads and streams. A total of 138 thin sections from the Winterville Formation, 101 from the Frenchville and Aroostook River Formations, and 9 from tuffs of the otherwise very calcareous rocks of the Jemtland Formation were studied. Thin sections of rare graywacke from the Madawaska Lake, New Sweden, and Spragueville Formations and a few sections of mafic rocks from the Dockendorff Group (Boucot and others, 1964) were also examined.

Petrographic Analyses

Identification of minerals was based on optical properties that could be determined in thin section. Plagioclase compositions in the igneous rocks were determined by the Michel-Lévy method. In a few thin sections, plagioclase compositions were checked on a universal stage. X-ray diffraction patterns were used where possible to corroborate the identification of fine-grained minerals in some samples. Analcime in three sections was analyzed by electron microprobe to determine its Ca/Na ratios.

It is very difficult to define equilibrium mineral assemblages in low-grade metamorphic rocks. Some workers insist that the metamorphic phases be in physical contact and show no signs of replacement. Zen (1974) suggested that the stable-appearing phases present in a 1-mm-diameter field of view (80×) of a thin section may be considered in equilibrium with each other. In general, Zen's criteria for defining equilibrium assemblages was used in this study.

METAMORPHISM

We found mineral assemblages characteristic of the prehnite-pumpellyite facies in 56 of 138 localities of the Winterville Formation, 41 of 101 localities of the Aroostook River and Frenchville Formations, and 5 samples from a 20-m section of aquagene tuff in the Jemtland Formation. The assemblage information is plotted in Figure 3. The rock types in which these assemblages occur include spilitic flows, diabase, tuff, breccia, and minor graywacke in the Winterville Formation and lithic graywacke in the Aroostook River and Frenchville Formations. No mineral assemblages indicative of a specific metamorphic grade were found in the Dockendorff Group or the Madawaska Lake, New Sweden, and Spragueville Formations.

Texture

The primary igneous or sedimentary textures of the assemblage-bearing samples are obvious both in thin section and in hand specimen. None of the samples has a cleavage fabric. The metamorphic minerals occur as veins, amygdules, and partial or complete replacements of original crystals in the crystalline igneous rocks. They occur in a similar manner in both the clasts and matrix of the tuff, breccia, and graywacke. In general, there is no relationship of metamorphic mineral growth to sedimentary fabric, with the exception of bedding-plane orientation of pumpellyite in the tuff of the Jemtland Formation. There is, however, a general increase in grain size and abundance of metamorphic minerals with increasing grade in the igneous rocks of the Winterville Formation. No such variation with grade is observed in graywacke assemblages.

Mineralogy

Albite. Albite is the dominant plagioclase in the igneous and sedimentary rocks of the area. It is not generally possible to determine precise An content by optical methods because of the fine grain size and lack of good twinning.

Albite has the following four distinct modes of occurrence in volcanic rocks: (1) Most commonly, albite occurs in the matrix of partially recrystallized rocks as fine-grained, poorly twinned, clouded, sheaflike aggregates associated with equally fine-grained aggregates of sphene. (2) Albite also occurs as albitized calcic plagioclase. In this habit it is characteristically cloudy in appearance because of abundant fluid inclusions. Such crystals of plagioclase are commonly also partially replaced by calcite and prehnite or, less commonly, by pumpellyite, chlorite, and white mica. (3) In some recrystallized pyroclastic rocks, albite forms long, relatively clear individual crystals, which are usually poorly twinned and typically bent or broken. (4) Coarse albite is present as amygdule fillings and veins in pyroclastic rocks. This albite is characteristically clear and finely twinned and has an An content of less than 5 percent.

In the graywacke and conglomerate, albite occurs in polymineralic and monomineralic clasts. There are no certain criteria for determining whether these clasts were albitized before or after deposition. Albite more rarely occurs as small rims and overgrowths on larger plagioclase clasts; this albite is assumed to be postdepositional.

Chlorite. Chlorite is present in almost every thin section examined. It most commonly forms irregular, fibrous, and subradial mats in the fine-grained matrix of pyroclastic rocks. In some samples, it is an alteration product of augite or biotite. Chlorite is a common inclusion in plagioclase. In the graywacke, chlorite occurs as small masses in the recrystallized argillaceous matrix, as clasts, and more rarely as discrete plates.

At least two distinct types of chlorite have been recognized. The more common chlorite is typically pale green, slightly pleochroic, optically negative, and length slow, and it exhibits anomalously blue (and rarely purple or brown) interference colors. This chlorite is inferred to have a relatively high $Fe/(Fe + Mg)$ ratio according to Albee's (1962) classification and is probably penninite. The other variety of chlorite is yellowish green, has higher relief than the more common chlorite, and has normal interference colors. The two varieties of chlorite have been observed in intimate intergrowths in the same thin sections.

Bishop (1972) also has reported two varieties of chlorite in rocks of the prehnite-pumpellyite facies in the Otago schist of New Zealand. He found that the type of chlorite varies with different rock types and with relative grade within the facies. No such correlations between chlorite types and metamorphic grade or rock type can be made for the occurrences in northern Maine.

Analcime. Analcime, always associated with prehnite, has been observed in thin sections from five localities in the Winterville Formation. Optical identifications were confirmed by x-ray diffraction and electron microprobe. Electron-microprobe analysis shows that the analcime from three localities (19, 13, and 16 of Figure 3) has an average of 55 percent SiO_2, 22 percent Al_2O_3, 13 percent Na_2O, 0.3 percent CaO, and 0.1 percent K_2O. These data indicate that the mineral is nearly pure Na-analcime. Analcime occurs as amygdule fillings, in veins with irregular and gradational margins (Fig. 4A), and as pseudomorphs after plagioclase. In one sample from locality 13, analcime is sufficiently abundant to yield a prominent pattern in whole-rock x-ray diffraction.

Prehnite. Prehnite has been observed in 40 of the igneous rocks of the Winterville Formation. It is commonly a replacement mineral after plagioclase (Fig. 4A) and is present in veins and as vesicle fillings (Figs. 4A, 4B). It is associated with analcime, calcite, quartz, and pumpellyite in veins and also occurs as small inclusions in albitized plagioclase. In the Aroostook River and Frenchville graywackes, prehnite most commonly forms scattered anhedral crystals in the matrix, and in one sample (locality 104) prehnite cements detrital grains (Fig. 5A). Prehnite has not been observed in unequivocal clasts. Prehnite has been identified in 16 percent of the Silurian graywacke samples, in which it is usually a minor constituent. Prehnite is abundant in only two outcrops of graywacke (localities 90 and 104, Fig. 3), which are near the base of the Frenchville Formation.

Pumpellyite. Pumpellyite occurs in both the igneous and sedimentary rocks as brightly pleochroic (α, γ colorless; β = bright green) crystals with anomalous yellowish-brown interference colors. The deep-green color probably reflects a relatively high iron content (Coombs, 1953). In the volcanic rocks of the Winterville Formation, relatively coarse grained pumpellyite, which has been observed in 19 of the localities, forms prisms and subradial aggregates in veins with prehnite or in cavities with epidote and chlorite (Fig. 5B). Finer grained pumpellyite appears as overgrowths on plagioclase crystals that are also partially replaced by calcite and as small crystals in chlorite patches. In many of the devitrified tuffs of the Jemtland Formation, pumpellyite is concentrated in thin mats that follow the bedding planes (Fig. 5C).

Extremely fine grained fibrous pumpellyite occurs as rounded, fine-grained, monomineralic aggregates in graywacke in all the metamorphic zones. It is difficult to distinguish whether the pumpellyite is detrital, a replacement of a clast, or recrystallized from the matrix. Because the aggregates of pumpellyite are delicate and because they occur with other indisputable metamorphic minerals, they are here interpreted to have been formed in place; a part of one of these aggregates is shown in Figure 6C. Pumpellyite also occurs in the matrix of graywacke at locality 90 in fanlike clusters (Fig. 5D). Pumpellyite has been observed in 18 percent of the specimens from the Silurian graywacke and conglomerate.

Actinolite. In the area south of Big Machias Lake, which was studied by Coombs and others (1970), actinolite (commonly with pumpellyite) replaces augite in six samples (Fig. 6A) and forms fine-grained acicular crystals in the matrix of three of the six actinolite-bearing samples. (Typically, very dark green, pleochroic, stubby crystals of actinolite are present as overgrowths of oriented crystals with their long axes perpendicular to the edge of augite crystals; see Fig. 6B.) Actinolite has been identified in only one sample from outside the Big Machias Lake area (locality 20, Fig. 3). Actinolite has not been observed in any of the sedimentary rocks.

Epidote Group. Two members of the epidote group have been observed. The more common variety is pale yellowish-green, slightly pleochroic, highly birefringent, iron-rich epidote. Epidote has been observed in seven of the thin sections from the Winterville Formation as aggregates in the matrix and amygdules of pyroclastic rocks (Fig. 5B). At locality 12 (Fig. 3), near Ashland, epidote lines amygdules containing calcite and (or) chlorite. Epidote also occurs in 25 percent of the graywackes of the Frenchville Formation, where it forms monomineralic polycrystalline aggregates or intergrowths with quartz.

The other epidote-group mineral, clinozoisite, has been observed in a few samples of graywacke from the Frenchville Formation near Ashland. The clinozoisite is colorless and shows anomalously blue interference colors, inclined extinction, and positive sign. It occurs as overgrowths on pumpellyite aggregates (Fig. 6C) and as independent euhedral crystals in the matrix of these sandstones.

Other Metamorphic Minerals. Quartz, white mica, sphene, hematite, and calcite are present in variable amounts in all rock types. Quartz, a relatively minor constituent in most of the igneous rocks, is commonly present in interstices and veins and, rarely, in amygdules. In the graywacke, secondary quartz forms small veinlets and minute overgrowths on detrital quartz grains.

Very fine grained white mica is a ubiquitous alteration product of feldspar. Sphene, usually mixed with leucoxene, occurs as small, brownish nodular aggregates in matrix material of nearly every thin section. Hematite is a common alteration product of mafic minerals; in some samples it is difficult to determine whether the hematite is a weathering product or is of metamorphic origin. Skeletal ilmenite forms geometric intergrowths with sphene in diabase.

Samples containing abundant calcite occur sporadically throughout the area (Fig. 3). Calcite, commonly with deformed twin planes, is present in amygdules and veins (Fig. 4B) as replacements of plagioclase and forms much of the matrix of many rocks.

Relict or Detrital Minerals. Relict calcic plagioclase is difficult to recognize. Phenocrysts of calcic plagioclase are present in flows and shallow intrusive rocks

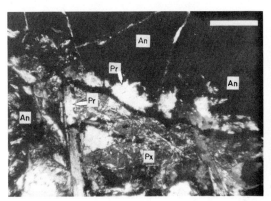 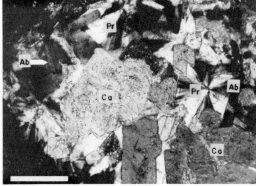

Figure 4. Photomicrographs using crossed polarized light of samples from the prehnite-analcime zone; A, Analcime (An) and prehnite (Pr) in a vein from locality 22, Winterville Formation; prehnite has partially replaced plagioclase laths, and analcime occurs in interstices in the rock; bar is 0.5 mm. B, Prehnite (Pr), calcite (Ca), and albite (Ab) fill a vesicle in a breccia from locality 16, Winterville Formation; bar is 0.2 mm.

of the Winterville Formation. Extinction-angle measurements on rare suitable grains indicate that this calcic plagioclase is labradoritic. Locally, the calcic plagioclase phenocrysts contain blebs of chlorite, white mica, and calcite.

Fresh augite occurs in both the volcanic rocks and graywacke. In the crystalline volcanic rocks, it poikilitically encloses albitized plagioclase; in the recrystallized tuff of the Winterville Formation, augite forms stringy, subvariolitic aggregates. Augite occurs as irregular, subrounded to angular detrital fragments in graywacke and conglomerate.

Potassium feldspar, quartz, muscovite, and biotite are present as detrital minerals in the Aroostook River and Frenchville Formations. The potassium feldspar is usually orthoclase, commonly in micrographic intergrowths with quartz, and is generally partially altered to white mica. The micas are bent and frayed fragments "sandwiched" between other grains.

Metamorphic Zoning

Systematic variation in the areal distribution of the observed assemblages plotted in Figure 3 enables the definition of two distinct mineralogic zones: (1) a prehnite-

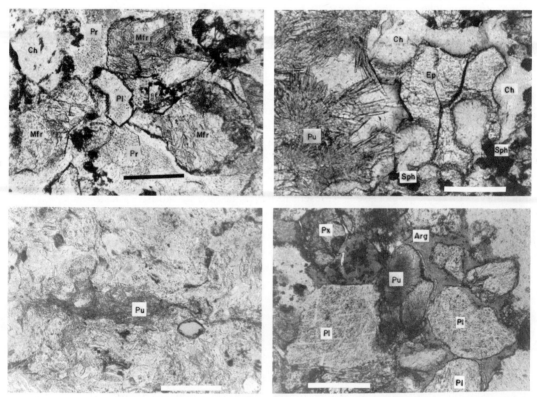

Figure 5. Photomicrographs using plane polarized light of samples from the pumpellyite-epidote-actinolite zone; bar is 0.2 mm long. A, Prehnite (Pr) in the matrix of graywacke from locality 104, Frenchville Formation; Mfr = mafic rock fragments; Pl = plagioclase. B, Pumpellyite (Pu), chlorite (Ch), epidote (Ep), and sphene (Sph) filling a vesicle in a sample from locality 55, Winterville Formation. C, Fine-grained mat of pumpellyite (Pu) in tuff from locality 105, Jemtland Formation; mat is elongated parallel to bedding. D, Sheaflike clusters of pumpellyite (Pu) in the matrix of lithic graywacke from locality 90, Frenchville Formation; Pl = plagioclase; Px = pyroxene; Arg = argillaceous matrix (recrystallized).

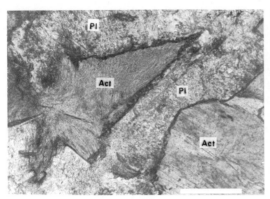
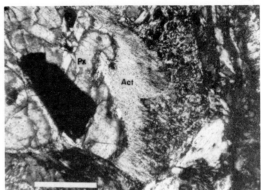
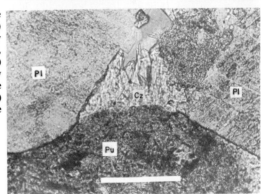

Figure 6. Photomicrographs of samples from the pumpellyite-epidote-actinolite zone. A, Actinolite (Act) pseudomorphs after pyroxene; plagioclase (Pl) is highly altered to white mica; plane polarized light; locality 10, Winterville Formation; bar is 0.2 mm. B, Actinolite (Act) overgrowths on augite (Px); crossed polarized light; locality 57, Winterville Formation; bar is 0.2 mm. C, Clinozoisite (Cz) overgrowths on a very fine grained pumpellyite (Pu) aggregate; plane polarized light; locality 88, Frenchville Formation; bar is 0.1 mm.

analcime zone in the northwest and (2) a pumpellyite-epidote-actinolite zone in the southeast. An intermediate prehnite-pumpellyite zone is possibly present between these two zones. The intermediate zone has prehnite-pumpellyite assemblages but lacks analcime, epidote, and actinolite, the index minerals of the prehnite-analcime and pumpellyite-epidote-actinolite zones. Unfortunately, the number of samples in the inferred prehnite-pumpellyite zone is too few and their distribution insufficiently widespread to establish the extent of the zone with confidence.

Metamorphic grade increases to the east and southeast, and the zone boundaries appear to cross the contacts of formations and the axes of the major folds. The prehnite-analcime zone thus far has been established only in rocks of the Winterville Formation. The pumpellyite-epidote-actinolite zone includes rocks of the Winterville, Aroostook River, Frenchville, and Jemtland Formations. The prehnite-pumpellyite zone as mapped in Figure 3 is recognized only in rocks of the Winterville and Frenchville Formations. The Winterville Formation is therefore the best indicator of change in metamorphic grade, because it contains rocks of similar composition in all three zones.

The metamorphic minerals in each zone are shown in Figure 7. This figure emphasizes that the metamorphic minerals present in each zone are similar and that the zones are essentially defined by three minerals—analcime, epidote, and actinolite.

Prehnite-Analcime Zone. The prehnite-analcime zone occurs in volcanic rocks and graywacke of the Winterville Formation in the northwestern part of the study area (Fig. 3). The zone is defined by the pair analcime-prehnite. In addition to albite, quartz, sphene, and (or) white mica, the following partial assemblages are observed in this zone: analcime-prehnite-chlorite-calcite, analcime-prehnite-chlorite, prehnite-chlorite-calcite, prehnite-chlorite, prehnite-pumpellyite-chlorite-calcite, pumpellyite-chlorite, and chlorite-calcite. Neither epidote or actinolite has been observed in this zone.

Prehnite-Pumpellyite Zone. A prehnite-pumpellyite zone appears to be present in the central part of the study area (Fig. 3). This zone contains metamorphic assemblages similar to those of the prehnite-analcime zone, but it lacks analcime. Its existence is supported by the fact that the area of analcime-bearing rocks is limited to the northwestern part of the chevron-shaped belt of volcanic rocks of the Winterville Formation. Prehnite-pumpellyite assemblages are also observed in graywacke of the Frenchville Formation along the northwestern flank of the Ashland synclinorium. The southern limit of the prehnite-pumpellyite zone is defined by the first appearance of epidote and (or) actinolite. The observed assemblages are as follows: prehnite-chlorite, prehnite-chlorite-calcite, prehnite-chlorite-pumpellyite-calcite, prehnite-chlorite-pumpellyite, pumpellyite-chlorite, and chlorite-calcite, together with albite, quartz, white mica, hematite, and (or) sphene.

Pumpellyite-Epidote-Actinolite Zone. This zone is defined by the appearance of epidote-pumpellyite and actinolite-pumpellyite pairs. Actinolite has been observed only in the volcanic rocks of the Winterville Formation in the southwestern part of the area and not in any of the graywacke units. Epidote, on the other hand, is abundant in both the graywacke and volcanic rocks of the area. Epidote is therefore the best index mineral of the zone. The following are the observed assemblages of this zone: prehnite-pumpellyite-chlorite, prehnite-chlorite, prehnite-chlorite-clinozoisite, prehnite-calcite, pumpellyite-chlorite, pumpellyite-chlorite-clinozoisite, pumpellyite-calcite, epidote-pumpellyite-chlorite, epidote-prehnite-chlorite, epidote-chlorite, epidote-calcite, epidote-prehnite-calcite, actinolite-epidote-chlorite, actinolite-pumpellyite-chlorite, actinolite-chlorite, actinolite-prehnite, and actinolite-prehnite-pumpellyite, along with variable amounts of albite, quartz, sphene, and white mica.

DISCUSSION AND CONCLUSIONS

Metamorphic Facies

Coombs (1960, p. 339) defined the prehnite-pumpellyite metagraywacke facies as including "those assemblages produced under physical conditions in which the following are commonly formed: quartz-prehnite-chlorite or quartz-albite-pumpellyite-chlorite, without zeolite and without the characteristic minerals of the glaucophane schist facies, jadeite, or lawsonite." Prehnite and pumpellyite appear at the expense of Ca-zeolites such as laumontite, the index mineral of the zeolite facies; albite replaces analcime and quartz. Coombs (1960) further noted that two subfacies or zones may be distinguished within the prehnite-pumpellyite facies: a quartz-prehnite zone and an unnamed higher grade zone in which prehnite-pumpellyite assemblages might include some sphene, epidote, stilpnomelane, and actinolite.

As more areas of prehnite-pumpellyite facies metamorphism were reported, it became clear that the combinations of minerals in the metamorphic assemblages varied in each metamorphic terrane. As stated in Coombs's (1960) definition, zeolites do not occur in rocks of the prehnite-pumpellyite facies in New Zealand. Seki and others (1969) found zeolites and analcime to be stable in the lowest part of the prehnite-pumpellyite facies in the Tanzawa Mountains of Japan. Therefore, the analcime-prehnite zone of central Aroostook County may still be considered part of the prehnite-pumpellyite facies.

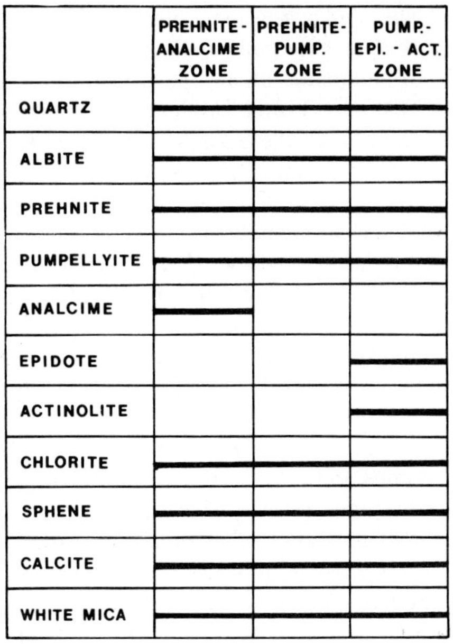

Figure 7. Distribution of metamorphic minerals relative to metamorphic grade in central Aroostook County, Maine.

	PREHNITE-ANALCIME ZONE	PREHNITE-PUMP. ZONE	PUMP.-EPI.-ACT. ZONE
QUARTZ	▬▬▬	▬▬▬	▬▬▬
ALBITE	▬▬▬	▬▬▬	▬▬▬
PREHNITE	▬▬▬	▬▬▬	
PUMPELLYITE	▬▬▬	▬▬▬	▬▬▬
ANALCIME	▬▬▬		
EPIDOTE			▬▬▬
ACTINOLITE			▬▬▬
CHLORITE	▬▬▬	▬▬▬	▬▬▬
SPHENE	▬▬▬	▬▬▬	▬▬▬
CALCITE	▬▬▬	▬▬▬	▬▬▬
WHITE MICA	▬▬▬	▬▬▬	▬▬▬

The assemblages of our intermediate prehnite-pumpellyite zone are typical of the prehnite-pumpellyite facies (Coombs and others, 1959; Seki and others, 1969; Smith, 1969; and others).

On the basis of a survey of several regions of prehnite-pumpellyite-facies metamorphism, Hashimoto (1966) divided the prehnite-pumpellyite metagraywacke facies of Coombs (1960) into a lower grade prehnite-pumpellyite facies and a higher grade pumpellyite-actinolite schist facies. Hashimoto and Saito (1970) and Hashimoto (1972) emphasized that the formation of actinolite can take place in at least two ways—that is, by the replacement of augite or by the decomposition of prehnite and (or) pumpellyite. They concluded on the basis of field occurrences that the replacement of augite by actinolite can proceed at lower pressure and temperature than the decomposition of prehnite and pumpellyite. Hashimoto (1966, 1972) stressed that only actinolite that is formed at the expense of other metamorphic minerals defines an actinolite isograd. Hashimoto implied that actinolite after augite is not sufficient to define the transition into the pumpellyite-actinolite schist facies.

Most of the actinolite occurring in the rocks of northern Maine is in the form of overgrowths and partial replacements of augite. At three localities within a small area (Fig. 3, localities 10, 45, 57), actinolite needles occur in the matrix of igneous rocks; this actinolite may have replaced either fine-grained augite or metamorphic minerals. These three occurrences of matrix actinolite are not here considered to be sufficient to define a discrete area of the pumpellyite-actinolite schist facies.

The first appearance of clinozoisite (and actinolite), coupled with the disappearance of pumpellyite, defines the beginning of the greenschist facies in some parts of the world (Bishop, 1972). Bishop, however, noted that in the Otago schist belt of New Zealand there is overlap between the occurrences of pumpellyite and clinozoisite. In the vicinity of Ashland, clinozoisite occurs with pumpellyite in a few thin sections of graywacke of the Frenchville Formation. These clinozoisite-bearing samples plus the actinolite-bearing volcanic rocks in the southwest part of the study area may therefore be considered as representing the highest grade of the prehnite-pumpellyite facies.

In summary, all of the assemblages of the three mineralogic zones in central Aroostook County, Maine, are compatible with the prehnite-pumpellyite facies. Furthermore, the zones seem to span the lowest to highest grades of the facies when compared to similar metamorphic terranes elsewhere.

Metamorphic Reactions and Conditions of Metamorphism

A number of reactions have been suggested for the observed changes in mineralogy in regions of low metamorphic grade. Most of the reactions are based on textural relations coupled with analyses of the phases involved.

The most likely reaction for the disappearance of analcime at the boundary of the prehnite-analcime zone is as follows: albite + water = analcime + quartz. This reaction is well known in subgreenschist-facies metamorphism (Coombs, 1954; Coombs and others, 1959; Hay, 1966; Seki and others, 1969; Miyashiro and Shido, 1970; Liou, 1971). A careful search of the analcime-bearing thin sections in this study has shown that apparently no quartz coexists with analcime, although the analcime pseudomorphs after plagioclase are surely of metamorphic origin. If one assumes that analcime formed by the above reaction and that most of the SiO_2 was dissolved in the pore fluid, then the analcime may represent a disequilibrium partial assemblage.

Many reactions have been suggested for the formation of prehnite and pumpellyite in low-grade metamorphic rocks; these reactions may also apply to the assemblages in northern Maine (for example, Seki and others, 1969; Smith, 1969; Bishop, 1972). Our work, together with that of Coombs and others (1970), indicates that the four-phase partial assemblage of prehnite-pumpellyite-chlorite-calcite is widespread in northern Maine. To account for the assemblage, Coombs and others (1970) suggested that a buffered reaction—23 prehnite + 2 chlorite + $4H_2O$ + $6CO_2$ = 10 pumpellyite + 6 calcite + 15 quartz—may have taken place. This reaction appears to have occurred in all three of the zones defined in this study.

Most of the epidote in the pumpellyite-epidote-antinolite zone is Fe rich and is not associated with actinolite. Reactions can be written that produce epidote at the expense of pumpellyite (Hashimoto, 1972). Where clinozoisite occurs as an overgrowth on pumpellyite (as in Fig. 6C), the following reaction is suggested: 15 pumpellyite + $14CO_2$ = 23 clinozoisite + 14 calcite + 12 quartz + 3 chlorite + $29H_2O$ (Hashimoto, 1972).

Most of the actinolite in the pumpellyite-epidote-actinolite zone replaces augite and occurs independently of epidote. Hashimoto and Saito (1970) proposed the following reaction for the replacement of pyroxene by actinolite in regions of low-grade metamorphism: 5 diopside + H_2O + $3CO_2$ = actinolite = 3 calcite + 2 quartz. Because this reaction can occur at temperatures lower than other reactions in which actinolite is produced at the expense of prehnite or pumpellyite (Hashimoto, 1972), actinolite that is formed by the alteration of augite can stably coexist with prehnite and pumpellyite, as it appears to in the pumpellyite-epidote-actinolite zone of central Aroostook County.

The association of acicular crystals of actinolite and epidote in a few samples from the Big Machias Lake area, reported by Coombs and others (1970), caused Hashimoto (1972) to suggest that those minerals may have formed by a reaction of prehnite, chlorite, and quartz. Although this reaction is possible, we have found no textural evidence for it.

The thermodynamic role of CO_2 in low-grade metamorphic reactions is not completely understood. Albee and Zen (1969) and Coombs and others (1970) have interpreted the chemical potential of CO_2 as being externally controlled. They demonstrated that a high μ_{CO_2}/μ_{H_2O} ratio will suppress the formation of Ca-Al silicates in metamorphic rocks and will favor clay-carbonate assemblages. Albee and Zen also showed that assemblages typical of the zeolite and prehnite-pumpellyite facies can be formed at the same total pressure and temperature by varying the μ_{CO_2}/μ_{H_2O} ratio.

Zen (1974) analyzed the assemblages of several localities of prehnite- and (or) pumpellyite-bearing rocks by the method of Schreinemakers in two ways. In the first treatment, Zen considered calcite as an inert phase (CO_2 is an initial-value component); he found that most of the assemblages of six phases common in his rocks were divariant. In the second treatment, calcite was considered an excess phase (CO_2 is therefore a boundary-value component), and he found that the same assemblages must be univariant. Because it is unlikely that the commonly occurring assemblages are univariant, Zen tentatively concluded that μ_{CO_2} at a given set of external conditions (T, P, μ_{H_2O}) was determined by the bulk composition of the rock, including the initial amount of CO_2.

Zen cautioned that the consideration of CO_2 as an initial-value component does not mean that the amount of CO_2 does not change after the onset of metamorphism; he implied that it might change during metamorphism. Calcite is a pervasive vug filling and vein mineral, and it cannot be determined whether all calcite in a given rock formed at the same time.

The low-grade metamorphic rocks of the western part of the area of Figure 3 are found in many localities with calcite-chlorite assemblages (with or without quartz, albite, white mica, and sphene) intermingled with those containing Ca-Al silicate assemblages. Coombs and others (1970), taking CO_2 as a boundary-value component, suggested that the calcite-chlorite assemblages may be the result of local metasomatic exchange reactions in which the components of prehnite, pumpellyite, and epidote are removed from one rock and concentrated in another rock. An alternative explanation for the calcite-chlorite assemblages that can be inferred from Zen (1974) is to consider CO_2 an initial-value component that varied from rock to rock. This would explain the existence of calcite-chlorite, calcite–chlorite–Ca-Al silicate, and Ca-Al silicate assemblages as a result of local differences in bulk (rock plus fluid) compositions.

East of the Frenchville and Winterville exposures (Fig. 3), limited observations in this study, together with reports by Williams and Gregory (1900) and Boucot and others (1964), suggest the predominance of the chlorite-calcite assemblage in both graywacke and volcanic rocks. In the east, the Ordovician and Silurian rocks are predominately calcareous slate with interstratified limestone and calcareous graywacke. The absence of mineral assemblages indicative of the prehnite-pumpellyite facies in these widespread and thick sequences may be the result of a generally higher μ_{CO_2}/μ_{H_2O} ratio in the fluid phase during metamorphism than was present in the west. For a similar reason, Albee and Zen (1969) suggested the possible suppression of minerals of the prehnite-pumpellyite facies along the western margin of the northern Appalachian fold belt in Vermont. The predominance of andesitic to rhyolitic compositions in the volcanic rocks of the Dockendorff Group probably was an additional factor preventing diagnostic Ca-Al silicate assemblages from developing.

The zonation of the prehnite-pumpellyite facies that we have observed can be interpreted as resulting from regional variations in temperature, pressure, and (or) μ_{CO_2}/μ_{H_2O} ratios. Assessment of the roles of these variables must await detailed chemical analysis of the important phases.

Seki (1969) divided the subgreenschist metamorphic facies into a facies series based on the pressure conditions during metamorphism. He used the variations in metamorphic mineral zonation as his criteria for differentiating pressure conditions. The central Kii Peninsula of the Sanbagawa belt of Japan and the Alpine belt of New Zealand are his type examples of "intermediate pressure"-type metamorphism. These two areas have a narrow zone of pumpellyite-actinolite assemblages occurring between the prehnite-pumpellyite and actinolite-greenschist zones. The "high-pressure" type (Kanto Mountains of the Sanbagawa belt, Japan) has a wide zone of pumpellyite-actinolite assemblages (without prehnite and with lawsonite); the "low-pressure" type (Tanzawa Mountains, Japan) has no pumpellyite-actinolite assemblages at all. In the Tanzawa Mountains, analcime and the other zeolites, wairakite and yugawaralite, persist in the prehnite-pumpellyite zone.

The occurrence in northern Maine of the prehnite-pumpellyite facies probably could be classified as Seki's (1969) intermediate-pressure metamorphism, because it is characterized by mixed prehnite-pumpellyite and actinolite-epidote assemblages in the highest grade metamorphic zone in the area. The metamorphic assemblages of the prehnite-analcime zone also show some similarity to the assemblages of the Tanzawa Mountains, but they lack any minerals diagnostic of high temperature-low pressure metamorphism such as wairakite.

Age of Metamorphism

Previous authors have attributed the metamorphism of northern Maine rocks to the Acadian orogeny (Boucot and others, 1964; Pavlides and others, 1964; Hall, 1970). This assignment has been based largely on the presence in pelitic rocks of Middle Ordovician to Early Devonian age of a single well-developed cleavage. Thus far, unfortunately, no mineral assemblages completely definitive as to metamorphic grade have been observed in the pelitic rocks of the region.

The absence of tectonic fabrics in the volcanic rocks and sandstones investigated in this study prevents a clear association of the alteration with a particular period of deformation. The data presently available relevant to the ages of rocks containing diagnostic mineral assemblages are as follows: (1) rocks of Middle Ordovician to late Silurian age were affected; (2) Lower Devonian volcanic rocks in at least the southwestern part of the study were affected (Coombs and others, 1970); (3) mafic detrital fragments, derived from the Ordovician Winterville Formation, and the matrix of Silurian lithic sandstone contain metamorphic minerals, but we have not observed any cases where metamorphic minerals cross a clast-matrix boundary; (4) dikes that contain analcime near Mapleton intrude deformed Silurian rocks (Boucot and others, 1964; Williams and Gregory, 1900); (5) mafic dikes containing prehnite, pumpellyite, actinolite, chlorite, and epidote in the northwest part of the Howe Brook quadrangle south of Mapleton, briefly described by Pavlides (1973) and Zen (1974), intrude Lower Devonian slate; and (6) the Middle Devonian Mapleton Sandstone, which unconformably overlies near-vertical Lower Devonian rocks, although broadly folded and faulted, is not metamorphosed.

Dikes containing minerals compatible with the prehnite-pumpellyite facies complicate the age picture somewhat. Teschenite dikes similar to those near Mapleton and considered to be post-Acadian are present in the Moose River synclinorium in western Maine (Boucot, 1969; Boucot and others, 1959), where they cut Lower Devonian rocks of low greenschist grade. Elsewhere in New England, rare prehnite- and zeolite-bearing veins are found cutting high-grade rocks. Considering the general increase in grade to the southeast as shown in Figure 3, it appears that the Mapleton teschenite dikes also intrude rocks of a metamorphic grade higher than the stability range of analcime. We therefore conclude that the analcime in the teschenite dikes is probably younger than the mineral parageneses described in this paper. Pavlides (1973) considered the prehnite- and pumpellyite-bearing dikes in the Howe Brook quadrangle to be of Devonian age and to reflect the regional metamorphism, but he did not give details of their relationships to cleavage. With our current information, these dikes are difficult to relate to the rocks examined in our study, but if they were regionally metamorphosed, their mineral assemblages and location are at least consistent with our data (see Zen, 1974).

The coincidence of metamorphic grade in Ordovician and Lower Silurian rocks in the Ashland area and the alteration of Upper Silurian tuff and Lower Devonian mafic volcanic rocks elsewhere point to post-Early Devonian metamorphism. The unmetamorphosed Middle Devonian Mapleton Sandstone (Fig. 1) suggests a minimum age for the alteration. We therefore assign the metamorphism to the Acadian orogeny. It is possible that the Ordovician volcanic rocks were earlier metamorphosed to the prehnite-pumpellyite facies by the Taconian deformation (Ashgillian-early Llandoverian) and that this previous alteration is reflected in the mafic lithic fragments of Frenchville Formation sandstone. It is as yet not possible to "filter out" the effects of Devonian metamorphism and establish a Taconian alteration.

ACKNOWLEDGMENTS

James W. Skehan, S.J., Priscilla Perkins, and J. Christopher Hepburn read various drafts of the manuscript. We appreciate the critical reviews by Sharon Bachinski, Gary Boone, and Robert Moench, which greatly improved the manuscript. Ely Mencher lent 57 thin sections of the Winterville Formation, and Robert Horodyski lent 39 sections of the Winterville Formation, which in large part made this study possible. E-an Zen confirmed some mineral identifications at an early stage of this project and kindly provided an English copy of Albee and Zen (1969).

REFERENCES CITED

Albee, A. L., 1962, Relationships between the mineral association, chemical composition, and physical properties of the chlorite series: Am. Mineralogist, v. 47, p. 851-870.
——1968, Metamorphic zones in northern Vermont, in Zen, E-an, White, W. S., Hadley, J. B., and Thompson, J. B., eds., Studies of Appalachian geology: Northern and Maritime: New York, Interscience Pubs., Inc., p. 329-341.
Albee, A. L., and Zen, E-an, 1969, Dependence of the zeolite facies on the chemical potentials of CO_2 and H_2O, in Zharikov, V. A., ed., D. S. Korzhinsky Festschrift Volume: Moscow, p. 249-259 (in Russian).
Bishop, D. G., 1972, Progressive metamorphism from prehnite-pumpellyite to greenschist facies in the Dansey Pass area, Otago, New Zealand: Geol. Soc. America Bull., v. 83, p. 3177-3198.
Boone, G. M., 1970, The Fish River Lake Formation and its environments of deposition: Maine Geol. Survey Bull. 23, p. 27-41.
Boucot, A. J., 1969, Geology of the Moose River and Roach River synclinoria, northwestern Maine: Maine Geol. Survey Bull. 21, 117 p.
Boucot, A. J., Harper, C., and Rhea, K., 1959, Geology of the Beck Pond area, township 3-range 5, Somerset County, Maine: Maine Geol. Survey Spec. Geol. Studies Ser., no. 1, 33 p.
Boucot, A. J., Field, M. T., Fletcher, R., Forbes, W. H., Naylor, R. S., and Pavlides, L., 1964, Reconnaissance bedrock geology of the Presque Isle quadrangle, Maine: Maine Geol. Survey Quad. Mapping Ser., no. 2, 123 p.
Cady, W. M., 1968, The lateral transition from the miogeosyncline to the eugeosynclinal zone in northwestern New England and adjacent Quebec, in Zen, E-an, White, W. S., Hadley, J. B., and Thompson, J. B., eds., Studies of Appalachian geology: Northern and Maritime: New York, Interscience Pubs., Inc., p. 151-161.
——1969, Regional tectonic synthesis of northwestern New England and adjacent Quebec: Geol. Soc. America Mem. 120, 181 p.
Coombs, D. S., 1953, The pumpellyite mineral series: Mineralog. Mag., v. 30, p. 113-135.
——1954, The nature and alteration of some Triassic sediments from Southland, New Zealand: Roy. Soc. New Zealand Trans., v. 82, p. 65-109.
——1960, Lower grade mineral facies in New Zealand: Internat. Geol. Cong., 21st, Copenhagen 1960, Rept., pt. xiii, p. 339-351.
Coombs, D. S., Ellis, A. J., Fyfe, W. S., and Taylor, N. M., 1959, The zeolite facies, with comments on the interpretation of hydrothermal synthesis: Geochim. et Cosmochim. Acta, v. 17, p. 53-107.
Coombs, D. S., Horodyski, R. J., and Naylor, R. S., 1970, Occurrence of prehnite-pumpellyite facies metamorphism in northern Maine: Am. Jour. Sci., v. 268, p. 142-156.
Doyle, R. G., 1967, Preliminary geologic map of Maine: Maine Geol. Survey, scale 1:500,000.
Hall, B. A., 1970, Stratigraphy of the southern end of the Munsungun anticlinorium, Maine: Maine Geol. Survey Bull. 22, 63 p.
Hashimoto, M., 1966, On the prehnite-pumpellyite metagreywacke facies: Geol. Soc. Japan Jour., v. 72, p. 253-265 (in Japanese with English summary).
——1972, Reactions producing actinolite in basic metamorphic rocks: Lithos, v. 5, p. 19-31.
Hashimoto, M., and Saito, Y., 1970, Metamorphism in the Tamba Plateau, Kyoto Prefecture: Geol. Soc. Japan Jour., v. 76, p. 1-6.
Hay, R. L., 1966, Zeolites and zeolitic reactions in sedimentary rocks: Geol. Soc. America Spec. Paper 85, 130 p.
Horodyski, R. J., 1968, Bedrock geology of portions of the Fish River Lake, Winterville, Greenlaw, and Mooseleuk Lake quadrangles, Aroostook County, Maine [M.S. thesis]: Cambridge, Massachusetts Inst. Technology, 192 p.
Liou, J. G., 1971, Analcime equilibria: Lithos, v. 4, p. 389-402.
Miyashiro, A., and Shido, F., 1970, Progressive metamorphism in zeolite assemblages: Lithos, v. 3, p. 251-260.
Mossman, D., and Bachinski, D., 1972, Zeolite facies metamorphism in the Silurian-Devonian fold belt of northeastern New Brunswick: Canadian Jour. Earth Sci., v. 9, p. 1703-1709.

Naylor, R. S., 1968, Origin and regional relationships of the core rocks of the Oliverian domes, *in* Zen, E-an, White, W. S., Hadley, J. B., and Thompson, J. B., eds., Studies in Appalachian geology: Northern and Maritime: New York, Interscience Pubs., Inc., p. 231-240.

Nitsch, K., 1971, Stabilitätsbeziehungen von Prehnit- und Pumpellyit-haltigen Paragenesen: Contr. Mineralogy and Petrology, v. 30, p. 240-260.

Papezik, V. S., 1972, Burial metamorphism of late Precambrian sediments near St. John's, Newfoundland: Canadian Jour. Earth Sci., v. 9, p. 1568-1572.

Pavlides, L., 1973, Geologic map of the Howe Brook quadrangle, Aroostook County, Maine: U.S. Geol. Survey Geol. Quad., Map GQ-1094.

Pavlides, L., Mencher, E., Naylor, R. S., and Boucot, A. J., 1964, Tectonic features of eastern Aroostook County, Maine: U.S. Geol. Survey Prof. Paper 501-C, p. C28-C38.

Pavlides, L., Boucot, A. J., and Skidmore, W. B., 1968, Stratigraphic evidence for the Taconic orogeny in the northern Appalachians, *in* Zen, E-an, White, W. S., Hadley, J. B., and Thompson, J. B., eds., Studies of Appalachian geology: Northern and Maritime: New York, Interscience Pubs., Inc., p. 61-82.

Richter, D. A., 1973, Prehnite-pumpellyite facies metamorphism in central Aroostook County, Maine [M.S. thesis]: Chestnut Hill, Mass., Boston College, 70 p.

Rosenfeld, J. L., 1968, Garnet rotations due to the major Paleozoic deformations in southeast Vermont, *in* Zen, E-an, White, W. S., Hadley, J. B., and Thompson, J. B., eds., Studies of Appalachian geology: Northern and Maritime: New York, Interscience Pubs., Inc., p. 185-202.

Roy, D. C., 1970, The Silurian of northeastern Aroostook County, Maine [Ph.D. thesis]: Cambridge, Massachusetts Inst. Technology, 483 p.

———1973, The provenance and tectonic setting of the Frenchville Formation, northeastern Maine: Geol. Soc. America Abs. with Programs, v. 5, p. 214.

Roy, D. C., and Mencher, E., 1976, Ordovician and Silurian stratigraphy of northeastern Aroostook County, Maine, *in* Page, L. R., ed., Contributions to the stratigraphy of New England: Geol. Soc. America Mem. 148 (in press).

Seki, Y., 1969, Facies series in low grade metamorphism: Geol. Soc. Japan Jour., v. 75, p. 255-266.

Seki, Y., Oki, Y., Mikami, K., and Okumura, K., 1969, Metamorphism in the Tanzawa Mountains, central Japan: Japanese Assoc. Mineralogists, Petrologists and Econ. Geologists Jour., v. 61, p. 1-29, 50-75.

Smith, R. E., 1969, Zones of progressive regional burial metamorphism in part of the Tasman geosyncline, eastern Australia: Jour. Petrology, v. 10, p. 144-163.

Thompson, J. B., and Norton, S. A., 1968, Paleozoic regional metamorphism in New England and adjacent areas, *in* Zen, E-an, White, W. S., Hadley, J. B., and Thompson, J. B., eds., Studies of Appalachian geology: Northern and Maritime: New York, Interscience Pubs., Inc., p. 319-327.

Williams, H. S., and Gregory, H. E., 1900, Contributions to the geology of Maine: U.S. Geol. Survey Bull. 165, 212 p.

Zen, E-an, 1974, Prehnite- and pumpellyite-bearing mineral assemblages, west side of the Appalachian metamorphic belt, Pennsylvania to Newfoundland: Jour. Petrology, v. 15, p. 197-242.

MANUSCRIPT RECEIVED BY THE SOCIETY JUNE 14, 1974
REVISED MANUSCRIPT RECEIVED FEBRUARY 7, 1975
MANUSCRIPT ACCEPTED FEBRUARY 20, 1975

Printed in U.S.A.

Geological Society of America
Memoir 146
© 1976

Stratigraphic Relationships on the Southeast Limb of the Merrimack Synclinorium in Central and West-Central Maine

KOST A. PANKIWSKYJ
Department of Geology and Geophysics
University of Hawaii
Honolulu, Hawaii 96822

ALLAN LUDMAN
Department of Earth and Environmental Sciences
Queens College
Flushing, New York 11367

JOHN R. GRIFFIN
Department of Geology
University of California, Davis
Davis, California 95616

W.B.N. BERRY
Department of Paleontology
University of California, Berkeley
Berkeley, California 94720

ABSTRACT

Detailed mapping on the southeast limb of the Merrimack synclinorium in central and west-central Maine has revealed a complex lithofacies pattern in Silurian and Silurian or Devonian eugeosynclinal metasedimentary rocks. The mapping includes tracing of stratigraphic units from the highly metamorphosed terraine of western Maine into the slightly recrystallized rocks of the central Maine slate belt. Ages of the Silurian part of the section and faunal confirmation of the lithofacies interpretations come from 25 new graptolite localities.

The lithofacies relationships indicate filling of a now greatly compressed sedimentary trough from both western (metamorphic, plutonic, and volcanic) and eastern (dominantly volcanic) sources from at least late Llandoverian through early Ludlovian time. From post-early Ludlovian through early Devonian(?) time, thick sequences of metasandstone and flyschlike materials were derived from indeterminate sources and blanketed the synclinorium and much of New England. The total section, dominated by metagraywacke and metashale, is more than 6 km thick.

INTRODUCTION

The Merrimack synclinorium is a major structural feature that can be traced from southern Connecticut (Dixon and Lundgren, 1968) through central Massachusetts (Rodgers, 1970) and southern New Hampshire (Billings, 1956) into Maine (Osberg and others, 1968). It contains Silurian and Early Devonian rocks in its core and Cambrian and Ordovician or older rocks on its flanks. In Maine, where the synclinorium attains a width of over 150 km, stratigraphy of the northwest limb has been established by Moench and Boudette (1970).

The axial region and southeast limb have been less well known. In 1968, Osberg described stratigraphic relationships in the Waterville-Vassalboro area, basing ages of two units on graptolites from two localities first reported by Perkins (1924) and Perkins and Smith (1925). Other work on the southeast limb was less detailed, including that of Barker (1961), Warner (1967), Warner and Pankiwskyj (1965), Caldwell (unpub. data), and Pankiwskyj (1964). Much of the pioneering geologic mapping in central Maine was carried out by students at Boston University, stimulated and coordinated by C. W. Wolfe (Borns, 1959; Cariani, 1958; Furlong, 1961; Glidden, 1963; and Skapinski, 1961).

This paper presents (1) a compilation of recent mapping on the southeast limb of the synclinorium (Figs. 1, 2), most of it carried out by Pankiwskyj, Ludman, and Griffin; (2) a stratigraphic summary of much of central and west-central Maine, based largely on 25 new graptolite localities and including detailed evidence for major lithofacies changes; and (3) some inferences about regional tectonism.

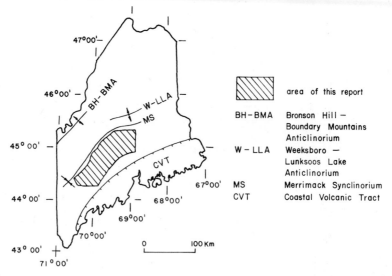

Figure 1: Index map of Maine showing the location of the area of this report and related major structural features.

GEOLOGIC SETTING

The Merrimack synclinorium in central Maine is bounded on the northwest by the Bronson Hill-Boundary Mountain and Weeksboro-Lunksoos Lake anticlinoria, and on the southeast by a complex region composed of metasedimentary and metavolcanic rocks of early Paleozoic (and Precambrian?) age (Fig. 1). Stratigraphy in these bordering areas and relationships with rocks of the synclinorium have been described by Green and Guidotti (1968), Hussey (1968), Pavlides and others (1964), and Gates (1969). The axial trace of the synclinorium can be mapped across the Bryant Pond and Dixfield quadrangles into the Kingfield, Anson, Bingham, and Greenville quadrangles, and on into incompletely known terrain to the east. The region described in this paper lies entirely to the southeast of the synclinorium axis and extends to within 5 km of the metavolcanic rocks exposed in the coastal volcanic tract.

The southeast limb of the synclinorium in central Maine lies in what Rodgers (1970, p. 127) called the central Maine slate belt. This limb can be conveniently divided into a central syncline (the Currier Hill syncline of Ludman, 1969), bounded on the southeast by dominantly anticlinal structures (the Waterville anticlinorium of Rodgers, 1970, p. 128), and on the northwest by several anticlines and synclines broken by major faults (see Fig. 2, especially cross-sections). In most of the region, folds and faults trend northeast. A northwest-trending warp deflects regional trends to nearly due east in the northeast part of the region, and locally aberrant northwest-trending fold axes are prominently developed in the Buckfield quadrangle in the extreme southwest corner of the area. A detailed analysis of the multiphase deformation recorded in the rocks in the map area is beyond the scope of this paper. Interested readers are referred to Griffin (1973), Ludman (1969), Pankiwskyj (1964), Warner (1967), and Osberg (1968).

Metamorphic grade in the northeastern half of Figure 2 is low, generally biotite- and chlorite-grade, but intensity of recrystallization increases toward the southwest, where sillimanite-grade rocks are extensively exposed. A number of stocks and batholiths of the Early Devonian New Hampshire Plutonic Series intrude the stratified rocks and have well-developed contact aureoles in the lower grade terrane.

STRATIGRAPHY

General Statement

The low intensity of metamorphism in the eastern half of the area shown in Figure 2 accounts for the excellent preservation of a wide variety of primary sedimentary structures, graptolites, crinoid columnals, and trace fossils. Facing indicators are abundant in most units and provide valuable stratigraphic and structural data. Metagraywacke and metapelite make up most of the section, interrupted infrequently by distinctive metaconglomerate, metalimestone, and carbonaceous metapelite. Extreme care had to be taken in using these lithologies as marker horizons, as each of them is found at several stratigraphic levels within the section.

The strongly recrystallized rocks in the southwestern part of the area shown in Figure 2 are nonfossiliferous and exhibit fewer primary sedimentary features, but a sequence has been established that can be traced into the lower grade rocks. Facies changes that are demonstrable in the low-grade terrane are interpreted but not confirmed in the high-grade region. All changes are consistent, however, with the regional interpretation to be detailed in this paper.

Inferred predeformation relationships between the stratigraphic units described below are illustrated in Figure 2. Complete stratigraphic sequences for low-grade terranes southeast and northwest of the Currier Hill syncline are given in Tables 1 and 2, and for a high-grade area, in Table 3. Four new formation names and one revised designation are summarized in Appendix 1, and a compilation of all fossil data is presented in Appendix 2.

Silurian System

Mayflower Hill Formation (Smh). The oldest rocks shown in Figure 2 east of the Currier Hill syncline are the thick-bedded metasandstone and metapelite of the Mayflower Hill Formation (Osberg, 1968, p. 8). These rocks form the cores of three doubly plunging anticlines, two in the Waterville-Vassalboro area and a third, truncated by the southern margin of Figure 2, in the Augusta quadrangle. The rocks consist of typically poorly sorted, delicately cross-laminated calcareous metagraywacke and subordinate gray metapelite. Graded bedding is well developed, and a metagraywacke:metapelite ratio of approximately 8:1 is characteristic of the formation. A single horizon of sulfidic, carbonaceous metapelite 30 to 60 m thick (Smhg) occurs within a few metres of the upper contact. Berry (Osberg, 1968, p. 32) suggested a late Llandoverian age for graptolites collected from the Mayflower Hill Formation in the Waterville quadrangle, and this age is assigned to the formation (Osberg, 1968, p. 29).

Waterville Formation (Swe, Sww). The Waterville Formation, as redefined by Osberg (1968, p. 11), consists of thin-bedded (eastern facies, Swe) and thick-bedded (western facies, Sww) metagraywacke, metasiltstone, and metapelite. The transition zone between the facies has been removed by erosion in the Waterville-Vassalboro area from the crest of the westernmost anticline in which the Mayflower Hill Formation is exposed (Fig. 2).

The western facies consists of thick-bedded, well-graded metagraywacke and metapelite very similar to those of the Mayflower Hill Formation. Several metalimestone horizons (Swl) are found in the western facies, but only the thickest is shown in Figure 2. The eastern facies, in contrast, is composed of cyclically and

TABLE 1. STRATIGRAPHY OF THE SECTION SOUTHEAST OF THE CURRIER HILL SYNCLINE

Formation	Age	Map symbol	Lithology	Thickness (m)
Vassalboro Formation	Devonian or Silurian	DSv	Massively bedded weakly calcareous metasandstone and minor interbeds of massive dark metapelite	800+
Waterville Formation	Silurian (probably Wenlockian or Ludlovian)	Swe	Eastern facies: thinly laminated metasiltstone and metapelite	1,000
		Sww	Western facies: massively bedded graywacke metasandstone and subordinate massive or thinly laminated metapelite	1,000
		Swl	Metalimestone member: interbedded metalimestone, calcareous metasiltstone, and metapelite	100-300
		Swg	Black phyllite member: sulfidic and graphitic rusty-weathering metasiltstone and metapelite	>30
		Swp	"Pittsfield member": thinly interlaminated maroon and green metasiltstone and metapelite	150(?)
Mayflower Hill Formation	Silurian (probably late Llandoverian)	Smh	Massive bedded prominently graded bedded graywacke metasandstone and minor metapelite	300+
		Smhg	Black phyllite member: sulfidic and graphitic rusty-weathering metasiltstone and metapelite	>30

Note: Adapted after Osberg (1968) and Griffin (1973).

TABLE 2. STRATIGRAPHY OF THE SECTION NORTHWEST OF THE CURRIER HILL SYNCLINE

Formation	Age	Map symbol	Lithology	Thickness (m)
Solon Formation	Early Devonian	Ds	Massive silvery gray metapelite; rhythmically graded bedded metasiltstone and silvery gray metapelite	2,000
Fall Brook Formation	Late Silurian or Early Devonian	DSf	Massive thick bedded weakly calcareous metasandstone and minor massive dark metapelite; regularly graded bedded metasandstone and metapelite	1,000
Parkman Hill Formation	Silurian (probably early Ludlovian)	Sph	Sulfide rich, rusty weathering metasandstone, quartz-rich granule metaconglomerate, thinly laminated metasiltstone and metapelite	0-700
Sangerville Formation	Silurian (probably late Llandoverian to early Ludlovian)	Ss	Thick bedded fine- to coarse-grained weakly calcareous graywacke metasandstone in graded units with massive or thinly laminated metapelite	2,000+
		Ssl Ssll	Metalimestone member and lower metalimestone member: bedded micritic metalimestone and calcareous metasiltstone and metapelite	150±
		Ssc Sscu	Granule metaconglomerate member and upper granule metaconglomerate member: polymictic granule metaconglomerate in addition to all rock types listed under (Ss)	200-250 each
		Ssd	"Dover member": thin-bedded graywacke metasandstone and metapelite	150?
Anasagunticook Formation	Silurian (possibly late Wenlockian to early Ludlovian)	Ss	Graded bedded metasandstone and metapelite; massive metapelite; thinly laminated metasiltstone and metasiltstone and metapelite	1,000
		Sal	Metalimestone member: bedded micritic metalimestone, calcareous metasiltstone, and metapelite	100±

thinly interlaminated ("pinstriped") metasiltstone and metapelite, with a single prominent metalimestone horizon (Swl) near the middle of the formation. Discontinuous carbonaceous metapelite lenses (Swg) occur below the metalimestone and also at the upper contact with the Vassalboro Formation. Only the latter occurrence is shown in Figure 2. To the northeast, in the Pittsfield, Stetson, and Boyd Lake quadrangles, Griffin (1973) has mapped a series of interbedded maroon and green metasiltstones and metapelites at the top of the eastern facies. These rocks have been informally termed the Pittsfield member (Swp) and are so designated in Figure 2.

Osberg (1968, p. 32, 33) proposed an age of Wenlockian through Ludlovian for the Waterville Formation on the basis of graptolites collected in the eastern facies and the position of the formation (above the Mayflower Hill Formation). New ages for graptolites from the Stetson quadrangle (App. 2) are in agreement with his suggestion. Nondiagnostic trace fossils similar to *Nereites ichnofacies* of Seilacher (1964) have been collected at several localities in the Pittsfield member.

Sangerville Formation (Ss; new name). The oldest rocks exposed west of the axis of the Currier Hill syncline are here designated as the Sangerville Formation. The Sangerville Formation is composed for the most part of poorly to moderately sorted, variably bedded (5 cm to 1 m), typically well-graded metagraywacke and metapelite. Cross-laminations and convolute laminations are common, and sedimentary breccias and soft-sediment (slump) folds are present in many exposures. The proportions of metagraywacke to metapelite are also highly variable, ranging from 10:1 to 1:1. A wide variety of lithic fragments in the metagraywacke is cemented by a calcareous and (or) argillaceous matrix that often includes a ferroan carbonate (both ankerite and siderite have been observed) in outcrops in the chlorite zone.

Three distinctive rock types form horizons and lenses of varying extent within the Sangerville Formation; they are mapped as separate lithologic members. Polymictic granule metaconglomerate (Ssc, Sscu) contains clasts ranging from 2 mm to 5 cm in longest dimension. Clasts identified in the metaconglomerate include quartz, feldspar, quartzite, chert, volcanic fragments ranging from felsic to mafic

TABLE 3. STRATIGRAPHY OF THE SECTION IN THE BUCKFIELD QUADRANGLE

Formation	Age	Map symbol	Lithology	Thickness (m)
Thompson Mountain Formation	Silurian (probably early Ludlovian)	Stm	Sulfidic, rusty-weathering mica schist, muscovite schist, and quartzite	500
Anasagunticook Formation	Silurian (possibly late Wenlockian to early Ludlovian)	Sa	Coarse-grained spangled muscovite-sillimanite-feldspar-biotite-garnet schist and migmatized gneiss; well-bedded where least metamorphosed	750+
Moody Brook Formation	Silurian (possibly late Wenlockian)	Sm	Sillimanite-biotite-muscovite schist, locally slightly rusty-weathering	200+
Berry Ledge Formation	Silurian (possibly Wenlockian)	Sb	Marble and interbedded calc-silicate granofels with biotite-quartz granofels	70
Noyes Mountain Formation	Silurian (possibly Wenlockian)	Sn	Sillimanite-biotite-muscovite-garnet schist with zones and beds of biotitic quartzite	250+
Patch Mountain Formation	Silurian (possibly Wenlockian)	Sp	Marble and (or) calc-silicate granofels interbedded with biotite-quartz granofels and (or) biotitic quartzite	300+
Turner Formation	Silurian	St	Sillimanite-biotite-muscovite-garnet schist with many zones and beds of biotitic quartzite and biotitic metagraywacke	500+

Note: Modified after Warner and Pankiwskyj (1965).

in composition, slate and schist chips, and fragments of hypabyssal igneous rock. Blue quartz is abundant in some outcrops. Both massive homogeneous and well-graded bedding habits have been observed. Metalimestone members (Ssl, Ssll) consist of thinly intercalated micritic metalimestone and noncalcareous metasiltstone, metagraywacke, and metapelite, commonly referred to as "ribbon rock". A carbonaceous metapelite member crops out in small lenses which could not be shown at the scale of Figure 2. This metapelite is commonly sulfidic and limonitic and is interbedded with sulfidic metasandstone and metasiltstone. All but one of the graptolite localities in the Sangerville Formation are in this member.

All three minor members are commonly, but not always, associated in the field. Each occurs at several levels within the formation. A reconstructed section showing interpreted relationships within the Sangerville Formation is given in Figure 2. Another member, informally termed the Dover member (Ssd), is recognized in the Guilford, Dover-Foxcroft, and Boyd Lake quadrangles, and is characterized by uncommonly thin-bedded metasandstone and metapelite in 1:1 proportion.

Also included within the Sangerville Formation is a weakly calcareous biotitic metasandstone informally named the Ludden Brook formation by Warner and Pankiwskyj (1965). This name has now been abandoned.

Thirteen graptolite localities have been found in the Sangerville Formation in a five-quadrangle area (App. 2). The age suggested for the formation in the type area, where it is overlain by the Fall Brook Formation, is late Llandoverian through early Ludlovian. To the west, in the Skowhegan, Kingsbury, and Anson quadrangles, where the Sangerville Formation is overlain by the Parkman Hill Formation, the suggested range is late Llandoverian through middle Wenlockian.

Anasagunticook Formation (Sa; revised name). The upper part of the Sangerville Formation (above the middle metalimestone member) interfingers to the southwest in the Anson, Farmington, and Dixfield quadrangles with rocks of the Anasagunticook Formation. As redefined here, the Anasagunticook Formation consists of well-graded, typically strongly cross-laminated and convolute laminated metasandstone and metapelite in 4:1 to 1:1 ratios. Bedding ranges in thickness from 4 to 25 cm, but thinly interlaminated metasiltstone and metapelite also occur in sets 2 to 3 cm thick. Metalimestone (Sal) and polymictic granule metaconglomerate (Sacu) members are differentiated.

No fossils have been found in the Anasagunticook Formation, but a Wenlockian age is suggested because of its interfingering relationship with the upper part of the Sangerville Formation in the Farmington, Anson, and Skowhegan quadrangles.

Moody Brook (Sm), Berry Ledge (Sb), Noyes Mountain (Sn), Patch Mountain (Sp), and Turner (Stm) Formations. All but one of these units were first described by Guidotti (1965) in sillimanite-grade terrane in the Bryant Pond quadrangle (Fig. 2). The Turner Formation was described by Warner and Pankiwskyj (1965) from the Buckfield Quadrangle. The Noyes Mountain and Moody Brook Formations, sillimanite schist with subordinate quartzite and locally rusty sillimanite schist, respectively, can be traced into the Anasagunticook Formation. The distinction is based on a decrease in the amount of muscovite and feldspar and an increase in biotite and sillimanite from northeast to southwest, but this trend may be in part caused by increasing metamorphic grade toward the southwest.

In the Buckfield quadrangle, the calc-silicate granofels of the Patch Mountain and Berry Ledge Formations can be traced into the Ssl member of the Sangerville Formation and the Sal member of the Anasagunticook Formation, respectively. One belt of Sangerville Formation metagraywacke (Ss) can be traced into the interbedded sillimanite schist and biotitic metagraywacke of the Turner Formation.

Because of these relationships with the Sangerville and Anasagunticook Formations, a Silurian (late Llandoverian to Wenlockian) age is suggested for these units (see Table 3).

Parkman Hill Formation (Sph; new name). The Parkman Hill Formation consists of typically sulfidic, rusty-weathering rocks that lie above the Sangerville and Anasagunticook Formations in the central and southwestern parts of the map area. Common rock types include sulfidic carbonaceous metapelite, quartz-rich granule metaconglomerate and metasandstone, and thinly laminated fissile metasiltstone and metapelite. Some nonsulfidic varieties are also present but are minor, and metalimestone occurs near the top of the formation in the Kingsbury and Guilford quadrangles. The coarser grained rock types are most common in the northeast, less abundant in the southwest, and generally absent to the southeast in the Currier Hill syncline. The entire formation is absent in the northeasternmost part of the map area and farther south as well. Parkman Hill lithologies interfinger with and are replaced by Sangerville types in the western half of the Guilford quadrangle and the eastern part of the Kingsbury quadrangle.

Eight graptolite localities have been discovered in the Parkman Hill Formation in four quadrangles (see App. 2). An age of middle Wenlockian to early Ludlovian is proposed for the formation, on the basis of the ages of these graptolites.

Thompson Mountain Formation (Stm). Rusty-weathering sulfidic schist and quartzite that lie above the Anasagunticook and Sangerville Formations in the Buckfield quadrangle were named the Thompson Mountain Formation by Warner and Pankiwskyj (1965). Because of its stratigraphic position, the Thompson Mountain Formation is considered to be the equivalent of the Parkman Hill Formation, and is thus of probable middle Wenlockian to early Ludlovian age.

Silurian or Devonian(?)

Vassalboro Formation (DSv). Massive metasandstone that lies above the Waterville Formation in the Waterville-Vassalboro area was first called the Vassalboro

Sandstone (Perkins and Smith, 1925), then renamed the Vassalboro Formation (Fisher, 1941). As described by Osberg (1968, p. 22), the Vassalboro Formation consists for the most part of variably (8 cm to >3 m) but generally thick-bedded metasandstone with abundant but subordinate metapelite (metasandstone:metapelite ratio approximately 7:3). Calc-silicate lenses are developed at high metamorphic grades. Carbonaceous metapelite was described as occurring at two horizons within the Vassalboro Formation in the type locality (Osberg, 1968, p. 22). Griffin (1973) has collected graptolites from two localities in carbonaceous metapelite at the base of the formation in the Stetson quadrangle (App. 2). The poor preservation of these fossils limits their usefulness; a Silurian age is suggested.

Fall Brook Formation (DSf; new name). The Fall Brook Formation lies above the Parkman Hill, Thompson Mountain, and Sangerville Formations in the area northwest of the Currier Hill syncline. It is composed for the most part of massively bedded (average thickness 50 cm), well-sorted, weakly calcareous metasandstone with minor metapelite beds and partings. Calc-silicate beds and pods are locally abundant. Thinly (1 to 3 cm) and regularly interbedded metasandstone and muscovitic metapelite occur at several levels within the formation, but apparently not near the base. These rocks are often strongly cross-laminated. The Fall Brook Formation grades upward into the Solon Formation, and the uppermost Fall Brook rocks are well-graded and poorly sorted metasandstone and metapelite. Carbonaceous metapelite is present in the Fall Brook Formation in the keel of the Currier Hill syncline.

No fossils have been found in the Fall Brook Formation, but a Silurian or Devonian age is proposed because of its position above the fossiliferous Parkman Hill Formation and below fossiliferous rocks correlated with the Solon Formation.

Devonian(?) System

Solon Formation (Ds; new name). The youngest unit in the map area, the Solon Formation, lies conformably above a 70-m-thick transition zone at the top of the Fall Brook Formation. The Solon Formation is composed of massive gray metapelite with minor metasiltstone and metasandstone, and rhythmically interbedded laminated metasiltstone and metapelite in distinctive graded sets. A lenticular polymictic granule metaconglomerate has been found in a few localities at the base of the formation.

Poorly preserved brachiopods collected in Solon-like rocks in the Sebec and Sebec Lake quadrangles suggest an Early Devonian age for the Solon Formation (Espenshade and Boudette, 1967).

REGIONAL CORRELATION

Stratigraphic correlation across the Currier Hill syncline by direct physical tracing of formations has only been accomplished in the northeast corner of the Livermore quadrangle. Difficulty in correlating is caused by major differences in stratigraphic sequences on either side of the axis of the Currier Hill syncline, specifically, the absence of the distinctive Parkman Hill Formation in the northeast and southeast parts of the map area, and the different stratigraphic positions (in some instances, total absence) of distinctive thinly interlaminated metasiltstone and metapelite (similar to the eastern facies of the Waterville Formation). New faunal and field evidence presented above and in Figure 2 confirms the existence of lithofacies changes across the Currier Hill syncline that resolve the uncertainties.

Osberg (1968) suggested an equivalence of the eastern and western facies of the Waterville Formation but had little faunal or physical evidence. Paleontologic evidence (App. 2) now indicates the equivalence of the Sangerville Formation with the Mayflower Hill Formation and the eastern facies of the Waterville Formation. In the Livermore quadrangle, the western facies of the Waterville Formation can be traced into an outcrop band of the Sangerville Formation. Direct physical evidence for the equivalence of the two facies of the Waterville Formation was suspected by Glidden (1963). It is now confirmed by mapping in the north-central part of the Pittsfield quadrangle. There, the contact between the eastern and western facies crosses the middle metalimestone member (Fig. 2). The time-trangressive nature of this facies contact is also shown schematically in the stratigraphic column of Figure 2.

Fossil evidence also confirms the equivalence of the Parkman Hill Formation with the upper part of the Sangerville Formation. Near the border between the Kingsbury and Guilford quadrangles, Parkman Hill and Sangerville lithologies interfinger, and the Parkman Hill lithology disappears toward the central part of the Guilford quadrangle. Also, the Parkman Hill (and the correlative Thompson Mountain) Formation appears to thin from northwest to southeast, until small lenticular bodies of sulfidic carbonaceous metasedimentary rocks at the appropriate stratigraphic level in the Sangerville and Waterville Formations are all that remain.

On both sides of the Currier Hill syncline, then, the Silurian section is dominated by two sedimentary facies types: a graywacke metasandstone-minor metapelite facies (Mayflower Hill Formation and Waterville Formation western facies east of the axis, and Sangerville Formation to the west), and a thinly interlaminated metasiltstone-metapelite facies (Waterville Formation eastern facies east of the axis, and Anasagunticook Formation and Dover member of the Sangerville Formation to the northwest). In any given area, the metagraywacke-minor metapelite facies is always low in the section and is replaced by the metasiltstone-metapelite facies at some higher level. The position of this change differs throughout the region. It is low in the section in the Waterville-Vassalboro area, where it lies at the Mayflower Hill-Waterville contact; higher in the central part of the Farmington quadrangle, where it lies above the median metalimestone member of the Sangerville Formation; and highest in the northeast, where the metagraywacke-minor metapelite facies was continuously deposited throughout the entire period of Sangerville deposition.

These data, and the general eastward-fining nature of the Llandoverian and Wenlockian clastic rocks, indicate that the source for the metasedimentary rocks was to the northwest, north, and perhaps northeast of the central part of the region. The coarse metaconglomerate at the base of the Rangeley Formation in the Rangeley quadrangle (Osberg and others, 1968, p. 245) attests to the presence of a local source which was most likely the Somerset Island of Boucot and others (1964), also referred to as the Somerset geanticline by Cady (1969).

Correlations with the western limb of the synclinorium proposed by Osberg and others (1968), particularly those between the Rangeley-Phillips and Waterville-Vassalboro sections, are consistent with the regional relationships suggested above. Sediment was derived from a pre-Silurian metamorphic-plutonic-volcanic terrain and transported downslope toward the east and southeast. Proximal (Rangeley and Perry Mountain Formations), intermediate (Sangerville Formation and Waterville Formation western facies), and distal (Waterville Formation eastern facies) facies are well represented.

Recent mapping by Griffin in the Stetson and Bangor quadrangles (Ludman and others, 1972) adds data on an eastern source area to these relationships. An unnamed unit at the stratigraphic position of the eastern facies of the Waterville

Formation exhibits thicker bedding and increased grain size along with an increased proportion of the coarser clastic material. This unit contains a markedly greater proportion of volcaniclastic debris than any of the formations described above, and some chlorite-rich metapelites in it may have been ash-fall tuffs. This suggests a second source area to the southeast, presumably the Silurian coastal volcanic belt of Berry and Boucot (1970, p. 56, 57). The uncertainty at this time about the relationships between the rocks of the Merrimack synclinorium and those of the coastal volcanic tract gives high priority to mapping in southeastern Maine.

SILURIAN AND EARLY DEVONIAN EVOLUTION OF THE MERRIMACK SYNCLINORIUM

Sedimentation recorded on the southeast limb of the Merrimack synclinorium was continuous from at least late Llandoverian through Early Devonian time, and perhaps from Ordovician through Early Devonian, since the map area is on strike with the zone shown by Pavlides and others (1964) in which the Taconic orogeny caused no interruption in sedimentation. The geologic evolution of the original sedimentary trough during the time span represented by the formations described above may be divided into two distinct phases.

During the early phase (late Llandoverian through early Ludlovian), the trough received sediment from western and eastern source areas, with these materials interfingering in the northeasternmost part of the map area. The debris derived from the western source was produced by rapid weathering and erosion, as indicated by the general immaturity of the sediment, characterized by poor sorting, wide variety of polymineralic lithic fragments, and high feldspar and mica content. Reduction of the western source is indicated by decreased grain size and bedding thickness and by increased abundance of metapelite in the metasiltstone-metapelite facies. Toward the end of Wenlockian time and continuing at least into the early Ludlovian, the western source had been reduced to the extent that chemical weathering had become more thorough. This is shown by a resulting increase in the quartz:feldspar ratio and a decrease in the amount of argillaceous matrix in metasandstone and metaconglomerate of the Parkman Hill Formation. Simultaneously, euxinic conditions became widespread in the northwestern part of the map area and extended as far west as the Rangeley quadrangle (Smalls Falls Formation), close to the Somerset geanticline.

The later phase began in the Late Silurian (post–early Ludlovian), when sediment encroached on and gradually covered the Somerset geanticline, producing a section of shallow-water deposits 60 m thick in that area (Boucot, 1961; Boone and others, 1970). A marked change resulted in the rocks of the Merrimack synclinorium. A thick blanket of homogeneous fine-grained calcareous metasandstone with some metapelite (Madrid, Fall Brook, Vassalboro, and Berwick Formations) covered most of the synclinorium in Maine. No significant systematic variation in grain size is apparent in these rocks, and their source is unknown. The Solon Formation, which lies gradationally above the Fall Brook Formation, is part of what Boucot (1970) described as the most widespread sedimentary unit in the northern Appalachians. Variously referred to regionally as the Seboomook Formation, Littleton Formation, Gile Mountain Formation, Compton Formation, Temiscouata Formation, Bolton Formation, St. Francis Group, and Fortin Series, this thick sequence of flyschlike sedimentary rocks blankets the Silurian section throughout much of the northern Appalachians.

Most of the preceding history concerns the western source area and its effect on the synclinorium. Little mention has been made of the role of the proposed eastern volcanic source simply because little is known about its precise history and relationships with the synclinorium sequence. As presently exposed, the original trough was grossly asymmetrical, with the western side being much wider than the eastern. The asymmetry may have been primary or tectonically induced. Silurian volcanic rocks now exposed along the Maine coast may have served as the eastern source, but major faults now separate them from the synclinorium sequence. Mapping now in progress will, we hope, document the relationships between these two rock successions.

Finally, it must be remembered that the present distribution of the metasedimentary rocks is partially the result of a considerable telescoping brought about by the Acadian orogeny. Estimates of crustal shortening to as little as one-fifth of original width have been suggested for the isoclinally folded region described in this paper (Rodgers, 1971). With the recognition of the major faults along the northwestern margin of the map area, this figure may have to be revised.

ACKNOWLEDGMENTS

Most of the field work summarized above was sponsored by the Maine Geological Survey, and we are grateful to Robert Doyle, Arthur M. Hussey, and Walter Anderson of the Survey for their support and suggestions. The geologic map (Fig. 2) incorporates the work of many geologists (see Introduction and References), modified to varying degrees in this paper.

We have benefited in all phases of our work from valuable discussions with several colleagues, especially G. McG. Boone, A. J. Boucot, E. L. Boudette, D. W. Caldwell, B. Keith, R. H. Moench, P. E. Osberg, J. A. Raabe, and J. L. Warner.

Finally, we wish to acknowledge our debt to C. W. Wolfe, whose inspiration led students to break the trail that we have followed.

APPENDIX 1. DESCRIPTION OF NEW AND REVISED STRATIGRAPHIC UNITS

Anasagunticook Formation (revised name)

Type Locality. The type locality of the Anasagunticook Formation is along the east shoreline and adjacent hills of Anasagunticook Lake in the northeastern part of the Buckfield quadrangle.

Lithology. The Anasagunticook Formation consists of typically well-bedded metasandstone and thinly laminated metasiltstone and metapelite. Thicknesses of whole beds range from 4 cm to 25 cm and the ratio of metasandstone:metasiltstone + metapelite ranges from 4:1 to 1:1. The metasandstone is characteristically strongly cross-laminated and convolute laminated. In addition, there are entire sections composed of thinly laminated metasiltstone and metapelite. A bedded member composed of micritic metalimestone, calcareous metasiltstone, and metapelite is mapped about midway in the formation. A polymictic granule metaconglomerate, commonly associated with a thin micritic metalimestone, is developed locally at the top.

Thickness. The thickness of the Anasagunticook Formation has been estimated at about 1,000 m. The median metalimestone member is 50 to 100 m thick, and the metaconglomerate member is as much as 40 m thick.

Age. On the basis of its correlation with the fossiliferous upper part of the Sangerville Formation, the age of the Anasagunticook Formation is estimated as possibly Wenlockian.

Sangerville Formation (new name)

Type Area. The type area of the Sangerville Formation is along Maine State Highway 23 between North Dexter and Sangerville in Sangerville Township, in the eastern third of the Guilford quadrangle.

Lithology. The Sangerville Formation consists of typically well-bedded, weakly calcareous metagraywacke, metasiltstone, and noncalcareous metapelite in graded units 5 cm to 1 m thick. Proportion of metagraywacke and metasiltstone to metapelite ranges from 10:1 to 1:1 in most instances, but some horizons are entirely metapelite or metagraywacke. Metalimestone members consisting of thinly interbedded micritic metalimestone, metasiltstone, and metapelite are mapped near the middle of the formation and at several levels in the lower half of the formation. Polymictic granule metaconglomerate members occur discontinuously at the top of the formation and at several levels within the formation, often associated with the metalimestone. Metaconglomerate occurs in massive homogeneous beds and also in well-graded units with metasiltstone or laminated metapelite tops. Carbonaceous metapelite occurs in lenses throughout the formation.

Thickness. The thickness of the exposed portion of the Sangerville Formation (the lower contact is not exposed) is more than 2,000 m. Metalimestone members are approximately 150 m thick, and metaconglomerate is 200 to 250 m thick.

Age. On the basis of ages from 13 graptolite localities, the age of the Sangerville Formation is estimated as late Llandoverian through early Ludlovian in the type area, where it lies below the Fall Brook Formation, and as late Llandoverian through middle Wenlockian to the west, where it lies below the Parkman Hill Formation.

Parkman Hill Formation (new name)

Type Locality. The type locality of the Parkman Hill Formation is on the south and southeast slopes of the hill of this name in the northeast part of the Anson quadrangle. The best outcrops are exposed between the elevations of 800 and 940 ft.

Lithology. The Parkman Hill Formation is dominantly composed of sulfidic rusty-weathering rocks of the following types: quartz-rich granule metaconglomerate, medium- and fine-grained quartzose metasandstone, thinly laminated metasiltstone, and metapelite; also, non-sulfidic calcareous fine-grained metasandstone and local micritic metalimestone and metasiltstone. The thinly laminated metasiltstone and metapelite forms thick sections but is also found in the graded tops of beds that contain coarser grained metasandstone or granule metaconglomerate at the base. The thickness of individual beds ranges from 1 cm to several metres, and the ratio of coarser grained rocks to metasiltstone and metapelite ranges from over 10:1 to 1:2.

Thickness. The thickness of the Parkman Hill Formation at the type locality is between 200 and 300 m. It increases up to as much as 700 m to the southwest, in the Farmington and Buckfield quadrangles, and pinches out to the northeast, in the Guilford and Dover-Foxcroft quadrangles.

Age. On the basis of graptolites from 8 localities, the Parkman Hill Formation is dated as middle Wenlockian to early Ludlovian.

Fall Brook Formation (new name)

Type Locality. The type locality of the Fall Brook Formation is in Fall Brook in the village of Solon, in the northern part of the Anson quadrangle. The best outcrops are between the elevations of 320 and 400 ft.

Lithology. Most of the formation is composed of massively bedded rocks. Sequences of thinly, rhythmically bedded rocks are found at several levels, but not near the base. Three distinct lithologic types are recognized: (1) Typically massively bedded metasandstones grading into minor dark metapelite. The thickness of beds averages 30 to 60 cm but many beds exceed 1.5 m. (2) Rhythmically graded, bedded laminated metasiltstone and metapelite. The thickness of beds ranges from 1 to 3 cm, with the metasiltstone and metapelite parts of approximately equal thickness. (3) A 70-m section at the very top of the formation showing a gradational change from massive metasandstone to thinly, rhythmically interbedded metasiltstone and metapelite.

Thickness. The thickness of the Fall Brook Formation is estimated to be between 600 and 1,000 m.

Age. No fossils have been recognized in the Fall Brook Formation. Because it lies above the Parkman Hill Formation, which has been dated to be as young as early Ludlovian, and because it lies below the Solon Formation, which is considered to be Early Devonian(?), the age of the Fall Brook Formation is believed to be Late Silurian (post-early Ludlovian) to Early Devonian(?).

Solon Formation (new name)

Type Locality. The type locality of the Solon Formation is in the township of Solon, in the north and northeast parts of the Anson quadrangle, on the tops and slopes of Whipple and French Hills.

Lithology. The Solon Formation is composed of massive silvery gray metapelite; rhythmically interbedded, thinly laminated metasiltstone and silvery gray metapelite; sections of massive metasandstone; and several horizons of polymictic granule metaconglomerate near the base. In the interbedded metasiltstone and metapelite sections, the metasiltstone layers are typically 5 mm to 2 cm thick, and the metasiltstone to metapelite ratio ranges from 1:4 to 3:2.

Thickness. The top of the Solon Formation has been removed by erosion in all the areas shown on the map. The thickness of the preserved section is estimated to be more than 2,000 m.

Age. On the basis of ages of brachiopods in the Sebec and Sebec Lake quadrangles, and also because of its correlation with the fossiliferous Seboomook Formation, the age of the Solon Formation is considered to be Early Devonian(?).

APPENDIX 2. NEW FOSSIL LOCALITIES AND FAUNAL IDENTIFICATIONS

Sangerville Formation

F7A51. Location: Anson quadrangle; 1.5 mi north of Route 43 on east side of Shaddagee Road.
Fauna: *Monograptus* sp. Fragments of possible cyrtograptid cladia with hooked thecae.
Age: Silurian; if the fragments are cyrtograptids, late Llandoverian through Wenlockian.

F8A29. Location: Skowhegan quadrangle; north side of Route 150 just west of jeep trail to Peeks Hill.
Fauna: *Monograptus* sp. (plain thecae); *Monograptus* sp. (probable hooked thecae); possibly distorted specimens of *M. flemingii* group.
Age: Silurian, possibly Wenlockian. If *M. flemingii* is represented, age is Wenlockian.

F9B32. Location: Kingsbury quadrangle; east side of road 1.75 mi north of Wellington, just north of house on west side.
Fauna: *Monograptus* sp. (of the *M. dubius* group); *Monograptus* sp. (very similar to *M. flemingii* [Salter]); *Monograptus* sp. (hooked thecae); *Monograptus* sp.
Age: Silurian, probably Wenlockian.

F71C55. Location: Kingsbury quadrangle; 100 ft west of dirt road, 0.8 mi south of Burdin Corner.
Fauna: *Cyrtograptus*(?) sp.; *Monograptus* sp.
Age: Silurian, possibly late Llandoverian through Wenlockian.

F71D36. Location: Kingsbury quadrangle; in Higgins Brook, 0.2 mi south of junction with tributary that flows through Wellington.
Fauna: *Cyrtograptus* sp. (probably *C. perneri*); *Monograptus flemingii*(?); *Monograptus pseudodubius*; *Monograptus* sp.
Age: Middle Wenlockian.

F106. Location: Dover-Foxcroft quadrangle; in ditch on north side of Route 16, 300 ft west of Sebec Corner, 7 mi east-northeast of Dover-Foxcroft.
Fauna: possible monograptid.
Age: Silurian to Early Devonian.

F165. Location: Dover-Foxcroft quadrangle; south end of long roadcut on Route 15, 1 mi south of South Dover.
Fauna: unidentifiable graptolite scraps.
Age: Ordovician to Early Devonian.

F391. Location: Dover-Foxcroft quadrangle; 2.5 mi east-southeast of Dover-Foxcroft in pit off south side of road that parallels Piscataquis River on south bank.
Fauna: *Monograptus* sp. (*M. dubius* type); *Monograptus* sp. (*M. colonus?*); *Monograptus* sp. (slender fragments with plain thecae; fragments curved).
Age: Late Llandoverian through Ludlovian; if *M. colonus* is present, possibly early Ludlovian.
Second collection:
Fauna: *Monograptus* sp. (*M. dubius* group); *Monograptus* sp. (probably *M. flemingii* group).
Age: Silurian, probably Wenlockian.

F7-435. Location: Dover-Foxcroft quadrangle; 1.75 mi south of Dover-Foxcroft in dirt road 500 ft east of Route 7, on strike with F391.
Fauna: *Monograptus* sp. (*M. dubius* type).
Age: Late Llandoverian through Ludlovian.

F8-155. Location: Dover-Foxcroft quadrangle; small pit 50 ft west of the East Dover–South Dover road, 1.8 mi south-southwest of East Dover.
Fauna: *Monograptus* sp. (*M. dubius* type); *Monograptus* sp. (fragment has thecae with shape and spines similar to those in *M. chimaera*).
Age: Silurian, possibly early Ludlovian.

F6-60. Location: Guilford quadrangle; 2 mi southwest of Dover-Foxcroft on east side of

Maine Central Railroad track, 50 ft north of dirt road that goes from Route 7 to East Sangerville.

Fauna: *Monograptus* sp. (*M. dubius* type); *Monograptus* sp. (possibly *M. colonus* type); *Monograptus* sp. (long, slender fragments).

Age: Late Llandoverian through Ludlovian, possibly early Ludlovian.

F7-506. Location: Guilford quadrangle; 3.5 mi northwest of Dexter on east side of Route 23.

Fauna: *Monograptus praedubius*(?); *Cyrtograptus*(?) sp.; *Monograptus* sp. (*M. priodon–M. flemingii* group, possibly *M. flemingii*).

Age: Late Llandoverian to Wenlockian, probably Wenlockian.

F6-746. Location: Dover-Foxcroft quadrangle; outcrop on south bank of Piscataquis River, at southeast abutment of bridge at east end of town of Dover-Foxcroft (in metalimestone member).

Fauna: Dendroid graptolite scraps?

Age: Ordovician to Early Devonian.

Waterville Formation

F9-430. Location: Stetson quadrangle; pit on south side of east-west dirt road. Pit is 6,000 ft south-southeast of South Exeter and 2,000 ft west of intersection with Exeter-Stetson road. (Top of Waterville Formation.)

Fauna: *Monograptus* sp. (*M. dubius* type); *Monograptus* sp. (thecae are hooked in manner similar to *M. priodon*).

Age: Silurian, probably late Llandoverian to Wenlockian.

F9-454. Location: Stetson quadrangle; pit on south side of east-west dirt road at town line between Corinna and Exeter. Pit is 14,000 ft southwest of South Exeter and 12,000 ft west of intersection of road with the Exeter-Stetson road.

Fauna: Unidentifiable graptolite scraps.

Age: Ordovician to Early Devonian.

Parkman Hill Formation

F9D4. Location: Guilford quadrangle; in dirt road, 0.19 mi northeast of crest of Crow Hill.

Fauna: *Monograptus bohemicus* (Barrande); *Monograptus* cf. *M. dubius* (Suess); *Monograptus* sp. (*M. nilssoni* [Barrande]?); *Monograptus* sp. (with uncinate thecae?).

Age: Early Ludlovian

F9A70. Location: Kingsbury quadrangle; a few hundred feet south of Route 154, at Trout Pond, on a small knoll.

Fauna: Vague impressions of graptolites.

Age: Ordovician or Silurian.

F9B41. Location: Kingsbury quadrangle; top of unnamed hill, elevation 1,192 ft, 1.25 mi west of Huff Corner.

Fauna: Vague impressions of graptolites; may be monograptids.

Age: Possibly Silurian.

F9B39. Location: Kingsbury quadrangle; west side of road from Kinsgbury to Huff Corner, a few feet north of junction with side road to southwest, on strike with F9B41.

Fauna: *Monograptus* sp. (possibly *M. vomerinus* type).

Age: Silurian; possibly in the span of late Llandoverian–Wenlockian. Rhabdosome size suggests large member of *M. vomerinus* group, and if so, age would most likely be Wenlockian.

F70A37. Location: Kingsbury quadrangle; west side of Route 151, 15 ft south of Central Maine Power Company pole 217, 2 mi south of Brighton.

Fauna: *Monograptus* sp. (*M. priodon* [Bronn]?); *Monograptus* sp.; possible cyrtograptid fragments.

Age: Silurian; most likely late Llandoverian or Wenlockian.

F8-127. Location: Pittsfield quadrangle; south side of Route 154 at a crossroad 0.7 mi west of Route 152.

Fauna: *Monograptus* sp. (distal fragment of rhabdosome; thecal form and overlap are suggestive of thecae in the distal part of rhabdosomes of the *M. colonus* group).

Age: Silurian; if the specimens are indeed members of the *M. colonus* group, or even *M. dubius* or *M. ludensis* types, the age is probably early Ludlovian.

F71B20. Location: Kingsbury quadrangle; at bend in dirt road 0.4 mi east of junction of road with Route 154, approximately 1 mi east of Trout Pond.

Fauna: *Monograptus* sp. (*M. dubius* group); *Monograptus* sp. (*M. uncinatus?*).

Age: Silurian, possibly Ludlovian.

F7C51. Location: Skowhegan quadrangle; north side of road leading east from major road, approximately 1.5 mi north of West Athens.

Fauna: *Monograptus* sp. (tubular thecae) similar to those in *M. dubius* group); *Monograptus* sp. (hooked thecae, possibly *M. flemingii*).

Age: Silurian, possibly Wenlockian.

Vassalboro Formation

F6-462. Location: Stetson quadrangle; 0.25 mi north of Stetson, in pit on west side of Stetson-Exeter road.

Fauna: *Monograptus* sp. (*M. dubius* type); *Monograptus* sp. (long slender fragments with plain thecae).

Age: Late Llandoverian through Ludlovian.

F8-191. Location: Stetson quadrangle; bulldozed hill 0.25 mi northeast of road intersection at center of Stetson.

Fauna: Vague linear features suggestive of graptolites.

Age: Ordovician to Early Devonian.

REFERENCES CITED

Barker, D. S., 1961, Hallowell granite and associated rocks, south-central Maine [PhD. thesis]: Princeton, N.J., Princeton Univ., 240 p.

Berry, W.B.N., and Boucot, A. J., 1970, Correlation of the North American Silurian rocks: Geol. Soc. America Spec. Paper 102, 289 p.

Billings, M. P., 1956, The geology of New Hampshire, Pt. 2: Bedrock geology: Concord, New Hampshire State Plan. and Devel. Comm., 203 p.

Boone, G. M., Boudette, E. L., and Moench, R. H., 1970, Bedrock geology of the Rangeley Lakes-Dead River Basin region, western Maine, in Boone, G. M., ed., Guidebook for field trips in the Rangeley Lakes-Dead River Basin region, western Maine: New England Intercollegiate Geol. Field Conf., 62d Ann. Mtg., p. 1-24.

Borns, H. W., 1959, Geology of the Skowhegan quadrangle, Maine [PhD. thesis]: Boston, Boston Univ., 183 p.

Boucot, A. J., 1961, Stratigraphy of the Moose River synclinorium, Maine: U.S. Geol. Survey Bull. 1111-E, p. 153-188.

——1970, Devonian slate problems in the northern Appalachians: Maine Geol. Survey Bull. 23, p. 42-48.

Boucot, A. J., Field, M. T., Fletcher, R., Forbes, W. H., Naylor, R. S., and Pavlides, L., 1964, Reconnaissance bedrock geology of the Presque Isle quadrangle, Maine: Maine Geol. Survey Quad. Mapping Ser. no. 2, 123 p.

Cady, W. M., 1969, Regional tectonic synthesis of northwestern New England and adjacent Quebec: Geol. Soc. America Mem. 120, 180 p.

Cariani, A. R., 1958, The geology of the Anson quadrangle, Maine [PhD. thesis]: Boston, Boston Univ., 177 p.

Dixon, H. R., and Lundgren, W. L., 1968, Structure of eastern Connecticut, in Zen, E-an, and others, eds., Studies of Appalachian geology, northern and maritime: New York, Wiley Interscience Pubs., Inc., p. 219-229.

Espenshade, G. H., and Boudette, E. L., 1967, Geology and petrology of the Greenville quadrangle, Piscataquis and Somerset counties, Maine: U.S. Geol. Survey Bull. 1241-F, 60 p.

Fisher, L. W., 1941, Structure and metamorphism of Lewiston, Maine, region: Geol. Soc. America Bull., v. 52, p. 107-160.

Furlong, I. E., 1961, The geology of the Farmington quadrangle, Maine [PhD. thesis]: Boston, Boston Univ., 162 p.

Gates, O., 1969, Lower Silurian-Lower Devonian volcanic rocks of New England coast and southern New Brunswick, in Kay, M., ed., North Atlantic—geology and continental drift: Am Assoc. Petroleum Geologists Mem. 12, p. 484-503.

Glidden, P. E., 1963, Geology of the Pittsfield quadrangle, Maine [PhD. thesis]: Boston, Boston Univ., 256 p.

Green, J. C., and Guidotti, C. V., 1968, The Boundary Mountains anticlinorium in northern New Hampshire and northwestern Maine, in Zen, E-an, and others, eds., Studies of Appalachian geology, northern and maritime: New York, Wiley Interscience Pubs., Inc., p. 255-266.

Griffin, J. R., 1973, A structural study of the Silurian metasediments of central Maine [PhD. thesis]: Riverside, Univ. California at Riverside, 157 p.

Guidotti, C. V., 1965, Geology of the Bryant Pond quadrangle, Maine: Maine Geol. Survey Quad. Mapping Ser. no. 3, 116 p.

Hussey, A. M., II, 1968, Stratigraphy and structure of southwestern Maine, in Zen, E-an, and others, eds., Studies of Appalachian geology, northern and maritime: New York, Wiley Interscience Pubs., Inc., p. 291-301.

Ludman, A., 1969, Structure and stratigraphy of Silurian rocks in the Skowhegan area, south-central Maine: Geol. Soc. American, Abs. with Programs for 1969, Pt. 1, (Northeastern Sec.), p. 37-38.

Ludman, A., Griffin, J., and Lindsley-Griffin, N., 1972, Sedimentary facies relationships in the Siluro-Devonian slate belt of central Maine: Geol. Soc. America, Abs. with Programs (Northeastern Sec.), v. 4, no. 1, p. 28.

Moench, R. H., and Boudette, E. L., 1970, Stratigraphy of the northwest limb of the Merrimack synclinorium in the Kennebago Lake, Rangeley, and Phillips quadrangles, western Maine, *in* Boone, G. M., ed., Guidebook for field trips in the Rangeley lakes–Dead River basin region, western Maine: New England Intercoll. Geol. Field Conf., 62d Ann. Mtg., p. A1-1–A1-25.

Osberg, P. H., 1968, Stratigraphy, structural geology, and metamorphism of the Waterville-Vassalboro area, Maine: Maine Geol. Survey Bull. 20, 64 p.

Osberg, P. H., Moench, R. H., and Warner, J., 1968, Stratigraphy of the Merrimack synclinorium in west-central Maine, *in* Zen, E-an, and others, eds., Studies of Appalachian geology, northern and maritime: New York, Wiley Interscience Pubs., Inc., p. 241–253.

Pankiwskyj, K., 1964, Geology of the Dixfield quadrangle, Maine [PhD. thesis]: Cambridge, Mass., Harvard Univ., 224 p.

Pavlides, L., Mencher, E., Naylor, R. S., and Boucot, A. J., 1964, Outline of the stratigraphic and tectonic features of northern Maine: U.S. Geol. Survey Prof. Paper 501-C, p. 28–38.

Perkins, E. H., 1924, A new graptolite locality in central Maine: Am. Jour. Sci., v. 208, p. 223–227.

Perkins, E. H., and Smith, E.S.C., 1925, Contributions to the geology of Maine, No. 1; A geological section from the Kennebec River to Penobscot Bay: Am. Jour. Sci., v. 209, p. 204–228.

Rodgers, John, 1970, The tectonics of the Appalachians: New York, Wiley Interscience Pubs., Inc., 271 p.

——1971, The Taconic orogeny; 1970 presidential address, Geol. Soc. America: Geol. Soc. America Bull., v. 82, p. 1141–1178.

Seilacher, A., 1964, Biogenic sedimentary structures, *in* Imbrie, J., and Newell, N. D., eds., Approaches to paleoecology: New York, John Wiley & Sons, Inc., p. 296–316.

Skapinski, S. A., 1961, Geology of the Kingfield quadrangle, Maine [PhD. thesis]: Boston, Boston Univ., 266 p.

Warner, J. L., 1967, Geology of the Buckfield quadrangle, Maine [PhD. thesis]: Cambridge, Mass., Harvard Univ., 230 p.

Warner, J. L., and Pankiwskyj, K., 1965, Geology of the Buckfield and Dixfield quadrangles in northwestern Maine, *in* Hussey, A. M., II, ed., Guidebook to 57th Ann. Mtg: New England Intercoll. Geol. Conf., Trip K, p. 103–118.

MANUSCRIPT RECEIVED BY THE SOCIETY APRIL 29, 1974
REVISED MANUSCRIPT RECEIVED AUGUST 22, 1974
MANUSCRIPT ACCEPTED AUGUST 26, 1974

Structural Evolution of the White Mountain Magma Series

CARLETON A. CHAPMAN
University of Illinois
Urbana, Illinois 61801

ABSTRACT

The shape, arrangement, and great differences in age of the intrusive complexes of the White Mountain magma series (of Jurassic to Cretaceous age) indicate an origin from discrete magma chambers, probably formed within the upper mantle. Initially basaltic, the melt of these closed chambers soon acquired a capping of granitic magma, formed by selective melting, as the chambers ascended through the crust. When the polymagmatic chambers reached the upper part of the crust, the granitic cap was fully developed and the underlying melt had become syenitic through assimilation and differentiation. As cumulate material settled, the chamber floor was built upward, and the volume of melt was greatly reduced.

Ring dikes and stocks formed a complicated sequence after the chambers reached roughly their diametral distance from the Earth's surface. Some ring dikes may have formed along paraboloidal fractures, but the steep dips of the ring dikes and evidence of repeated subsidence imply formation along cylindrical fractures. A new theory of the mechanism of ring dike formation, involving cone fracturing followed by stoping and subsidence, is explained. It accounts for the later interior stock so common in ring complexes.

The model of a floored chamber with coexisting syenitic and granitic melts explains why the subsiding block stops, why mafic ring dikes are rare, why granite is found more commonly in stocks and syenite more commonly in ring dikes, why quartz syenite cuts granitic rock, why rhyolite appears at the surface before quartz syenite forms in associated ring dikes, and why rhyolite dominates the associated volcanic rocks. The model should remind us that petrographic similarity is not an infallible criterion of age, and it suggests the probability that certain rocks of the White Mountain magma series are products of mixed magmas.

INTRODUCTION

This paper presents a broad, unifying hypothesis that accounts for the structural evolution of the White Mountain magma series (of Jurassic to Cretaceous age) of New England. The hypothesis deals with the process from the origin of the magma at depth to the final stage of solidification at or near the Earth's surface. Although it deals with evolutionary changes largely from a structural point of view, the hypothesis is compatible with the requirements imposed by the theories of petrological evolution. The treatment of the earlier stages of evolution is naturally highly speculative, but some direct observations made at the Earth's surface do support this view.

The literature on the White Mountain magma series is voluminous, but only a brief description of this igneous rock association is necessary here. The series is composed of an association of mildly alkalic plutonic and volcanic rocks clustered in central complexes that form a broad belt across New Hampshire (Fig. 1). Granite is the dominant rock type; it occurs largely in stocklike plutons and much less commonly in ring dikes. Syenite and quartz syenite are common in ring dikes but less common in stocks. Gabbro and diorite are uncommon and are usually restricted to plug- or funnel-like bodies. Other plutonic rocks are rare. Volumetrically the volcanic rocks (Moat Volcanics) are rare, but locally they constitute thick sequences preserved in areas of cauldron subsidence. Volcanic flows, tuff, and breccia are dominantly rhyolitic, with subordinate amounts of andesitic and basaltic material. Trachytic varieties are rare.

The close chemical, petrologic, and structural similarities among certain of the posttectonic, mildly alkalic igneous complexes of New Hampshire long ago suggested a common magma source. This idea was clearly expressed by Billings (1945, p. 55): "The tectonic history of the White Mountain magma series is controlled by the magmatic differentiation of the main underlying reservoir. The remarkable uniformity of the White Mountain magma series throughout New Hampshire implies that a single reservoir underlay much of the state. The petrographic peculiarities found in the smaller complexes, such as Red Hill and Mount Tripyramid, imply that these rocks evolved in small cupolas rising from the main reservoir . . ." However, Billings (1942, p. 91) recognized an alternative hypothesis when he stated, "A single master reservoir may not have existed; rather, each complex may have been underlain by a circular, steep-walled reservoir 15 to 35 km high." Nevertheless, at that time he felt that the remarkable petrographic similarities supported the first hypothesis.

As new investigations revealed additional complexes of the "White Mountain type" at more remote localities within the state and in neighboring parts of southwestern Maine and southeastern Vermont, it became necessary, on the basis of Billings' hypothesis, to postulate a universal reservoir of correspondingly greater dimensions.

But the assumption of a magma reservoir sufficiently extensive to supply igneous complexes within an area about 290 × 225 km raises a serious question about the stability of the shallow roof zone—postulated (Billings, 1942) to be perhaps 8 to 15 km thick—above such a flat-topped magma reservoir. It is unlikely that a roof zone of this thickness, spanning a magma reservoir ranging in width from 225 to 290 km, would maintain its integrity. Extensive foundering might not be expected in the early stages of differentiation, but repeated fracturing of the roof rocks should be recorded by the presence of numerous sets or swarms of steep dikes whose compositions reflect successive changes in the chemical nature of the differentiating magma. Partly because of this objection, I visualized (Chapman,

1963, 1967, 1968) small, discrete magma reservoirs.

Geochemical age dating by numerous workers has raised another objection to Billings' hypothesis. In an especially pertinent paper on the geochemical ages of these rocks, Foland and others (1971, p. 327) stated, "Because of the above constraints, it is suggested that one magma chamber was not sufficient to supply all the 'White Mountain' rocks in spite of their similarities. A single magma chamber model where magmatic differentiation controlled the compositions of the different phases would require that all early phases predate the Conway granite. However, the monzonite from the Mount Pawtuckaway complex (120 m.y.), where no granite occurs, is clearly younger than some Conway granites and the monzonite associated with them." Here the term "chamber" is used in the same sense as the word "reservoir." Foland and others continued, "The data suggest that two or more magma chambers existed and that a generally similar process operated in each one but at different times in the period of 70 m.y." Although I agree with both these statements, they are in my opinion most conservatively expressed.

STRUCTURAL CONTROL AND DISCRETE MAGMA CHAMBERS

When Billings (1956) published his geologic map of New Hampshire, I was impressed by the striking reticulate or netlike arrangement of the roughly circular igneous complexes (Fig. 1). This pattern suggested a strong structural control, as if the complexes were located at the intersections of two sets of fracture zones running parallel to the principal lines of the net (roughly due east and south-southeast). If the complexes represented cupolas above an extensive shallow-roofed magma reservoir, then any controlling fractures must have existed in the crust above the reservoir—that is, in the upper part of the crust, perhaps 5 to 15 km from the surface. Because no extensive reticulate fracture or dike patterns had been observed at the present surface, it seemed reasonable to assume that controlling fractures, if they did exist, were deep-seated, perhaps near the base of the crust. In any case, the supposition of a structural control seemed a logical one.

The hypothesis of structural control and formation of discrete magma chambers as applied to the White Mountain magma series has already been presented with illustrations (Chapman, 1967, 1968) and need only be outlined here. I postulated that movements produced, in the upper part of the mantle and (or) lower part of the crust, two sets of steeply inclined, sheetlike zones of deformation along which basaltic magma formed and (or) accumulated. Repeated movements served to produce or accumulate new melt and move existing magma to locations of lower pressure potential. Upward migration and collection of magma took place most rapidly along the more active zones and particularly at intersections of fracture zones. Discrete magma chambers thus developed. These were closed, equilibrium forms of roughly ellipsoidal shape; they were stable under the high confining pressures. The chambers migrated upward perhaps largely by the processes of plastic yielding of the wall rock and overhead stoping. As melting and stoping became effective near the chamber roof, xenoliths and newly crystallized material collected on the floor.

The hypothesis of structural control receives further support from the shapes of many plutons and complexes of the White Mountain magma series. A number of individual plutons (Fig. 1) are elongate or elliptical in ground plan. The elongate form of such plutons may have been derived from the underlying magma chamber, whose outline in turn reflects the control of a principal fracture zone.

Much more significant is the elongation of many complexes parallel to a lattice line. Examples of south-southeasterly elongation are found in the Belknap Mountain complex and in the unnamed complex about 40 km southeast of the Ossipee Mountain complex; examples of easterly elongation are seen in the Ascutney Mountain, Pliny, and Percy complexes (Fig. 1). In ground plan, elongation appears to represent a series of successive intrusions that overlapped one another somewhat both in time and space. This phenomenon, described from many parts of the world, has long been recognized in central complexes, where it is referred to as "migration of centers of intrusion."

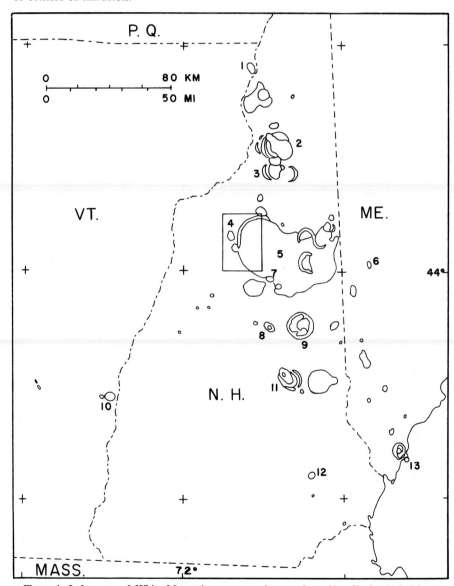

Figure 1. Index map of White Mountain magma series, northern New England. 1, Mount. Monadnock pluton; 2, Percy complex; 3, Pliny complex; 4, Franconia quadrangle; 5, White Mountain batholith; 6, Pleasant Mountain pluton; 7, Tripyramid complex; 8, Red Hill complex; 9, Ossipee Mountain complex; 10, Ascutney Mountain complex; 11, Belknap Mountain complex; 12, Pawtuckaway complex; 13, Cape Neddick complex.

My explanation of this phenomenon is perhaps most convincingly demonstrated by Turner's (1963) study of the Sara Fier complex of northern Nigeria. I interpret Turner's findings to indicate that the series of five successive and contiguous to overlapping intrusions formed from upward-moving, discrete magma chambers. These chambers were disposed parallel to a lattice line in ground plan but were staggered en echelon in the vertical dimension. This arrangement permitted intrusions to arrive on line at the present surface level at different times. I believe Turner's data constitute a compelling argument for the existence of structurally controlled, discrete magma chambers.

POLYMAGMATIC CHAMBERS

The concept of a simple rock series formed by differentiation of a basaltic melt and leading from gabbroic through dioritic, monzodioritic, syenitic, and quartz syenitic to granitic rock types was found by early workers to be a quantitatively inadequate explanation for the White Mountain magma series. For example, it failed to account for the large amount of quartz-rich rocks. By assuming that considerable assimilation accompanied the differentiation process, Chapman and Williams (1935) minimized this problem. More recently I showed (Chapman, 1966a) that by assuming two coexisting magmas, the problem of quartz-rich rocks may be more satisfactorily resolved and additional objections raised by monomagmatic theory may be removed.

The two magmas, basaltic and granitic, could have formed at nearly the same time but at different levels in the earth, or the granitic magma could have been formed by selective melting of crustal material by the hot basaltic magma. Mainly because of its high viscosity and low density, much felsic melt would escape mixing and rise as liquid globs to the top of the magma chamber. It should not be inferred that the failure of these two implicated magmas to homogenize implies liquid immiscibility.

Once a thin layer of granitic melt had formed above the mafic melt, it would develop its own pattern of thermal convection, which would operate in harmony with that established in the mafic melt beneath (Fig. 2A). Heat from the cooling mafic melt would be transferred upward through the granitic magma by tandem convection, and the process of melting the roof rocks could continue. The granitic melt, it should be noted, would possess considerable superheat for some time because its temperature would be maintained at roughly that of the mafic magma below.

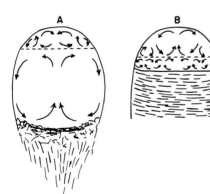

Figure 2. Hypothetical polymagmatic chamber showing convection patterns. A, granitic magma above basaltic magma in young chamber; B, granitic magma above syenitic magma resting on cumulate material (long dashes) in old chamber.

Presumably while the cap of granitic melt was still forming, the basaltic melt was crystallizing, and the heavy minerals that formed early collected, with the xenolithic material, at the base of the magma chamber and built up on the floor. I agree with Holmes (1931) that in time the lower magma in such a system would evolve to one of syenitic composition, passing through such intermediate stages as dioritic, monzodioritic, and monzonitic. So, as the lower melt in our two-magma chamber continued to crystallize, its volume was reduced, and melting of and reaction with xenoliths, which were fed to it from the still-active granitic melt above, became progressively less important. Melting by the granitic liquid, when deprived of its superheat, became negligible, but stoping remained active in the more brittle and easily fractured sialic crust. Great quantities of xenolithic material accumulated on the floor of the magma chamber. As convection within the magma layers became weaker due to changes of temperature, composition, and viscosity, a fairly good stratification of melts, granitic above syenitic, was maintained (Fig. 2B).

As we shall see, this concept of coexisting felsic (granitic and syenitic) magmas in the late stages of the magma chambers does explain, among other things, certain variable sequences of eruption, certain compositional variations within apparently simple intrusive bodies, and the reason why syenite generally forms ring dikes whereas granite tends to form stocklike plutons.

Actually, we know little of the magma chamber during its early history, but fortunately, a more complete record of late-stage evolution is well preserved and available for study in the rocks of the upper crust. It is with this record that we will now be most concerned. The declining stage of the magma chamber is characterized by (1) severe fracturing or shattering of the roof rocks, (2) extensive and intensive stoping by an active but rapidly cooling granitic magma, and (3) emplacement, at high levels in the crust or at the surface, of a variety of rock melts in an equally great variety of geometric forms. These phenomena will now be considered in detail.

INTEGRITY OF THE ROOF ROCKS

As the magma chamber comes close to the Earth's surface, the roof rocks become highly susceptible to rupture and disintegration. Such conditions might be expected to arise at about the time that the thickness of the roof-rock layer equals the diameter of the underlying magma chamber. Disruptive forces in the overlying rocks could be caused either by uplift or depression of the chamber roof.

A frequently cited cause of increased chamber pressure, which could affect the roof rocks, is rising gas pressure, particularly in the late stages of the melt's crystallization. Rather sudden and intense increases of pressure, effecting great bursting pressures in the magma chamber, could result from tectonic adjustments in the crust. Sharp reductions in chamber pressure could also be induced tectonically. Reduction in roof support could come about when rapid expulsion of magma, caused by gas accumulation, partially empties the chamber. I believe that discrete magma chambers completely surrounded by crystalline rock would be highly sensitive to such crustal movements as crowding, flexing, and faulting.

Anderson's theory (Bailey and others, 1924; Anderson, 1936) of rupture in crustal rocks above certain magma reservoirs has been widely accepted as an explanation for the origin of ring dikes and cone sheets. A number of recent studies, however, have questioned the validity of Anderson's analysis, and Roberts (1970) has discussed the subject in detail. Opinions still vary widely; it will suffice here, therefore, to speak in general terms. We will consider that when the pressure in the magma chamber is increased and the superjacent rocks are pushed upward, a series of curved fractures (Anderson's cone fractures) may develop. Where such fractures are injected with magma, cone sheets will form.

Should the chamber pressure be reduced, arch fractures roughly parallel to the chamber roof and normal to the direction of maximum tensile stress may develop; but due to the archlike fashion in which they span the magma chamber, they would not normally open and permit the roof to collapse. Where interrupted by favorably oriented cross fractures, however, arch-shaped shells of roof rock might be freed to settle into the magma chamber.

Also under conditions of reduced chamber pressure, a third type of fracture, paraboloidal fracture, may form. Where sufficiently extensive, such a fracture would isolate a huge, roughly paraboloidal block, and the outward inclination of the fracture would enable the block to settle into the melt. This large, heavy block should settle to the floor of the felsic magma chamber. With concomitant upward displacement of the melt, the magma chamber will advance upward rapidly, in one fell swoop, by what we may call paraboloidal stoping (Fig. 3A). It seems unlikely, however, that a paraboloidal fracture would extend high enough to completely isolate the block as Anderson indicated. As soon as it starts to propagate upward, the lower part of the paraboloidal fracture provides favorable conditions for tensional release in the roof and consequent opening of one or more broadly curved arch fractures. The growing paraboloidal fracture is therefore expected to terminate upward in a subhorizontal arch fracture. Thus a bluntly truncated,

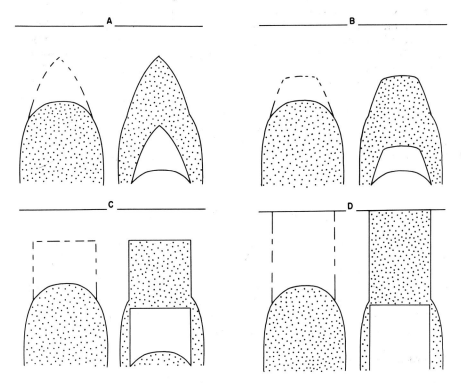

Figure 3. Special types of high-level stoping. A, paraboloidal fracture and stoping; B, the same with truncated paraboloidal block; C, cylindrical fracture and stoping forming underground cauldron; D, the same forming surface cauldron.

paraboloidal block should be isolated (Fig. 3B). Evidence for this type of termination will be considered later in connection with ring dikes.

Without paraboloidal fracture or some comparable structure, arch fracturing is not likely, but in combination with such features, extensive opening of arch fractures could lead to arch stoping. And so we might expect to find all gradations between arch stoping and paraboloidal stoping. This type of subsidence is considered by Locke (1926) and Wisser (1927) to have operated over certain ore bodies and mining slopes, and Billings (1945) suggested it as a possible stoping mechanism in New Hampshire. It seems to me that the huge curved inclusions in the syenite plutons in Vermont at Ascutney Mountain (Chapman and Chapman, 1940) and Mount Monadnock (Chapman, 1954) may now be better interpreted as archlike slabs fallen to the floor of closed magma chambers. The large inclusions of volcanic rocks in the syenite body at Pleasant Mountain, Maine (Jenks, 1934), are arranged in a roughly arclike fashion nearly concentric with the pluton. I interpret these also to represent huge slabs formed by arch fracturing showered down upon the chamber floor.

Detailed studies of subsidence and collapse structures formed above mining stopes and certain ore bodies throw much light on the subject of fracture formation in rocks above magma chambers. Here another type of ring fracture appears to control the shape of the subsided roof block. For example, over some of the stopes and ore bodies at Bisbee, Arizona (Wisser, 1927), and in the Pilares pipe in Sonora, Mexico (Locke, 1926), a cylindrical block, somewhat elongate vertically, has subsided. Each block appears to be bounded by a vertical, cylindrical fracture zone whose horizontal cross section corresponds approximately in size and shape to that of the cavity above which the subsidence occurred. The cylindrical fractures are believed to become surfaces of shear that permit downward displacement. These observations support the concept, proposed by Clough and others (1909) to explain the ring intrusions at Glen Coe and Ben Nevis, Scotland, that a cylindrical block might develop through subsidence above a magma chamber (Fig. 3C, 3D). For the Scottish area, however, no similarity was expressed between the shape of the horizontal cross-sectional area of the subsided block and that of the underlying magma reservoir. After Anderson's theory of the formation of ring dikes was proposed (Bailey and others, 1924), the hypothesis of a cylindrical block almost fell into disuse. The idea was revived, with considerable modification (Billings, 1945), and was applied to the White Mountain magma series.

ORIGIN OF THE RING DIKES

Billings (1943, 1945) presented detailed descriptive summaries of the ring dikes and related igneous features of the White Mountain magma series and gave a critical account of the various theories of ring-dike origin. Even after nearly three decades, these annular bodies are still puzzling structural features (Roberts, 1970). It is known that some ring dikes are clearly related to cauldron subsidence, whereas others may have formed independent of such subsidence. But there are still many questions that prevailing theories of ring-dike genesis fail to answer.

Paraboloidal Ring Dikes

Slight subsidence of a paraboloidal block, accompanied by flowage of magma into the dilated ring fracture, could form a ring dike. A common objection to this hypothesis is that it does not explain why the block stops subsiding. Where the melt is felsic, the denser block is likely to settle deeply into the magma reservoir. It is commonly concluded that this type of subsidence should lead to a stocklike pluton and not a ring dike (Fig. 3A).

The model of a closed or floored magma chamber has supplied two possible answers to this question (Chapman, 1966a). First, the subsiding block comes to rest on the floor of the chamber or on a thick cover of xenolithic material spread over that floor (Fig. 4A). Second, the block is stopped by a constructed floor at the magma-magma interface (Fig. 4B). Either mechanism keeps the block partly within the cupola so a ring dike can form.

A more serious problem is posed, however, by the fact that paraboloidal ring dikes should dip outward and in many cases at relatively low angles. Paraboloidal fractures extended upward should form isolated blocks with gently inclined conical tops. Subsidence of such blocks should create very broad ring dikes with thickness to radius ratios commonly as much as 0.5 or so (Fig. 4C). But such features are not observed in the New Hampshire area. The ring dikes of the White Mountain magma series are steep—almost vertical—and, as Billings (1945) has indicated, are not satisfactorily explained by the hypothesis of subsidence of a paraboloidal block. Therefore, paraboloidal fractures controlling ring dikes are probably truncated by subhorizontal fractures (arch fractures), as already indicated (Fig. 4D). Subsided

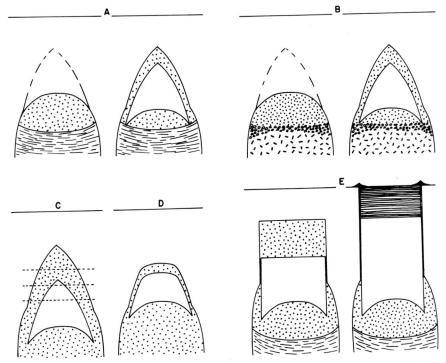

Figure 4. Types of ring dikes. A, paraboloidal ring dike with subsided block resting on chamber floor; B, the same with block resting on constructed floor of xenoliths frozen in along felsic-mafic melt interface; C, paraboloidal fracture may lead to formation of a stock at high erosion level (uppermost dashed line) or very broad ring dikes at lower levels (two lower dashed lines); D, truncated paraboloidal fracture leads to steep narrow ring dike and subhorizontal sheet-like cap; E, cylindrical ring dikes showing subsurface and surface cauldrons with cylindrical block arrested above chamber floor. Long dashes = cumulate material in chamber; dots = felsic magma; dashes = mafic magma; open symbols = constructed floor of xenoliths frozen in along interface; solid lines = volcanic rocks in surface cauldron.

blocks with partial or complete subhorizontal tops are common in numerous areas, for example, Northern Ireland (Richey, 1927; Emeleus, 1955) and Nigeria (Jacobson and others, 1958). These tops pass rather abruptly to steep boundaries near the margins of the subsided blocks. This suggests an arch fracture (subhorizontal) that truncates a steep ring fracture and not a single, continuous paraboloidal fracture. Fairly complete paraboloidal fractures, therefore, do not appear to bound ring dikes in the New Hampshire region, but they may have controlled the boundaries of some felsic stocklike plutons. At least some such contacts are known to dip as low as 45° in the Percy complex (Chapman, 1948).

Cylindrical Ring Dikes

Formation of ring dikes by subsidence of a cylindrical block is a concept that overcomes certain objections raised by the paraboloidal-fracture hypothesis (Fig. 4E). It explains why the dip of the ring dikes of the New Hampshire region is almost vertical, and why the isolated block may undergo repeated subsidence and arrest, a phenomenon apparently common in ring dike areas. An imperfectly formed cylindrical fracture could lead to repeated lodging of the block, thereby interrupting subsidence.

For formation of the dike we must postulate a cylindrical fracture zone produced perhaps as the cylindrical block shears downward. Billings (1945) suggested that a similar fracture zone might form by a sudden upward push of the magma on its roof. Although I agree in general with this concept, I will explain in the next section a different version, which seems to me more satisfactory. Space for the cylindrical dike filling may be provided by displacement of rock rubble upward (flushing out) or downward (settling or stoping). Again, space might be created along the fracture through radial contraction of the cylinder due to sag in the block. It is doubtful if forcefully intruded melt is important in crowding back the walls of the fracture zone.

If a cylindrical fracture or fracture zone reached the surface, subsidence would lead to the formation of a surface cauldron, and the cauldron would fill with upward-rushing melt as rapidly as the block subsided. The upward-streaming magma would likely loosen and carry with it xenoliths formed by brecciation along the fracture zone. This action would clear away much rubble and widen the fissure zone for emplacement of a ring dike.

Whether the cylindrical zone is formed by a sudden upward push of the magma on its roof or by a reduction in chamber pressure is difficult to say. Experience with small-scale examples in mining stopes and ore bodies indicates that cylindrical blocks do form and subside directly under the control of gravity. As Billings (1945) pointed out, the scheme of cylindrical subsidence applies well to some of the most conspicuous ring dikes in New Hampshire, especially those involving porphyritic quartz syenite (for example, in the Ossipee Mountain complex and the northeastern part of the Franconia quadrangle).

Ring Dikes Formed by Cone Fracturing and Subsidence

Some time ago I outlined a model that I think more suitably accounts for many of the ring dikes (Chapman, 1966b); it is explained in detail in Figure 5. For simplification, only one felsic magma is illustrated. It should be noted that this hypothesis differs considerably from the one proposed by Reynolds (1956), which also involves the formation of cone fractures.

My model not only explains why block subsidence is limited (by depth of magma chamber; see Fig. 5F) and how some ring intrusions may form with little or no subsidence (Fig. 5A), but it also accounts for the repeated subsidence observed in so many localities (for example, Belknap Mountain, Percy and Pliny complexes, and Franconia quadrangle). As parts of the obstructing peripheral shoulder are trimmed back and as the melt encroaches upon the block margin by stoping along the ring fracture (Fig. 5B, 5C), subsidence may be renewed from time to time. In some instances a stable condition may not be reached until the block comes to rest on the chamber floor or on the blocky debris that covers the floor (Fig. 5F).

The model also explains why the ring dikes of the White Mountain magma series are steep or vertical. The apparent inward dip of 78° for the Devil's Slide ring dike in the Percy complex (Chapman, 1948) is in agreement with the cone fracture hypothesis.

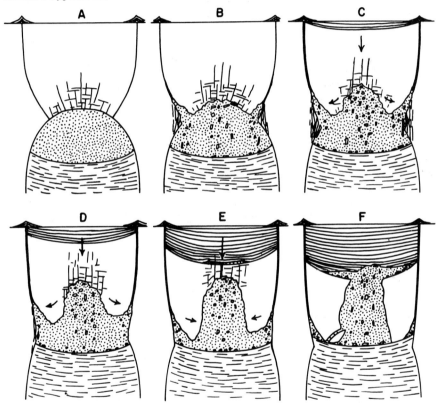

Figure 5. Ring dike formed by cone fracturing and subsidence. A, Major cone fracture extends to surface, permitting volcanism; minor steep cone fractures, steep radial joints, and subhorizontal arch fractures form at base of conical block. B, Stoping initiated where fractures abound, at center of block base and along lower part of major cone fracture; spalling of shoulder at lower wall of cone fracture begins. C, Continued piecemeal stoping and spalling of shoulder decreases constriction beneath block, permitting interrupted subsidence, sagging, fracturing, and volcanism. D, Continued retreat of peripheral shoulder and upward stoping along block core promote further subsidence, fracturing, and sagging of weakened block and filling of surface cauldron with volcanic rock. E, Peripheral shoulder is trimmed back, permitting further subsidence and volcanism; central stope nearly penetrates original block; some magma reaches volcanic rocks at base of cauldron. F, Block stops at chamber floor and is penetrated by central stope: crystallizing magma forms interior stock within ring dike. Dots = felsic magma; solid lines = volcanic rocks; long dashes = cumulate material in chamber.

The steep radial and tangential (conical) fracture pattern assumed to have formed in the base of the conical block when the major cone fracture formed (Fig. 5A) is exemplified by the spider-web joint pattern of the Cape Neddick complex (Gaudette and Chapman, 1964) of southwestern Maine (Fig. 1). Here steep radial and tangential joints were formed by a slight doming action that accompanied explosive eruption and development of cone sheets (Hussey, 1962).

The spider-web pattern of jointing is assumed by my model to be reactivated from time to time during subsidence of the block (Fig. 5C, 5D, 5E). In the Mount Monadnock complex, Chapman (1954) attributed the steep radial-tangential joint system to subsidence and sag during emplacement of the complex. Although the data are not numerous, the dikes at Mount Monadnock also appear to be radial and tangential and were thought by Chapman to have been emplaced along the joints. In an earlier study of the Pawtuckaway Complex, Roy and Freedman (1944) placed no structural interpretation on the distribution and attitude of the associated dikes, but their large-scale map shows a spider-web pattern.

Undoubtedly the dikes, which cover a wide range of composition in each of these two areas, formed over an extended period of time. The long period of formation indicates that dilation of old and creation of new radial and tangential fractures was repeated a number of times in these complexes, perhaps in response to repeated upward thrust as well as to repeated subsidence. Similar joint and dike patterns undoubtedly occur in other complexes, but data are not sufficient to verify their presence.

Large-scale adjustments within the subsiding block are to be expected in certain complexes, particularly during the late stages, by dilation of or slip along large radial fractures. These processes should result in the formation of steep radial faults and dikes. Kingsley (1931) has reported numerous faults and dikes displaying a roughly radial map pattern in the subsided block of the Ossipee Mountain complex. The large dikes appear as inward-projecting offshoots of the great encircling ring dike.

Reduction of the peripheral shoulder and encroachment of magma along the block margin (Fig. 5B, 5C) would most likely proceed at an uneven rate around the cone fracture, effecting a more pronounced subsidence along one side of the

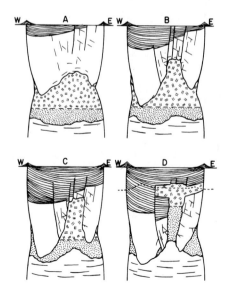

Figure 6. Hypothetical east-west cross sections of Ossipee Mountain complex. A, Cone fracture forms above magma chamber, and rhyolitic volcanic rocks accumulate in surface cauldron as block subsides differentially. B, Continued differential sag and tilting crowds granitic magma to eastern side of chamber where it stopes upward along weakened zones; rhyolitic rocks fill surface cauldron. C, More uniform subsidence of block drives both granitic and syenitic magma up along ring fracture, and central stope nearly penetrates original conical block. D, Partially disintegrated block comes to rest on xenolithic piles on chamber floor as granitic melt invades and spreads out in eastern part of cauldron to form off-centered interior stock. Double dashes = granitic magma; dots = syenitic magma; solid lines = volcanic rocks; long dashes = layered crystallized rock; dashed line in D = present land surface.

block than along the other. Tilted and differentially down-sagged blocks therefore should be common in the ring complexes. Perhaps the clearest example of such a feature is in the Ossipee Mountain complex, for which Kingsley (1931) showed that a much thicker sequence of the Moat Volcanics is preserved within the western part of the subsided area than within the eastern part. This implies that the top of the original block has been tilted to the west, probably due to actual rotation and (or) differential sag of the block itself.

This hypothesis of cone fracturing and subsidence explains the formation of what has been called the interior stock (Fig. 5E, 5F), which is found in many of the complexes. These bodies are usually small, are composed of granitic rock, and seem to have formed late. I believe they represent silicic melt temporarily trapped beneath the block, which under favorable circumstances may eventually penetrate the block itself and intrude the volcanic rocks resting on the floor of the surface cauldron (Fig. 5F). A fine example of this is found in the Ossipee Mountain complex (Kingsley, 1931).

One might expect from my model that pronounced tilt or differential sag of the subsiding block would displace or confine to one side of the chamber the capping granitic magma below. The central stope should therefore be displaced opposite to the direction of apparent tilting. This is precisely the relationship found in the Ossipee Mountain complex. My interpretation of the late structural history of this complex is illustrated in Figure 6.

Another expected consequence of tilting and differential sagging is the formation of partial or incomplete ring dikes. Billings (1943, 1945) has already pointed out that this type of ring dike is the one most commonly represented by the White Mountain magma series. Although the type of tilt-sag action considered here is not necessarily a requisite for partial ring dike formation, there is evidence that the two phenomena have accompanied one another in a number of complexes.

In the Percy complex (Chapman, 1948), large syenitic ring dikes are found on the west, whereas the late granitic interior stock occurs on the east. This polarization of granitic and syenitic rock suggests an effective westward tilting of the subsiding country rock. My interpretation is presented in somewhat simplified form in Figure 7. As the Cape Horn, Parks Brook, and Pilot Range fractures were fed from the syenitic magma layer in the underlying chamber (Fig. 7A, 7B, 7C), the granitic melt was displaced eastward to appear later as a rather broad "stock" at the eastern margin of the ring complex (Fig. 7D).

In ground plan the Pliny complex (Chapman, 1942) appears to be composed of two roughly circular subcomplexes disposed along an easterly line and slightly overlapping (poorly shown in Fig. 1), as if formed around two main centers of intrusion (two separate magma chambers). In the eastern subcomplex is the Crescent Range ring dike. This partial ring dike outlines roughly the eastern half of a circle that bounds the subcomplex. At the western point of this circle, diametrically opposite the center of the ring dike, is a small stock of granite. Presumably, subsidence along the eastern half of a circular ring fracture here created a crescent-shaped opening for the partial ring dike. Consequent tilt or differential sag of the block displaced to the west the remaining granitic melt in the chamber and caused the interior stock to appear diametrically opposite the ring dike. Other examples of this type of polarized intrusion may be found in the Belknap Mountain and Ascutney Mountain complexes and in the northeastern part of the Franconia quadrangle. Polarized intrusions therefore may constitute evidence of tilting and (or) differential sagging of the subsiding block.

The kind of tilting and differential sagging discussed above would probably be less readily accomplished if the subsiding block were cylindrical instead of conical,

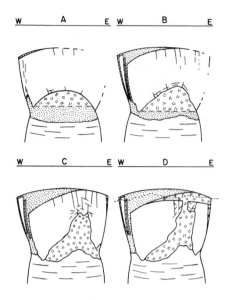

Figure 7. Hypothetical east-west cross sections of Percy complex. A, Incomplete cone fracture initiated; rapid stoping along west side permits differential subsidence; cone fracture terminates upward in subhorizontal arch fracture and Cape Horn ring dike is initiated. B, Subsidence and fracturing are renewed to form Parks Brook ring dike to east and to feed subsurface cauldron with syenitic magma; granitic magma is crowded to east and stopes upward into tilted block. C, West side of tilted block comes to rest, and syenitic melt in enlarged subsurface cauldron crystallizes to form Pilot Range syenite mass; ring fracture gapes along eastern side and fills with granitic melt (hastingsite-riebeckite granite); interior stope is extended upward. D, Interior stope penetrates original conical block now freed to settle along eastern side; granitic melt enters reopened subsurface cauldron and crystallizes to form off-centered interior stock of Conway Granite. Numerous small ring dikes of Conway Granite around the complex were probably formed at this time. Double dashes = granitic magma; dots = syenitic magma; long dashes = layered crystallized rock; heavy straight line = former land surface; dashed line in D = present land surface.

because of the absence of a supporting peripheral shoulder and the higher symmetry of the cylindrical form. For the paraboloidal block, the probability seems even less.

MOAT VOLCANICS

The Moat Volcanics is considered (Billings, 1956) to represent the extrusive phase of the White Mountain magma series. The rocks are primarily rhyolitic with considerable andesitic and basaltic material locally. Trachytic types are clearly subordinate. Flows, tuff, and breccia are represented; measured sections are as thick as nearly 3,660 m. It is not known how much of the volcanic section has been removed by erosion. The formation is now represented mainly by a few large patches confined to the concave side of prominent ring dikes of porphyritic quartz syenite. These patches seem to represent remnants of a much more extensive formation preserved within areas of cauldron subsidence (Noble and Billings, 1967). Although volcanic rocks are cut by and included in the more abundant plutonic rock types of the region, both the volcanic and plutonic rocks were probably derived from the same melts and, therefore, probably overlap considerably in age. The thickest sections of the Moat Volcanics are found near the White Mountain batholith, where intrusive activity was the most intense. Far from the batholith, volcanic material appears only as inclusions in the plutonic rocks, and it may have been restricted originally to the vicinity of the small complexes.

The great abundance of rhyolitic material in the Moat Volcanics supports my contention that granitic melt was relatively abundant in the magma chambers when they reached their near-surface positions. The paucity of trachyte, on the other hand, indicates a lower percentage of trachytic melt and (or) a failure of the ring fracture to tap this melt in the bottom of the chamber. The type of melt fed to the ring fracture will depend on the types and relative levels of melts in the chamber (Fig. 8A, 8B). Thus, different types of melts may alternate repeatedly or may even be erupted simultaneously from the same fracture zone.

When a ring fracture developed before the lower melt had reached the syenitic stage, we might expect basalt or andesite to appear in the lower part of the volcanic assemblage; these rocks might even accompany or alternate with rhyolitic ones, as in the Ossipee Mountain complex (Kingsley, 1931). It seems likely, however, that in the Ossipee Mountains and around the White Mountain batholith, where eruptive centers were relatively closely spaced, a volcanic sequence in one locality would have been derived from a number of different, neighboring chambers. This would be particularly true for the finer tuffaceous rocks and for the lower parts of sequences. But once subsidence started in a given complex, the volcanic rocks to accumulate subsequently in the surface cauldron would be largely of local derivation.

If we assume granitic melt in the upper part of the chamber and syenitic melt below, we can understand why rhyolite could be erupted to the surface before quartz syenite magma crystallized in the associated ring dike. Such a sequence, perhaps common in the White Mountain magma series, is certainly not what we would expect if the series were derived from a monomagmatic cupola or chamber.

PROBLEMS OF THE SYENITE-GRANITE ASSOCIATION

Further evidence for the coexistence of syenitic and granitic magma is found in the nature of the relationship between syenite and granite in the various complexes. Once completely freed, the roof block would settle through the layer of granitic magma and into the syenitic melt beneath, coming to rest on the floor of the magma chamber or on the jumbled xenolithic layer accumulated on the chamber floor. The granitic melt would tend to collect in the region newly evacuated above the block, whereas syenitic melt would tend to surround the block. If subsequent erosion were slight, only a stocklike mass of granite would be exposed. Deeper erosion, however, would likely eliminate the granite body and expose a ring dike of syenite (Fig. 8C). This explains the strong tendency in the New Hampshire area for syenitic rocks to form ring dikes and for granitic rocks to concentrate in stocks (Billings, 1945) I proposed the concept of floored magma chambers

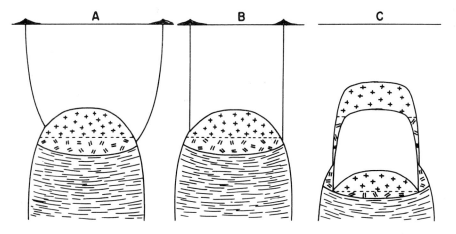

Figure 8. Relation of ring fracture to felsic melts. A, Steep cone fracture taps granitic melt at one point and syenitic melt at another; B, Same relation involving cylindrical fracture; C, Granitic melt collects above subsided block to form stock-like mass, and syenitic melt collects along ring fracture to form ring dike. Crosses = granitic melt; double dashes = syenitic melt; long dashes = cumulate material in chamber.

for central complexes (Chapman, 1966) largely because of this syenite-granite relationship.

If the volume of syenitic melt were insufficient to fill the ring fracture completely, then the upper part of the ring dike would be granitic. An abundance of syenitic melt could cause an overflow into the cauldron above the block, and, depending on the level of erosion, either a granitic or a syenitic stocklike pluton would subsequently be exposed. If much xenolithic rubble had accumulated on the floor, the subsiding block would be stopped at a higher level so syenitic melt could not enter the fracture zone. Thus a granitic ring dike might be formed. If the ring fracture should extend low enough to tap the syenitic melt, the interface between the two melts in the chamber would be lowered until it reached the fracture, and then granitic melt could enter the opening. Moreover, if the lower part of the fracture zone tapped the different melts at different locations around the chamber, both liquids could be fed in (Fig. 8A, 8B). This could lead to mixing and hybridization and (or) to development of a complex dike, in part granitic, in part syenitic.

It seems likely that more or less mixing may already have begun in some chambers at the magma-magma interface. If so, this could lead to the formation of quartz syenitic rocks. Syenitic melt might become trapped in the interstices of xenolithic material accumulating near the chamber floor until finally only a thin layer of syenitic melt was left standing above the rubble. Convection within such a thin layer would become weak, and this melt would be drawn into the stronger convecting system of overlying granitic magma. More or less homogenization could thereby create large quantities of quartz syenitic magma.

Evidence for the mixing of granitic and syenitic material may be found in the nature of the porphyritic quartz syenite that forms many ring dikes. The rock of the Ossipee Mountain ring dike (Kingsley, 1931), for example, is highly variable both in texture and mineralogical composition. It commonly carries a small percentage of quartz phenocrysts with rounded and embayed outlines. Wilson (1969) explained this apparent resorption by pressure change during dike injection. Although this may be the best interpretation, I suggest that quartz may have crystallized from the granitic melt and settled into or become mixed with syenitic or quartz syenitic melt before or during intrusion upward into the ring fracture. In any event, it seems to me that an extremely detailed field and laboratory study of one or more of these porphyritic quartz syenite bodies would be highly rewarding.

Strong evidence for coexisting syenitic and granitic melts is found in the Franconia quadrangle, where Williams and Billings (1938) described relationships that are extremely difficult to reconcile with the theory of a simple magma series running from gabbro to granite. Here porphyritic quartz syenite of the Mount. Garfield ring dike cuts granite porphyry of the Franconia Ridge ring dike. The boldness with which the porphyritic quartz syenite ring dike doubly transects and cuts out the granite porphyry ring dike at one location and sends a hooklike tongue about 0.8 km wide into the granite porphyry dike at another is very conspicuous, even on the 1:62,500-scale map. It leaves no doubt about which unit is the younger. The two ring dikes run roughly parallel and close to one another for about 15 km and appear to have been controlled by roughly the same major ring fracture zone. Here I believe we can make an excellent case for successive intrusions along the general fracture zone, first of a granitic melt and then of a quartz syenitic one. A polymagmatic chamber therefore seems to be the best explanation.

The important point here is that with coexisting melts the order of eruption may vary, and the mixing of melts and their products is not only a likely but perhaps a common occurrence. Many relationships outlined here may have counterparts in situations where granitic and dioritic or monzonitic melts are involved.

There are many other phenomena explained by this model of closed polymagmatic chambers, but space does not permit discussion of them here.

SUMMARY

The hypothesis of independently derived, closed magma pockets as a source for the central complexes of the White Mountain magma series is preferred to the concept of a group of cupolas fed from a common or universal underlying magma reservoir. In addition to eliminating numerous structural and petrologic problems, it negates the problem of crustal instability above an extensive magma layer. The marked similarities between the rocks of the various and widely separated complexes may be explained by the uniformity and consistency with which the processes of selective melting, reaction, and crystallization operate within the outer shell of the Earth.

The idea of small independent magma chambers is consistent with recent geochemical age-dating studies, which show wide ranges in age for rocks of the same petrographic types even within adjoining areas.

The hypothesis finds support in the reticulate arrangement of the large plutons and complexes, which suggests deep structural control. Apparently the magma collected at nodal points in a system of intersecting fracture zones in the upper mantle or base of the crust. Magma pockets were rapidly transformed to diapiric shapes, and ellipsoidal masses became detached from the reservoir rock. These discrete chambers of relatively light, superheated basaltic magma moved upward through the crust, aided by plastic flowage, melting, and stoping of the adjacent rock.

Not all magma pockets formed at one time, and rates of upward migration differed among neighboring magma chambers. This led to the overlapping of intrusions and complexes near the surface, a phenomenon commonly referred to as "migration of centers of intrusion."

Selective melting in the crust caused large volumes of granitic magma to collect in the upper parts of magma chambers. The light, viscous felsic melt, continually supplied with heat from the mafic magma beneath, remained active and augmented itself by further selective melting. By the time a polymagmatic chamber had reached a high level in the crust, the mafic melt had assimilated much material and had crystallized sufficiently to evolve through the dioritic and monzonitic stages to a syenitic magma.

Ring dikes formed by several methods. Some may have involved subsidence of steep-sided paraboloidal blocks, truncated above by subhorizontal fractures. Many perhaps involved subsidence of cylindrical blocks bounded by zones of brecciation along which magma was emplaced. This mechanism explains why the ring dikes of the White Mountain magma series are generally very steep, and it finds support in similar mechanisms known to have operated over mining stopes and certain ore bodies.

It is proposed that some ring structures formed as a consequence of steep ring fractures (cone fractures) produced by increased chamber pressure. Extensive stoping along the lower, inwardly inclined parts of a cone fracture would eventually remove the supporting shoulder that encircled the magma chamber, and would permit the block to subside. This also could lead to the formation of a steep ring dike.

A steep-sided block would likely subside intermittently because of temporary wedging along its walls. This would explain the episodes of repeated subsidence apparent in many ring complexes. Where the space created along a ring fracture was too wide for wedging, as in the case of paraboloidal fractures, the subsiding block would come to rest on the floor or on the xenolithic carpet on the floor of the magma chamber.

Differential settling or tilting of the subsiding block is noted in a number of the complexes. This apparently has led to an asymmetrical distribution or polarization of rock types, with the older units erupting along the side of greatest subsidence.

Many igneous bodies exposed as stocks at the present surface may represent magma-filled cauldrons above subsided blocks. Deeper erosion at these localities would in time destroy the stocks and expose the ring dikes below. In polymagmatic chambers, the lighter granitic melt would tend to fill the cauldron above the subsiding block before syenitic melt could enter the fracture. This explains why granite is found more frequently in stocks and syenite in ring dikes of the White Mountain magma series.

Coexistence of syenitic and granitic melts indicates that mixing may have occurred in certain areas. Mixing of the two melts within the magma chamber might create quartz syenitic material. Partial mixing could be accomplished as early crystals in the granitic melt settled into syenitic magma below. Mixing could also take place if the ring fracture tapped the magma chamber at sufficiently different levels to admit both liquid phases. Such mixing may account for the variable nature of some ring dikes. Strong evidence for coexisting magmas is found in the Franconia quadrangle, where a granite porphyry ring dike is clearly cut and transected by a porphyritic quartz syenite ring dike.

The great abundance of rhyolitic rock and the paucity of trachytic material in the Moat Volcanics accords well with the concept of polymagmatic chambers. This model also explains why rhyolitic material apparently erupted to the surface before quartz syenite formed in the associated ring dike.

ACKNOWLEDGMENT

An earlier version of this paper was reviewed by M. P. Billings and H. B. Gaudette.

REFERENCES CITED

Anderson, E. M., 1936, The dynamics of the formation of cone sheets, ring dikes, and cauldron subsidences: Royal Soc. Edinburgh Proc., v. 56, pt. 2, p. 128-157.
Bailey, E. B., Clough, C. T., Wright, W. B., Richey, J. E., Wilson, G. V., and others, 1924, The Tertiary and post-Tertiary geology of Mull, Lock Aline and Oban: Scotland Geol. Survey Mem., 445 p.
Billings, M. P., 1942, Geology of the central area of the Ossipee Mountains, New Hampshire, earthquakes: Seismol. Soc. America Bull., v. 32, p. 83-92.
―――1943, Ring dikes and their origin: New York Acad. Sci. Trans., ser. 2, v. 5, p. 131-144.
―――1945, Mechanics of igneous intrusion in New Hampshire: Am. Jour. Sci., v. 243-A, Daly Volume, p. 40-68.
―――1956, The geology of New Hampshire, Pt. II: Bedrock geology: Concord, New Hampshire, New Hampshire Planning and Devel. Comm., 203 p.
Chapman, C. A., 1963, Structural control on magmatic central complexes in central New England [abs.]: Geol. Soc. America Spec. Paper 73, p. 128.
―――1966a, Paucity of mafic ring dikes―evidence for floored polymagmatic chambers: Am. Jour. Sci., v. 264, p. 66-77.
―――1966b, Origin of igneous central complexes and formation of ring dikes: [abs.]: Geol. Soc. America Spec. Paper 87, p. 31.
―――1967, Magmatic central complexes and tectonic evolution of certain orogenic belts *in* Etages tectoniques, Colloque de Neuchâtel, Neuchâtel Univ. Inst. Geol., 1966: Neuchâtel, Switzerland, La Baconnière, p. 41-52.
―――1968, A comparison of the Maine coastal plutons and the magmatic central complexes of New Hampshire, *in* Zen, E-an, White, W. S., Hadley, J. B., and Thompson, J. B., Jr., eds., Studies in Appalachian geology: Northern and Maritime: New York, Interscience Pubs., Inc., p. 385-596.
Chapman, R. W., 1942, Ring structures of the Pliny region, New Hampshire: Geol. Soc. America Bull., v. 53, p. 1533-1568.
―――1948, Petrology and structure of the Percy quadrangle, New Hampshire: Geol. Soc. America Bull., v. 59, p. 1059-1100.
―――1954, Criteria for mode of emplacement of the alkaline stock at Mount Monadnock, Vermont: Geol. Soc. America, Bull. v. 65, p. 97-114.
Chapman, R. W., and Chapman, C. A., 1940, Cauldron subsidence at Ascutney Mountain, Vermont: Geol. Soc. America Bull., v. 51, p. 191-212.
Chapman, R. W., and Williams, C. R., 1935, Evolution of the White Mountain magma series: Am. Mineralogist, v. 20, p. 502-530.
Clough, C. T., Maufe, H. B., and Bailey, E. B., 1909, The cauldron-subsidence of Glen Coe, and the associated igneous phenomena: Geol. Soc. London Quart. Jour., v. 65, p. 611-678.
Emeleus, C. H., 1955, The granites of the Western Mourne Mountains, County Down: Royal Dublin Soc. Sci. Proc., v. 27, p. 35-50.
Foland, K. A., Quinn, A. W., and Giletti, B. J., 1971, K-Ar and Rb-Sr Jurassic and Cretaceous ages for intrusives of the White Mountain magma series, northern New England: Am. Jour. Sci., v. 270, p. 321-330.
Gaudette, H. E. and Chapman, C. A., 1964, Web joint pattern of the Cape Neddick gabbro complex, southwestern Maine: Illinois State Acad. Sci. Trans., v. 57, p. 203-207.
Holmes, Arthur, 1931, The problem of the association of acid and basic rocks in central complexes: Geol. Mag., v. 68, p. 241-255.
Hussey, A. M., Jr., 1962, The geology of southern York County, Maine: Maine Geol. Survey Spec. Studies Ser., no. 4, 67 p.
Jacobson, R.R.E., MacLeod, N. W., and Black, Russell, 1958, Ring-complexes in the Younger Granite province of Northern Nigeria: Gel. Soc. London Mem., no. 1, 72 p.
Jenks, W. F., 1934, Petrology of the alkaline stock at Pleasant Mountain, Maine: Am. Jour. Sci., v. 28, p. 321-340.
Kingsley, Louise, 1931, Cauldron subsidence of the Ossipee Mountains: Am. Jour. Sci., v. 22, p. 139-169.

Locke, Augustus, 1926, The formation of certain ore bodies by mineralization stoping: Econ. Geology, v. 21, p. 431-453.

Noble, D. C. and Billings, M. P., 1967, Pyroclastic rocks of the White Mountain magma series, New Hampshire: Nature, v. 216, p. 906-907.

Reynolds, D. L., 1956, Calderas and ring-complexes: Nederlandsch Geol.-Mijnbouw. Genoot. Verh., Geol. Ser., v. 16, p. 355-379.

Richey, J. E., 1927, The structural relations of the Mourne granites: Geol. Soc. London Quart. Jour., v. 83, p. 653-688.

Roberts, J. L., 1970, The intrusion of magma into brittle rocks, *in* Newall, Geoffrey and Rast, Nicholas, eds., Mechanisms of igneous intrusion: Liverpool, Gallery Press, p. 287-338.

Roy, C. J., and Freedman, Jacob, 1944, Petrology of the Pawtuckaway Mountains, New Hampshire: Geol. Soc. America Bull., v. 55, p. 905-920.

Turner, D. C., 1963, Ring-structures in the Sara-Fier Younger Granite complex, northern Nigeria: Geol. Soc. London Quart. Jour., v. 119, p. 345-366.

Williams, C. R., and Billings, M. P., 1938, Petrology and structure of the Franconia quadrangle, New Hampshire: Geol. Soc. America Bull., v. 49, p. 1011-1044.

Wilson, J. R., 1969, The geology of the Ossipee Lake quadrangle, New Hampshire: Concord, New Hampshire, New Hampshire Dept. Resources and Econ. Devel., 116 p.

Wisser, Edward, 1927, Oxidation subsidence at Bisbee, Arizona: Econ. Geology, v. 22, p. 761-790.

MANUSCRIPT RECEIVED BY THE SOCIETY MARCH 29, 1974
REVISED MANUSCRIPT RECEIVED SEPTEMBER 26, 1974
MANUSCRIPT ACCEPTED OCTOBER 8, 1974

Printed in U.S.A.

Gravity Models and Mode of Emplacement of the New Hampshire Plutonic Series

Dennis L. Nielson*
Dartmouth College
Hanover, New Hampshire 03755

Russell G. Clark
Albion College
Albion, Michigan 49224

John B. Lyons
Dartmouth College
Hanover, New Hampshire 03755

Evan J. Englund
Phelps Dodge Corporation
Morenci, Arizona 85540

AND

David J. Borns
Dartmouth College
Hanover, New Hampshire 03755

ABSTRACT

A gravity network of 700 stations over an area of approximately 4,500 km^2 in central and southern New Hampshire, and mapped intrusive contacts and density determinations for rocks of the New Hampshire Plutonic Series, were used to deduce the structural relations of these Acadian-age intrusives. The bodies occur principally as subhorizontal sheets no thicker than 2.5 km, some of them once probably continuous or semicontinuous over much of the area examined.

---

*Present address: The Anaconda Company, Salt Lake City, Utah 84116.

Forcible injection of the earliest members of the series, the Bethlehem Gneiss and Kinsman Quartz Monzonite, occurred during a cycle of nappe development. Next, biotite quartz monzonite and the Spaulding Quartz Diorite were forcibly emplaced during a cycle of predominantly horizontal compression, but the magmas were channeled beneath and along the margins of the earlier intrusives.

Posttectonic Concord Granite was irrupted after the temperature of the region had cooled to the point where brittle fracture was possible, and the intrusive mechanism may have involved forcible injection, cauldron stoping and possible displacement of roof zones.

Elsewhere in New England the only Acadian-age plutons which have vertical dimensions greater than 2 or 3 km are those which occur outside the sillimanite zone of regional metamorphism. We suggest that the rheidity and the pre-existing foliation of the metasediments dictated the formation of thin plutonic sheets at the level of the infrastructure, and that a necessary condition for the formation of a thick stock or batholith is emplacement in the brittle superstructure.

INTRODUCTION

The Acadian-age New Hampshire Plutonic Series (Billings, 1956, p. 53) is widespread throughout northern New England. In the central and southwestern portions of New Hampshire, the series consists of four principal members. Arranged in order of decreasing age these are (1) Bethlehem Gneiss, strongly foliated biotite granodiorite to quartz monzonite; (2) Kinsman Quartz Monzonite, moderately foliated garnet-biotite granodiorite and quartz monzonite characterized by megacrysts of potash feldspar up to 13 cm in maximum dimension; (3) Spaulding Quartz Diorite and associated biotite quartz monzonite, weakly to moderately foliated hornblende gabbro, biotite quartz diorite and quartz monzonite, with hypersthene and garnet sporadically present; and (4) Concord Granite, light-colored, nonfoliated to weakly foliated biotite-muscovite granodiorite, quartz monzonite and granite. Although the age relations of Bethlehem and Kinsman rocks are somewhat uncertain, intensity of foliation is a rough guide to relative age within the New Hampshire Plutonic Series. Within the area which we have examined (Figs. 1, 2, 3) all the plutons lie within the regional sillimanite isograd, and there is no readily detectable contact metamorphism.

Structural relations within some parts of the New Hampshire Plutonic Series have long been known or suspected. Billings (1956, p. 125-128) interpreted the geologic relations of the Bethlehem Gneiss to indicate that it is a synorogenic sheet, originally subhorizontal but now folded. Kruger and Linehan (1941) had shown by seismic refraction methods that the westernmost body of Bethlehem Gneiss, the Bellows Falls pluton (Fig. 3), is a floored sheet no more than 1 km thick. In agreement with a suggestion by Billings (1956, p. 128), the reinterpretation of the geology of western New Hampshire by Thompson and others (1968, Pl. 15, fig. 1b) shows the Bellows Falls pluton as a western extension of the Mount Clough pluton, detached from it by erosion. In their cross-section the Bethlehem Gneiss is sandwiched between two west-facing nappes which root an uncertain distance (>25 km) to the east.

Structural interpretations for the remaining members of the New Hampshire Plutonic Series are less certain. Typical cross-sections (Billings, 1956, Pl. 1) show the plutons extending indefinitely to depth. An exception is the Ashuelot pluton of Kinsman Quartz Monzonite in southwestern New Hampshire (Fig. 3), which Billings (1956, p. 128) has suggested may be analogous to the Bellows Falls pluton, and which Thompson and others (1968, Pl. 15, fig. 1b) depict as lying between the same two nappes which enclose the Bellows Falls pluton.

Hamilton and Myers (1967) have proposed that rocks of the New Hampshire Plutonic Series represent the remnants of a major batholith, now largely eroded away. We will present evidence to show that the plutons were injected as thin sheets either at the same time as or after the formation of nappes during the Acadian orogeny. There is evidence also that the Bethlehem Gneiss and Kinsman Quartz Monzonite are, at least locally, parts of composite sheets which may at one time have been nearly coextensive over much of the 5,000 km^2 of central and southwestern New Hampshire.

Previous gravity studies covering all or portions of the area outlined by Figure 1 have been published by Bean (1953), Joyner (1963) and Kane and others (1972).

Recent interpretations on gravity-structural relations for the New Hampshire Plutonic Series have been published by Clark and Englund (1972) and by Nielson and others (1973). In this paper the evidence is set forth in detail.

GRAVITY MEASUREMENTS AND REDUCTIONS

In order to investigate the three-dimensional geometry of the New Hampshire Plutonic Series, we established a gravity network of approximately 700 stations, covering all or parts of the following 11 quadrangles (Figs. 1, 2): Holderness (Englund, 1971, 1974), Cardigan (Fowler-Billings, 1942), Penacook, Mount Kearsarge, Sunapee (Chapman, 1952), Concord (Vernon, 1971), Lovewell Mountain (Heald, 1950), Peterborough (Greene, 1970), Monadnock (Fowler-Billings, 1949), and Hillsboro quadrangle (Nielson, 1974). The Mount Kearsarge and Penacook quadrangles were mapped in reconnaissance by Lyons and Clark (unpub. data).

The metasedimentary rocks which surround the plutons are generally pelitic schists and metaquartzites with minor amounts of calc-silicate rocks. We have listed these schists as Devonian Littleton Formation on Figures 4 and 5, although recent mapping in the area suggests that older Paleozoic rocks may also be present.

Gravity readings were made at bench marks or road intersections of known elevation using a Worden Gravimeter, model 133. The base stations of our survey have been tied by four loops to Joyner's (1963) gravity station at Farmington, New Hampshire, with the results indicated in Table 1.

At each of the 700 gravity stations the gravimeter was read at least three times until readings were within 0.06 mgal of one another. Averaged readings were then adjusted for drift, and a computer program was used to compute latitude, free-air, and Bouguer corrections. For the Bouguer reduction a slab density of 2.67 was assumed. Because all gravity stations lie within one degree of latitude, a uniform gravity gradient was assumed (Dobrin, 1960, p. 76) and the differential of the

TABLE 1. GRAVITY BASE STATIONS, CENTRAL AND SOUTHWESTERN NEW HAMPSHIRE

Location	Absolute gravity (mgal)
Farmington Town Hall (Joyner, 1963)	980,455.3
New London Town Hall	980,376.7 ± 0.1
Warner, Pillsbury Library	980,430.3 ± 0.1
Concord City Hall	980,427.5 ± 0.1
Bristol, cannon pediment	980,452.2 ± 0.1
Franklin Post Office	980,451.4 ± 0.1

Note: All stations U.S. Geological Survey bench marks.

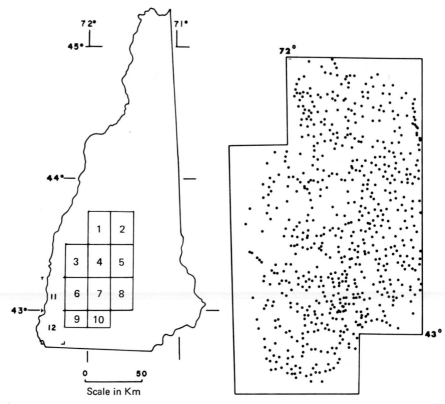

Figure 1. Index map. Map of New Hampshire showing area covered by gravity survey map. Numbers refer to quadrangles. 1, Cardigan; 2, Holderness; 3, Sunapee; 4, Mount Kearsarge; 5, Penacook; 6, Lovewell Mountain; 7, Hillsboro; 8, Concord; 9, Monadnock; 10, Peterborough; 11, Bellows Falls; 12, Keene.

Figure 2. Location of gravity stations within area outlined in Figure 1.

International Gravity Formula was used to make the latitude correction. Terrain corrections were computed for approximately 75 stations; for the remainder, the terrain correction is less than 0.2 mgal.

The gravity observations thus corrected were plotted on a Bouguer anomaly map (Fig. 4) with an arbitrary scale of milligals; the absolute Bouguer anomaly may be found by subtracting approximately 50 mgal. The original map was contoured on a 1-mgal scale, but for ease of display we show a 2-mgal interval in Figure 4. The anomalies have been superimposed on a geologic map showing the distribution of the plutonic rocks.

Several correlations are apparent from Figure 4. The large Cardigan pluton of Kinsman Quartz Monzonite is defined by a series of broad gravity lows with a maximum amplitude of 8 mgal. The biotite quartz monzonite and Spaulding Quartz Diorite complexes are characterized by 2- to 3-mgal lows. Concord Granite plutons are also indicated by gravity lows, the largest (in the Concord quadrangle) with an amplitude of 6 mgal. Note that this low is offset to the west with respect to the outcrop pattern of the granite. A strong gravity high of 17 mgal is shown along the southwestern side of Figure 4. It is centered on metasedimentary rocks, but far exceeds the broad 3- to 5-mgal highs characterizing the metasedimentary terrane elsewhere.

GRAVITY MODELS

Table 2 displays the density measurements and averaged densities that we adopted in calculating the gravity profiles of Figure 5. It is the lateral density contrasts between adjacent rock bodies which are significant in these profiles, and in all cases we can account for the gravity contrasts within the upper 3 km of the crust.

The Bouguer anomaly profiles of Figure 5 are taken from a 1-mgal contour map and have been corrected for an eastward regional gravity gradient of +0.36 mgal/km (taken from a map of Kane and others, 1972); therefore the profiles of Figure 5 are residual Bouguer anomaly profiles. Profiles I-I', V-V', and VII-VII', however, have not been corrected in this fashion. The eastern end of I-I' is under the local influence of a strong northward negative gravity gradient toward the White Mountain batholith, V-V' lies in an area where the gravity gradient is flat, and the western end of VII-VII' is under the influence of a strong local positive anomaly (Kane and others, 1972). Attempting to account for these local effects by assuming that they are of upper crustal origin involves a subjectivtity that

TABLE 2. DENSITY DETERMINATIONS, NEW HAMPSHIRE PLUTONIC SERIES AND METASEDIMENTARY ROCKS

Formation	Number of Samples (This paper)		Number of Samples (Clark, 1973)		Number of Samples (Joyner, 1963)		Number of Samples (Bean, 1963)		Adopted value	Density contrast
Bethlehem Gneiss (Bg)							63	2.67 to 2.88	2.73	+0.06
Kinsman Quartz Monzonite (Kqm)	6	2.68 to 2.75	33	2.59 to 2.95	16	2.65 to 2.80	11	2.68 to 2.74	2.72	+0.05
Biotite quartz monzonite (bqm)	8	2.60 to 2.69							2.66	−0.01
Concord Granite (Co)	12	2.60 to 2.67	7	2.60 to 2.69	54	2.59 to 2.77	21	2.61 to 2.75	2.64	−0.03
Pelitic schist (D1)	15	2.76 to 2.95	69	2.67 to 3.04	88	2.68 to 2.93	76	2.70 to 2.96	2.82	+0.15
Calc-silicate schist (D1)	11	2.70 to 2.92							2.82	+0.15
Oliverian plutonic series					15	2.58 to 2.68			2.63	
Ammonoosuc Volcanics					17	2.70 2.91			2.80	

makes the residual Bouguer profiles of questionable validity as compared with the simple Bouguer profiles.

The computer-calculated gravity profiles are based on a two-dimensional model by Talwani and others (1959). The model is valid for rock bodies longer than they are wide, but yields thicknesses slightly less than would be obtained from a three-dimensional model. When the calculated and observed Bouguer anomaly profiles show a close fit, as in Figure 5, the geometric-geologic solution is reasonable,

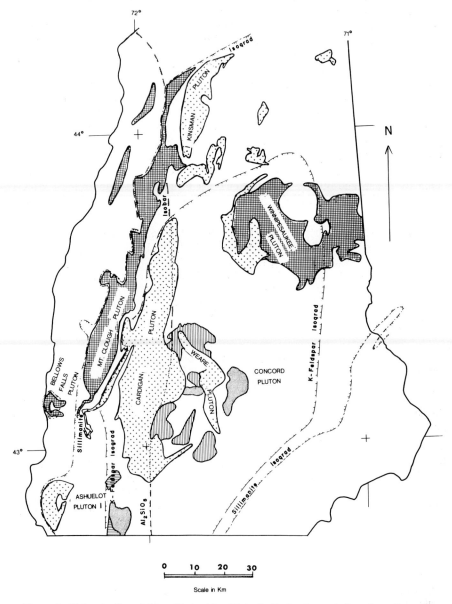

Figure 3. Major bodies of New Hampshire Plutonic Series in central and southern New Hampshire. Sillimanite and K-feldspar isograds and Al_2SiO_5 isobar after Thompson and Norton (1968). Grid pattern is Bethlehem Gneiss; crosses, Kinsman Quartz Monzonite; horizontal lines, Spaulding Quartz Diorite; vertical lines, biotite quartz monzonite; dots, Concord Granite.

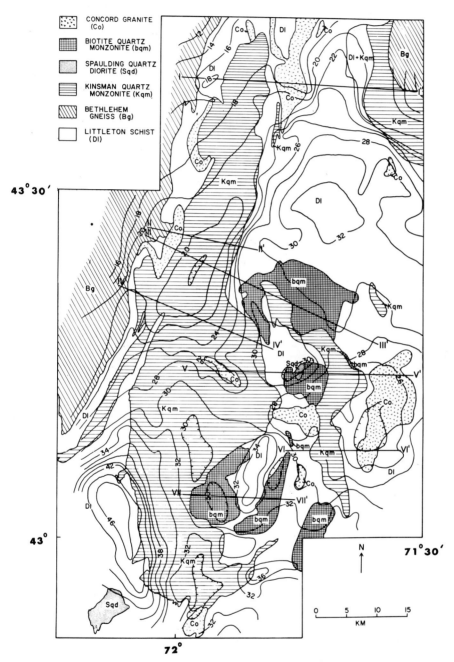

Figure 4. Bouguer anomaly map for area outlined in Figure 1. Contour interval is 2 mgal. Roman numerals refer to profiles displayed in Figure 5.

although only one of a spectrum of possibilities. However, other constraints such as measured densities, structures, and mapped distributions of the geologic formations severely limit additional possible solutions, and for most of the cross-sections we have constructed the number of alternatives is relatively few.

Several features of profile I-I′ typify much of the geology of central New

Hampshire. Most of the low Bouguer anomalies are explicable by assuming a sheet-like form for the individual plutons, with no individual body exceeding a 2.5-km thickness and Kinsman rocks typically displaying a saucer shape, with inward-dipping contacts.

As noted earlier, the Winnipesaukee pluton at the eastern end of this profile is under the influence of a strong northward regional gravity gradient toward the White Mountain batholith (Kane and others, 1972). In order to account for the strong negative gradient at the eastern end of our profile I-I', we have inserted a wedge of Concord Granite at depth; we suspect, however, that a combination of the northward regional gravity gradient and an intrusion of the Conway Granite of the White Mountain batholith at depth are more likely solutions. Englund (1971, p. 84) made a correction for the local strong negative gravity gradient toward the White Mountains, and on this basis calculated a maximum thickness of 3.5 km for the Winnipesaukee pluton at the eastern end of profile I-I'.

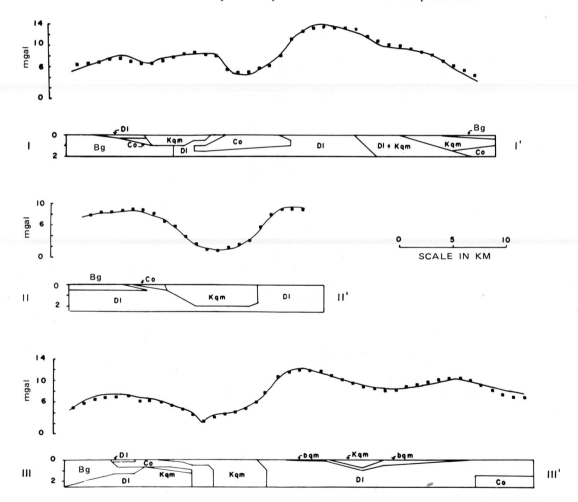

Figure 5. Residual Bouguer anomaly profiles (solid lines) and computed Bouguer anomalies (square dots) based on models shown below each profile. Locations of profiles I-I' through VII-VII' shown on Figure 4. Profile VIII-VIII' is a NW-SE section across the Ashuelot pluton (Fig. 3); Op is Partridge Formation, Oam is Ammonoosuc Volcanics.

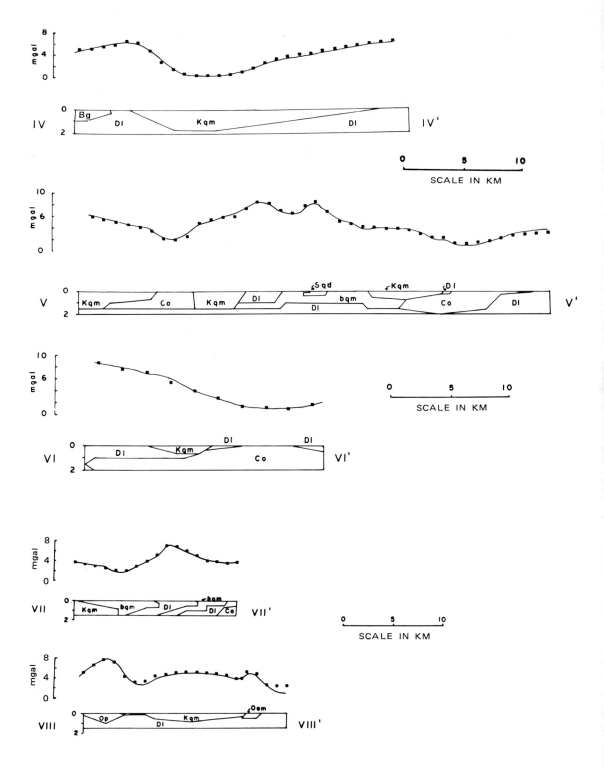

Figure 5. *(Continued)*

Rocks immediately west of the Winnipesaukee pluton consist of a migmatitic complex of Kinsman and metapelitic rocks. The model suggests a wider migmatite zone at depth, but there is latitude for several variations in this part of the cross-section.

Geologic and gravity constraints also bear strongly on the interpretation of the western end of Section I-I'. Exposed contacts indicate that the Bethlehem, Littleton, and Kinsman formations dip easterly. To accommodate the gravity profile here it is necessary to insert a wedge of Concord Granite beneath the Littleton, the justification being that the granite is exposed a short distance to the south of the line of the profile. The Kinsman–wall rock contacts dip towards the center of the pluton, and the Cardigan pluton here and throughout the other sections is bowl-shaped in cross-section.

A more complete picture of the Cardigan pluton (sections II-II', III-III', and IV-IV') shows that the western margin of the body maintains an easterly dip, but the eastern margin steepens and becomes overturned at about mid-latitude; to the south (section V-V') it resumes a westerly dip. The maximum thickness of the Cardigan pluton is 2.5 km (section III-III'), but the body shallows northward and southward, so that the entire mass is like an elongated irregular platter or half cigar.

The small Weare pluton to the east has the same general three-dimensional shape except that it has a maximum thickness of 1 km (sections III-III', V-V', VI-VI'), and its contacts dip inward more regularly. The close approach of the Kinsman and Weare plutons on the geologic map (Fig. 4) suggests that they were once coextensive and have been separated by erosion across an antiformal arch. We are hesitant about postulating a tie between the Cardigan and Winnipesaukee plutons, but there is a remote geometric possibility of this (section I-I'). Toward the west, Kinsman rocks apparently lie structurally above the Bethlehem, but in the Winnipesaukee pluton the reverse relation holds. However, the recent geologic map of the Bronson Hill anticlinorium (Thompson and others, 1968; this paper, Fig. 3) shows Kinsman and Bethlehem formations tracing into one another in southwestern New Hampshire; either two separate sheets of these formations locally interfinger, or their relative positions here are due to structural convolutions. In any case, although the total number of sheets is uncertain they must be at approximately the same structural level in all of our profiles.

Both geologic structure sections of the Bethlehem Gneiss (Billings, 1956, Pl. 1; Thompson and others, 1968, Pl. 15, Fig. 1b) and gravity data imply that this formation has a maximum thickness of about 3 km. Because roof rocks are not exposed in the Weare, Cardigan, and Winnipesaukee plutons, estimating Kinsman thicknesses is more difficult.

One place where gravity and structural data allow an estimate of total thickness of the Kinsman sheet is in the Ashuelot pluton of southwestern New Hampshire (Fig. 3; section VIII-VIII' of Fig. 5). As shown by Moore's (1949) geologic map, the Ashuelot pluton is bordered by granitic Oliverian dome rock along most of the northern, eastern, and southwestern margins, by the Ordovician Partridge Formation (a metapelite) on the west, and by the Ordovician Ammonoosuc Volcanics along a part of the eastern margin. A roof pendant of Partridge apparently sits on top of the eastern part of the pluton (Thompson and others, 1968, p. 214). An east-west gravity traverse was made across the Ashuelot pluton (Fig. 3). The gravity profile and model are shown in section VIII-VIII'; in calculating the model we used Joyner's (1963) density values for the Oliverian and Ammonoosuc formations, and assumed a Partridge density (2.67) equivalent to that of pelitic schist (D1) in Table 2. According to this model, Kinsman rocks of the pluton form a flat sheet 0.8 km in maximum thickness. Thompson and others' (1968) interpretation of the Ashuelot pluton places two extremely attenuated nappes between the Kinsman and Oliverian rocks of our section, and another nappe above the Kinsman. If two nappes are continuous beneath the Ashuelot pluton (a possibility not ruled out by the gravity data) and if the Partridge roof pendant belongs to an overlying nappe, the average thickness of Kinsman is less than our gravity model implies. The Ashuelot pluton can then be interpreted as a detached portion of the south end of the Cardigan pluton brought westerly as a sandwich between the same two nappes that enclose the Bellows Falls and Mount Clough plutons of Bethlehem Gneiss (Fig. 3). In this sense Billings' (1956, p. 128) suggestion that the Ashuelot pluton may be a displaced westerly portion of the Cardigan pluton seems to be substantiated.

A more detailed but unpublished gravity map covering the area of the Ashuelot pluton has been completed by students at the University of Massachusetts (Peter Robinson, oral comm.). Models from this map suggest a somewhat greater thickness for the Ashuelot pluton than the 0.8 km shown in section VIII-VIII'.

The 17-mgal high in the southwestern part of Figure 4 lies on the southwest flank of the Cardigan pluton in an area underlain by pelitic schist. As noted by Kane and others (1972), this high is along the strike of a similar high in Massachusetts and parallels a north-south high associated with the Green Mountain anticlinorium to the west. Diment (1968) attributed the Green Mountain high to an upwarp of dense rocks from below the Conrad discontinuity, and suggested a similar origin for the parallel Massachusetts and New Hampshire highs. Another alternative explanation could be a buried mafic intrusion. Gabbro of the New Hampshire Plutonic Series is present but rare within the area covered by this gravity survey, but Acadian-age gabbro is common along-strike in Maine (Doyle, 1967), and we cannot rule out the possibility that large masses of gabbro are present at depth in New Hampshire. The 17-mgal anomaly has a maximum gradient of 4 mgal/km. Using the maximum-depth formula of Bancroft (1960), we calculated that the top of the perturbing mass is at a depth of 1.4 km. If the density contrast between this mass and the overlying schist is 0.16 gm/cc (a likely value for the density difference between gabbro and pelitic schist) the minimum thickness of the gabbro body is 2.5 km. If the perturbing mass is Spaulding Quartz Diorite, a possibility suggested by map relations of Figure 4, its top would remain at a depth of 1.4 km but its thickness would increase to 2.8 km.

REGIONAL TECTONIC-INTRUSIVE RELATIONS

Englund (1971) has demonstrated three cycles of Acadian folding in the area between the Cardigan and Winnipesaukee plutons. F_1 is recumbent, F_2 is a folding about northwesterly axes, and F_3 a folding about northeasterly axes. F_1 and F_3 are the dominant fold directions, and the recumbent folds are mappable at least as far southeast as the Hillsboro quadrangle (Fig. 1; Nielson, 1974).

Bethlehem Gneiss displays a strong foliation (S_1) which is related to F_1. The Kinsman Quartz Monzonite displays S_1 and S_3, implying that initial emplacement took place during recumbent folding and that later folding superimposed a regional north-northeasterly fabric. We take these structural observations, along with the gravity interpretations we have presented, to indicate a tectonic control on the emplacement of Bethlehem and Kinsman plutons. We conclude that both formations were injected as thin sheets, originally not much thicker than their presently indicated maximum thicknesses of 2 to 3 km.

Like the Bethlehem and Kinsman formations, the late or posttectonic biotite quartz monzonite, Spaulding Quartz Diorite and Concord Granite complexes tend to form sheets, but probably for different reasons. On section III-III' the biotite quartz monzonite is shown as a thin sheet, partly because of field geologic relations. This sheet apparently spreads out laterally beneath the Kinsman Quartz Monzonite of the older Weare pluton. The biotite quartz monzonite–Spaulding pluton of section V-V', which is continuous at depth with that of section III-III', suggests lateral forcible injection, partly beneath the Weare and partly beneath the Cardigan pluton. The missing roof rock may have been entirely Kinsman Quartz Monzonite belonging to a now-eroded connection between the Cardigan and Weare plutons, as we have suggested earlier.

The westernmost biotite quartz monzonite intrusive of profile VII-VII' was irrupted into its present position along the contact between the Kinsman and the metasediments. It spread along the upper contact of the Kinsman toward the west and into the metasediments toward the east, taking on the shape of a flat mushroom.

The eastern pluton of profile VII-VII' is known from field mapping to occupy the core of an F_3 antiform, overturned toward the east. The pluton has a maximum thickness of 1 km and was injected into the axial region of the fold, possibly during F_3 folding. The general similarity between the shapes of the upper and lower walls of all the biotite quartz monzonite bodies clearly implies forcible injection as the mechanism of igneous intrusion, coinciding in time with F_3 folding.

Posttectonic Concord Granite plutons are generally similar in that many of them widen or are asymmetric at depth (sections I-I', III-III', V-V', VI-VI'). Structural control by other members of the New Hampshire Plutonic Series is not obvious, and in many ways these plutons remain an unsolved problem. Although xenoliths are not unknown, they are generally uncommon, and the three-dimensional shapes of the bodies lack the simple geometric relations which imply proof of forcible injection by the wedging apart of the wall rocks. The tendency to spread laterally is also puzzling; our suggestion is that it partly reflects emplacement along an S_1 or S_3 structural grain. The Concord Granite at the type locality is apparently the largest massif of this formation; a large portion of the pluton remains unexposed.

MAGMA GENERATION AND EMPLACEMENT

The gravity study of the New Hampshire Plutonic Series has shown that plutons of the four principal members of the series form intrusive sheets which extend to depths of no more than 2.5 km. The gravity anomalies over these plutons have amplitudes of no more than 8 mgal. The regional Bouguer gravity map of New England, contoured on a 5-mgal interval (Kane and others, 1972), shows few local anomalies as large as 10 mgal over New Hampshire Series plutons in southern New Hampshire and southwestern Maine, suggesting that all these plutons are thin sheets. For example, a recent study by Bothner (1974) has shown that the Exeter Diorite of southeastern New Hampshire has a maximum thickness of 3 km. However, in northern Maine and eastern Vermont strong local anomalies are associated with some of the Acadian-age felsic plutons, indicating that these bodies are much thicker. The Barre Granite of northeastern Vermont, with an anomaly of 20 mgal, is probably coeval with the Concord Granite, considering their structural and petrographic similarities. Diment (1968, p. 408) has shown that the anomaly over the Barre Granite can be explained by the presence of a mass having the shape of an upright truncated cone and extending to a depth of 7 km, where it seems to be floored. Several similarly large negative anomalies occur in northern

Vermont and northwestern Maine, suggesting that underlying plutons are probably at least as thick as the Barre Granite. By contrast, the large Acadian-age Sebago pluton of southwestern Maine causes almost no perturbation in the regional gravity gradient, and must therefore be very thin.

The dividing line between these two contrasting intrusive styles is approximately delineated by the sillimanite isograd as drawn by Thompson and Norton (1968) and Morgan (1972). The thicker plutons always lie in the low-temperature region. Kane and others (1972) also noted this correlation between thin plutonic sheets and the sillimanite isograd. Thus the shapes of the plutons are probably functions of varying temperatures and relative rheologic differences between the intrusives and their wall rocks, with folding of the rocks being an additional modifying factor. The zone of plastic flowage is marked by thin sheets; the zone of brittle fracture, by thick or deep plutons.

None of our gravity profiles allows a feeder conduit for any of the intrusives. However, if the feeder is less than 1 km in width and at a depth greater than 2.5 km its presence would not be detectable on the profiles. About all we can state with confidence is that the feeders, if present, must be quite narrow, or that the conduits may have been necked off after passage of the magma, as in the scale-model experiments of Grout (1945) and Ramberg (1970). An alternative explanation, for which we have no direct evidence, is that the density of the plutons increases with depth, the density contrast thereby being eliminated.

None of the plutonic rocks we have studied in the field show evidence of having been generated in situ. Flecky gneisses within the area covered by the gravity survey might appear to be anatectic melts, but Nielson (1973) has shown them to be products of metamorphic diffusion and differentiation. Some peculiar garnet-rich rocks in central New Hampshire (Clark, 1972, p. 97–142; 1973) are explicable as residua or "restites" (Mehnert, 1968, p. 119) of anatexis, but such rocks are relatively rare, and the petrochemistry of pelitic rocks elsewhere in central New Hampshire (Clark, 1972) suggests no additional anatectic residua. On this basis we conclude that the volume of anatectic melt generated from schists at the presently exposed tectonic level was small, if not nonexistent, and cannot reasonably be supposed capable of accounting for the observed volume of the New Hampshire Plutonic Series.

Thompson and others (1968, p. 208) have proposed that a potential source for the Bethlehem-Kinsman rocks could be anatexis of a hypothetical pile of Early Devonian felsic volcanics within the Littleton Formation of central New Hampshire. They pointed to the analogue of the Piscataquis volcanic belt of north-central Maine, where some felsic volcanic piles locally attain thicknesses of at least 3,400 m (Rankin, 1968, p. 359). It is difficult to disprove this hypothesis completely, but three stumbling blocks are obvious: (1) recognizable felsic volcanics within the Littleton Formation elsewhere in New Hampshire are thin, and there is little or no evidence that they thicken toward central New Hampshire; (2) the problem of the heat source for anatexis is unexplained: it is true that regional isograds rise to a broad K-feldspar or second-sillimanite plateau in central New Hampshire (Fig. 3), but the isograds cross the Al_2SiO_5 triple-point isobar, and this geometry suggests heating by New Hampshire Plutonic Series magmas rather than anatexis; and (3) although Bethlehem-Kinsman rocks are broadly concordant with their wall rocks, they are also sharply discordant locally and crosscut formational boundaries. There is no doubt therefore that the plutonic rocks were mobile intrusive magmas, and have a present distribution more or less independent of depth of burial as measured by the alumino-silicate triple-point isobar.

The facts reviewed lead us to seek a deeper source for the anatectic melts, possibly somewhere near the crust-mantle interface as petrologic, geochemical, and isotopic evidence so strongly implies for the Mesozoic batholiths of California (Presnall and Bateman, 1973; Kistler and Petermann, 1973; Doe and Delevaux, 1973). We believe that the magmas rose in three distinct pulses, each of these during different tectonic events and all within a relatively short time span. Naylor (1971) has shown that the time available for Early Devonian sedimentation, folding, regional metamorphism, and intrusion is less than 30 m.y.

According to the scenario of Bird and Dewey (1970) and Naylor (1971), the onset of continental collision and of the Acadian orogeny of central New Hampshire resulted first in development of nappes and consequently in a probable increase in cover thickness. Whereas much of the overfolding was toward the west, transport directions were irregular. In the Hillsboro quadrangle (Fig. 1), transport was northeasterly, resulting in northwest-trending fold axes. This same pattern is recognizable in the Holderness quadrangle to the north, in the area mapped by Englund (1971), but in southwestern New Hampshire and also in north-central Massachusetts (Thompson and others, 1968, Pl. 15, Fig. 1b), some of the transport was both easterly and westerly. In eastern Connecticut (Dixon and Lundgren, 1968) it was easterly.

According to Presnall and Bateman (1973), the generation of magma near the crust-mantle interface may be due to a combination of conductive heat from the mantle and radioactive heat in the crust, possibly aided by upward transport of andesitic or basaltic magma generated along a deeper-lying subduction zone. Fractional crystallization of resultant magma ensues. Almandine-rich garnet phenocrysts in the Kinsman give a clue to the site of this fractional crystallization. Experimental work by Green and Ringwood (1968) on similar almandine-rich phenocrysts in calc-alkaline rocks suggests a pressure level of between 9 and 18 kb and $P_{H_2O} < P_{load}$; this would correspond to conditions at the lower crust–upper mantle interface. Garnets generated under these conditions should break down to pyroxene and plagioclase at crustal levels. They have survived, particularly in Kinsman rocks, because of extremely low oxygen fugacities which are indicated mineralogically by the presence of accessory graphite (Hsü, 1968). The Ordovician Borrowdale volcanic rocks of northern England (Fitton, 1972) and the Early Devonian felsic ash flows of the Piscataquis belt in northern Maine (Rankin, 1968) likewise carry almandine-rich garnet phenocrysts, suggesting a similar depth of origin for those magmas.

A combination of density differences between quartz monzonite–granodiorite melts and enclosing lower crustal wall rocks, along with tectonic squeezing, would provide the buoyant force for lifting the Bethlehem-Kinsman magmas from their source site into a higher tectonic level, which in this case is close to that of the Al_2SiO_5 triple-point isobar (Fig. 3). In accordance with model experiments by Ramberg (1970), we infer that the magmas kneaded their way upward through the infrastructure into a zone of active tectonic transport (nappe formation) where they were channeled into or between the still-forming recumbent folds. Continued deformation was enhanced both because of the heating and weakening effect of the hot magmas (which also raised metamorphic isograds to the second sillimanite plateau), and because of their lubricating effect within the sliding structures. A possible effect of the nappe formation, however, was the eventual detachment of the plutons from their feeders, which were then pinched off and closed down, leaving Kinsman-Bethlehem sheets (or a single sheet) with a strong S_1 foliation.

A period of magmatic quiescence followed, during which the plutonic rocks solidified and the enclosing metamorphic rocks also gradually cooled. However, the rocks were subjected to continuing compression, which resulted in north- and northeast-trending folds and the superposition of an S_3 foliation. The rise of a series of remobilized Oliverian gneiss domes along the Bronson Hill anticlinorium of western New Hampshire at this time (Thompson and others, 1968) protected much of the Bethlehem Gneiss from D_3 compression, but east of the anticlinorium D_3 effects were strong.

At this stage in tectonic history another period of intrusion occurred, apparently because magmatism was still active at the crust-mantle interface (garnet is sparingly present in the second-cycle intrusives) and because the combination of density differences and tectonic squeezing were sufficient to raise the biotite quartz monzonitic and quartz dioritic magmas. During this intrusive cycle it is apparent that (1) igneous activity was more restricted and the volume of magma less; (2) magmas were guided along pre-existing structures, a favored site being beneath sheets of brittle Kinsman, angular fragments of which are enclosed within biotite quartz monzonite; and (3) folding was still active because some biotite quartz monzonite (section VII-VII') was injected into axial regions of S_3 folds, and because S_3 foliation is locally strong in rocks of this series. Wall rocks were still sufficiently warm to deform plastically.

For the youngest member of the New Hampshire Plutonic Series, the Concord Granite, field relations and foliation patterns (parallel to intrusive contacts, rather than to S_3) indicate a posttectonic origin. In Vermont several comparable granite stocks (Barre Granite, for example) superimpose a local sillimanite contact thermal aureole across the regional metamorphic patterns. The site of origin of the muscovite-biotite granodiorite, quartz monzonite, and granite magmas is not revealed by their mineralogy, but because they show petrographic features gradational into those of the biotite quartz monzonite series, a similar source region is possible. By the time of intrusion of the Concord Granite, however, the crust at the presently exposed level had become more brittle, and we postulate that magmatic emplacement was accomplished by a process of brittle fracturing of the crust, ascent of the magma along these fractures or along pre-existing zones of structural weakness, cauldron stoping, or forcing of overlying crustal blocks toward the surface.

If New Hampshire ever had a batholith, it is likely to have been several kilometers above the present structure level. Deeply extending stocks or batholiths seem to need a brittle fracture zone for their emplacement, and hence are found in the superstructure of an orogen. In the infrastructure, on the other hand, lateral intrusive directions are taken, partly because this is the structural grain and partly because vertically extending zones of weakness are closed off by flowage. The combination of lateral forcible intrusions and high superincumbent load would also explain why, as in the New Hampshire Plutonic Series, vertical lifting of roof rocks was relatively ineffectual, and plutons were restricted to thicknesses on the order of 2.5 km.

The idea that there may be a close causal relationship between the shapes of syntectonic intrusions and their positions relative to the regional isograd pattern has been noted by others. Perhaps the best documented case is that of the Adirondacks, in which Buddington (1959) long ago observed that within the catazone (our infrastructure) plutons are invariably thin phacolithic sheets.

ACKNOWLEDGMENTS

W. S. Bothner and D. S. Hodge have been very helpful in their thoughtful and constructive editorial reviews. D. R. Nielson and C. W. Montgomery made some of the gravity readings in the Hillsboro quadrangle and also computed a gravity profile which is not shown on Figures 4 or 5, but which is in agreement with our results. This study was supported in part by National Science Foundation Grant GA-35078 to Lyons, and in part by research grants from the Geological Society of America and the Society of Sigma Xi to Clark, Englund, and Nielson. These research funds are gratefully acknowledged.

REFERENCES CITED

Bancroft, A. M., 1960, Gravity anomalies over a buried step: Jour. Geophys. Research, v. 65, p. 1630-1631.
Bean, R. J., 1953, Relations of gravity anomalies to the geology of central Vermont and New Hampshire; Geol. Soc. America Bull., v. 64, p. 509-538.
Billings, M. P., 1956, Bedrock geology, in the geology of New Hampshire, Pt. 2: Concord, New Hampshire Planning and Devel. Comm., 203 p.
Bird, J. M., and Dewey, J. F., 1970, Lithosphere plate-continental margin tectonics and the evolution of the Appalachian orogen: Geol. Soc. America Bull., v. 81, p. 1031-1059.
Bothner, W. A., 1974, Gravity study of the Exeter pluton, southeastern New Hampshire: Geol. Soc. America Bull., v. 85, p. 51-56.
Buddington, A. F., 1959, Granite emplacement with special reference to North America: Geol. Soc. America Bull., v. 70, p. 671-784.
Chapman, C. A., 1952, Structure and petrology of the Sunapee quadrangle, New Hampshire: Geol. Soc. America Bull., v. 63, p. 381-425.
Clark, R. G., 1972, Petrology and structure of the Kinsman Quartz Monzonite and some related rocks [Ph.D. thesis]: Hanover, N.H., Dartmouth College, 255 p.
——1973, The occurrence and origin of some almandine-rich rocks in central New Hampshire: Geol. Soc. America, Abs. with Programs (Ann. Mtg.), v. 5, no. 2, p. 147-148.
Clark, R. G., and Englund, E., 1972, Gravity evidence for shallow-floored plutons in New England: (EOS) Amr. Geophys. Union Trans., v. 53, no. 4, p. 342.
Diment, W. H., 1968, Gravity anomalies in northwestern New England, in Zen, E-an, White, W. S., Hadley, J. B., and Thompson, J. B., Jr., eds., Studies of Appalachian geology, northern and maritime: New York, Interscience Pubs., Inc., p. 399-413.
Dixon, H. R. and Lundgren, L. L., 1968, Structure of eastern Connecticut, in Zen, E-an, White, W. S., Hadley, J. B., and Thompson, J. B., Jr., eds., Studies of Appalachian geology, northern and maritime: New York, Interscience Pubs., Inc., p. 219-234.
Dobrin, M. B., 1960, Introduction to geophysical prospecting: New York, McGraw-Hill Book Co., 435 p.
Doe, B. R., and Delevaux, M. H., 1973, Variation in lead isotope composition in Mesozoic granitic rocks of California: A preliminary investigation: Geol. Soc. America Bull., v. 84, p. 3513-3526.
Doyle, R. G., 1967, Preliminary geologic map of Maine: Augusta, Maine Geological Survey.
Englund, E. J., 1971, Geology of the Holderness quadrangle, in Lyons, J. B., and Stewart, G. W., eds., Guidebook for field trips in New Hampshire and contiguous areas: New England Intercollegiate Geol. Conf., New Hampshire 1971, p. 78-87.
——1974, The bedrock geology of the Holderness quadrangle, New Hampshire [Ph.D. thesis]: Hanover, N.H., Dartmouth College, 144 p.
Fitton, J. G., 1972, The genetic significance of almandine-pyrope phenocrysts in the calc-alkaline Borrowdale volcanic group, northern England: Contr. Minerology and Petrology, v. 38, p. 231-248.
Fowler-Billings, K., 1942, Geologic map of the Cardigan quadrangle, New Hampshire: Geol. Soc. America Bull., v. 53, p. 177-178.

Fowler-Billings, K., 1949, Geology of the Monadnock region of New Hampshire: Geol. Soc. America Bull., v. 60, p. 1249-1280.
Greene, R., 1970, The geology of the Peterborough quadrangle, New Hampshire: New Hampshire Dept. Resources and Econ. Devel., Bull., no. 4, 88 p.
Green, T. H., and Ringwood, A. E., 1968, Origin of garnet phenocrysts in calc-alkaline rocks: Contr. Mineralogy and Petrology, v. 18, p. 163-174.
Grout, F. F., 1945, Scale models and structures related to batholiths: Am. Jour. Sci., v. 243A, p. 260-287.
Hamilton, W., and Myers, W. B., 1967, The nature of batholiths: U.S. Geol. Survey Prof. Paper 554-C, p. C1-C30.
Heald, M. T., 1950, Structure and petrology of the Lovewell Mountain quadrangle, New Hampshire: Geol. Soc. America Bull., v. 61, p. 43-89.
Hsü, L. C., 1968, Selected phase relationships in the system Al-Mn-Fe-Si-O-H: A model for garnet equilibria: Jour. Petrology, v. 9, p. 40-88.
Joyner, W. B., 1963, Gravity in north-central New England: Geol. Soc. America Bull., v. 74, p. 831-858.
Kane, M. F., Simmons, G., Diment, W. H., Fitzpatrick, M. M., Joyner, W. B., and Bromery, R. W., 1972, Bouguer gravity and generalized geologic map of New England and adjoining areas: U.S. Geological Survey Geophys. Inv. Map GP-839.
Kistler, R. W., and Petermann, Z. C., 1973, Variations in Sr, Rb, K, Na, and initial Sr^{87}/Sr^{86} in Mesozoic granitic rocks and intruded wall rocks in central California: Geol. Soc. America Bull., v. 84, p. 3489-3511.
Kruger, F. C., and Linehan, D., 1941, Seismic studies of floored intrusives in western New Hampshire: Geol. Soc. America Bull., v. 52, p. 633-648.
Mehnert, K., 1968, Migmatites and the origin of granitic rocks: Amsterdam, Elsevier Pub. Co., 393 p.
Moore, G. E., 1949, Structure and metamorphism of the Keene-Brattleboro area, New Hampshire-Vermont: Geol. Soc. America Bull., v. 60, p. 1613-1670.
Morgan, B. A., 1972, Metamorphic map of the Appalachians: U.S. Geol. Survey Misc. Geol. Inv. Map I-724.
Naylor, R. S., 1971, Acadian orogeny: An abrupt and brief event: Science, v. 172, p. 558-559.
Nielson, D. L., 1974, The structure and petrology of the Hillsboro quadrangle, New Hampshire [Ph.D. thesis]: Hanover, N.H., Dartmouth College, 254 p.
Nielson, D. L., Lyons, J. B., and Clark, R. G., 1973, Gravity and structural interpretations of the mode of emplacement of the New Hampshire Plutonic Series: Geol. Soc. America, Abs. with Programs (Ann. Mtg.), v. 5, no. 7, p. 750.
Nielson, D. R., 1973, Formation of flecky gneiss by metamorphic differentiation: Geol. Soc. America, Abs. with Programs (Ann. Mtg.), v. 5, no. 7, p. 750.
Presnall, D. C., and Bateman, P. C., 1973, Fusion relations in the system $NaAlSi_3O_8$-$CaAl_2Si_2O_8$-$KAlSi_3O_8$-SiO_2-H_2O and the generation of granitic magma in the Sierra Nevada batholith: Geol. Soc. America Bull., v. 84, p. 3181-3202.
Ramberg, H., 1970, Model studies in relation to intrusion of plutonic bodies, in Newell, G. and Rast, N., eds., Mechanism of igneous intrusion: Liverpool, Gallery Press, p. 261-286.
Rankin, D. W., 1968, Volcanism related to tectonism in the Piscataquis volcanic belt, an island arc of Early Devonian age in north-central Maine, in Zen, E-an, White, W. S., Hadley, J. B., and Thompson, J. B., Jr., eds., Studies of Appalachian geology, northern and maritime: New York, Interscience Pubs., Inc., p. 355-378.
Talwani, M., Worzel, J. L. and Landsman, M., 1959, Rapid gravity computations for two-dimensional bodies with application to the Mendocino submarine fracture zone: Jour. Geophys. Research, v. 64, p. 49-59.
Thompson, J. B., Jr. and Norton, S., 1968, Paleozoic regional metamorphism in New England and adjacent areas, in Zen, E-an, White, W. S., Hadley, J. B., and Thompson, J. B., Jr., eds., Studies of Appalachian geology, northern and maritime: New York, Interscience Pubs., Inc., p. 319-328.

Thompson, J. B., Jr., Robinson, P., Clifford, T. N., and Trask, N. J., 1968, Nappes and gneiss domes in west-central New England, *in* Zen, E-an, White, W. S., Hadley, J. B., and Thompson, J. B., Jr., eds., Studies of Appalachian geology, northern and maritime: New York, Interscience Pubs., Inc., p. 203-218.

Vernon, W. W., 1971, Geology of the Concord quadrangle, *in* Lyons, J. B., and Stewart, G. W., eds., Guidebook for field trips in central New Hampshire and contiguous areas: New England Intercollegiate Geol. Conf., New Hampshire 1971, p. 118-125.

MANUSCRIPT RECEIVED BY THE SOCIETY MARCH 11, 1974
REVISED MANUSCRIPT RECEIVED JULY 24, 1974
MANUSCRIPT ACCEPTED AUGUST 26, 1974

Printed in U.S.A.

Geological Society of America
Memoir 146
©1976

Nickeliferous Pyrrhotite Deposits, Knox County, Southeastern Maine

George D. Rainville
Department of Geology
Boston University
Boston, Massachusetts 02215

AND

Won C. Park
Ledgemont Laboratory
Kennecott Copper Corporation
Lexington, Massachusetts 02173

ABSTRACT

Nickeliferous ores (pyrrhotite, pentlandite, and chalcopyrite) in the Harriman peridotite and Warren gabbro-diorite of Knox County, southeastern Maine, occur as lenticular bodies in the Ordovician Penobscot Formation. The ore bodies were emplaced in Late Ordovician time, most likely synchronously with the Taconic orogeny; they were later modified by Acadian dynamothermal and Permian thermal metamorphism.

The Harriman body is a feldspathic lherzolite peridotite with limited serpentinization in the central zone. The amount of serpentine, amphiboles, talc, and carbonate minerals increases markedly away from the central zone, thus defining a crude zoning pattern. Sulfides occur interstitially or, less commonly, as thin massive layers. The Warren body is composed predominantly of andesine-labradorite, hornblende, cummingtonite, and biotite. Sulfides occur interstitially, as irregular disseminations, and in massive segregation bodies.

The textural intergrowths of sulfides are generally simple, although locally they are more complex as a result of metamorphism. The interstitial nature and macroscale and microscale segregation drops of the sulfides indicate an orthomagmatic origin. This is supported by the high combined concentrations of nickel, cobalt, and copper in pyrrhotite (approximately 0.7 wt percent) and more than 2 wt percent cobalt plus copper in all varieties of pentlandite. These values are consistent with values for similar minerals in deposits of presumably early magmatic origin, such as those of Sudbury, Canada; Pecenga, USSR; and Insizwa, South West Africa.

Lack of primary pyrite, as well as higher than stoichiometric metal to sulfur ratios of the major sulfides, indicates relatively low sulfur fugacities. These features, along with the virtual absence of primary magnetite and the presence of as much as several volume percent of graphite, distinguish the Harriman and Warren deposits from other nickeliferous pyrrhotite deposits. The occurrence of chromian ulvospinels somewhat similar to those from lunar samples, the dominant ilmenite-spinel assemblage, and the absence of hematite and primary magnetite indicate that low oxygen fugacities prevailed during the development of the ore deposits.

INTRODUCTION

The Warren (now referred to as area 5) and Harriman (formerly Union) deposits are located in the Seaboard Lowland Section of the New England province as delineated by Thornbury (1965). The deposits are located in central Knox County, southeastern Maine, approximately 110 km northeast of Portland, Maine (Fig. 1).

The Harriman peridotite was first described by Bastin (1908a, 1908b); it was later mapped in some detail by Houston (1956), who described both the opaque and nonopaque mineralogy. Although the Warren deposit lies only 3 km southwest of the Harriman, it was not discovered until 1965 owing to the lack of exposure of mafic rocks in the vicinity of the ore body. (Joint evaluation of the ore properties is now under way under the auspices of Hanna Mining Company and Basic Incorporated, both based in Cleveland, Ohio.) The Warren ore body contains an estimated 10 million tons of ore with 0.99 wt percent nickel plus 0.49 wt percent copper (Mining Engineering, 1972). Several reports by R. S. Young (unpub. data) and D. G. Rogich and others (unpub. data) have been prepared with special emphasis on the economic aspects of these ore bodies. The purpose of this study is to describe and discuss the opaque and nonopaque mineralogy, sulfide-silicate relationships, trace-element content of the major sulfides, and the paragenetic history of the deposits. Several interesting mineralogical problems that arose during the course of this study are discussed in some detail. More comprehensive data and discussion of these mineralogical problems are found in Rainville (1972). It is hoped that a better understanding of the Harriman and Warren ore bodies will provide a basis for further exploration and evaluation of similar alpine-type peridotite-gabbro sequences throughout the New England region.

METHODS OF INVESTIGATION

Surface, pit, and core samples were obtained for examination. Approximately 150 polished sections, 6 polished thin sections, and 100 thin sections were studied. Modal analyses of 23 selected polished sections were made for the major sulfides, oxides, and graphite. An average of more than 3,000 points per section were counted. Electron microprobe analyses, including semiquantitative scanning photography and quantitative analyses by point, step, and linear scanning, were used to investigate pyrrhotite, pentlandite, chalcopyrite, mackinawite, silicates, and oxides. The ARL Electron Microprobe at Heidelberg University and Model MAC-5 electron microprobe at Massachusetts Institute of Technology were utilized. Reference standards for the sulfides were synthetic FeS, Co- and Ni-bearing FeS, and Ookiep chalcopyrite obtained from the Geophysical Laboratory, Washington, D.C. Supplementary analyses of the sulfides and oxides were accomplished by x-ray diffraction and laser microprobe analyses at Jarrel Ash Company, Waltham, Massachusetts. The polished-section collections of P. Ramdohr of Heidelberg University and Boston University were utilized for comparison with other nickeliferous pyrrhotite deposits and in opaque-mineral identification. A technique employing a magnetite colloidal

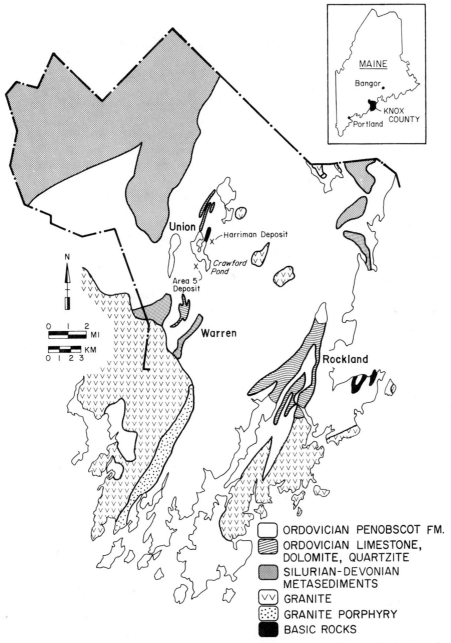

Figure 1. Location and generalized geologic map of Knox County, Maine, with location of Harriman (Union) and Warren deposits. Modified after map by the Sewall Company, Old Town, Maine (Knox County Regional Planning Commission, 1963).

solution was used to supplement optical and x-ray identification of monoclinic pyrrhotite.

GENERAL GEOLOGY

The Harriman and Warren deposits were emplaced in the areally extensive Penobscot Formation of Middle to Late Ordovician age (Boucot and others, 1972). The restriction of these mafic bodies and several others of the area to the Penobscot Formation suggests a Late Ordovician age for the mafic bodies. Amos (1963) stated that similar isolated mafic bodies are aligned in southeastern Maine in three northeast-trending belts, to which the Harriman and Warren deposits may be genetically related. Cheney (1965) indicated that the mafic intrusive rocks of Knox County were emplaced prior to the Taconic orogeny, but Amos (1963) postulated an Early Devonian age. An Ordovician age is more consistent with the age of numerous mafic to ultramafic intrusive rocks throughout the Appalachian belt and with the restriction of the Harriman, Warren, and related mafic bodies to pre-Silurian metasedimentary rocks.

The Penobscot Formation contains quartz, biotite, sillimanite, and muscovite as the dominant minerals. The formation is, however, highly variable lithologically and is locally gneissic in character, with albite and orthoclase as primary constituents. Almandine garnet, graphite, and sulfides are locally abundant. Marble and quartzite occur in the vicinity of the Harriman deposit and have been correlated by Bickel (1971) with the Silurian Appleton Formation. This agrees well with the work of Houston (1956), who demonstrated that the marble-quartzite sequence is stratigraphically younger than the Penobscot Formation. Locally, quartz monzonite, granite, and simple granite pegmatite occur within the schist and mafic units. Although some of the silicic intrusive rocks clearly show crosscutting relationships, it must

TABLE 1. CHEMICAL COMPOSITION OF SILICATES AND OXIDES

	Warren deposit				Harriman deposit									
	Plagioclase (An_{65})	Hornblende	Biotite	Ilmenite	Olivine (Fo_{76})	Bronzite (En_{80})	Hornblende*	Hornblende†	Plagioclase	Biotite	Ilmenite	Dark spinel phase	Light spinel phase	Titanium-bearing chrome spinel
SiO_2	50.22	48.90	35.54	0.00	38.47	54.87	41.68	42.04	52.01	38.75	0.00	0.35	0.40	0.43
Al_2O_3	31.84	7.10	16.61	0.11	0.03	1.63	13.87	17.82	31.03	17.59	0.13	38.80	4.49	4.62
FeO	0.10	13.22	16.92	44.83	21.92	13.00	8.17	6.30	0.08	7.27	43.03	31.96	68.97	55.20
MgO	0.01	14.44	13.33	0.39	39.59	29.19	14.98	16.54	0.11	20.27	1.66	7.49	1.16	1.59
CaO	13.16	11.08	0.02	..	0.02	0.23	11.14	11.42	13.81	0.00	..	..	..	..
Na_2O	3.91	0.60	0.16	..	0.01	0.05	2.85	3.02	3.82	1.11	..	..	..	..
K_2O	0.02	0.17	7.66	..	0.00	0.00	0.73	0.33	0.02	7.75	..	..	..	..
TiO_2	0.01	0.73	2.15	51.98	0.01	0.02	3.03	0.08	0.03	3.14	52.86	0.20	3.82	13.86
MnO	0.02	0.37	0.22	1.66	0.35	0.37	0.18	0.17	0.01	0.07	1.74	0.00	0.11	0.32
Cr_2O_3	0.03	0.15	0.21	0.01	0.03	0.02	0.32	0.01	0.02	0.63	0.00	17.32	13.13	16.64
NiO	..	..	..	0.02	0.07	0.02	0.11	0.03	0.07	0.05	0.01	0.06	0.18	0.27
CoO	..	..	..	0.10	..	..	..	..	..	..	0.18	0.12	0.41	0.38
ZnO	..	..	..	0.02	..	..	..	..	..	..	0.03	0.98	0.50	0.47
V_2O_3	..	..	..	0.23	..	..	..	..	..	..	0.03	..	..	..
Total	99.32	96.76	92.82	99.42	100.50	99.40	97.06	97.75	101.01	96.68	99.67	97.28	93.17	93.78

Note: All values are average of 7 or more analyses. Values corrected after the Bence-Albee methods utilizing the PDP-11 digital computer, Massachusetts Institute of Technology.
*Formed as a reaction rim between feldspar and pyrrhotite.
†Associated with olivine and bronzite as part of corona sequence.

be emphasized that contacts between the Penobscot Formation and silicic intrusive rocks are often clearly gradational and support granitization.

The area of the Harriman and Warren deposits is a zone of high-grade metamorphism, with the lower sillimanite and sillimanite–potassium feldspar zones dominant. The principal schistosity of the area strikes north-northeast and variously dips from 45° to nearly vertical. At least two separate schistosities have developed with recognizable bedding approximately parallel to the earlier schistosity. Folding is extremely tight, with axial surfaces dipping steeply to the east and southeast. Folds generally plunge moderately north-northeast. Normal faults, with displacements of as much as 30 m, were recognized.

The ore-bearing Harriman and Warren mafic bodies are conformable to the principal schistosity. D. G. Rogich and others (unpub. data) indicated that the stratigraphic thickness of the Warren body exceeds 120 m. The Harriman body is exposed at the surface in isolated scattered outcrops, whereas massive sulfide zones of the Warren body occur within several metres of the surface on the western side of that body. The sulfide ores are restricted almost entirely to gabbro and peridotite, although minor ore minerals are present in Penobscot schist in the immediate vicinity of these mafic-ultramafic intrusive rocks. In the Warren deposit, sulfides occur in massive zones often several metres thick and are disseminated in varying amounts throughout the gabbro. Only thin massive zones, seldom more than 0.5 m in width, occur in the Harriman peridotite. The peridotite ore is generally interstitial, with sulfides constituting about 25 volume percent of the rock.

SILICATE PETROGRAPHY

Harriman Feldspathic Lherzolite Peridotite

The presence of magnetite-bearing olivine, anhedral plagioclase, and interstitial sulfides gives the rock its unusual appearance. It has been studied petrographically by Bastin (1908a, 1908b) and Houston (1956). The relatively unaltered central zone of the peridotite was studied in more detail for this report, as the altered peridotite is generally depleted in sulfide minerals, and these constitute only a few percent of the rock.

Electron microprobe data for silicates and oxides are given in Table 1.

Olivine (Fo_{76}) is the major mineral in the unaltered Harriman peridotite; Bastin (1908a, p. 12) and Houston (1956, p. 63) estimated that it constitutes 60 vol percent of the rock. This percentage decreases rapidly toward the margin of the peridotite body. The olivine grains are generally round, with grain size varying from several tens of micrometres to several millimetres in diameter. Nickel (NiO, 0.07 wt percent) cobalt, and manganese (MnO, 0.35 wt percent) were detected in trace amounts in olivine by electron microprobe. The olivine grains are not highly shattered but are crosscut by numerous fractures that contain serpentine. Serpentine also borders olivine grains. Many of the serpentine veinlets enclose magnetite veinlets; this is partly responsible for the extremely dark color of the peridotite.

Pyroxene is generally subhedral and interstitial to olivine but occasionally partially or totally encloses olivine. The primary pyroxenes are bronzite (En_{80}) and very minor augite. Bronzite surrounds olivine in thin bands seldom exceeding 100 μm; it grades into light-brown hornblende in contact with plagioclase. A similar multilayered corona was described in detail by Spry (1969, p. 105). This was recognized by Bastin (1908b) and probably represents equilibration of the iron and magnesium-rich minerals with the sodium- and calcium-enriched residual silicate liquid. A similar light-brown hornblende occurs as a reaction product of feldspar and pyrrhotite. The only significant chemical difference between the two hornblendes is the several percent TiO_2 contained in the type formed as a pyrrhotite-feldspar reaction product (Table 1). Some augite is partially uralitized.

Plagioclase is generally anhedral and interstitial to olivine. Its average composition is An_{67}, as determined by electron microprobe analysis. Reaction zones between feldspar and sulfides consist of brown hornblende. Cathodoluminescent studies of these feldspars show light-blue luminescent colors. This is indicative of Ti^{+4} activation and a low Fe^{+3}/Fe^{+2} ratio (Mariano and others, 1973). The plagioclase shows minor saussuritization.

Hornblende is formed by uralitization of pyroxene and as a reaction product, as noted above. Minor amounts of biotite and chlorite are present in the unaltered peridotite. Biotite is the result of primary crystallization, whereas chlorite (penninite) is the result of the alteration of hornblende or biotite.

Serpentine constitutes a few modal percent of the unaltered peridotite. Antigorite borders and crosscuts olivine grains and is associated with small disseminated grains of a second variety of serpentine.

Garnet was not recognized by earlier investigators, but uvarovite has tentatively been identified with the aid of G. Van der Kaaden of Heidelberg University. Minor scattered grains of carbonate occur. The vein-filling carbonate is calcite, but cathodoluminescent studies confirm the presence of dolomite in small scattered grains. The spinel of the deposit tends to be opaque or semiopaque; it occurs as scattered round grains generally less than 100 μm in diameter and is generally chromiferous. Houston (1956, p. 64) indicated that the altered peridotite consists largely of serpentine, tremolite-actinolite, anthophyllite, talc, chlorite, hornblende, quartz, carbonates, and very minor amounts of the sulfides and oxides characteristic of the unaltered peridotite.

Warren Gabbro-Diorite

The Warren body consists primarily of rocks of gabbroic composition that have undergone fairly extensive alteration. Locally, more mafic and silicic rocks are present, but they show no systematic distribution or relationship to the gabbroic phases.

The presence of Crawford Pond between the Harriman and Warren bodies makes any gradational nature of the mafic-ultramafic bodies very difficult to ascertain. Drill core from an island in the pond between the two ore bodies shows a hybrid rock (a few percent olivine), which indicates some type of gradation within the mafic–ultramafic complex.

Augite is present in minor amounts and is extensively uralitized to hornblende and cummingtonite.

Plagioclase (An_{41-75}) constitutes the dominant mineral of the body and shows varying degrees of saussuritization. The plagioclase shows some minor alignment and reverse zoning.

Cummingtonite occurs as radiating aggregates of prismatic crystals commonly with irregular crystal boundaries. Cummingtonite is almost invariably intergrown with hornblende, where both minerals are present, and they appear to be contemporaneous. The use of the hornblende-cummingtonite pair (Eskola, 1950) as a metamorphic indicator is not firmly established and may represent nonequilibration (Vernon, 1962), as do many of the sulfide phases.

Hornblende is generally more abundant than cummingtonite, but the latter dominates in some sections. Amphiboles are next in abundance to feldspar. The presence of some primary hornblende is inferred but difficult to ascertain because of the large amount of textural modification. The ophitic texture of the gabbro is recognizable in some instances, but a blastophitic texture is common.

Biotite is present in amounts exceeding several modal percent. The first generation consists of fine-grained biotite with subparallel alignment. The second generation is much coarser biotite that is randomly oriented and frequently disrupts and crosscuts earlier formed silicates and sulfides. The latter biotite appears to be a product of recrystallization.

Quartz grains are generally small and show undulatory extinction. Quartz increases in the vicinity of pegmatite dikes and silicic intrusive rocks within the gabbro. The myrmekitic quartz intergrown with plagioclase and the amphiboles seems to confirm the observation of Bickel (1971) that quartz is liberated during the uralitization process by the following reaction: clinopyroxene = hornblende + biotite + quartz. Quartz that later filled veins is either related to associated silicic intrusive rocks or is locally derived by segregation.

Chlorite is present as an alteration product of biotite and the amphiboles. Minor amounts of olivine, garnet, apatite, and quartz are also present.

SULFIDE MINERALOGY

In light of the economic potential and our particular interests, considerable emphasis was placed on the opaque mineralogy of these deposits. Quantitative and qualitative evaluations of the minerals have been made, and a paragenetic sequence has been established.

Pyrrhotite

Pyrrhotite is the most abundant sulfide; it constitutes an average of 92 wt percent of the total sulfides. Pyrrhotite occurs in massive zones where it is the chief constituent, in disseminated grains throughout the mafic and ultramafic rocks, interstitially to olivine grains, in veinlike masses in granite pegmatite, and within the schist bordering and enclosed by the ore bodies.

Pyrrhotite is typically anhedral with a recognizable basal parting. There are several textural varieties of pyrrhotite which fall predominantly into the following categories: (1) simple intergrowths with pentlandite and chalcopyrite with gently curved or rectilinear boundaries (Fig. 2A); (2) large anhedral masses in the silicates, with or without minor amounts of associated chalcopyrite and pentlandite, commonly replaced by the growth of later formed hydrous silicates (Fig. 2B); (3) intersitital grains between earlier formed orthosilicates (especially in the peridotite); (4) complex intergrowths with chalcopyrite and pentlandite, including replacement of pyrrhotite by chalcopyrite and exsolution flames and rosettes of pentlandite; (5) fracture fillings crosscutting the silicates (Fig. 2C); and (6) early segregation droplets and myrmekitic intergrowths with the orthosilicates (Fig. 2D).

The pyrrhotite in these deposits is strongly anisotropic, but it lacks the strong pleochroism often characteristic of pyrrhotite. About 90 percent of the pyrrhotite is optically homogeneous, but careful examination under oil immersion shows strong inhomogeneity in the remainder of the pyrrhotite which is characterized by slight differences in reflectivity (Fig. 3). The use of a colloidal magnetite solution (Einaudi, 1968) to determine if these areas represent monoclinic pyrrhotite yielded negative results. The microscopic inhomogeneity of pyrrhotite may simply reflect variations in trace elements such as nickel, as suggested by Dennen (1943, p. 46). The more regular pyrrhotite exsolution lamellae, however, are identified as monoclinic pyrrhotite by colloidal magnetite solution (Fig. 3). These distinct lamellae constitute 30 percent of the Harriman pyrrhotite and are virtually absent from the Warren pyrrhotite.

Microprobe analyses of pyrrhotite from the ore deposits are given in Table 2. Several x-ray diffraction analyses of pyrrhotites confirm that hexagonal ($2A$ and $5C$ structural type) pyrrhotite constitutes virtually the entire pyrrhotite content. This is compatible with the low magnetic readings and optical observation. Using the d_{102} methods of Arnold and Reichen (1962) and Arnold (1966), the average atomic percent of metals of six pyrrhotites is 47.70 (Table 3).

The average microprobe atomic percent of metals of 46.76 corresponds fairly well with the value of 47.70 obtained by x-ray diffraction. The high nickel and cobalt substitution is indicative of magmatic pyrrhotites as described by Cambel and Jarkovsky (1969). Copper is the only other trace element of any significance in pyrrhotite, but it rarely exceeds several hundredths of a percent by weight. The amount of cobalt substitution in pyrrhotite is fairly constant (ranging from 0.06 to 0.16 wt percent, averaging 0.11 wt percent of 11 pyrrhotites) throughout the deposits (Table 2). There is a possibility that some of the cobalt values may

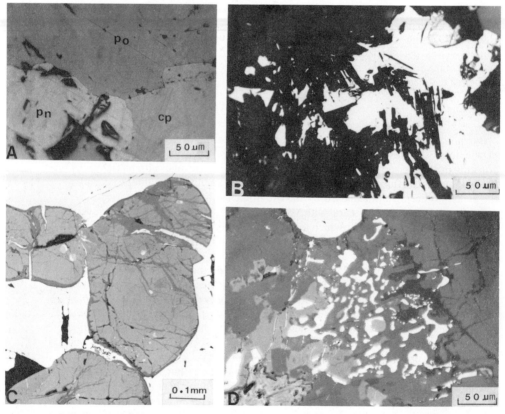

Figure 2. Pyrrhotite textures. A, simple grain-boundary relationships of pyrrhotite (po)–pentlandite (pn)–chalcopyrite (cp); Warren deposit. B, pyrrhotite (white) replaced by hydrosilicates; Warren deposit. C, olivine grains (light gray) surrounded and filled by pyrrhotite (white); dark gray coronas and veinlets are serpentine and magnetite, only in veinlets; note small round primary pyrrhotite-spinel grains within olivine; Harriman deposit. D, pyrrhotite (white) myrmekitically intergrown with olivine (dark gray) and orthopyroxene (light gray); graphite at extreme bottom left; Harriman deposit.

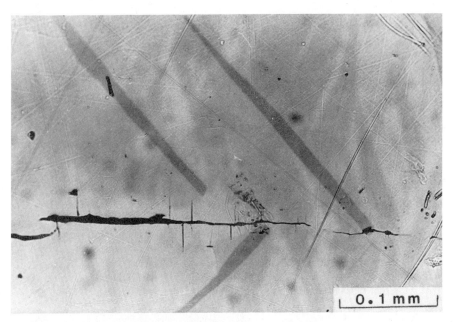

Figure 3. Optically inhomogeneous hexagonal pyrrhotite (light to medium gray) with exsolution lamellae (dark gray) of monoclinic pyrrhotite. Harriman deposit.

TABLE 2. ELECTRON-MICROPROBE ANALYSIS OF MAJOR SULFIDES OF THE HARRIMAN AND WARREN DEPOSITS, SOUTHEASTERN MAINE

	Average wt %	Standard deviation	Range
	Pyrrhotite*		
Fe	59.2	0.9	58.25–61.31
Ni	0.51	0.15	0.14–0.68
Co	0.11	0.03	0.06–0.16
Cu	0.07	0.06	0.00–0.21
S	39.1	1.6	36.21–41.80
Total	98.99		
	Pentlandite†		
Fe	32.1	1.2	30.81–34.09
Ni	34.3	1.8	32.30–36.80
Co	2.6	1.6	1.09–5.60
Cu	0.22	0.07	0.09–0.35
S	31.6	0.5	30.73–32.53
Total	100.82		
	Chalcopyrite§		
Fe	31.1	0.9	30.89–31.58
Cu	33.8	0.2	33.57–34.09
Ni	0.15	0.11	0.03–0.29
Co	0.05	0.08	0.00–0.12
S	34.6	0.9	33.19–35.30
Total	99.70		

*Average of 11 analyses.
†Average of 10 analyses.
§Average of 4 analyses.

328 George D. Rainville and Won C. Park

TABLE 3. X-RAY DIFFRACTION DATA OF HARRIMAN AND WARREN PYRRHOTITES

Deposit	d_{102}	Metals* (atomic %)	Structural variety
Harriman	2.072	47.85	Hexagonal
Harriman	2.072	47.85	Hexagonal
Harriman	2.071	47.75	Hexagonal > monoclinic
Warren	2.071	47.75	Hexagonal
Warren	2.070	47.66	Hexagonal
Warren	2.067	47.39	Hexagonal

*Calculated from the method of Arnold and Reichen (1962).

be contributed by FeK$_\beta$ radiation, but no dilution analyses were performed. Careful scanning has shown, however, that the contribution of FeK$_\beta$ radiation is extremely small.

The amount and distribution of nickel in pyrrhotite is interesting genetically as well as economically. Figure 4 shows scans of various elements in pyrrhotite associated with serpentine and olivine. A linear scan across the hexagonal and monoclinic pyrrhotite shows that nickel is concentrated in the monoclinic phase (0.31 wt percent) as compared to the hexagonal phase (0.07 wt percent). The increase

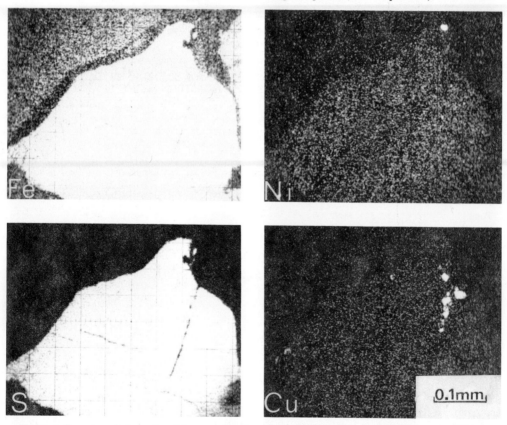

Figure 4. Scanning photographs of hexagonal and monoclinic pyrrhotite in olivine. Ni, Co, and Cu are the only trace elements present (Co is less than Cu). A broad lamella of monoclinic pyrrhotite is associated with the host hexagonal phase. Average amount of Ni in the hexagonal phase is 0.07 wt percent; in the monoclinic phase, 0.31 wt percent.

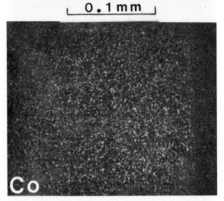

Figure 5. Scanning photographs of Ni and Co in pyrrhotite. Co distribution is uniform, but Ni distribution does not conform to recognizable optical boundaries. Crystallographically oriented distribution of Ni. Warren deposit.

in nickel is abrupt and not gradational. The amount of nickel within the monoclinic and hexagonal phases shows almost no variation from the average nickel value in each phase. Other monoclinic-hexagonal assemblages were studied by step and linear scanning techniques, but these failed to prove a consistent relationship between nickel partitioning and the type of pyrrhotite. Blatt (1972) concluded that nickel tends to enrich in the hexagonal over the coexisting monoclinic phase. This conclusion cannot be supported for our deposits, and in some instances the reverse has been found. For most of the inhomogeneous pyrrhotite masses, there was no consistent relationship with optical inhomogeneity. This conflicts with the conclusion of Scholtz (1936), who believed that a small amount of variation in nickel content is responsible for the inhomogeneous pyrrhotite phases. Although the nickel distribution in pyrrhotite does not necessarily correspond to optical variations, such a relationship is often observed. Figure 5 shows a remarkable oriented distribution of nickel in pyrrhotite that crosscuts recognizable optical boundaries.

Some correlation between an increase in nickel substitution in pyrrhotite at depth and a sympathetic decrease in the percentage of pentlandite appears to exist. A similar increase in nickel content with depth was recognized by Hawley and Nichol (1961) at the Campbell-Chibougamu deposit of Quebec.

Pentlandite

Pentlandite is the second most abundant sulfide in both ore bodies. It occurs as (1) large granular masses sometimes exceeding 1 mm in diameter; (2) narrow veins within pyrrhotite and between pyrrhotite grains; (3) subhedral grains bordering or enclosed by pyrrhotite; (4) flames or rosettes exsolved from pyrrhotite; (5) intimate intergrowths with chalcopyrite, in places resulting from replacement of pentlandite by chalcopyrite; and (6) subround or irregular grains in mosaic intergrowths with chalcopyrite and pyrrhotite (Figs. 6A to 6D).

Pentlandite comprises an average of 5 wt percent of the total sulfides. Elemental microprobe analyses (Fe, Ni, Co, Cu, S) of pentlandite are given in Table 2. W. A. Hockings and V. L. Doane (unpub. data) obtained a value of 34 ± 2 wt percent nickel in pentlandite, which is in agreement with our analyses. The percentage of cobalt in granular pentlandite is consistently higher than in coexisting flame pentlandites. The increase in cobalt corresponds to a sympathetic decrease in iron content. The average cobalt value of 2.6 wt percent is higher than the average obtained by Harris and Nickel (1972, p. 866) for similar pyrrhotite-pentlandite assemblages. The range of values from the Harriman and Warren deposits (1.09 to 5.60), however, is comparable with their range of values (0.0 to 6.5). The amount of cobalt in pyrrhotite and pentlandite accounts for virtually the entire cobalt assay. This is compatible with the absence of any primary cobalt minerals.

The development of subhedral and large massive granular pentlandite led Houston (1956) to speculate on the direct crystallization of this mineral from a magma in these deposits. Naldrett and others (1967) and several investigators since have recognized that all varieties of pentlandite are the result of the breakdown of (Fe, Ni)S solid solution. Only in extremely Ni-rich sulfide deposits is primary pentlandite formation probable. The sulfide iron to nickel ratio exceeds 25:1 in the Harriman and Warren bodies, and it has not been proved experimentally that primary pentlandite could have formed from a magma with this ratio.

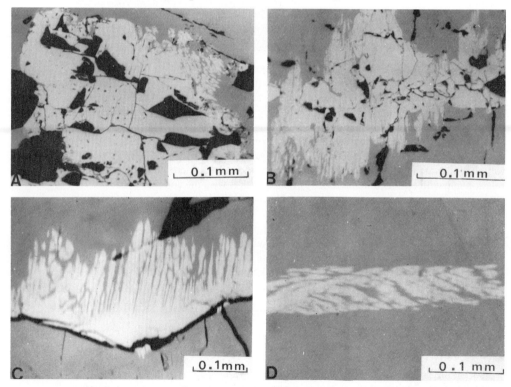

Figure 6. Pentlandite textures. A, granular pentlandite (light gray) with flamelike border at upper right in pyrrhotite (gray); Warren deposit. B, vein pentlandite with flamelike borders in pyrrhotite; Warren deposit. C, pentlandite exsolved along fracture in pyrrhotite; Warren deposit. D, large pentlandite flame exsolved from pyrrhotite; Warren deposit.

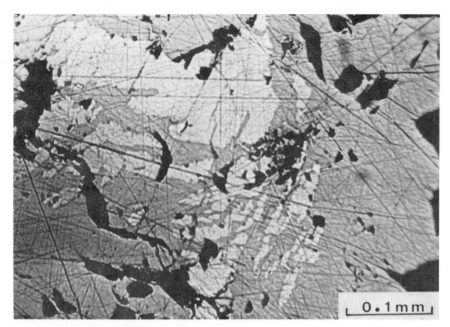

Figure 7. Pentlandite (grayish white) remaining as "islands" within replacment chalcopyrite. Warren deposit.

The variations in iron and cobalt contents from granular to flame pentlandites may reflect a more complex exsolution history than is expected from the phase relations in the experimental Fe-Ni-S system. The commonly observed replacement of granular pentlandite by chalcopyrite is rare for flame pentlandite. This indicates some intervening period before the exsolution of flame pentlandite, which may in part have been during metamorphism of the ore bodies. Both types of pentlandite are exsolution products. The textures of flames and rosettes indicate exsolution, and even granular pentlandite, which tends toward idiomorphic development, frequently "flames out" into the adjacent pyrrhotite, which lends credence to its exsolution origin (Fig. 6A).

An average pentlandite composition of $(Ni, Fe, Co, Cu)_{9.8}S_8$ indicates a sulfur deficiency. The metal to sulfur ratios range between 9.49:8 and 10.04:8 for ten pentlandites. The compatibility of a sulfur-deficient sulfide assemblage with coexisting sulfur-deficient pentlandite was suggested by Harris and Nickel (1972). Their observations seem valid in light of our data. The metal to sulfur ratio of 10.04:8.0 is higher than any that they reported and may reflect an error in sulfur probe analyses, but it is compatible with the relatively low sulfur fugacities postulated for our deposits.

Chalcopyrite

Chalcopyrite is the third most abundant sulfide and constitutes 3 wt percent of all sulfides. It occurs as simple intergrowths with granular pentlandite in pyrrhotite, irregular grains in pentlandite, isolated grains within the silicates, fracture fillings, and early segregation droplets. Some chalcopyrite, along with granular pentlandite, appears to be an unmixing product of pyrrhotite. Figure 7 demonstrates incomplete pentlandite replacement with isolated islands of pentlandite remaining in chalcopyrite.

Chalcopyrite appears to be the latest major sulfide formed. The textural relations determining its paragenetic position may in part result from mobilization during metamorphism. Very minor amounts of substitution by nickel, cobalt, and chromium in chalcopyrite have been detected. Average analyses are given in Table 2.

Mackinawite

In spite of its extreme pleochroism and anisotropism, mackinawite has commonly remained undetected in microscopic investigations of ores because of its extremely small grain size. Investigations of several nickeliferous pyrrhotite ores throughout the world show that this sulfide is common, although decidedly minor, in these deposits.

In the Harriman and Warren deposits, mackinawite is restricted to occurrences within chalcopyrite or at its borders. It occurs in spindlelike forms and wormlets and as irregular and subrounded grains. As there has been much confusion in the literature regarding the identification of mackinawite, we employed the electron microprobe to determine chemical composition, because the optical properties of mackinawite merge imperceptibly with those of valleriite with the substitution of trace elements.

The average analyses of the Harriman and Warren mackinawites show large

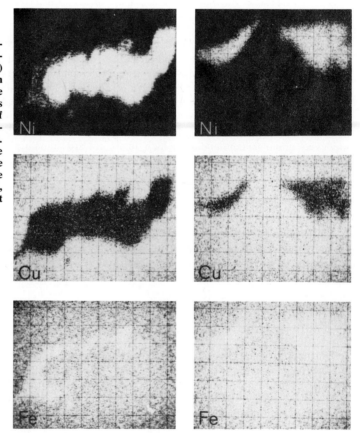

Figure 8. Scanning photographs of mackinawites (outlined by light area in Fe scans) in chalcopyrite. Ni distribution is uniform throughout the mackinawite grains, whereas Cu is abundant at the edges of mackinawite grains and deficient in the central portions. Quantitative analysis of these grains gives an average Ni value of 6.0 wt percent and an average Cu value of 4.0 wt percent, ranging from 1 to 10 wt percent Cu. Warren deposit.

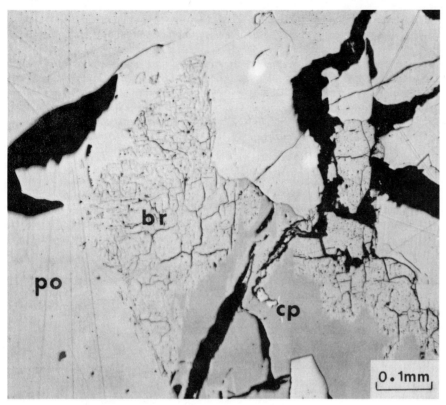

Figure 9. Bravoite (br) altered from pentlandite; po = pyrrhotite; cp = chalcopyrite. Warren deposit.

amounts of nickel and copper (6.0 and 4.01 wt percent, respectively) and very little cobalt. This high copper content is anomalous compared to values cited in the literature. Careful examination of the scanning photographs of mackinawite (Fig. 8) shows that whereas nickel distribution is uniform, copper is concentrated at the grain boundaries. Values vary from less than 1 percent Cu at the central part of a mackinawite grain to more than 10 percent at the edge of the grain. Since the electron beam penetrates to a depth of 3 to 5 μm, there is a possibility that the larger copper values obtained at the edge of mackinawite may reflect underlying chalcopyrite. The constancy of the relationship of increased copper values at the margins was found in all of the several grains analyzed. Step scanning also shows a constant increase from the central to the outer portions of mackinawite grains, which makes a diffusion interpretation more likely.

Bravoite

Bravoite occurs only as an alteration product of pentlandite. Bravoite formation generally starts along fractures in pentlandite and proceeds in some instances to a stage of nearly complete alteration of pentlandite (Fig. 9). The initial stages of bravoite development commonly appear as thin blotches in pentlandite. When no granular or flame pentlandites have been bravoitized, bravoite formation is restricted to pentlandite grains enclosed in chalcopyrite. This pentlandite-chalcopyrite association is common and may reflect a slightly different genesis for these pentlandites, probably of double exsolution of pentlandite and chalcopyrite out of monosulfide solid solution.

Bravoite in the Warren ore body disappears below a depth of about 50 m, which indicates that it is a product of supergene alteration. Enough samples from the Harriman body were not available to establish the same relation for this ore body. A low temperature of formation is compatible with the experimental results of Clark and Kullerud (1963), who showed that bravoite is stable only below 137°C.

Pyrite

Pyrite is very rare in both ore bodies. It occurs as veinlike masses; it is porous and characterized by a reflectivity lower than is common for pyrite. The mineral occurs entirely as late fracture filling. The lack of pyrite formation during any of the primary stages of ore development indicates rather low sulfur fugacities. The authors examined a few other basic bodies of the same age in the Knox County area and found alteration of pyrrhotite to pyrite, marcasite, and an "intermediate product" (Ramdohr, 1969, p. 592-598) quite common. This is an oxidation reaction as evidenced by the other mineralogic changes that occurred contemporaneously, such as ilmenite breaking down to rutile and hematite.

Molybdenite

Molybdenite is minor but widespread throughout the deposits. It occurs as laths and irregular grains and is in close association with graphite and pyrrhotite. The paragenetic position of this mineral is uncertain, but most of the molybdenite appears to have formed early—contemporaneously with pyrrhotite. Some is well developed in pegmatites within the ore body, so that two generations may be likely.

Sphalerite

Sphalerite is very minor, and its paragenetic position is uncertain. It occurs as fine aggregates of irregular sphalerite grains intergrown with the early silicates. In view of the comparable amount of Zn (105 ppm), Cu (87 ppm), and Ni (130 ppm) in mafic rocks, it is surprising that sphalerite is not even an important minor constituent (Turekian and Wedepohl, 1961).

Marcasite

Marcasite is not common in these deposits. It is found only in those few polished sections containing pyrite, in which it comprises small irregular grains within the pyrite veins. The lack of alteration of pyrrhotite to marcasite and an intermediate product is unusual in view of the common occurrence of these minerals in pyrrhotite deposits. As previously noted, the paucity of marcasite, as well as that of pyrite, indicate a relatively low sulfur fugacity.

Other Sulfides

Other sulfides that have rarely been observed include bornite, cubanite, galena (R. S. Young, unpub. data), and gersdorffite. Niccolite is locally abundant in pegmatites and is commonly rimmed by rammelsbergite, which demonstrates a response to local changes in arsenic composition of the residual ore fluids. The chalcopyrite in association with these arsenic minerals contains many exsolution starlets of sphalerite, which distinguish it from the chalcopyrite of the primary ore.

Other Minerals

The platinoids are extremely rare and are enclosed by pyrrhotite and pentlandite grains. Most grains are only a few micrometres in size and show good idiomorphic development. Most grains are isotropic, but one mineral displayed strong anisotropism. These grains were compared to several cobalt and platinoid minerals from the collection of P. Ramdohr at Heidelberg University. None of the isotropic minerals was identified except sperrylite ($PtAs_2$), but the anisotropic mineral has been identified as niggliite (PtTe-PtSn), because it compares favorably to samples of that mineral from South African deposits.

OXIDES

Ilmenite is the dominant primary oxide in the Harriman and Warren deposits. The combined average of the primary oxides (ilmenite, rutile, magnetite, and spinel) from the deposits is 0.6 vol percent, of which ilmenite constitutes over 90 percent. Several different size and textural categories into which ilmenite may be classified are as follows, in decreasing order of abundance: (1) several hundred micrometre- to several millimetre-sized irregular grains; (2) a few micrometre- to several hundred micrometre-sized grains of varying shape, although generally round to elliptical; (3) oriented needles within biotite; (4) minor irregular grains associated with spinels, either within or bordering grains; and (5) ilmenite-magnetite exsolution intergrowths.

The near absence of *primary* magnetite from the Warren and Harriman bodies is an unusual feature of these deposits. We have found no similar type of deposit described in the literature where magnetite was not at least a well-distributed minor constituent. In most deposits, in fact, magnetite clearly predominates over ilmenite. It is also evident that the ilmenite of this deposit, although generally strongly anisotropic, does not display as strong a pleochroism as is characteristic of ilmenite in general. The ilmenite is of at least two distinct generations (categories 1 and 2 and category 3), in addition to the ilmenite associated as an exsolution product of spinel.

Rutile

Rutile occurs as a decomposition product of ilmenite and as fine needles in quartz. Although much of the decomposition of ilmenite to rutile appears rather irregular, there is a tendency for rutile to form along grain boundaries, within fractures, or along crystallographic directions of ilmenite (Fig. 10). A myrmekitic intergrowth of chalcopyrite and rutile is commonly observed in altered ilmenite grains. Chalcopyrite is always intergrown with rutile in these instances and is never contained within ilmenite in this association (Fig. 10). It is not certain whether the alteration of ilmenite to rutile occurred during the last stages of chalcopyrite deposition or during metamorphism when chalcopyrite was a mobile phase.

Ramdohr (1969) indicated that the breakdown of ilmenite to rutile with accompanying sulfide formation is attributable to hydrothermal alteration. No generation of ilmenite has escaped rutilization. Some rutile is associated with a breakdown

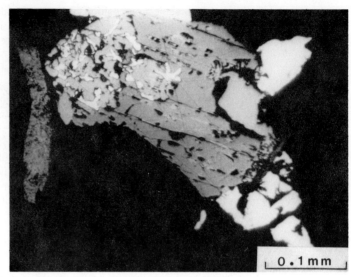

Figure 10. Ilmenite (dark gray) altered to rutile (lighter gray), which is myrmekitically intergrown with chalcopyrite (white). Lighter gray in ilmenite is rutile formed along preferred crystallographic directions in ilmenite. Elongate grain to the left is graphite; white areas surrounding ilmenite are pyrrhotites. Warren deposit.

product that is dark brownish gray in reflected light. This mineral was described as "replacement rutile" by Ramdohr (1969) and is relatively common in nickeliferous pyrrhotite deposits such as Nysteen and Askim in Norway.

Minor amounts of rutile formed as isolated grains within silicates; the rutile appears to represent the total decomposition of ilmenite. Minor epitaxial rutile needles in quartz are especially well displayed in pegmatite associated with the ore bodies.

Magnetite

Magnetite is decidedly minor in the Warren deposit, whereas it is common as a secondary breakdown product of olivine in the Harriman deposit. Very minor magnetite occurs also in the Harriman body as an early-formed oxide. The much more common secondary magnetite occurs as veinlets that are several micrometres to several tens of micrometres in width and that border and crosscut olivine and form the central part of serpentine veinlets. Fractures in the sulfides are filled with secondary magnetite and serpentine, which indicates that serpentinization postdated ore formation. In the Harriman deposit, some exsolved pentlandites flame out from magnetite-serpentine-filled fractures. It is difficult to estimate how much magnetite is actually primary because of several textural variations, but the early magnetite that can be identified with certainty is very minor.

The few grains of magnetite recognized from the Warren body show slightly lower reflectivity than normal magnetite and electron-microprobe analyses show them to be titanomagnetites. Magnetite-ilmentite or magnetite-hematite exsolution textures are virtually absent.

Spinels

The Harriman and Warren bodies contain genetically significant and complex early oxides, which rarely exceed a few tenths of a modal percent in any section. Most of the complex oxides are chrome-rich spinels intergrown with spinels of varying composition. In the Harriman peridotite, the spinels generally occur as dropletlike grains that seldom exceed 100 μm in diameter, whereas the spinels from the Warren gabbro occur as several hundred micrometre-sized round grains or as subhedral to euhedral grains. A striking feature of these spinels is their optical complexity. Some of the grains demonstrate symmetrical zoning patterns (Fig. 11A) or, more often, a mixture of several optically distinct phases within a single grain (Fig. 11B).

Oriented rods and blebs of ilmenite are found as exsolution products of these spinels, and generally, the ilmenite crosscuts the complex, optically distinct phases (Fig. 11C). Less commonly, the ilmenite associated with spinel occurs as distinct boundary grains (Fig. 11D). A discussion of spinels in this deposit can be found in a previous report (Rainville and Park, 1973).

Spinels of similar optical complexity have been reported from Giant Nickel Mine, Hope, British Columbia, by Muir and Naldrett (1973), but these are generally low in titanium. Unmixing of a single-phase grain is a likely explanation of most of the Harriman and Warren spinel grains, but some differences in phases simply represent a change in the composition of the magma. The "woven-looking" spinels described by Muir and Naldrett (1973) have also been found; these represent minute exsolution growths of another spinel.

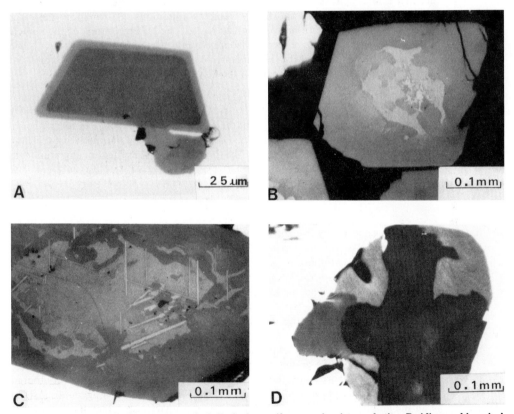

Figure 11. A, idiomorphic chrome spinel displaying uniform zoning in pyrrhotite. B, idiomorphic spinel grains in silicate; small light-gray blebs in central part are exsolved ilmenite; spinel displays complex optical inhomogeneity. C, spinel grain in pyrrhotite with pronounced optical inhomogeneity; exsolution rods and blebs of ilmenite (light gray) oriented along three preferred crystallographic directions. D, two-phase spinel grain (almost black and gray), with small grain of ilmenite (dark gray) at left border, in pyrrhotite (white). All from Warren deposit.

The excellent elemental zoning often found in spinels can be seen in Figure 12. Iron is the only element common in all the spinels, but elemental scanning of several spinels proved that chromium is a major constituent in most of the spinel phases. Magnesium, aluminum, and titanium are variously present as major or minor constituents.

Most of the Harriman spinels consist of two distinct phases. The dark phase (reflected light) is an aluminum-rich chrome spinel; the light phase is an iron-rich chrome spinel containing several percent titanium. A third phase, found in association with only a small number of spinels, is exceptionally rich in titanium. All three phases are evident in the scanning photographs of Figure 13. Average analyses of the three spinel phases for several grains is given in Table 1. The third phase, a chromian ulvospinel, is of special interest because the only terrestrial occurrences reported are from the Bushveld (Frankel, 1942; Cameron and Glover, 1971). A series of these chromian ulvospinels have recently been described from lunar samples (Haggerty and other, 1970). There is still some uncertainty as to whether there is complete or limited solid solution between ulvospinel (Fe_2TiO_4) and chromite ($FeCr_2O_4$) (Taylor and others, 1971; Kushiro and others, 1970). There is little doubt, however, that true chromian ulvospinels do occur in lunar samples. The fact that similar spinels are formed in the Harriman and Warren deposits indicates a titanium-rich paragenesis and highly reducing conditions of formation (the lunar spinels are associated with native iron and nickel). Fe^{+3}/Fe^{+2} ratios of the spinels are needed for a definitive comparison between these and lunar spinels. The dominance of ilmenite in the Harriman and Warren deposits is comparable to lunar assemblages in which ilmenite is the dominant oxide.

Hematite

Hematite is nearly absent throughout the deposits. It is probably a minor associate of rutile in the breakdown of ilmenite, but this has not been definitely established. The only occurrence is a few grains of the three-phase assemblage, magnetite-ilmenite-hematite.

GRAPHITE

The occurence of significant amounts of graphite in this deposit distinguishes it from other nickeliferous pyrrhotite deposits, except Kotalahti, Finland (Haapala, 1969), and Duluth, Minnesota. As much as 6.6 modal percent of graphite is found in a single section of ore-bearing gabbro, although a few sections show no appreciable amounts of graphite. The minimum average value of graphite from 23 sections is 1 modal percent. This is an extremely high amount, but it is clear that the highest percentages are at the margins of the mafic intrusive rocks bordering the Penobscot Formation.

Graphite occurs as botryoidal masses, individual laths, and more commonly, as a mass of stringerlike grains. Frequently it is associated with shattered sulfides, and it fills in and surrounds these sulfides. Some of the graphite has contoured itself to the sulfide grains and is locally folded (Fig. 14). The paragenetic position of all the graphites, with the exception of minor primary botryoidal grains, appears to be late in the development of the ore body. Graphite in the Penobscot Formation tends to align itself along the dominant schistosity.

It is clear that much of the graphite in the Harriman and Warren bodies resulted from assimilation of carbonaceous material from the Penobscot Formation. The preservation of material probably indicates that graphitization was well established before assimilation. Miyashiro (1964) noted the importance of graphite in determining the oxidation state and thus the oxide assemblage of a particular rock. In studying various metamorphic assemblages, he noted that ilmenite-magnetite is the dominant assemblage in graphite-bearing rocks and hematite-magnetite ± ilmenite is the dominant assemblage in graphite-free rocks. Miyashiro further noted that the control of oxygen fugacity by graphite is a local phenomenon, and sequential beds may demonstrate alternating reducing-oxidizing conditions, even though graphite is in close proximity in the adjoining rock. This may explain why many mafic bodies associated with graphitic rocks show little evidence of reducing conditions, because graphite was not assimilated by the igneous body and, therefore, a reducing environment was restricted to the associated graphite-bearing sediment.

MODAL ANALYSES

Modal analyses of polished sections from two drill holes (EDA-2 and EDA-11) and sections from other drill holes were completed for the major sulfides, combined oxides (excluding secondary magnetite), and graphite (Table 4). Noteworthy is the large amount of graphite and the irregular but general pattern of an increasing ratio of pyrrhotite to pentlandite plus chalcopyrite with depth (Fig. 15). Most of the deep drill-hole data from additional sections in Table 4 also show greatly increased ratios of pyrrhotite to pentlandite plus chalcopyrite.

The average sulfide percentages (converted to 100 wt percent total sulfide) for

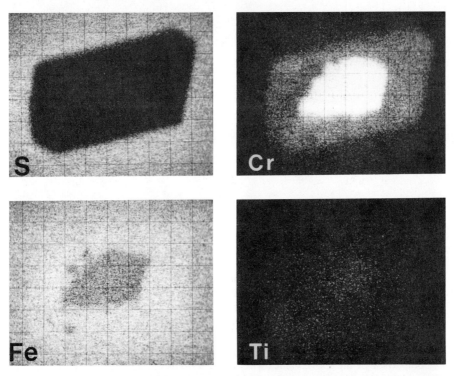

Figure 12. Idiomorphic spinel grain within pyrrhotite (spinel is black and pyrrhotite is white in scanning photograph for S). Chromium content decreases away from core, and iron increases. Trace amounts of titanium in chrome-rich central zone. Warren deposit.

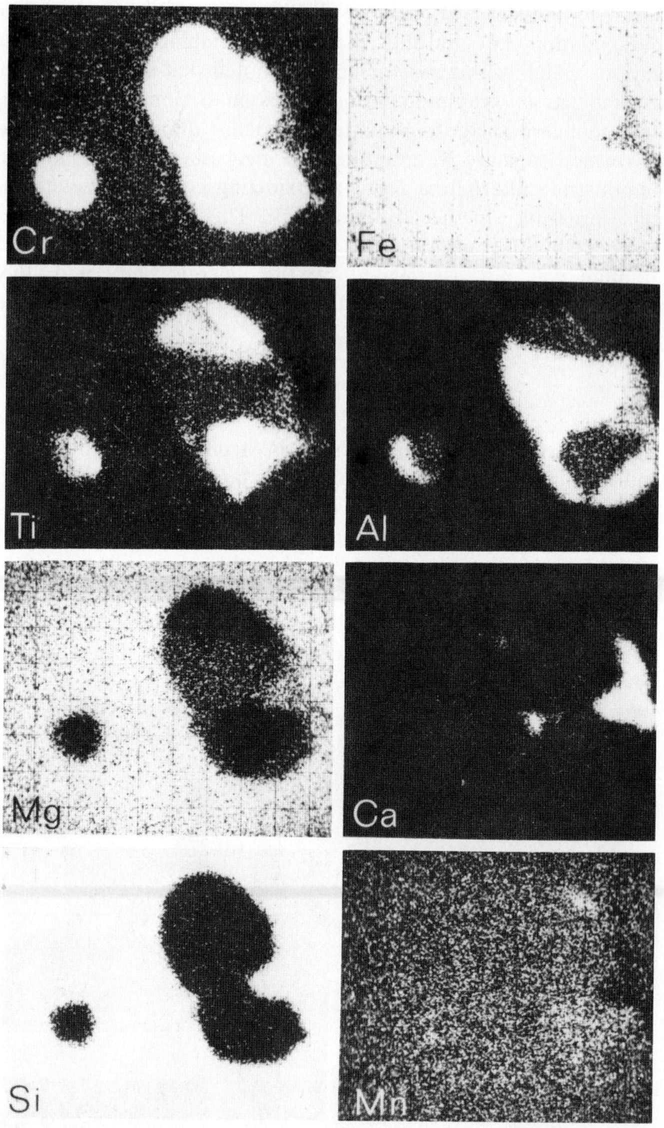

Figure 13. Scanning photographs of spinel in olivine (spinel is black and olivine is white in scanning photograph for Si). Chromium and iron are present in all the spinel phases. The recognizable phases are Cr-Fe-Al-Mg spinel, Cr-Fc-Al-Ti, Cr-Fe-Ti-Al (chromian ulvospinel), Fe-Ti (ilmenite), Ca-Si-Al-Mg-Ti (hornblende), and Si-Fe-Mg-Mn (olivine). The Cr-Fe-Ti-Al phase is a counterpart to the lunar spinels. Harriman deposit.

the 17 sections from drill holes EDA-2 and EDA-11 are pyrrhotite, 92 wt percent; pentlandite, 5.0 wt percent; and chalcopyrite, 3.0 wt percent. Combining the electron-probe data of pyrrhotite and pentlandite, stoichiometric chalcopyrite, and the point-count data from drill holes EDA-2 and EDA-11 gives a fairly accurate estimate of the total sulfide composition, as follows: Fe, 57.5; Ni, 2.2; Cu, 1.1; Co, 0.2; and S, 39.0 wt percent.

Comparison of the above values with the Cu-Fe-Ni-S system (Yund and Kullerud, 1966; Naldrett and others, 1967; Craig and Kullerud, 1969) indicates that all Ni, Cu, and Co were incorporated in monosulfide solid solution, at least well above 600°C.

PARAGENETIC SEQUENCE OF CRYSTALLIZATION

Figures 16 and 17 show the paragenetic sequence of crystallization for the Harriman and Warren bodies, respectively. They are treated separately because of their mineralogic and paragenetic variations and differences. The pyrrhotite-chalcopyrite-sphalerite-niccolite-rammelsbergite assemblage is not included in the paragenetic sequence, because these minerals correspond to a minor localized pegmatitic paragenetic sequence. The paragenetic sequence as shown in Figures 16 and 17 is fairly straightforward but needs some clarification.

The orthomagmatic stage is the main crystallization stage in which most silicates, as well as a significant amount of sulfides, crystallize. The temperature range of this stage varies with composition and water content but probably exceeded a temperature of 1000°C in the case of both ore bodies, as indicated by the presence of high-temperature spinels.

The hydrothermal stage corresponds to the subsequent cooling history of the mafic body. This term *does not* refer to any substantial introduction of material from outside, but rather to the autohydrothermal stages of the ore body itself. It is unlikely that the ore body was exposed to erosion before metamorphism and subsequent Acadian intrusion; rather, it probably remained deeply buried and retained temperatures of a few hundred degrees Celsius. It is impossible, therefore, to define a lower temperature limit for the hydrothermal stage.

The last stage, designated as metamorphic and supergene, involves all the subsequent changes that occurred after the emplacement of the ore body. This

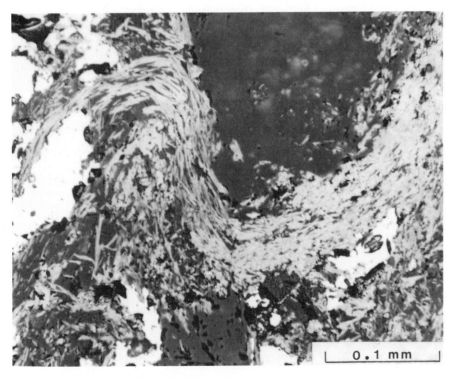

Figure 14. Folded graphites (light gray) in silicates (dark gray) with associated pyrrhotite (white). Minor ilmenites (barely visible) within graphite layer. Warren deposit.

means that the pattern of decreasing temperature for the first two stages is no longer followed, and the last stage involves temporal fluctuations of temperature in response to contact and regional metamorphism. The last stage, then, in the order of events occurring, is regional metamorphism of the ore body, intrusion of the Acadian silicic rocks, later retrograde metamorphism, and subsequent cooling to the present day, accompanied by supergene alteration.

SUMMARY AND CONCLUSIONS

Although Acadian dynamothermal and Permian thermal metamorphism (Zartman and others, 1970) have modified the ore bodies, primary igneous textures and the early paragenetic position of the sulfides are still evident. Results of this study and some genetic implications are summarized below.

The interstitial character of the sulfides and limited replacement of all but the early orthotectic silicates, as well as limited fracture filling, indicate an early magmatic origin of the deposits. The fact that the Warren deposit sulfides occur commonly in massive zones, as contrasted with the interstitial nature of the Harriman peridotite sulfides, most probably reflects viscosity differences that prevented segregation of the immiscible sulfide-silicate liquids.

High trace-element substitution in pyrrhotite (0.7 percent, combined Ni, Co, and Cu) is indicative of sulfides of magmatic origin. Detailed analyses of several thousand pyrrhotites by Cambel and Jarkovsky (1969) demonstrate that deposits of presumed high temperature of formation show consistently higher substitution of nickel and cobalt in pyrrhotite than those of hydrothermal or metamorphic origin. The substitution of several weight percent cobalt in pentlandite is also generally consistent with a magmatic origin, comparable to such deposits as Sudbury, Pecenga, Sohland an der Spree, Ookiep, and many others (Cambel and Jarkovsky, 1969, p. 70-71).

Partitioning of nickel and cobalt in pyrrhotite between the hexagonal and exsolved monoclinic phases cannot be demonstrated for the majority of samples in the

TABLE 4. POINT-COUNT DATA FROM HARRIMAN AND WARREN DEPOSITS

	I (vol %)	II (vol %)	III (vol %)
Silicates	61.9	41.9	57.1
Oxides*	0.4	0.7	0.7
Graphite	0.2	1.5	1.1
Total sulfides	37.5	56.3	41.0
Points counted	33,870	30,344	19,471
Sulfides taken as 100 wt percent			
Pyrrhotite	93.8	90.7	95.8
Pentlandite	4.4	5.3	3.3
Chalcopyrite	1.8	4.0	0.9

Note: I. Drill-core data from depth range of 2 to 66 m below the surface. Average of 8 sections, drill hole EDA-2, west-central part of Warren deposit. II. Drill-core data from depth range of 15 to 55 m below the surface. Average of 9 sections, drill hole EDA-11, southwestern part of Warren deposit. III. Drill-core data from scattered areas throughout the Harriman and Warren deposits, mostly from the base of the intrusive rocks. Average of 6 sections.

*Dominantly ilmenite (greater than 90 percent); also spinel, rutile, and magnetite, in order of abundance.

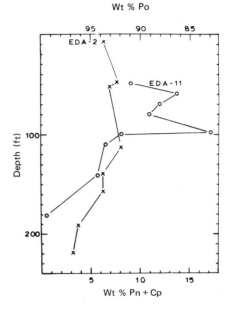

Figure 15. Variation of weight percent of pyrrhotite versus weight percent of chalcopyrite plus pentlandite with depth for two drill holes, EDA-2 and EDA-11. Each point value represents over 3,000 point counts. This shows a pronounced increase of pyrrhotite at the expense of chalcopyrite and pentlandite with depth. Warren deposit.

deposit; however, when observed, nickel is shown to concentrate in the monoclinic phase. Cobalt distribution in pyrrhotite is uniform within a single grain. Nickel distribution may vary considerably within a single pyrrhotite grain and sometimes shows preferred concentration along crystallographic directions in the pyrrhotite. There is often not a good correlation between optical inhomogeneity in the pyrrhotite and increased nickel content.

The ratio of pyrrhotite to pentlandite plus chalcopyrite has been demonstrated to increase with depth in the Warren ore deposit. There is some quantitative evidence indicating that the decrease in pentlandite and chalcopyrite is balanced by a sympathetic increase of nickel, copper, and cobalt in pyrrhotite, but enough data is not available to establish definitively this trend.

The variations in cobalt content between granular and flame pentlandite point to a rather complex exsolution history for this mineral. The presence of two separate generations of pentlandite exsolution is a phenomenon not predictable by experimental investigation of the Fe-Cu-Ni-S system. The time factor is especially important when dealing with problems of exsolution of pentlandite.

The probable ability of mackinawite to accept significant metal substitution by solid-state diffusion after exsolution from within chalcopyrite restricts the value of nickel + cobalt + copper content as an indicator of the thermal stability of this mineral. We agree with the work of Schot and others (1972), which indicates that the amount of these elements substituted into mackinawite may only be used as a general indicator of the increased temperature of thermal stability in natural assemblages.

The absence of any significant amounts of early-formed pyrite, or of marcasite alteration of pyrrhotite, is indicative of continuously low sulfur fugacities. This assumption is supported by the high metal to sulfur ratio in pyrrhotites obtained by x-ray diffraction analyses (47.7 percent metals) and the high metal to sulfur ratio (9.8:8.0) obtained from pentlandite microprobe data. In the absence of externally introduced sulfur, there must be another mechanism for controlling sulfur fugacity. Sulfur fugacity may be effectively increased by oxidation. Taylor's (1970) experi-

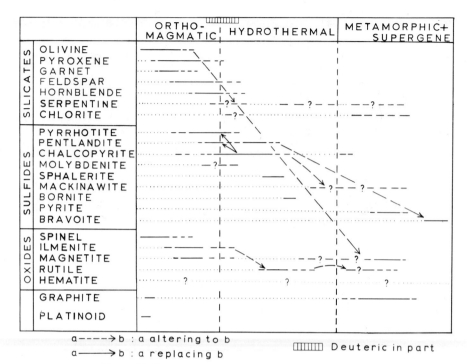

Figure 16. Paragenetic sequence of the Harriman deposit.

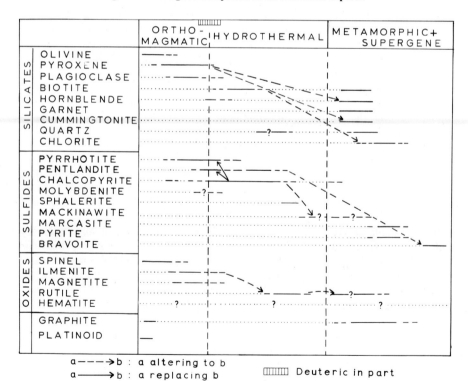

Figure 17. Paragenetic sequence of the Warren deposit.

ments demonstrated that pyrrhotite may be converted to pyrite as oxygen pressure is increased. The effective decrease in activity of iron by oxide formation results in increased sulfur fugacity. It is likely then, in the Harriman and Warren deposits, that both low sulfur and oxygen fugacities prevented pyrite-marcasite formation.

The role of graphite in these deposits appears to be critical. Graphite can produce carbon dioxide by oxidation, which Miyashiro (1964) noted is an effective buffer to increased oxygen fugacity. However, in the presence of graphite and under a wide range of carbon dioxide pressures (10 to 10^3 atm), the equilibrium oxygen pressure is so low (10^{-15} to 10^{-35} atm) that graphite is a strong reducing agent. If a low oxygen pressure is maintained, then an increase in sulfur pressure is unlikely unless sulfur is derived from some external source; consequently, FeS_2 is unlikely to form. The constancy of reducing conditions throughout the history of the deposit indicates the assimilation of large amounts of carbonaceous material by the mafic magma.

The occurrence of ilmenite-magnetite or ilmenite alone—either of these with no associated hematite—is compatible to the graphite-bearing metamorphic assemblages observed by Miyashiro (1964). This further indicates the importance of graphite in controlling oxygen fugacity in the deposit.

Electron-microprobe analyses of several spinels from the Harriman and Warren deposits reveal minor amounts of a chromian ulvospinel phase. These have been previously recognized only from Bushveld and lunar samples. The lunar spinels formed in an environment that was highly reducing, as evidenced by their association with native iron, so that the occurrence of this type of spinel in the Harriman and Warren deposits adds another clue pointing to conditions more reducing than are characteristic of most magmatic nickel sulfide deposits. The dominance of ilmenite to the virtual exclusion of primary iron oxides also bears a strong resemblance to lunar assemblages where ilmenite predominates.

ACKNOWLEDGMENTS

We thank the Knox Mining Company, Rockland, Maine, for samples and sections supplied for this study and their kind cooperation and assistance in the field. We also thank Professor G. C. Amstutz of Heidelberg University for extending the use of his institute's research facilities during our stay in Heidelberg in the summer of 1971. We benefited from discussions and counseling with Professors Paul Ramdohr, J. Ottemann, and G. Van der Kaaden, all of the University of Heidelberg. Erik H. Schot, then of Heidelberg, did some of the electron-microprobe analyses and provided valuable discussions on mackinawite problems. A. H. Brownlow of Boston University offered useful suggestions on ore genesis. L. A. Taylor of the University of Tennessee and Ulrich Petersen of Harvard University read the first draft of the manuscript and gave valuable suggestions. Joseph Kowalik and Steven Kurz of Boston University aided us in x-ray and electron-microprobe analyses. A research grant from the Boston University Graduate School (grant GRS-285-GL) for the cost of polishing ore samples is acknowledged.

REFERENCES CITED

Amos, D. H., 1963, Petrology and age of plutonic rocks, extreme southeastern Maine: Geol. Soc. America Bull., v. 74, p. 169-194.

Arnold, R. G., 1966, Mixtures of hexagonal and monoclinic pyrrhotite and the measurement of the metal content of pyrrhotite by x-ray diffraction: Am. Mineralogist, v. 51, p. 1221-1227.

Arnold, R. G., and Reichen, L. E., 1962, Measurement of metal content of naturally occurring, metal-deficient, hexagonal pyrrhotite by an x-ray spacing method: Am. Mineralogist, v. 47, p. 105-111.

Bastin, E. S., 1908a, Rockland folio: U.S. Geol. Survey Geol. Atlas, folio 149.

———1908b, A pyrrhotitic peridotite from Knox County, Maine: Jour. Geology, v. 16, p. 124-139.

Bickel, C. F., 1971, Bedrock geology of the Belfast quadrangle, Maine [Ph.D. dissert.]: Cambridge, Mass., Harvard Univ., 341 p.

Blatt, A. P., 1972, Nickel distribution in hexagonal and monoclinic pyrrhotite: Canadian Mineralogist, v. 11, p. 892-897.

Boucot, A. J., Brookins, D., Forbes, W., and Guidotti, C. V., 1972, Staurolite zone Caradoc (Middle-Late Ordovician) age, Old World Province brachiopods from Penobscot Bay, Maine: Geol. Soc. America Bull., v. 83, p. 1953-1960.

Cambel, B., and Jarkovsky, J., 1969, Geochemistry of pyrrhotite of various genetic types: Bratislava, Czechoslovakia, Bratislava Univ. Press, 333 p.

Cameron, E. N., and Glover, D., 1971, Unusual titanium-chromium spinels from the eastern Bushveld Complex: Geol. Soc. America Abs. with Programs, v. 3, no. 7, p. 522.

Cheney, E. S., 1965, Reconnaissance and economic geology of the northwestern Knox County marble belt: Maine Geol. Survey Bull. 19, 32 p.

Clark, L. A., and Kullerud, G., 1963, The sulfur rich portion of the Fe-Ni-S system: Econ. Geology, v. 60, p. 853-885.

Craig, J. R., and Kullerud, G., 1969, Phase relations in the Cu-Fe-Ni-S system and their application to magmatic ore deposits, in Wilson, H.D.W., ed., Magmatic ore deposits: Econ. Geology Mon. 4, p. 344-358.

Dennen, W. H., 1943, A nickel deposit near Dracut, Massachusetts: Econ. Geology, v. 38, p. 25-55.

Einaudi, M. T., 1968, Pyrrhotite-pyrite-sphalerite relations at Cero de Pasco, Peru [Ph.D. dissert.]: Cambridge, Harvard Univ., 381 p.

Eskola, P., 1950, Paragenesis of cummingtonite and hornblende from Muuruvesti, Finland: Am. Mineralogist, v. 35, p. 728-734.

Frankel, J. J., 1942, Chrome-bearing magmatic rocks from the eastern Bushveld: South African Jour. Sci., v. 38, p. 152-157.

Haapala, P. S., 1969, Fennoscandian nickel deposits, in Wilson, H.D.W., ed., Magmatic ore deposits: Econ. Geology Mon. 4, p. 262-275.

Haggerty, S. E., Boyd, F. R., Bell, P. M., Finger, L. W., and Bryan, W. B., 1970, Iron-titanium oxides and olivine from 10020 and 10071: Science, v. 167, p. 613-615.

Harris, D. C., and Nickel, E. H., 1972, Pentlandite compositions and associations in some mineral deposits: Canadian Mineralogist, v. 11, p. 861-878.

Hawley, J. E., and Nichol, I., 1961, Trace elements in pyrite, pyrrhotite and chalcopyrite of different ores: Econ. Geology, v. 56, p. 467-487.

Houston, R. S., 1956, Genetic study of some pyrrhotite deposits of Maine and New Brunswick: Maine Geol. Survey Bull. 14, 117 p.

Knox County Regional Planning Commission, 1963, Mineral deposits map, Knox County, Maine: James W. Sewall Co.

Kushiro, I., Nakamura, Y., and Akimoto, S., 1970, Crystallization of Cr-Ti spinel solid solutions in an Apollo 12 rock, and source rock of magmas of Apollo 12 rocks [abs.]: EOS (Am. Geophys. Union Trans.), v. 51, p. 585.

Mariano, A. N., Ito, J., and Ring, P. J., 1973, Cathodoluminescence of plagioclase feldspars: Geol. Soc. America Abs. with Programs, v. 5, no. 7, p. 726.

Mining Engineering, 1972, March issue, p. 8.
Miyashiro, A., 1964, Oxidation and reduction in the earth's crust with special reference to the role of graphite: Geochim. et Cosmochim. Acta, v. 28, p. 717-729.
Muir, J. E., and Naldrett, A. J., 1973, A natural occurrence of two-phase chromium-bearing spinels: Canadian Mineralogist, v. 11, p. 930-939.
Naldrett, A. J., Craig, J. R., and Kullerud, G., 1967, The central portion of the Fe-Ni-S system and its bearing on pentlandite exsolution in iron-nickel sulfide ores: Econ. Geology, v. 62, p. 826-847.
Rainville, G. D., 1972, Geology and ore mineralogy of the Harriman-Warren nickeliferous pyrrhotite deposits, Knox County, Maine [M.S. thesis]: Boston, Mass., Boston College, 243 p.
Rainville, G. D., and Park, W. C., 1973, Ti-V-Cr spinels from nickeliferous pyrrhotite deposits, Knox County, Maine: Geol. Soc. America Abs. with Programs, v. 5, no. 2, p. 209.
Ramdohr, P., 1969, The ore minerals and their intergrowths: New York, Pergamon Press, 1174 p.
Scholtz, D. L., 1936, The magmatic nickeliferous ore deposits of East Griqualand and Pondoland: Geol. Soc. South Africa Trans. and Proc., v. 39, p. 81-210.
Schot, E. H., Ottemann, J., and Omenetto, P., 1972, Some new observations on mackinawite and valleriite: Soc. Italiana Mineralogia i Petrographia Rend., v. 28, p. 241-295.
Spry, A., 1969, Metamorphic textures: Oxford, England, Pergamon Press, 350 p.
Taylor, L. A., 1970, Oxidation of pyrrhotites and the formation of anomalous pyrrhotite: Carnegie Inst. Washington Year Book, no. 70, p. 287-289.
Taylor, L. A., Kullerud, G., and Bryan, W. B., 1971, Opaque mineralogy and textural features of Apollo 12 samples and a comparison with Apollo 11 rocks, in Proceedings of the Second Lunar Science Conference, Vol. 1: Cambridge, Mass., M.I.T. Press, p. 855-871.
Thornbury, W. D., 1965, Regional geomorphology of the United States: New York, John Wiley & Sons, Inc., 609 p.
Turekian, K. K., and Wedepohl, K. H., 1961, Distribution of the elements in some major units of the earth's crust: Geol. Soc. America Bull., v. 72, p. 175-191.
Vernon, R. H., 1962, Co-existing cummingtonite and hornblende in an amphibolite from Duchess, Queensland, Australia: Am. Mineralogist, v. 47, p. 360-370.
Yund, R. A., and Kullerud, G., 1966, Thermal stability of assemblages in the Cu-Fe-S system: Jour. Petrology, v. 7, p. 454-488.
Zartman, R. E., Hurley, P. M., Krueger, H. W., and Giletti, B. J., 1970, A Permian disturbance of K-Ar radiometric ages in New England: Its occurrence and cause: Geol. Soc. America Bull., v. 81, p. 3359-3374.

MANUSCRIPT RECEIVED BY THE SOCIETY APRIL 15, 1974
REVISED MANUSCRIPT RECEIVED FEBRUARY 6, 1975
MANUSCRIPT ACCEPTED FEBRUARY 20, 1975

Printed in U.S.A.

Author Index

Abbott, M. L., 150, 152-154, 156, 158, 160, 177
Abu-moustafa, A. A., 36, 39, 62, 68
Akimoto, S., 346
Albee, A. L., 242, 249, 257, 260
Allen, R., 46, 55, 57-58, 69
Alvord, D. C., 35, 39, 68, 111, 113, 123
Amos, D. H., 322, 346
Anderson, E. M., 286, 299
Andrews, H. N., Jr., 170, 177
Appleman, E. E., 105, 123
Arber, E. A. N., 190, 196
Arnold, R. G., 326, 328, 346
Artis, E. T., 152, 160, 162, 177
Atlas, E., 148, 150, 152, 154, 158, 160, 163

Baadsgaard, H., 100
Bachinski, D., 240, 260
Bailey, E. B., 286, 288, 299
Bancroft, A. M., 311, 316
Barghoorn, E. S., 68, 123, 178, 196
Barker, D. S., 264, 279
Barth, T. F. W., 84, 100, 119-120, 123
Barton, G. N., 182, 185-186, 196
Bascom, F., 128, 140
Bastin, E. S., 320, 323, 346
Bateman, P. C., 314, 317
Bean, R. J., 303, 305, 316
Bell, K. G., 35, 39, 68, 111, 123
Bell, P. M., 69, 346
Bell, W. A., 148, 150, 152, 154, 156, 158, 160, 162-164, 166, 170, 173, 177, 187-188, 190, 192-193, 196
Bentall, R., 179
Berry, W. B. N., 207, 236, 238, 266, 272, 279
Bickel, C. F., 322, 325, 346
Billings, M. P., 3, 6, 8, 11, 13-17, 25, 27, 29-30, 35, 68, 76, 88-89, 100, 103-104, 123, 126, 128, 140, 185, 196, 204-205, 211, 232, 236, 264, 279, 282-283, 288-290, 293-296, 299-300, 302, 310-311, 316
Bird, J. M., 314, 316
Bishop, D. G., 249, 255-256, 260

Black, R., 101, 299
Blatt, A. P., 329, 346
Boone, G. M., 204, 207, 210, 214-215, 224, 233, 236, 240, 243, 260, 272, 279
Borley, G. D., 85-86, 100
Borns, H. W., 264, 279
Bothner, W. A., 233-234, 236, 312, 316
Bottino, M. L., 95, 100, 128, 140, 230, 236
Boucot, A. J., 231, 236, 240, 242, 245, 248, 257-258, 260-261, 271-272, 279-280, 322, 346
Boudette, E. L., 207, 236-237, 264, 270, 279-280
Bowen, N. L., 82-83, 88, 90, 100, 102, 120, 124
Boyd, F. R., 346
Bromery, R. W., 205-206, 213-214, 233, 237, 317
Brongniart, A., 150, 152, 158, 160, 162-164, 166, 168, 170, 177
Brookins, D., 346
Bryan, W. B., 346-347
Buddington, A. F., 315-316
Buma, G., 74, 100
Burke, R. E., 10, 29, 183, 193, 196
Burr, H. T., 10, 29, 183, 193, 196

Cady, W. M., 242, 260, 271, 279
Caldwell, D. W., 10, 29
Cambel, B., 326, 342, 346
Cameron, B. W., 30, 178
Cameron, E. N., 338, 346
Cariani, A. R., 264
Cassidy, M. M., 30
Chapman, C. A., 89, 91, 95-96, 100, 282-283, 288-290, 292, 296, 299, 303, 316
Chapman, R. W., 85, 88, 285, 288, 290-291, 293, 299
Cheney, E. S., 322, 346
Christiansen, R. L., 135, 140
Chute, N. E., 7-8, 14, 21, 24, 29, 72-73, 100, 126, 136, 140, 182, 184-187, 193, 195-196
Clapp, C. H., 83, 100
Clarke, F. W., 88, 100
Clark, L. A., 334, 346
Clark, R. G., 237, 303, 305, 313, 316-317

349

Clarke, E. C., 6, 15–16, 29
Cleaves, A. B., 194, 197
Clifford, T., N., 38, 68
Clough, C. T., 288, 299
Cook, E. F., 130, 140
Coombs, D. S., 240, 242, 250, 254–258, 260
Corsin, P., 163, 177
Craig, J. R., 340, 346–347
Cridland, A. A., 194, 196
Crookall, R., 150, 152, 158, 163–164, 177, 188–190, 196
Crosby, I. B., 8, 29
Crosby, P., 117, 119–120, 123
Crosby, W. O., 6, 8, 15, 29, 36, 68, 73, 90, 100, 107, 111, 113, 123, 126, 128, 140, 182, 185–187, 196
Currier, L. W., 104, 115, 123

Daly, R. A., 46, 51, 68
Darrah, W. C., 152, 156, 158, 160, 162–164, 168, 173, 177, 189, 192–193, 196
Dawson, J. W., 148, 152, 156, 177
Delevaux, M. H., 314, 316
Dennen, W. H., 39, 68, 325, 346
Dewey, J. F., 314, 316
Diment, W. H., 237, 311–312, 316–317
Dixon, H. R., 264, 279, 314, 316
Dobrin, M. B., 303, 316
Dodge, W. W., 186, 196
Doe, B. R., 314, 316
Dott, R. H., Jr., 10, 29
Dowse, A. M., 7, 29, 194, 196
Doyle, R. G., 242, 260, 311, 316
Dunn, P. J., xii

Edwards, A. L., 117, 123
Einaudi, M. T., 325, 346
Ellis, A. J., 260
Elmore, P. L. D., 100, 236
Emeleus, C. H., 290, 299
Emerson, B. K., 7, 9, 26, 29, 32, 35, 68, 72, 74, 100, 103–104, 107, 111, 115, 123, 126, 140, 144, 177, 185, 187, 196
Englund, E. J., 303, 311, 314, 316
Enlows, H. E., 130, 132, 140
Ernst, W. G., 88, 90, 100
Eskola, P., 324, 346
Espenshade, G. H., 270, 279
Ettingshausen, C., 164, 177
Eugster, H. P., 90, 100
Evans, B. W., 225, 233, 236
Evans, H. T., Jr., 105, 123

Fabriès, P. J., 84–85, 87–88, 100
Fahlquist, F., 36, 68
Fairbairn, H. W., 94, 100–101, 140
Faul, H., 92–93, 96, 100–101, 121, 123, 229, 231, 236

Faure, G., 101
Field, M. T., 236, 260, 279
Filbert, C. R., 101
Finger, L. W., 346
Fischer, W. L., 36, 68
Fisher, I. S., 206, 208, 212, 214, 236
Fisher, L. W., 270, 279
Fisher, R. V., 135, 140
Fitton, J. G., 314, 316
Fitzpatrick, M. M., 237, 317
Fletcher, R., 236, 260
Foerste, A. F., 30, 179, 197
Foland, K. A., 92, 95, 100, 283, 299
Forbes, W. H., 236, 260, 279, 346
Fowler, W. C., 39, 68
Fowler-Billings, K., 303, 316–317
Frankel, J. J., 338, 346
Freedman, J., 292, 300
Frey, F. A., 100
Friend, P. F., 230, 236
Frondel, C., xii
Fullagar, P. D., 100, 140, 230, 236
Furlong, I. E., 264, 279
Fyfe, W. S., 260

Gates, O., 265, 279
Gaudette, H. E., 292, 299
Geinitz, H. B., 152, 154, 158, 160, 171, 173, 177
Germar, E. F., 158, 160, 170–171, 173, 177
Gilbert, M. C., 69
Giletti, B. J., 100, 102, 121, 124, 238, 299, 347
Glidden, P. E., 264, 271, 279
Glover, D., 338, 346
Goeppert, H. R., 163, 178
Goldich, S. S., 92–93, 100
Gore, R. Z., 36, 39, 68, 105, 107, 120–121, 123
Gothan, W., 150, 152, 154, 158, 162–164, 171, 173, 178
Green, J. C., 206, 208, 212–213, 236, 265, 279
Green, T. H., 314, 317
Greene, R., 303, 317
Gregory, H. E., 257–258, 261
Grew, E. S., 27, 29, 36, 39, 68, 104, 111, 120–121, 123, 178, 183, 196, 232
Griffin, J. R., 209, 224, 237–238, 264–267, 270–271, 279
Grout, F. F., 58, 68, 313, 317
Guidotti, C. V., 117, 123, 206–208, 210, 212–215, 217, 219, 221, 225–228, 233, 236, 238, 265, 269, 279, 346
Gutbier, A., 154, 156, 178

Haapala, P. S., 338, 346
Haffty, J., 140
Haggerty, S. E., 338, 346
Hall, B. A., 242, 260
Hamilton, W. B., 206, 231–234, 236, 303, 317
Handford, L. S., 100

Handwecker, D. C., 105, 123
Hansen, W. R., 32, 36, 39, 68, 104-105, 109, 111, 123
Harakal, J. E., 93, 101
Harper, C., 260
Harris, D. C., 330-331, 346
Hart, S. R., 91, 101
Harwood, D. S., 206-207, 236
Hashimoto, M., 240, 255-256, 260
Hawley, J. E., 329, 346
Hay, R. L., 256, 260
Heald, M. T., 303, 317
Heath, M. M. J., 36, 68, 100
Hechinger, L. A., 184, 194
Hedge, C. F., 140
Hepburn, J. C., 10, 30
Herd, H. H., 117, 123
Hildreth, C. T., 206, 208-210, 213, 216-217, 219-221, 225, 227, 237
Hills, E. S., 16, 29
Hitchcock, E., 143, 178, 186-187, 190, 196
Holdaway, M. J., 67-68
Holmes, A., 286, 299
Hon, R., 233, 237
Horodyski, R. J., 240, 245, 260
House, M. R., 230, 236
Houston, R. S., 320, 322-324, 330, 346
Hsu, L. C., 314, 317
Hurley, P. J., 121, 124
Hurley, P. M., 92-93, 100-102, 140, 347
Hussey, A. M., 211, 237, 265, 279, 292, 299

Ingamells, C. O., 69
Ito, J., 346

Jackson, C. T., 143, 178
Jacobson, R. R. E., 76, 80, 83, 88-89, 101, 290, 299
Jahns, R. H., 39, 69, 103-105, 123
Jarkovsky, J., 326, 342, 346
Jenks, W. F., 288, 299
Johannsen, A., 58, 69
Jongmans, W. J., 152, 178
Joyner, W. B., 90, 101, 237, 303, 305, 310, 317

Kaktins, U., 24, 29, 84, 89, 101
Kane, M. F., 205-206, 213-214, 233, 237, 303, 305, 308, 311-313, 317
Karner, F. R., 89, 101
Kaulfuss, F., 170, 178
Kay, M., 147, 178
Kaye, C., 9, 29
Kennedy, G. C., 139-140
Kerrick, D. M., 67, 69
Kidston, R., 152, 162, 170-171, 178
Kingsley, L., 292-293, 295-296, 299
Kistler, R. W., 314, 317
Knox, A. S., 3, 143, 178, 194, 196

Koenig, C., 158, 178
Krueger, H. W., 24, 29, 73, 92, 95, 101-102, 121, 124, 128, 140, 238, 347
Kruger, F. C., 302, 317
Kullerud, G., 334, 340, 346-347
Kushiro, I., 338, 346

Lacroix, A. V., 36, 69
La Forge, L., 6-7, 9-10, 29-30, 126, 140
Landsman, M., 317
Larsen, E., 39, 69
Laveine, J. P., 188-189, 196
Legarde, C. N., 36, 69
Lesquereux, L., 143, 148, 150, 152, 154, 158, 160, 163, 168, 178, 187, 190, 196
Lindsley-Griffin, N., 279
Linehan, D., 302, 317
Liou, J. G., 67, 69, 256, 260
Lipman, P. W., 135, 138, 140
Locke, A., 288, 300
Loomis, F. B., Jr., 29
Lord, G. S., 24, 29
Loughlin, G. F., 24, 29, 184, 196
Ludman, A., 209, 224, 237-238, 264-265, 271, 279
Lundgren, L. L., 314, 316
Lundgren, W. L., 264, 279
Luth, W. C., 66, 69
Lyons, J. B., 93, 101, 121, 123, 237, 317
Lyons, P. C., 10, 24, 26, 29-30, 72-76, 80-82, 84-86, 88, 91, 101, 128, 140, 143, 154, 156, 158, 163-164, 173-174, 178, 182-184, 186-187, 190, 193-195, 197

MacLeod, W. N., 101, 299
Mamay, S. H., 68, 111, 123, 178, 196
Mariano, A. N., 324, 346
Marmo, V., 120, 123
Marvin, R. F., 3, 24, 30, 35, 70, 73, 92, 94-95, 102, 128, 141
Maufe, H. B., 299
McCarter, R. S., 65, 70
McKinstry, H. E., 80, 82-83, 102
Mehnert, K. R., 120, 123, 313, 317
Mencher, E., 240-243, 245, 261, 280
Mikami, K., 261
Miller, F., 39, 69
Milton, D. J., 206-208, 212-213, 216, 237
Miyashiro, A., 256, 260, 339, 343, 347
Moench, R. H., xii, 206-210, 213, 215-221, 225-228, 231, 236-237, 264, 279-280
Moody, W. C., Jr., 212, 237
Moore, G. E., Jr., 73, 101, 310, 317
Moore, R. C., 178
Moorhouse, W. W., 58, 69
Morgan, B. A., 313, 317
Morris, J. E., 194, 196
Mossman, D., 240, 260
Muir, J. E., 337, 347

Author Index

Mutch, T. A., 178, 182, 186, 193–194, 197
Myers, W. B., 206, 231–234, 236, 302, 317

Naldrett, A. J., 330, 337, 340, 347
Naylor, R. S., 73, 102, 229, 231, 236–237, 261, 279–280, 314, 317
Nichol, I., 329, 346
Nichols, D. R., 92, 101
Nickel, E. H., 330–331, 346
Nielson, D. L., 224, 233–234, 237, 303, 311, 313, 317
Nier, A. O., 100
Nitsch, K., 240, 261
Noble, D. C., 83, 89, 101, 138–140, 294, 300
Nockold, S. R., 46, 55, 57–58, 69, 126, 140
Noé, A. C., 163, 178
Norton, S. A., 66, 69, 204–205, 232, 234, 238, 241–242, 245, 260–261, 313, 317

Oki, Y., 261
Okumura, K., 261
Oleksyshyn, J., 187, 197
Oliver, W. A., Jr., 144, 147, 179, 183–184, 187, 193–194, 197
Omenetto, P., 347
Orville, P. M., 117, 123
Osberg, P. H., 204, 206, 209, 224, 237, 264–267, 270–271, 280
Ottemann, J., 347

Page, L. R., 101, 109, 123
Palache, C., 85, 102
Pankiwskyj, K. A., 206, 208–209, 217, 238, 264–265, 268–269, 280
Papezik, V. S., 204, 261
Park, W. C., 337, 347
Parsons, I., 76, 80, 84, 90, 101
Pavlides, L., 224, 236, 238, 240, 242, 244, 258–261, 265, 272, 279–280
Peck, J. H., 36, 69
Peppers, R. A., 194, 197
Perkins, E. H., 146, 178, 194, 197, 264, 270, 280
Petermann, Z. C., 314, 317
Pettijohn, F. J., 51, 55, 65, 69
Pfefferkorn, H. W., 194, 197
Pinson, H. W., Jr., 100–101, 140
Pollard, M., 8, 10, 30, 183, 193, 197
Potonié, H., 162, 170, 173, 178
Presnall, D. C., 314, 317

Quinn, A. W., 73–74, 90, 92–93, 100–101, 144, 147, 179, 183–184, 187, 193–195, 197, 299

Rahm, D. A., 6, 8, 10, 16–17, 19, 29, 30
Rainville, G. D., 320, 337, 347
Ramberg, H., 313–314, 317
Ramdohr, P., 334–336, 347

Rankin, D. W., 313–314, 317
Ratté, J. C., 128, 139, 141
Ray, R. G., 101
Reesman, R. H., 92, 101
Rehmer, J. A., 10, 30
Reichen, L. E., 326, 328, 346
Remy, W., 150, 152, 154, 158, 162–164, 173, 178
Renard, A., 38, 69
Reynolds, D. L., 290, 300
Rhea, K., 260
Richardson, S. W., 66–67, 69
Richey, J. E., 290, 299–300
Richter, D. A., 244, 261
Ring, P. J., 346
Ringwood, A. E., 314, 317
Rittman, A., 138, 141
Roberts, J. L., 286, 288, 300
Robinson, F., 318
Rocci, G., 84–85, 87–88, 100
Rodgers, J., 93, 102, 204, 206, 231, 238, 264–265, 273, 280
Rosenfeld, J. L., 242, 261
Ross, C. S., 131, 138, 141
Round, E. M., 143, 179, 186–187, 197
Rouse, R. C., xii
Roy, C. J., 292, 300
Roy, D. C., 240–243, 245, 261

Saito, Y., 255–256, 260
Sayles, R. W., 10, 30
Schairer, J. R., 82, 100
Schenk, A., 154, 179
Schimper, W. P., 168, 179
Schlotheim, F. V., 153, 179
Scholtz, D. L., 329, 347
Schopf, J. M., 179
Schot, E. H., 343, 347
Scott, R., 138, 141
Seilacher, A., 267, 280
Seki, Y., 254–256, 258, 261
Self, S., 141
Seward, A. C., 183, 197
Seymour, W. L., 101
Shaler, N. S., 26, 30, 179, 182, 184–186, 194, 197
Shido, F., 256, 260
Shrock, R. R., 6, 30
Simmons, G. C., 237
Simonds, G., 317
Skapinski, S. A., 264, 280
Skehan, J. W., 36, 39, 62, 68–69, 113, 121, 123
Skidmore, W. B., 261
Smith, E. S. C., 264, 270, 280
Smith, R. E., 255–256, 261
Smith, R. L., 131, 138, 141
Sparks, R. S. J., 139, 141
Springer, G. H., 144, 179
Spry, A., 323, 347

Sriramadas, A., 104, 124
Stanley, D. J., 179, 185-186, 193-194, 197
Stern, T. W., 236
Sternberg, G. K., 148, 150, 152, 154, 160, 163-164, 168, 171, 179
Steven, T. A., 128, 139, 141
Stewart, D. B., 105, 115, 124
Stewart, G. W., 29
Stopes, M. C., 173, 179
Streckeisen, A. L., 212, 238
Stur, D. R. J., 170, 179
Suhr, N. H., 40, 69
Sundeen, D. A., 104, 124

Talwani, M., 306, 317
Taylor, L. A., 338, 343, 347
Taylor, N. M., 260
Teschemacher, J. E., 143, 179, 187, 197
Thomas, H. H., 100, 236
Thompson, J. B., Jr., 66, 69, 204-205, 232, 234, 238, 241-242, 261, 302-303, 310-311, 313-315, 317-318
Thornbury, W. D., 320, 347
Tierney, F. L., 6, 8, 16-17, 27, 29-30
Tiffney, B., 30, 178
Toulmin, P., III, 8, 30, 35, 69, 74, 84-85, 90, 94, 102
Towe, K. M., 144, 146, 179, 194, 197
Trask, N. J., 318
Turekian, K. K., 334, 347
Turner, D. C., 285, 300
Tuttle, C. L., 117, 123
Tuttle, O. F., 69, 83, 88, 90, 102, 120, 124
Tyrrell, G. W., 51

Unger, F., 171, 180

Van Houten, F. B., 194, 197
Vernon, R. H., 324, 347
Vernon, W. W., 303, 318

Walker, G. P., 141
Walker, T. R., 194, 197
Warner, J. L., 237, 264-265, 268-269, 280
Warren, C. H., 24, 30, 74, 80, 82-85, 102, 126, 128, 139, 141
Wedepohl, K. H., 334, 347
Weyland, H., 152, 164, 178
White, D., 148, 150, 154, 156, 158, 163, 166, 168, 180
White, W. S., 205, 217, 224, 238
Whitehead, W. L., 72, 80, 102
Whitehouse, U. G., 65, 70
Willard, B., 194, 197
Williams, C. R., 85, 88, 100, 199, 285, 296, 300
Williams, H., 135, 141
Williams, H. S., 257-258, 261
Wilmarth, M. G., 6, 11, 30
Wilson, G. V., 299
Wilson, J. R., 296, 300
Wilson, L. R., 179
Winkler, H. G. F., 66, 70
Wisser, E., 288, 300
Wolfe, C. W., xiii-xv, 72, 75, 84-85, 87-89, 91, 101-102, 215, 238
Wones, D. R., 90, 100
Woodworth, J. B., 30, 179-181, 186-187, 197
Worzel, J. L., 317
Wright, T. L., 105, 115, 117, 124
Wright, W. B., 299

York, D., 230, 238
Yund, R. A., 340, 347

Zartman, R. E., 3, 24, 30, 35, 70, 73, 91-95, 102, 121, 124, 128, 141, 206, 229, 231, 238, 342, 347
Zeiller, R., 148, 154, 160, 162-163, 166, 180
Zen, E., 240, 248, 257-261
Zenker, F. C., 154, 156, 180
Zwart, H. J., 222, 225, 238

Subject Index

Acadian. *See* Devonian
Acadian orogeny, 206
  age, 259
  anatexis, 122
  catalysis, 122
  Chelmsford granite, 121-122
  Clinton facies, 121-122
  compression, 206
  crustal shortening, 273
  Maritime Canada, 146
  Merrimack synclinorium, 206
  metamorphism, 239-240, 258
  New England, 146, 242
  plate tectonics, 314
  pulses, 104, 121-122
  sheet development, 303
  time interval, 231, 314
Acmite, 88. *See also* Aegirine
Actinolite, 250-251
Aegirine, 24, 95. *See also* Acmite
Aenigmatite, 88
Age determinations. *See* Radiometric dating; geologic periods by name
Alabama, 168
Albee Formation, 207
Albite, 249
Alkalic granite
  areas, 80, 89
  petrogenesis, 72
  surface form, 89-90
  temperature of crystallization, 72
Alkalic granite, southern New England
  age, 72-73
  cathodoluminescence studies, 74
  chemistry, 72, 83-84, 96
  correlations, 72, 84, 89, 96
  feldspar, 74, 81, 84
  lithologic types, 72, 76, 90
  mineralogy, 76, 94
  origin, 72, 74, 89-90, 95-97
  radiometric dating, 73, 91-97
  spatial relations, 90, 95-97
Alkalic plutons, 74. *See also* Alkalic granite, southern New England; White Mountain Plutonic-Volcanic Series
Alkalic rocks. *See* Alkalic granite, southern New England; Blue Hill Granite Porphyry; Blue Hills volcanic complex; Quincy Granite; White Mountain Plutonic-Volcanic Series
Blue Hills igneous complex, 24-25, 126
  chemical analyses, 71
  Maine, coastal, 96
  northern New England, 205
  White Mountain Plutonic-Volcanic Series, 281-298
Alkali feldspar. *See* Perthite
  binary feldspar, 84
  binary minimum, 90
  chemistry, 71, 84-86, 90, 96-97. *See also* X-ray data
  system Ab-Or-H_2O, 88
  ternary minimum, 90
  thermal history, 103
  x-ray data, 84, 103, 107, 115-117
  zoning, 76, 90
Allanite, 107, 111, 114-115
Alleghenyan (Alleghenian), 143. *See also* Pennsylvanian
Allegheny Group, 158, 166
Allochthonous granite, 87, 89
Almandine, 64. *See also* Garnet
Aluminum silicate isobar, 232-233, 306, 313-314
Ammonoosuc Volcanics, 115, 207, 245, 308, 310
Amphibolite, 37, 58, 108
Amygdules, 7, 46-47, 246, 249, 251
Analcime, 240, 248
Anasagunticook Formation, 268-269, 274
Anatexis, 120-122, 313
Andalusite, 227, 233
Andover Granite, 37, 39
Animal fossils
  amphibian footprints, 194
  brachiopods, 270, 275
  crinoids, 265
  graptolites, 207, 274-278
  shelly fauna, 207. *See also* Brachiopods

Subject Index

Annite, 84, 87. See also Biotite; Rattlesnake pluton
Anorthoclase, 77
Anthophyllite, 324
Anticlines. See Boston basin; Maine, central, western
Anticlinorium
  Aroostook-Matapedia anticlinorium, 244
  Boundary Mountain anticlinorium, 207
  Bronson Hill anticlinorium, 310, 315
  Bronson Hill-Boundary Mountain anticlinorium, 205, 207, 242, 265
  Green Mountain anticlinorium, 311
  Munsungun anticlinorium, 265
  Pennington Mountain anticlinorium, 241-242, 244-245
  Rockingham anticlinorium, 205
  Weeksboro-Lunksoos Lake anticlinorium, 264-265
Antiformal arch, 310
Antiforms, 208, 217, 312
Antiform-synform pair, 218
Antigorite, 324
Aplite, 20, 115, 212, 229
Aporhyolite, 83, 125, 128, 139
Appalachian orogeny. See Appalachian revolution
Appalachian region, 153, 164, 240, 322
  plant fossils, 143, 154, 162
  stratigraphy, 158, 193, 195
Appalachian revolution, 5, 25-27, 147
  age, 26-27
Arch fractures, 287-289, 294
Arizona, 288
Arkose, 10, 24, 145, 147, 183
Aroostook County, Maine, 239-261
Aroostook-Matapedia anticlinorium, 244
Aroostook River Formation, 244-245, 248
Ascutney Mountain, 288
Ascutney Mountain complex, 284, 293
Ashland, Maine, 244-245, 251, 255, 259
Ashland synclinorium, 241-242, 254
Ashuelot pluton, 302-303, 308-311
Assemblage quartz-aenigmatite-acmite, 88
Attleboro, Massachusetts, 144
Attleboro syncline, 146
Aureole, 115, 204, 265, 315
Autochthonous granite, 87
Axiolitic structure, 125, 135
Ayer area. See Harvard-Ayer area, Massachusetts
Ayer Crystalline Complex, 103-104. See also Ayer Granodiorite; Clinton facies; Devens-Long Pond facies
  feldspar, 105, 115-120
  origin 115, 117, 120, 122
Ayer Granite, 104. See also Ayer Crystalline Complex; Ayer Granodiorite
Ayer Granodiorite, 37, 39, 104-105
  lithology, 103-105, 121

Ayer, Massachusetts, 103-104, 113, 115
Aziscohos Formation, 207

Barre Granite, 312-313, 315
Basal conglomerate, 174
Basins
  nonmarine, 25. See also Boston basin; Narragansett basin; Norfolk basin
  structural. See Boston basin; Narragansett basin; Norfolk basin
  topographic. See Boston basin; Narragansett basin; Norfolk basin
Batholiths, 206, 232-233, 265, 302-303
  White Mountain batholith, 284, 294-295
Bear Hill syncline, 208, 216
Bedrock tunnels, 5-6, 33, 37, 39, 113
Belknap Mountain Complex, 284, 291, 293
Bellingham Conglomerate, 174-175, 183, 194
Bellows Falls pluton, 302, 306, 311
Berry Ledge Formation, 268-269
Berwick Formation, 272
Bethlehem Gneiss, 302-303, 306, 310-311, 315
Bethlehem pluton, 311
Big Machias Lake region, Maine, 240, 242, 250, 256
Billings Hill Formation, 207, 217
Biotite
  chemistry, 71, 87
  Nashoba Formation, 36, 41, 50, 55, 60
  phase equilibria studies, 90
  Rattlesnake pluton, 70-71, 76, 87
  western Maine, 225
Black Mountain syncline, 208, 216, 218
Blackstone Series, 144
Blastophitic texture, 325
Blister collapse, 72, 89, 96
Blister hypothesis, 89
Bloody Bluff fault zone, 31, 33
Blueberry Mountain fault, 208, 216-219, 221
Blue Hill Granite Porphyry 24-25, 127-128, 194
  chemistry, 86, 139
  contact relations, 128, 132
  feldspar, 81, 86
  origin, 5, 25, 128
Blue Hills igneous complex, 125-127, 194-195. See also Blue Hills volcanic complex; Blue Hills volcanic rocks
  lithology, 24, 74, 125-126, 128
Blue Hills, Massachusetts, 5-6, 12, 21, 74, 125
  tectonics, 5-6, 25-26
Blue Hills volcanic complex, 5, 25, 126. See also Blue Hills volcanic rocks
  bedding, 24, 125
Blue Hills volcanic rocks. See Chickatawbut Road flow; Great Dome Trail flow; Hemenway Hill flow; Pine Hill flows; Pyritic volcanic rocks; Wampatuck Hill flow

banding, 127, 130, 135-136
chemistry, 125, 137-139
contact relations, 128, 132, 136
correlation, 7, 24, 88, 126
description, summary, 129
fine structures, 24, 134-136, 138
modal analyses, 128-129
origin, 125, 129, 132, 134-135, 137-139
radiometric dating, 128
stratigraphic relations, 6, 125, 127-129, 136, 139
thickness, 24
zonation, 125, 128, 130-134
Bolton Formation, 272
Bolton Gneiss, 32. See also Nashoba Formation
Bornite, 334
Borrowdale volcanic rocks, 314
Boston basin, 5-30
age, 5, 7, 182
anticlines, 7, 12, 15
correlation, 193
dikes, 20-21, 23, 27
folds, 12, 14-16
fossils, 6, 10, 183, 193
fractures, 27-28
hypabyssal rocks, 19-21
joints, 17-18, 27
major structures, 12
minor faulting, 7, 17-20, 22, 27-28
minor structures, 16-22, 28
regional structure, 15, 27
sedimentation, 25-26
stratigraphy, 6, 184
subsidence, 25-26
thrust faults, faulting, 6-7, 12-13, 15-16, 25, 27
Boston Bay Group, 7, 8, 11, 25. See also Cambridge Argillite; Roxbury Conglomerate
age, 5, 7-8, 11
formations, 5, 8, 14
kaolinization, 9
Boston College, 15, 122
Boston Harbor, 7, 12, 14-15
Boston University, 122, 182, 195, 264, 345
Boudinage, 219
Bouguer anomaly map, 304, 307. See also Gravity maps
Bouguer anomaly profiles, 305-311
Bouguer corrections, 303
Boundary Mountain anticlinorium, 207
Brachiopods, 270, 275
Braintree Argillite (Formation), 24-25, 184
Bravoite, 334
Breccia. See Volcanic rocks; Sedimentary structures
Brecciation, 12
Brimfield Schist, 103, 108, 113
Brimstone Mountain anticline, 207, 224
British Columbia, 337

Brittle fracture, 28, 302, 315
Bronson Hill anticlinorium, 310, 315
Bronson Hill-Boundary Mountain anticlinorium, 205, 207, 242, 265
Bronzite, 323, 333-334
Brookline. See Newton-Brookline
Brookline member, 9-10, 12, 25
Brown University, 195
Bunker Pond pluton, 212, 215-216
Bushveld deposit, 338, 345

Calc-alkaline rocks, 314. See also Ayer Crystalline Complex; Dedham quartz monzonite; New Hampshire Plutonic Series
Calc-silicate rocks
Maine, 207, 217, 269-270
Nashoba Formation, 31, 36, 51, 66-67
Cambrian
Blue Hills, Massachusetts 5, 24
Early
Hoppin Hill, Massachusetts, 144, 194
Weymouth Formation, 24
flanks of Merrimack synclinorium, Maine, 264
fossils, 7
Late
Fossils in quartzite pebbles, Narragansett basin, 26
Middle
Braintree Formation, 24
quartzite, Massachusetts, 24, 146
Quincy Granite, 73
unconformity, 7, 25
Wachusett-Marlborough Tunnel, 39
Cambridge Argillite, 5, 11-12, 15, 17
contact relations, 12, 15-16
facies change, 5, 9
lithology, 9-13, 16
minor structure, 11, 17
thickness, 5, 9, 11, 13
thrusting, 12, 14, 17
Cambridge, Massachusetts, 9
Canton Junction, Massachusetts, 181-182, 185, 187, 194-195
Cape Ann Granite
chemistry, 74, 83-85, 87, 97
feldspar, 76, 84-85, 97
gravity studies, 90
lithology, 74, 82, 94
mineralogy, 74, 82, 87-88, 97
radiometric dating, 72-73, 94, 97
Cape Ann, Massachusetts, 72, 87, 94-95
Cape Ann pluton, 74, 94. See also Cape Ann Granite
Cape Horn, New Hampshire, 293
Cape Neddick Complex, 284, 292
Carboniferous. See Mississippian; Pennsylvanian
Cardigan pluton, 304, 306-307, 310-312
Carrabassett Formation, 210

Castle Hill anticline, 242
Cataclasis, 111, 115, 122, 219, 226, 228
Cataclastic texture, 212
Catazone, 315
Cathodoluminescence studies
  alkali feldspar 74, 77, 81
  apatite, 81
  calcite, 81
  dolomite, 324
  feldspar, 77
  fluorite, 76, 81
  Massachusetts, 97
  plagioclase, 324
Cauldron subsidence, 72, 96, 282-298
Central anticline, 12-13, 15
Central United States, 154, 156
Centripetal drainage, 195
Chalcopyrite, 320, 325, 327, 331-332, 343-344
Chapman syncline, 242
Charles River syncline, 5, 12, 14-16, 19
Chelmsford granite, 104, 115, 120-122
Chelsea, Massachusetts, 9. 14
Cherry Hill granite, 94
Chert, 62, 240
Chickatawbut Road flow, 113, 129, 132-134, 138
  chemical analyses, 137-138
  petrography, 132-134
  "Chilled" contact, 135. *See also* Blue Hills volcanic rocks
Chlorite, 20, 107, 324-325. *See also* Metamorphic grade; Metamorphism, greenschist; Penninite
  Nashoba Formation, 37-38, 55-56
  northern Maine, 239, 245-246, 249, 258
  Rattlesnake pluton, 76-77
  western Maine, 212, 227, 231
Chloritization of biotite, 77, 228
Chloritoid, 242
Chromian ulvospinels, 320
C.I.P.W. norms, 40, 48-50, 56, 58, 61
City Tunnel, 8, 10, 15, 19-20
City Tunnel Extension, 8-10, 12-14, 16-17, 20
Classification, igneous rocks, 87, 212
Clasts
  Blue Hills volcanic rocks, 134
  Boston Bay Group, 8-11, 26
  flattened, 222
  Narragansett basin, 144, 146, 194-195
  Norfolk basin, 128, 195
  northern Maine, 245-247, 249-250
  Sagerville Formation, 267-268
  stretched, 216
  volcanic, 240
  Winterville Formation, 258
Cleavage, 258. *See also* Fracture cleavage; Maine, western

designation of, 216
Clifton Formation, 148, 162
Clinton facies, 103, 107-109. *See also* Ayer crystalline complex
  age, 103, 121-122
  contact relations, 111-112, 119-120
  faulting, 109, 111
  feldspar, 116-119, 122
  foliation, 104, 107, 109-111, 120-122
  lithology, 105, 107, 121-122
  modal analyses, 105, 109
  origin, 103, 109, 111, 119-122
  tectonic pulses, 109, 111
Clinton fault, 33
Clinton, Massachusetts, 33, 103, 107-108, 113
Clinton-Newbury fault zone, 33
Clough quartzite, 115
Coal
  meta-anthracite, 144-145
  Narragansett basin, 145, 147-148, 183, 186
  Worcester, Massachusetts, 182-183, 193
Coal basins, 163-164. *See also* Coalfields; Narragansett basin; Worcester basin
Coalfields, 145, 154, 160, 162, 164, 168, 170
Columbia Point, Massachusetts, 15
Compton Formation, 272
Concordant rocks, 120, 313
Concord Granite, 302, 304, 306, 308, 310, 312, 315
Cone fractures, 290-291, 294, 297
Cone sheets, 286-287, 292
Conglomerate. *See* Basal Conglomerate; Bellingham Conglomerate; Dighton Conglomerate; Pondville Conglomerate; Purgatory Conglomerate; Rangeley Conglomerate
Connecticut, 33, 204, 264, 314
Connecticut River valley, 92
Conrad discontinuity, 311
Contact halo, 76, 88
Conway Granite, 283, 294
Cordierite, 233
Coticule, 38-39, 56, 66
Crescent Range ring dike, 293
Cretaceous, 281-282
Cross-bedding. *See* Sedimentary structures
Crustal shortening, 25, 206, 273
Cubanite, 334
Cumberland Group, 150, 153, 156, 158, 166, 192
Cummingtonite, 319, 324
Cummingtonite-hornblende association, 324
Cupolas, 204, 214-215, 233
Currier Hill syncline, 265, 267, 269-271
Cylindrical fracture, 287

Decussate structure, 41
Dedham Granodiorite. *See* Dedham quartz monzonite

Dedham, Massachusetts, 14
Dedham quartz monzonite, 72, 82, 91, 184. *See also* Weymouth Granite
  age, 7-8, 72-73, 76, 144
  clasts, 26, 195
  contact relations, 7, 13, 15
  emplacement, 89, 96
Deer Island, Massachusetts, 14
Density determinations, 305, 310
Devens-Long Pond facies, 113, 120. *See also* Ayer Crystalline Complex
  age, 120, 122
  contact relations, 104, 115
  feldspar, 103, 105, 117, 119, 122
  lithology, 103, 105, 113-115, 119
  modal analyses, 113-116
  origin, 103, 120-122
  textural variation, 113-115, 122
Devil's Slide ring dike, 291
Devitrification, 130-131, 135-136, 138-139
Devonian. *See* Acadian orogeny
  age of folding, New Hampshire and Maine, 224, 311
  Boston basin, 183
  Callixylon genus, 183
  Clinton facies, 103, 122
  Dalhousie Group, New Brunswick, 240
  disconformity, Presque Isle region, 240
  Early
    Acadian orogeny, northern Maine, 242
    Carrabasset Formation, 210
    cleavage in pelitic rocks, 258
    Fall Brook Formation, 270, 275
    Littleton Formation, 210, 232, 303
    Madrid Formation, 209
    mafic rocks, southern Maine, 322
    Merrimack synclinorium, sedimentation, 204, 231, 263-264, 272
    Mooselookmeguntic pluton, 230
    Piscataquis volcanic belt, 314
    Seboomook Formation, 210, 275
    Solon Formation, 270, 275
    timing of events, 314
    upper sequence, western Maine, 209-210, 231
    volcanic rocks, New Hampshire, 313
    volcanic rocks, northern Maine, 240, 258-259
  felsic plutons, Maine, Vermont, 312
  fracture systems, southern New England, 97
  gabbros, Maine, 205, 218, 221, 311
  K-Ar dating, 72, 91-96
  "Late Devonian Plutonic Series," 89
  Lower
    cut by teschenite dikes, western Maine, 258
  Maine, 204
  metamorphism, Devens-Long Pond facies, 120
  metamorphism, southern Maine, 319, 342

Middle
  Aroostook County, Maine, 240
  Mapleton Sandstone, 259
  Mooselookmeguntic pluton, 230
  Newbury Volcanic Complex, 8
  New Hampshire Plutonic Series, 209, 301-302
  northern Maine, 241
  Oakdale Quartzite, 122
  Peabody Granite, 72
  plant fossils, Norfolk basin, 186
  plutonism, western Maine, 218
  prehnite- and pumpellyite-bearing dikes, northern Maine, 259
  Rattlesnake pluton, 72
  regional metamorphism, Massachusetts, 120
  Roxbury Conglomerate, 10
  Sebago pluton, 313
  silicic rocks, southern Maine, 342
  slaty cleavage, 224
  thickness of plutons, 302
  time scale, 230
  Vassalboro Formation, 266, 269
  Wachusett-Marlborough tunnel, 39
Diorite. *See* Salem Gabbro-Diorite; Straw Hill Diorite; Warren gabbro-diorite
Dorchester Bay Tunnel, 6, 15-16
Dorchester, Massachusetts, 13
Dorchester Member, 9-11, 15
Dorchester Tunnel, 5-6
Drag folds, 16, 130

Early Paleozoic, 144, 239-240, 265
Electron microprobe analyses, 320, 322
Electron microprobe scanning photographs, 332, 337-340
Electron microprobe studies, 248, 250, 323
Engineering geology, 6
Epeirogenic movement, 146
Epicule, 39
Epidote, 40, 60, 107, 246, 251, 256. *See also* Pistacite
Epieugeosyncline, 147
Eugeosynclinal sequence, 32-33, 206
Europe, 152, 181, 192-193
Eutaxic structure, 125, 130, 132-133, 136
Euxinic conditions, 272
Exsolution
  chalcopyrite-pyrrhotite, 331
  chalcopyrite-sphalerite, 335
  feldspar, 76, 80-81
  ilmenite-spinel, 337
  machinawite-chalcopyrite, 343
  magnetite-hematite, 336
  magnetite-ilmenite, 336
  pentlandite, 331, 343
  pyrrhotite, 326-327
  pyrrhotite-pentlandite, 325, 327, 331

Facies. *See* Sedimentary facies; Metamorphic facies
Fall Brook Formation, 268, 270, 272, 275
Farmington, New Hampshire, 303
Fault contacts, 103
Faults defined, 17
Fault zones. *See* faults by name
   Bloody Bluff fault zone, 31, 33
   Clinton-Newbury fault zone, 31, 33, 111, 113
   Rattlesnake Hill fault zone, 113
Fayalite, 74, 82
Feeder conduit, 313
Feldspar. *See* Alkali feldspar; Albite; Anorthoclase; Labradorite; Microcline; Orthoclase; Perthite; Sanidine
   geothermometer, 119-120
   homogenization, 105, 107, 117
   sampling, 105, 116
   staining, 105
   zonation, 76
Ferrohastingsite, 91
Ferrohastingsite granite, 71, 74, 82, 84-85, 91, 96. *See also* Cape Ann Granite; Peabody Granite
Ferrohastingsite syenite, 74
Finland, 338
Flow banding, 127
   Blue Hills volcanic rocks, 24, 135-136
   Nashoba Formation, 46-47, 57
Flow structure, 213
Flow zones, 134
Fluorine-containing minerals, 86
Fluorite, 76, 92
Fluxes, 74, 81, 83
Flysch, 264, 272
Folds. *See* Boston basin; Maine, central, northern, western; Merrimack synclinorium; Narragansett basin; folds by name
Foliation
   Ayer Crystalline Complex, 104, 107, 109-111, 121-122
   Nashoba Formation, 37, 41, 47, 50, 59
   New Hampshire Plutonic Series, 302, 311
   northern Maine, 244
   western Maine, 204, 213, 216-217, 219, 221-222, 225, 227-228
Forge Pond, 107
Forsterite, 37, 57, 59
Fort Devens Army Reservation, 113
Fortin Series, 272
Fossil localities, 276-278
Fossils. *See* Animal fossils; Plant fossils; Pseudofossils; Trace fossils
Fossil zones
   Linopteris obliqua Zone, 148, 154, 166
   Ptychocarpus unitus Zone, 164, 170
Foxboro, Massachusetts, 156, 158, 164, 173
Fracture cleavage, 224, 242, 244

Fracture system, southern New England, 97
Franconia Ridge ring dike, 296
Frenchville Formation, 244-245, 247-248, 251, 254-255, 259

Gabbro
   New Hampshire Plutonic Series, 311
   Salem Gabbro-Diorite, 127
   thickness, New Hampshire, 311
   Warren gabbro-diorite, 323-324
   western Maine, 205, 212, 218, 221
   White Mountain Plutonic-Volcanic Series, 282
Galena, 82, 334
Garnet
   associated with chlorite, 64
   Kinsman Quartz Monzonite, 314
   Nashoba Formation, 33, 36-38, 40-41, 43-45, 47, 50, 52-53, 55, 59-60, 63-65
   Penobscot Formation, 322
   Spaulding Quartz Diorite, 302
   Warren gabbro-diorite, 325
   western Maine, 225-226
Gaspé Peninsula, 204
Geophysical laboratory, 320
Geosynclinal stage, 96
Geosyncline. *See* Epieugeosyncline
Geothermometry, 58, 88, 119-120, 341
Gersdorffite, 334
Gettysburg College, 195
Gile Mountain Formation, 272
Gneiss
   Ayer Crystalline Complex, 103, 105, 115, 120
   Nashoba Formation, 32-33, 36-37, 44, 58
   New Hampshire, 313
   western Maine, 204, 219
Gneissic domes, 115
Gneissic fabric, 219
Graded bedding. *See* Sedimentary structures
Granite
   classification, 87
   flow structures, 228
   mesozonal, 212
   two-feldspar. *See* Subalkalic granite
   two-mica, 212-213, 221
Granite porphyry. *See* Blue Hill Granite Porphyry; Rattlesnake pluton
   defined, 76
   Franconia Ridge ring dike, 296, 298
   origin, 90
   Rattlesnake pluton, 71, 76-77
   southern Maine, 321
Granitic rocks. *See* Granite; Granodiorite; Katahdin pluton; Quartz diorite; Quartz monzonite
   depth of emplacement, 233
   topographic expression, 213
Granitic sheets
   New Hampshire, 224

western Maine, 204, 213-215, 219, 224, 233
Granitization, 323
Granodiorite, 218. *See also* Ayer Granodiorite; Dedham Granodiorite; Songo Granodiorite; Umbago Granodiorite
Granophyric texture, 130, 137
Granulation, 109
Granulite, 36, 51, 53, 56
Graphite
   Kinsman Quartz Monzonite, 314
   Nashoba Formation, 36-37
   nickeliferous deposits, southern Maine, 320, 322, 326, 338-339, 341, 344-345
   western Maine, 226
Graptolites. *See* Animal fossils
Gravity field, 205
Gravity lows, 214, 233
Gravity maps, 213, 304, 307
Gravity studies. *See* Bouguer anomaly
   Cape Ann Granite, 90
   New Hampshire, 301-308
   western Maine, 206
Graywacke
   central Maine, 265-269, 271
   chemical composition, 51, 55, 65
   Dighton Conglomerate, 186
   lithic graywack, modes, 247
   metamorphism, 240
   modal analysis, 185
   Narragansett basin, 145, 147
   northern Maine, 240, 245, 247-252, 254, 257
   southeastern Massachusetts, 183
Great Basin, 89
Great Cedar Swamp, 134
Great Dome Trail Flow, 129, 130
Greenland, 84
Greenlodge Formation, 184
Green Mountain anticlinorium, 311
Greenschist terrane, 203, 233
Greenstone, 144, 207
Greenvale Cove Formation, 207
Grossularite, 53, 66. *See also* Garnet
Grunerite, 79, 82

Harriman ore deposit, 319, 323, 325-342
   paragenetic sequence, 341-342, 344
Harriman peridotite, 319-320, 323-324
Hastingsite-riebeckite granite, 294
Harvard-Ayer area, Massachusetts, 106
Harvard Conglomerate, 103, 111
Harvard-Littleton-Ayer area, 107
Harvard, Massachusetts, 103-104, 109
Harvard University, 33, 148, 195, 345
Heidelberg University, 320-321, 345
Hematite, 338
Hemenway Hill flow, 128-129, 136-137
Hiatus, 181, 195

Highlandcroft Plutonic Series, 209, 211-212
Hildreths Formation, 210, 217
Hingham anticline, 12
Hingham Bay, 15
Hingham, Massachusetts, 9, 10, 25. *See also* Nantasket-Hingham
Hoppin Hill.formation, 184
Hoppin Hill granite, 144. *See also* Dedham quartz monzonite
Hornblende, 324
Hornblende-actinolite ultrafic rock, 33
Hornfels, 225, 233
Hot spot, 93
Houghs Neck anticline, 12
Houlton oroflex, 224
Hull, Massachusetts, 15
Hutchinson Brook anticline, 217-218
Hyde Park, Massachusetts, 14
Hyde Park syncline, 12
Hydrologic program, 33
Hydrothermal alteration, 335
Hydrothermal stage, 341
Hypabyssal dikes, 212, 231
Hypabyssal intrusives, 19-21
Hypersolvus granite, 72. *See also* Alkalic granite
Hypersthene, 302

Igneous complexes, 284-285, 293, 297. *See also* Alkalic granite, southern New England; White Mountain Plutonic-Volcanic Series
Igneous rocks. *See* Volcanic rocks; plutonic rocks by name
   classification, 58, 87, 212
Ignimbrites, 138-139. *See also* Volcanic rocks, ash flows
Ilmenite-sphene intergrowth, 251
Ilmenite-spinel assemblage, 320
Imbricate structural belt, 113
Insizwa, South West Africa, 320
Intermontane basins, 146, 193, 195
Isograd
   actinolite, 255
   biotite, Maine, 241
   dips, western Maine, 226
   K-feldspar, New Hampshire, 306, 313
   New Hampshire, 314
   regional, 315
   sillimanite
      Maine, 204, 214-215, 218, 224-225, 227
      New England, 313
      New Hampshire, 302, 306, 313
   sillimanite-potash feldspar, Massachusetts, 32, 66
   staurolite, western Maine, 225, 227
Isotopic dating, 213. *See* Radiometric dating

Jadeite, 254

Japan, 254, 258
Japan Analytical Chemistry Research Institute, 75
Jemtland Formation, 248-250
Johannsen's classification, 58
Joints, 17, 19, 81-82
Jura Mountains, 28
Jurassic, 281-282

Katahdin pluton, 205, 233
Kink bands, 216, 219, 226
Kinks in biotite, 41
Kinsman pluton, 306, 310-311
Kinsman Quartz Monzonite, 302-304, 306, 308, 310-311
Knox Mining Company, 345
Kyanite, 242

Labradorite, 252
Late Paleozoic
  Massachusetts, 5, 35, 126, 182, 193
  thermal event, western Maine, 229-231
Laterization, 45, 51
Late-stage fluids, 71
Lawsonite, 258
Leucoxene, 135
Lexington pluton, 205, 214, 233
Limestone
  central Maine, 266-269, 271, 274
  northern Maine, 257
Liquid immiscibility, 285, 319, 325, 336-337, 342
Lithic graywacke, 145. See also Graywacke
Lithic sandstone modes, 247
Lithofacies, 263, 270. See Sedimentary facies
Littleton Formation, 207, 272, 310
  age, 230
  correlations, 206-207, 210
  lithology, 205, 232
Little Zircon Mountain synform, 217
Llandoverian. See Silurian, Early
Logan Brook anticline, 208, 217
Lovejoy Mountain anticline, 217
Lowell, Massachusetts, 113
Lower Allegheny subgroup, 158
Ludden Brook Formation, 268
Ludlovian. See Silurian, Late
Lunar minerals, 320, 338, 340, 345
Lynn Volcanic Complex, 5, 8, 12, 14, 17
  age, 5, 7-8
  thrusting, 12, 14

Mackinawite, 320, 332-333, 343
Madawaska Lake Formation, 248
Madrid Formation, 207, 209, 216-217, 272
Magma
  alkalic, 72, 94
  andesitic, 314
  basaltic, 285-286, 292, 297-298, 314

  convection, 285-286, 296
  crystal settling, 281, 286
  devolitization, 138-139
  differentiation, 281-283, 285
  granitic, 219, 281, 292, 294-295, 298
  mixed magmas, 281, 285
  origin of, 234, 314
  peralkaline, 87, 89, 138
  quartz dioritic, 315
  quartz syenitic, 296
  stratification, 286
  subalkalic, 89
  superheat, 285, 297
  syenitic, 281, 286, 292, 294-297
Magma chamber. See Roof pendant
  fracturing, 286-287
  polymagmatic chambers, 281-298
Magmatic assimilation, 91, 96
Magmatic origin
  nickeliferous ores, 342
Magnetite, 135, 137, 335-336
Main Drainage Tunnel, 8, 10, 15-17
Maine, 33, 117, 204, 233, 242. See also Maine, central; Maine, coastal; Maine, northern; Maine, southern; Maine, western; Merrimack synclinorium
Maine, central. See Merrimack synclinorium, southeast limb
  mapping, 264
  Merrimack synclinorium, 204
  metamorphism, greenschist, 203, 242
  New Hampshire Plutonic Series, 265
  Piscataquis volcanic belt, 313
  slate belt, 263, 265
  Somerset geanticline, 242
  stratigraphy, 263, 274-278
Maine, coastal
  coastal plutons, 89, 97
  coastal volcanic tract (belt), 264-265, 272
  radiometric dating, 96
  volcanic rocks, 265, 272
Maine, northern
  Acadian orogeny, 258
  age of Acadian orogeny, 259
  Aroostook River Formation, 245
  cleavage, 244, 258-259
  deformation, periods of, 240
  faults, 244, 258
  folds, 244, 253, 258
  Frenchville Formation, 245, 248
  gravity anomalies, 312-313
  Jemtland Formation, 248-249
  lithology, summary, 243
  map, geologic, 241
  metamorphic assemblages, 244, 254
  metamorphic zones, 244
  metamorphism, 242, 254, 256, 258-259

mineralogy, 249–254. *See also* Actinolite; Albite; Analcime; Chlorite; Epidote; Prehnite; Pumpellyite
petrographic study, 242
Piscataquis volcanic belt, 314
Seboomook Formation, 210
stratigraphy, 241–242
Taconian deformation, 242
teschenite dikes, 258–259
Winterville Formation, 245
Maine, southern
  Cape Neddick complex, 292
  Exeter Diorite, 312
  gravity map, 312
  Harriman peridotite, 319–320, 323–324
  igneous complex, 282
  mapping, 272
  metamorphic grades, 242
  nickeliferous ores, 319–345
  plutonic sheets, 312
  Warren gabbro-diorite, 319
Maine, western. *See* Merrimack synclinorium
  boundinage, 219
  cataclasis, 219, 226, 228
  chronology of major events, 231, 264
  clasts, 222
  cleavage, fracture, 224
  cleavage, slaty, 224–225
  cleavage, slip, 203, 217–218, 222, 224, 226–227
  correlation with central Maine, 263
  deformations, 203, 215–218
  dikes, 228–229, 258
  doming, 203, 219, 231
  faults, 203, 215–216, 218
  folds, cross, 217, 219–221, 224
  folds, flexural, 203, 218
  folds, other, 203, 208, 217, 219, 222
  folds, recumbent, 203, 219, 222–224, 227
  foliations, 219–222, 225, 227
  fossils, 207
  granitic sheets, 219, 224
  kink bands, 226
  lower sequence, 206–207, 216
  maps, 206, 208
  metamorphic events, 224–228, 233
  metamorphic grades, 219, 224
  metamorphic zones
    sillimanite, 221
    staurolite, 221–222, 225–226
  metamorphism
    contact metamorphism, 203–204, 225
    general, 211, 227–228
    greenschist, 228, 231
    isograds, 211
    pressure conditions, 232, 234
    progressive, 225–227
    retrogressive, 219, 225–228
  middle sequence, 206–207, 216, 218
  migmitization, 219
  pegmatite, 218, 228. *See also* Whitecap pegmatite
  petrologic studies, 232, 234
  plutonic rocks, 204, 206, 211–215, 219, 221, 226, 231
  pre-Silurian rocks, 207
  radiometric dating, 208, 231
  ring dikes, 219
  schistosity, 203, 216–217, 221, 223–224, 228
  sillimanite terrane, 203–204, 206, 231–232, 234
  slate layer, 204
  stratigraphic sequences, 206–207, 209–210, 220
  structure, 204, 208, 219
  transition zone, 203–205
  Tumbledown antiform, 218–219
  upper sequence, 206–207, 209–210, 217–218, 231
Malden, Massachusetts, 9
Malden Tunnel, 8, 12, 14–15, 17, 20
Mantle, 283, 314–315
Mapleton, Maine, 258
Mapleton sandstone, 258–259
Maps, geologic. *See* specific geologic areas; for example, Boston basin, Merrimack synclinorium
Marble
  Blackstone Series, 144
  Hildreths Formation, 210
  Nashoba Formation, 31–32, 36, 51, 53, 57
  southern Maine, 322
Marcasite, 334
Maritime Canada, 143, 146, 173, 181, 193. *See also* Cumberland Group; Pictou Group; Riversdale Group; Clifton Formation; Minto Formation; New Brunswick; Nova Scotia
Marlboro Formation
  comparison with Nashoba Formation, 31–32, 57, 65
  lithologic types, 31, 39
Marlboro, Massachusetts, 32–33
Marlborough Township, 32
Maryland, 160
Massachusetts, 311, 314. *See also* Ayer Crystalline Complex; Boston basin; Blue Hills volcanic rocks; Narragansett basin; Nashoba Formation; Norfolk basin; Rattlesnake pluton
Massachusetts Bay, 146
Massapoag Lake granite, 75
Massive segregation bodies, 319
Massive sulfides, 319, 323
Masslite quarry, 147
Mattapan anticline, 12, 14
Mattapan, Massachusetts, 6, 14
Mattapan Volcanic Complex, 5, 12, 14
  age, 5, 7–8, 13–14
  correlation, 7–8, 126
  dikes, 8

fossils, 183
lithology, 7-8
stocks, 8
Mayflower Hill Formation, 266, 271
Medford, Massachusetts, 9
Melaphyre, 7, 20
Merrimack synclinorium, 33, 265. *See also* Maine, central, western
   age, 204, 264
   compression, 231
   correlation, regional, 209, 270-272
   described, 204, 215, 264
   fossiliferous rocks, southeast limb, 209
   fossil localities, 276-278
   greenschist terrane, western Maine, 203
   lithofacies, 263-264, 266-267, 270-271
   Lower Devonian(?), 209-210
   major events, western Maine, 231
   Massachusetts, 31
   metamorphism, 203-205, 211, 215-218
   metasedimentary rocks, 204
   northwest limb, 264. *See* Maine, western
   plutonic rocks, 104, 211-215, 219
   relation to Cambrian-Ordovician rocks, 264
   sedimentary facies. *See* Lithofacies
   sedimentation, 231, 272-273
   slate belt, 204
   slump deformation, 203
   source of sediments, 271-272
   southeast limb. *See* Maine, central
      deformation, 265
      metamorphism, 265
      stratigraphic markers, 265
      stratigraphy, 263-278
      structure, 265
   structures, Maine, 204, 217-218, 264, 272-273
   thickness of sediments, 206, 264
   transition zone, Maine, 203-204
   unconformity, 115
   volcanic rocks, 204, 272
   width, 264
Mesozoic, 183, 205, 212, 231, 314
Metagraywacke. *See* Graywacke
Metalimestone. *See* Limestone
Metaluminous granite, 83-84
Metamorphic assemblage. *See* Prehnite-Pumpellyite metamorphism
   chlorite-calcite, 254, 257
   disequilibrium, 256
   equilibrium, 248
   graphite assemblages, 345
   Nashoba Formation, 53, 66-67
   northern Maine, 244
   prehnite and pumpellyite assemblages, 254, 256
   sillimanite assemblage, western Maine, 227
   western Maine, 233

Metamorphic facies
   Barrovian type, 120
   glaucophane schist, 254
   greenschist facies, 120, 126, 202, 204, 225, 231, 255
   intermediate-pressure facies, Maine, 204
   Nashoba Formation, 32, 66
   prehnite-pumpellyite facies, 239-261
   sillimanite-potash feldspar, Maine, 203, 219
   zeolite facies, 240
Metamorphic grade
   biotite grade, central Maine, 265
   central Maine, 269
   chlorite grade, 183, 242, 265
   relation to folds, 253
   sillimanite grade
      Barre Granite, Vermont, 315
      Maine, 219, 265
   sillimanite + K-feldspar-grade, western Maine, 219
   staurolite grade
      Narragansett basin, 183
      western Maine, 215, 222, 225-227, 234
Metamorphic rocks, 32, 213. *See also* Gneiss; Granulite; Marble; Migmatite; Quartzite; Schist
Metamorphic zones
   chlorite zone, central Maine, 267
   garnet zone, Vermont 242
   Kyanite-chloritoid, 242
   prehnite and pumpellyite, 252-254
   sillimanite
      New Hampshire, 302
      southern Maine, 323
      western Maine, 219, 221, 224-227, 231, 234
   sillimanite + K-feldspar
      southern Maine, 323
      western Maine, 219, 226
   staurolite zone, western Maine, 221-222, 225-227, 234
Metamorphic zoning, 252-254
Metamorphism. *See* Isograds; Prehnite-pumpellyite metamorphism
   Acadian, 240
   Buchan type, Maine, 204
   contact. *See* Aureole
   contact, western Maine, 120, 204, 233, 302, 315, 342
   disequilibrium, 226
   downgrading, western Maine, 219
   greenschist, 126, 203-204, 225, 228, 231
   intermediate-pressure type, 240, 258
   lower Paleozoic rocks, Maine 239
   low-grade, 183, 256. *See also* Metamorphism, greenschist; Prehnite-Pumpellyite metamorphism

metamorphic events, western Maine, 233
metamorphic grade, texture, 249
Narragansett basin, 92
overprinting, 225, 242
pressure conditions, 234
progressive, 224, 226-227
regional Maine, 117, 224, 342
relation to deformation and plutonism, 224-228
relation to granitic rocks, western Maine, 218
retrograde, southern Maine, 342
retrogressive, western Maine, 219, 225-228
sillimanite metamorphism, 203, 215
subgreenschist metamorphism, 256, 258. *See also* Prehnite-Pumpellyite metamorphism
Taconian, Maine, 240
western Maine, 211
zonation, 225
Metapelite, 310
central Maine, 265-267, 269-272, 274-275
Metasedimentary rocks, 265, 275. See also Metapelite
Maine, central, 265-275
Maine, southern, 321
Maine, western, 203-234
Metasomatism, 77, 94, 120. *See also* Metamorphism
Metropolitan District Commission, 6, 33
Mexico, 288
Microcline, 64, 213. *See also* Alkali feldspar
Micrographic intergrowth, 76, 252
Microperthite granite, 87. *See also* Alkalic granite; Peralkaline granite
Middle Paleozoic, 35. *See also* Silurian; Devonian
Migmatite, 40, 66, 104, 310
western Maine, 204, 206, 211, 215, 217, 219, 226, 231
Migmatitic gneiss, 217
Migmatization, 219
Migration of centers of intrusion, 284-285, 297
Milton, Massachusetts, 11, 24
"Milton" Quartzite, 11-14, 25
Minnesota, 338
Minto Formation, 143, 148, 162, 166, 173
Miocene, 88, 89
Mississippian
Blue Hill Granite Porphyry, 126, 128
Boston basin, 183
fossils, 183, 186
Quincy Granite, 126, 128
Lynn Volcanic Complex, 5, 7-8
M.I.T., 40, 320
Moat Volcanics, 232, 282, 293-295, 298
Molybdenite, 334
Monongahela Group, 158. *See* Pennsylvanian
Monzonite, 283
Moody Brook Formation, 268-269
Mooselookmeguntic Lake, 210-211

Mooselookmeguntic pluton, 208, 225, 228, 233
emplacement, 204, 233
lithology, 212-213
metamorphic zones, 221, 224, 227, 233
radiometric dating, 229-230
relation to structures, 205, 216, 218, 221
shape, 204, 213-214
Moose River synclinorium, 258
Morien Group, 156, 160. *See* Pictou Group
Mount Abraham, 234
Mountain Pond syncline, 221
Mount Clough pluton, 302, 306, 311
Mount Garfield ring dike, 296
Mount Hope fault, 13
Mount Monadnock, 288
Mount Monadnock pluton (complex), 284, 292
Mount Pawtuckaway complex, 283
Mount Tripyramid complex, 282, 284
Munsungun anticlinorium, 242
Mylonite, 111
Myrmekite, 115, 121
Myrmekitic intergrowths, 325-326, 332, 335-336

Nahant, Massachusetts, 5
Nantasket-Hingham, 6
Nappes, 115, 302-303, 311, 314. *See also* Recumbent folds
Narragansett basin, 7, 26, 144-148, 194
age, 25, 92, 143-144, 173-174, 182-183, 195
amphibian footprints, 194
clasts, 144, 146, 194-195
coal, 145, 183, 186
correlation, 183, 193
facies changes, 175, 186
intrusion of magma, 147
Late Cambrian fossils, 26
lithology, 144, 147, 174
metamorphism, 92, 183
paleoenvironment, 193-195
plant fossils, 143-180, 182-183, 194
red beds, origin of 147
source of sediments, 26, 144, 146-147, 195
stratigraphy, 144-148, 184, 186
synclines, 146
unconformity, 144
volcanism, 144, 147
Narragansett Pier Granite, 92
Nashoba Formation, 31-67
age, 39
amygdules, 46-47
chemistry, 31, 39, 54
comparison with Marlboro Formation, 31-32, 57, 65
flow banding, 46-47, 57, 65
forsterite, 57, 59
garnet. *See* Garnet

layering, 46–47, 51
lithologic types, 31–32, 66
lithologic units, 32–33, 36–38
   biotite-quartz granulite, 36, 51
   biotite-rich gneiss and schist, 33, 38, 40, 42–46, 48–50, 54–55
   calc-silicate granulite, 36, 53, 56, 58
   coticule, 56, 64–66
   hornblende gneiss, schist, and amphibolite, 54–59
   marble, 57–58
   muscovite- and (or) chlorite-rich schist and gneiss, 54, 62–65
   quartz-feldspar-biotite gneiss and schist, 46–47, 50–54, 56
   quartzite and quartz schist, 59–60, 62, 64
metamorphism, 31–32, 53, 66–67
origin, 31, 58
postmetamorphic deformation, 64
P-T conditions, 32, 66–67
relict texture, 62
sillimanite, 50, 60, 63–65, 67
source rocks, 31–32, 45–46, 53, 58, 65
stratigraphy, 32
temperature of formation, 32, 66–67
thickness, 31, 39
unconformity, 31
volcanic origin, 31–32, 53, 62, 65–66
Nashua River Gorge fault, 113
Native iron, 345
Neponset River, 14
Nevada, 89
New Brunswick, 143, 156, 162–163, 166, 240. See also Pictou Group
Newburyport, Massachusetts, 113
Newbury Volcanic Complex, 8
New England, 234, 240, 242, 302, 312
Newfoundland, 240
New Hampshire, 85, 224, 229, 234, 283, 314. See also New Hampshire Plutonic Series; White Mountain Plutonic-Volcanic Series
  Ammonoosuc Formation, 245
  Ayer Granodiorite, 104
  folding, 224, 311, 314–315
  gravity studies, 214, 301, 303–304, 311–312
  Merrimack synclinorium, 33, 264
  plutons, 204, 224
  sillimanite terrane, 232, 234
New Hampshire Plutonic Series, 305–306, 313, 315. See also Bethlehem Gneiss; Concord Granite; Kinsman Quartz Monzonite; Spaulding Quartz Diorite
  age, 209, 265, 301–302, 312
  batholithic structure, 303
  central Maine, 265
  Clinton facies, 109
  emplacement, 234, 302, 312–315

  gravity data, 303–311
  lithology, 209, 212–213, 311
  nappe formation, 314
  radiometric dating, 204
  sheet structure, 301–303, 310, 312
  thickness of plutons, 302, 308, 312, 315
  western Maine, 211, 215
Newry mines, 213
New Sweden Formation, 248
Newton-Brookline, Massachusetts, 6
New York, 240
New Zealand, 240, 249, 254, 258
Niccolite, 334
Nickeliferous deposits, 319–345
Nigeria, 285
Nigerian granites, 83, 85, 88, 90
Nigerian minerals, 86–88
Niggliite, 335
Norfolk basin, 6–7, 144, 146, 184, 193. See also Pondville Conglomerate; Wamsutta Formation
  age, 6, 25–26, 182–183, 186–188, 192
  clasts, 195
  deposition, 6
  lithology, 24, 183, 195
  metamorphism, 183
  paleoenvironment, 193–195
  plant fossils, 181–182, 187–194. See also Plant fossils, Norfolk basin
  red beds, origin of, 194
  stratigraphy, 183, 185–186
Norfolk, Massachusetts, 185–186
Normal faults, 218, 323
North Attleboro, Massachusetts, 185
Northern Appalachians, 204, 224, 231, 257, 272
  metamorphism, 239–259
Northern border fault, 5, 12
North Metropolitan Relief Tunnel, 8, 14, 20, 27
North Scituate basin, 146, 174, 182–183
Norway, 336
Nova Scotia, 143, 150, 153–154, 158, 173
  plant fossils, 156, 166, 168, 170
Noyes Mountain Formation, 268–269

Oakdale Quartzite, 103, 108, 111, 115
  age, 120, 122
  contact relations, 111–112, 115
Obsidian, 131. See also Volcanic rocks, ash flows
Ohio, 160
Oliverian gneiss domes, 310, 315
Oliverian Plutonic Series, 310–311
Olivine, 323, 325–326, 328
Ookiep deposit, 342
Ophitic texture, 325
Ordovician
  Albee Formation, 207
  Ammonoosuc Volcanics, 310

Borrowdale volcanic rocks, 314
Cape Ann Granite, 72
Devens-Long Pond facies, 104, 120
Early
  Albee Formation, 207
  Aziscohos Formation, 207
  flanks of Merrimack synclinorium, Maine, 264
  Frenchville Formation, 245
  Highlandcroft Plutonic Series, 209, 212
  Late (Ashgillian)
    metasedimentary rocks, western Maine, 206
    nickeliferous ores, Knox County, Maine
    northern Maine, 242
    Quimby Formation, 207
    Taconian deformation, 259
    volcanic rocks, northern Maine, 245, 259
    Winterville Formation, 245
  mafic rocks, southern Maine, 322
  Merrimack synclinorium, 204
  metamorphic grade, northern Maine, 259
  Middle
    Ammonoosuc Formation, 207
    Aroostook County, Maine, 240
    black shale, western Maine, 207
    cleavage in pelitic rocks, northern Maine, 258
    Dixville Formation, 207
    Partridge Formation, 207
    Taconian orogeny, New England, 240
    volcanic rocks, graywacke, Maine, 240
  northern Maine, 241
  Partridge Formation, 310
  Pennington Mountain anticlinorium, 242
  Penobscot Formation, 319, 321–322
  Protolith of Devens-Long Pond facies, 104
  quartzite, Massachusetts, 146
  Quincy Granite, 24, 72
  radiometric dating, southern New England, 94
  sedimentation, Merrimack synclinorium, 231, 272
  southern Maine, 321
Orthoclase, 41, 47
Orthomagmatic stage, 341
Ossipee Mountain Complex, 284, 290, 292–293, 295
Ossipee Mountain ring dike, 296
Oxygen fugacity
  alkalic granites, 74, 86–87, 96–97
  nickeliferous deposits, 314, 320, 339–340, 345

Paleozoic, 32, 35, 183
"Palisade disturbance," 93
Paraboloidal fractures, 281, 287–290, 298
Parkman Hill Formation, 269, 271, 275
  correlation, 270–271
  lithology, 269, 272, 274–275
Parks Brook, 293
Partridge Formation, 207, 216, 308, 310–311
Patch Mountain Formation, 268–269

Pawtuckaway complex, 284, 292
Peabody Granite, 74, 84, 90, 94
  chemistry, 74, 83–84, 97
  feldspar, 74, 76, 84, 97
  radiometric dating, 72–73, 92, 94–95, 97
Peabody pluton. See Peabody Granite
Pecenga deposit, 342
Pecenga, USSR, 320
Pegmatite. See Whitecap pegmatite
  Cape Ann Granite, 82
  Clinton facies, 121
  Nashoba Formation, 32, 40
  Newry mines, 213
  Quincy Granite, 86
  Rattlesnake pluton, 80, 82–83, 88, 90–91
  southern Maine, 322, 325, 334
  western Maine, 213, 218, 221
Pegmatite dikes, 82, 115, 213, 228
Pennington Mountain anticlinorium, 241–242, 244–245
Penninite, 249. See also Chlorite
Pennsylvania, 160, 240
Pennsylvanian, 25, 92, 173, 187, 193. See also Allegheny Group; Conemaugh Group; Monongahela Group; Narragansett basin; Norfolk basin
  Allegheny Group, 158, 160, 162–163, 166, 173
  Boston basin, 5, 7, 183
  Conemaugh Group, 158, 160, 163, 173
  faulting, 147
  folding, 26, 147
  Harvard Conglomerate, 111
  Maritime Canada, 143
  metamorphism, 183
  Monongahela Group, 158, 160, 163, 173
  Narragansett basin, 73, 143, 154, 183, 195
  Norfolk basin, 6, 24, 181, 185, 192, 195
  paleoenvironment, 193–195
  sedimentation, 192
  stratigraphy, 146, 187, 193. See also Narragansett basin; Norfolk basin
  Worcester Phyllite, 39, 183
Penobscot Formation, 322–323, 338
  nickeliferous ores, 319–345
Pentlandite, 319–320, 325, 329–331, 343–344
  Chemical analyses, 327, 330–331
Pentlandite-chalcopyrite intergrowths, 329–331
Peralkaline granite, 83–84, 86, 96–97. See also Quincy granite; Rattlesnake pluton
Peralkaline rhyolite, 125, 138. See also Blue Hills volcanic rocks; Moat Volcanics
Peralkalinity, 83, 125
Peraluminous rocks, 80, 84
Percy complex, 284, 290–291, 293–294
Peridotite, 319, 323. See also Harriman peridotite
Peridotite-gabbro sequences, 320

Permian
　　disturbance, 92, 126
　　Dunkard group, 156
　　faulting, 121
　　fracture tectonics, 121
　　metamorphism, 121, 319, 342
　　plant fossils, 154, 163, 186
Perry Mountain Formation, 206–207, 271
Perthite, 116–119. *See also* Alkali feldspar
Phacolithic sheets, 315. *See also* Plutons, shape
Phenocrysts, 296
Phillips pluton, 205, 208, 211, 215
　　lithology, 212
　　metamorphic zoning, 225–226
Pictou Group, 143, 152–153, 156, 162–164, 173, 192
　　New Brunswick, 162–163, 187
　　Sydney coalfield, Nova Scotia, 154, 160, 162, 164, 170, 187
Pilot Range, New Hampshire, 293
Pine Hill flows, 128–129, 135–136
Pine Hill, Massachusetts, 125, 134
Piscataquis volcanic belt, 314
Pistacite, 67. *See also* Epidote
Pittsfield member, 267
Plainville beds, 173
Plainville, Massachusetts, 143, 147–148, 171, 175, 187
　　plant fossils, 143–172
Plant fossils
　　Annularia foliage, 153, 155
　　arthrophytes, 153, 155, 173
　　Asterophyllites foliage, 153, 155
　　Boston basin, 183
　　Cordaites genus, 183
　　Diplothmema genus, 170
　　Narragansett basin, 143–172
　　Norfolk basin, 181–197
　　Palmatopteris genus, 170
　　Sphenophyllum foliage, 153, 155
Plate tectonics, 206, 314
Pleasant Mountain, Maine, 288
Pleasant Mountain pluton, 284
Pliny complex, 284, 291, 293
Plumbago Mountain fault, 212, 216, 218
Plumbago pluton, 212–213, 218
Plutons. *See* Batholiths; Granitic sheets; Phacolithic sheets; Stocks
　　composite, 233
　　concordant, 204
　　depth of crystallization, 204
　　dips of contacts, 213
　　emplacement, 224, 234
　　origin of, 281–298
　　relation to metamorphism, 203, 218, 224–225
　　relation to structure, 217, 231, 313
　　shape, 89–90, 204, 212–215, 302–303, 310–311, 315

　　thickness, 90, 302
　　zoning, 212
Pondville Conglomerate, 183, 185
　　age, 174, 181, 195
　　contact relations, 128, 185
　　correlation, 195
　　lithology, 24, 144, 185, 195
　　paleoenvironment, 194
　　sedimentary structures, 185
　　thickness, 24, 185
Pondville Group. *See* Pondville Conglomerate
Pondville Station, 185–186
Porphyroblasts, 47, 113, 227
Portland, Maine, 320
Post-Acadian, 258. *See also* Late Paleozoic; Mississippian; Pennsylvanian; Permian
Postmetamorphic deformation, 41
Post-Pennsylvanian, 25
Posttectonic granite. *See* Concord Granite
Posttectonic rocks, 39, 312
Potassium analyses, 75
Precambrian, basement
　　Boston basin, 5, 7, 12, 14–15
　　central Maine, 265
　　Dedham quartz monzonite, 8, 15, 73
　　Massachusetts, 25, 35, 39, 144
　　Newfoundland, 240
　　Rhode Island, 39, 144
　　unconformity, 25
Prehnite-pumpellyite metamorphism, 248–259
　　age, 258–259
　　assemblage calcite-chlorite-Ca-Al silicate, 257
　　conditions of metamorphism, 255–258
　　metamorphic assemblages, 239
　　metamorphic zones, 239–240, 246, 254–255, 258
　　Triassic, New Zealand, 240
Pre-Mississippian, 92–93
Pre-Pennsylvanian, 126, 128
Pre-Silurian, 207, 322
Presque Isle, Maine, 244
Pressure-quench mechanism, 90
Primary features, 115. *See also* Sedimentary structures
Pseudofossils, 183
Pseudomorphs
　　actinolite after pyroxene, 246
　　analcime after plagioclase, 250, 256
　　chlorite after staurolite, 225
　　magnetite, 135, 137
　　muscovite after staurolite, 225, 227
Puklen complex, 84
Pumice, 125, 130, 132, 135–136
Pumpellyite. *See* Prehnite-pumpellyite metamorphism
Purgatory Conglomerate, 143, 146, 174, 194
Pyrite, 334
Pyritic volcanic rocks, 129, 134–135, 139
Pyroclastic eruption, 129

Pyroxene, 256
Pyrrhotite, 319-320, 322, 325-329, 342-344
Pyrrhotite-olivine association, 328

Quadrangles
  Maine
    Anson quadrangle, 265, 268-269, 274, 276
    Augusta quadrangle, 266
    Bangor quadrangle, 271
    Bethel quadrangle, 206, 210
    Bingham quadrangle, 265
    Boyd Lake quadrangle, 267-268
    Bryant Pond quadrangle, 206, 210, 265, 269
    Buckfield quadrangle, 265, 269, 274-275
    Dixfield quadrangle, 210, 217, 265, 268
    Dover-Foxcroft quadrangle 268, 275-277
    Farmington quadrangle, 268-269, 271, 275
    Greenville quadrangle, 265
    Guilford quadrangle, 268-269, 271, 274-277
    Howe Brook quadrangle, 258-259
    Kingfield quadrangle, 265
    Kingsbury quadrangle, 268-269, 271, 276-278
    Livermore quadrangle, 271
    Old Speck Mountain quadrangle, 206-207, 210, 212-213, 216, 227
    Oquossoc quadrangle, 210, 221, 227
    Phillips quadrangle, 206, 210, 214, 216, 225, 228
    Pierce Pond quadrangle, 224
    Pittsfield quadrangle, 267, 271, 278
    Rangeley quadrangle, 206, 210, 213-214, 216, 218, 225-226, 228, 271-272
    Rumford quadrangle, 206, 210, 213, 216, 218-219, 226-227
    Sebec Lake quadrangle, 270, 275
    Sebec quadrangle, 270, 275
    Skowhegan quadrangle, 268-269, 276, 278
    Stetson quadrangle, 267, 270-271, 277-278
    Waterville quadrangle, 266
  Massachusetts
    Ayer quadrangle, 36, 105-106
    Brockton quadrangle, 72, 144
    Clinton quadrangle, 36, 105
    Framingham quadrangle, 36
    Hudson quadrangle, 36, 105-106
    Mansfield quadrangle, 72, 186
    Marlboro quadrangle, 36
    Maynard quadrangle, 36
    Norwood quadrangle, 186
    Paxton quadrangle, 36
    Shrewsbury quadrangle, 36
    Sterling quadrangle, 36
    Worcester North quadrangle, 36
    Worcester South quadrangle, 36
  New Hampshire
    Bellows Falls quadrangle, 304
    Cardigan quadrangle, 303-304
    Concord quadrangle, 303-304
    Franconia quadrangle, 284, 290-292, 296, 298
    Hillsboro quadrangle, 303-304, 311, 314
    Holderness quadrangle, 303-304
    Keene quadrangle, 304
    Lovewell Mountain quadrangle, 303-304
    Monadnock quadrangle, 303-304
    Mount Kearsarge quadrangle, 303-304
    Penacook quadrangle, 303-304
    Percy quadrangle, 88
    Peterborough quadrangle, 303-304
    Sunapee quadrangle, 303-304
Quartz diorite, 82, 89, 115, 212, 221, 227. See also Spaulding Quartz Diorite
Quartzite, 108, 120, 144, 146, 183
  central Maine, 268-269
  Maine, 222, 322
  Nashoba Formation, 32, 37, 60, 62
  New Hampshire, 232, 303
Quartz monzonite. See Ayer Crystalline Complex; Dedham quartz monzonite; New Hampshire Plutonic Series
Quartz syenite, 281, 284, 296
Quebec, 240, 329
Quimby Formation, 207
Quincy Granite
  age, absolute, 24, 73, 94-95, 97, 128
  age, relative, 24, 73, 128
  chemistry, 74, 83-86, 139
  contact relations, 11, 14, 25, 128, 136
  correlation, 72, 88, 92
  lithology, 74, 86, 125, 136
  origin, 5, 25, 74, 84, 86, 96, 128
  phase equilibria studies, 88
  xenoliths, 128, 132
Quincy, Massachusetts, 72, 95
Quincy pluton, 71, 86, 89. See also Blue Hill Granite Porphyry; Blue Hills volcanic complex (rocks); Quincy Granite

Radiometric dating
  alkalic granite, southern New England, 73
  bimodal ages, 97
  Devonian granite, 89
  disturbance of age, 91-92, 94, 97, 230
  feldspars, correction factor, 92
  K-Ar dating, 72
    accuracy, 75
    argon retention in mica, 93
    biotite, Rhode Island, 93
    biotite, Scituate Granite Gneiss, 92
    hornblende, alkalic granite of southern New England, 94
  Maine, alkalic rocks, 96
  mica, 92
  Pennsylvanian rocks, Massachusetts, 92
  riebeckite, Quincy Granite, 95
  Rattlesnake pluton, 72, 91-99
  Triassic, Connecticut River, 92

western Maine, 229, 231
lead-alpha dating, western Maine, 212
Mount Pawtuckaway complex, 283
Pb/Pb age, Quincy Granite, 24
Rb-Sr dating
  loss of radiogenic Sr, Quincy Granite, 95
  New Hampshire, 229
  New Hampshire Plutonic Series, Maine, 204
  validity, Quincy Granite, 95
  western Maine, 229, 231
  whole-rock, 94
Rhode Island, alkalic granite, 92
riebeckite, 73, 91
U-Th-Pb, alkalic granite, southern New England, 94
White Mountain Plutonic-Volcanic Series, 283
zircon, 94
Rammelsbergite, 334
Rangeley Conglomerate, 216
Rangeley Formation, 207, 212, 216, 271
Rangeley Lake, 207, 216, 224
Rattlesnake Hill fault zone, 113
Rattlesnake Hill muscovite granite, 37
Rattlesnake Hill pluton, 33, 39. *Note:* Not to be confused with Rattlesnake pluton
Rattlesnake pluton, 71-97
  age of emplacement, 73
  alkalic granites, 72-73
  allochthonous granite, 87-88
  biotite, 84, 87-88
  blister collapse, 72, 89
  cathodoluminescence studies, 75, 81
  cauldron subsidence, 72, 91, 96
  chemical concentrations, 71, 82, 90-91
  chemical enrichment, 71, 84, 86
  coarse biotite granite, 72, 75-76, 78-79, 83-85, 88
  coarse riebeckite granite, 75, 78, 80
  feldspars, 77, 84-86, 90, 96-97
  ferrohastingsite granite, 82
  fine riebeckite granite, 79-81
    chemistry, 83-84, 91-92
    correlation, 72, 92
    feldspar, 80-81, 84
    varieties, 79-80
  granite porphyry, 71, 75-78, 85-86, 88, 96
  joints, 81
  lithologic types, 71, 96
  magmatic differentiation, 72, 82, 88, 91
  modal analyses, 72, 75, 78-79
  oxygen fugacity, 87, 97
  pegmatite, 74, 79-80, 82-83, 88
  pegmatite, chemical enrichment, 81, 90, 96
  pegmatite zone, 72
  petrogenesis, 72, 87-91, 96
  radiometric dating, 73, 91-99
  relative age, 73, 96

  riebeckite, 84-85, 87, 96, 98
  riebeckite-biotite granite, 82
  rock chemistry, 85
  staining of feldspar, 75, 77
  temperatures of crystallization, 86, 88, 96
  trachyte, 72, 77, 80, 83, 86, 90, 97
  vapor-phase transfer, 72, 91
  water-vapor pressure, 90
  water-vapor pressure-quench episode, 71
  xenoliths, 82
Recumbent folds
  New Hampshire, 224, 311, 314
  western Maine, 203, 219, 223-224, 227, 231
Red beds, 144, 147, 186, 194
Red Hill complex, 282, 284
Redington pluton, 205, 208, 212, 215, 225, 233
Reducing agent, 345
Reverse faults defined, 17
Rhode Island, 174. *See also* Narragansett basin; North Scituate basin; Woonsocket basin
  alkalic granite, 72-74, 85, 90, 92, 98
  Bristol, R. I., 144
  chemistry of riebeckite, 86
  Cumberland, R. I., 86, 90
  Diamond Hill, 93
  "hot spot," 93
  Metacom Granite Gneiss, 144
  Narragansett Pier Granite, 92
  Newport, R. I., 146
  pre-Mississippian rocks, 92-93
  radiometric dating, 92-93, 97
  Scituate Granite Gneiss, 92
  Westerly Granite, 92
Rhode Island Formation. *See* Plant fossils, Narragansett basin
  age, 143, 173-174, 183, 187, 195
  coal, 148
  paleoenvironment, 194
  plant fossils, 143-180
  plant fossils, new species, 143, 169-172
  sedimentary facies, 144, 175
Ribbon rock, 268
Riebeckite, 24, 71-72, 74, 81, 85, 87, 90-91
Riebeckite-aegirine granite, 73
Ring complexes, 89, 281-298
Ring dikes, 89, 219, 281, 296
  origin, 281-298
Ring fractures, 288, 291, 293-294, 296-297
Riversdale Group, 150, 152-153, 156, 158, 166, 192
Rockdale, 186
Rockingham anticlinorium, 205
Rockport granite, 74. *See also* Cape Ann Granite; Cape Ann pluton
Rockport, Massachusetts, 88
Rockport quarry, 94
Roof pendant, 129
Roslindale syncline, 12-13

Roxbury Conglomerate, 5, 10-11, 13, 15-16. *See also* Brookline Member; Dorchester Member; Squantum Member
  clasts, 8-10, 26
  contact relations, 14-15
  facies change, 5, 9
  fossils, 8, 10
  lithology, 9-11
  members, 5, 8
  ripple marks, 10
  thickness, 5, 9, 11, 15
Roxbury puddingstone. *See* Roxbury Conglomerate
Rumford pluton, 212, 221
Rutile, 335-336

Saar coal basin, 163
Saint Francis Group, 272
Salem anticline, 208, 218-219
Salem Gabbro-Diorite, 127
Salinic disturbance, 240
Sangerville Formation, 267-268, 271, 274
Sanidine, 130, 132-135, 137
Saprolitization, 31, 46, 58
Sara Fier complex, 285
Saussuritization, 228, 231, 324. *See also* Metamorphism, greenschist
Schist
  Blackstone Series, 144
  central Maine, 269
  Nashoba Formation, 32-33, 36-37, 44, 58, 60, 62
  western Maine, 204, 206, 210, 216-217, 219
Schistosity, 224, 323
  western Maine, 203, 216-218, 221, 223-224, 228
Schlieren, 122
Scituate Granite Gneiss, 92
Scotland, 288
Seaboard Lowland, 320
Sebago pluton, 204, 211, 213-215, 218-219, 313
Seboomook Formation, 210, 230, 272
Sedimentary facies. *See* Lithofacies
  Boston basin, 5, 9
  central Maine, 264, 266-267, 270-271
  currier Hill syncline, 270-271
  Narragansett basin, 195
  Wamsutta Formation, 144, 175, 186
Sedimentary structures
  breccia, 186, 193, 267
  central Maine, 265
  convolute lamination, 267, 274
  cross-bedding, cross-lamination, cross-stratification, 11, 146, 185, 195, 266-268, 270, 274
  cyclic bedding, 266-267
  graded bedding, 11, 185-186, 266
  mudcracks, 186, 193
  Norfolk basin, 185, 193

raindrop prints, 186, 193
ripple marks, 10, 194
sandstone dikes, 186, 193
scour and fill, 185-186
slump folds, 16, 46, 203, 267
Serpentine, 36, 319, 323, 328, 336
Shagg Pond Formation, 207
Shards, 125, 130, 132-135
Sharon, Massachusetts, 71-72
Sharon Syenite, 184
Shears, 18, 20, 22, 28
Shepleys Hill, 113
Silicified zone, 218
Sillimanite
  intergrown with biotite, 41, 50, 60
  Nashoba Formation, 31, 33, 36, 38, 41-45, 47, 52, 60, 63-65
  Penobscot Formation, 322
Sillimanite-clinozoisite association, 67
Sillimanite terrane (plateau) 203-206, 215, 231-232, 234, 314
Sills, 19
Silurian
  Anasagunticook Formation, 268-269
  Aroostook River Formation, 245
  Cape Ann Granite, 72
  central Maine, 268-269
  disconformity, Presque Isle region, 240
  Early (Llandoverian)
    Mayflower Hill Formation, 266
    northern Maine, 242
    Rangeley Formation, 207
    Sangerville Formation, 268, 274
    sedimentation, Merrimack synclinorium, 264, 272
    Taconian deformation, 259
    Taconian orogeny, 240
  Fall Brook Formation, 270
  flyschlike deposits, 272
  graptolites, Merrimack synclinorium, 263-266, 268-270, 274-278
  Late (Ludlovian)
    Ashland, Maine, 259
    Fall Brook Formation, 275
    Parkman Hill Formation, 269
    sedimentation, Merrimack synclinorium, 264
    Thompson Mountain Formation, 269
    tuff, northern Maine, 259
    Waterville Formation, 266-267
  Madrid Formation, 209
  Maine coastal volcanic belt, 272
  Mayflower Hill Formation, 266
  Merrimack synclinorium, central Maine, 263
  Middle (Wenlockian)
    Anasagunticook Formation, 269
    Parkman Hill Formation, 269, 275
    Sangerville Formation, 268

Thompson Mountain Formation, 269
Waterville Formation, 266-267
Newbury Volcanic Complex, 8
northern Maine, 241, 250
Oakdale Quartzite, 120, 122
Parkman Hill Formation, 269
Quincy Granite, 72
rocks cut by analcime-bearing dikes, northern Maine, 258
Sangerville Formation, 267-268, 274
Vassalboro Formation, 266, 269
Waterville Formation, 266-267
western Maine, 206
Slate belt, 263
Slaty cleavage, 17, 27, 216, 224-225, 232, 242
Slip cleavage, 203, 215-218, 222, 224, 226-227
Slump folds. *See* Sedimentary structures
Smalls Falls Formation, 206-207, 217, 272
Sodic granite, 84. *See also* Alkalic granite
Sohland an der Spree deposit, 342
Solon Formation, 270, 275
Somerset geanticline, 242, 271-272
Somerset Island, 271
Somerville, Massachusetts, 9
Songo Granodiorite, 212
Southeastern Massachusetts, 182-183
Spaulding pluton, 312
Spaulding Quartz Diorite, 302, 306, 311-312
Spectrographic analysis, 39-40
Sperrylite, 335
Sphalerite, 334
Sphene-magnetite association, 58
Spherulites, 129-135
Spider-web joints, 292
Spinel, 324, 336-338, 340
Spragueville Formation, 248
Squantum, Massachusetts, 15
Squantum Member, 5, 8-11, 14, 16
Squantum tillite. *See* Squantum Member
Standards, 39
Staurolite. *See* Pseudomorphs
Staining of feldspar, 74-75
Stephanian (Upper Pennsylvanian), 187
Stocks, 71, 265, 281-298, 302. *See also* Plutons
Stoping
New Hampshire Plutonic Series, 302, 315
Rattlesnake pluton, 71-72, 91
western Maine, 234
White Mountain Plutonic-Volcanic Series, 281, 283, 286-288, 290-292, 296-297
Stratigraphic markers, 265
Stratigraphy. *See* Boston basin; Narragansett basin; Norfolk basin; Maine; Merrimack synclinorium; Nashoba Formation
Straw Hollow Diorite, 103, 108, 111, 122
Strike-slip fault, 185
Stony Brook fault, 12-13, 185
Strontium analyses, 229

Subalkalic defined, 88
Subalkalic granite, 89. *See also* Ayer Crystalline Complex; Dedham quartz monzonite; Calc-alkaline rocks; New Hampshire Plutonic Series
Subduction zone, 314
Subgreenschist metamorphism. *See* Prehnite-pumpellyite metamorphism
Sudbury deposit, 320, 342
Sulfide mineralogy, 325-335
Sulfide ores, 323. *See also* Warren ore deposit
Sulfidic rocks. *See* Smalls Falls Formation
Sulfur fugacity, 320, 331, 334, 343, 345
Supergene alteration, 334, 341-342
Sydney coalfield, 154, 160, 162, 164, 170, 187
Syenite, 282, 288, 293, 295, 298. *See also* Ferrohastingsite syenite
Synclines. *See* Boston basin; Merrimack synclinorium; Narragansett basin
Synforms, 208, 217
Synintrusive blastesis, 104
Synorogenic sheet, 302
Syntectonic intrusions, 315
Syntectonic rocks, 37
System Ab-Or-SiO$_2$-H$_2$O, 120
System Fe-Cu-Ni-S, 343

Taconic orogeny, 240, 259, 319, 322
Tadmuck Brook Schist, 111. *See also* Brimfield Schist
Talc, 324
Taunton syncline, 146
Tectonic dewatering, 203
Temiscouata Formation, 272
Tennessee, 168
Ternary minimum, alkalic granite, 72
Teschenite dikes, 258
Thompson Mountain Formation, 207, 269
Thrusts, 12. *See also* Boston basin
Tillite. *See* Squantum Member
Tonalite. *See* Quartz diorite; Trondhjemite
Tory Hill syncline, 208, 218-219
Tourmaline, 114
Trace fossils, 265
Transverse fault, Stony Brook fault, 12
Tremolite-actinolite, 324
Triassic, 92-93, 240
Trondhjemite, 212. *See also* Quartz diorite
Tuff. *See* Volcanic rocks
Tumbledown antiform, 208, 217-219, 221, 225-226
Tunk Lake pluton, 89
Turner Formation, 268-269

Ultramafic complex, 324
Ultramafic rocks, 212. *See also* Harriman peridotite
Ulvospinel, 338
Umbagog Granodiorite, 212
Unconformity

Subject Index

Ammonoosuc Volcanics and Clough Quartzite, 115
angular, 240, 245
Blue Hills volcanic complex, 24
Cambrian-Precambrian, 25
Dedham quartz monzonite, 73
Harvard Conglomerate and Clinton facies, 111
Hoppin Hill, Massachusetts, 144
Oakdale Quartzite and Devens-Long Pond facies, 104, 115, 120
Pondville Conglomerate and pre-Pennsylvanian rocks, 144
Presque Isle region, 240
Taconic unconformity, western Maine, 207
Wachusett-Marlborough Tunnel, 31, 37
University of Heidelberg, 321, 324, 335, 345
University of Tennessee, 345
Upper Carboniferous, 147. *See also* Pennsylvanian
Upper Pottsville Group, 158, 162, 166
Uralite, 324
U.S.S.R., 320
Uvarovite, 324. *See also* Garnet

Vapor-phase transfer, 72, 91
Vassalboro Formation, 266, 269-270, 272
Vermont. *See* Barre Granite
  Appalachian fold belt, 257
  Ascutney Mountain, 288
  gravity anomalies, 312-313
  igneous complexes, 282
  pre-Silurian metamorphism, 242
Vesicular structures, 44
Vitroclastic texture, 136
Volcanic centers, 65
Volcanic neck, 135
Volcanic rocks. *See* Blue Hills volcanic rocks; Lynn Volcanic Complex; Maine, coastal; Mattapan Volcanic Complex; Piscataquis volcanic belt
  agglomerate, 135
  andesite, 7, 246, 257, 282, 294
  ash fall, 125, 272
  ash flows, 125-141
  basalt, 20, 55, 57-58, 144, 282, 294
  basalt-andesite-dacite-rhyolite association, 65
  bombs, 134, 138
  Boston basin, 193
  breccia, 10, 135-136, 282, 294
  coastal Maine, 272
  comendite, 83-84, 94, 138-139. *See also* Rhyolite, peralkaline
  dacite composition, 46, 55
  diabase, 246
  felsite, Mattapan Volcanic Complex, 7
  glass, 138
  lava flow. *See* Pine Hill flows, 135-136
  Narragansett basin, 183
  Nashoba Formation, 32
  Norfolk basin, 193
  Pillow lava, 245
  rhomb-porphyry, 74
  rhyolite, 24, 74, 125, 144, 257, 282, 294. *See also* Blue Hills volcanic rocks
  rhyolite, chemical analyses, 137
  rhyolite, peralkaline, 125
  splite, 245-246
  trachyte, 71, 77
  tuff
    ash-flow, 84
    Brookline Member, 10
    horizon, 248
    New Hampshire, 282, 293-295, 298
    northern Maine, 240, 245-246, 249-251, 258-259
    welded, 128, 130-131, 136. *See also* Blue Hills volcanic rocks
Volcanic vent, 135

Wachusett Aqueduct, 111, 113
Wachusett-Marlborough Tunnel, 33, 37, 39, 113. *See also* Nashoba Formation
Wachusett Reservoir, 113
Wairakite, 258
Waltham, 321
Wampatuck Hill flow, 128-132, 137-138
Wamsutta Formation, 147
  age, 24, 175
  amphibian footprints, 194
  sedimentary facies, 144, 175, 195
  stratigraphy, 183-186
Wamsutta Group. *See* Wamsutta Formation
Warren gabbro-diorite, 319, 324-325
Warren ore deposit
  bravoite occurrence, 334
  chemical composition of minerals, 322
  described, 319
  disseminated sulfides, 323
  massive segregation bodies, 319
  massive sulfides, 323
  mineralogy, 325-342
  modal data, 339, 342
  paragenetic sequence, 341-342, 344
  relation to schistosity, 323
  supergene alteration, 334
  thickness, 323
  tonnage of ore, 320
  variation of sulfides with depth, 343
Water-vapor quench episode, 96
Waterville anticlinorium, 265
Waterville Formation
  facies change, 271-272
  lithology, 266-267, 270, 272
  sedimentary facies, 266-267
Waterville-Vassalboro area, Maine, 264, 266, 269, 271
Weare pluton, 306, 310, 312

Weeksboro–Lunksoos Lake anticlinorium, 264–265
Welding, 130–132
Wenlockian. *See* Silurian, Middle
Westboro Quartzite (Formation), 194–195
West Central Maine. *See* Maine, central
Westerly Granite, 92
Westford, 107
Westphalian (Middle Pennsylvanian), 146. *See also* Pennsylvanian
West Roxbury Tunnel, 6, 8
West Virginia, 160
Weymouth Granite, 72. *See also* Dedham quartz monzonite
Weymouth, Massachusetts, 6
Whitecap pegmatite, 208, 213, 229–230, 235
White Mountain batholith, 284, 294–295, 305, 308
White Mountain Plutonic-Volcanic Series, 87–88, 281–300, 308
 age, 281
 magma source, 282
 Moat Volcanics, 232
 radiometric dating, 92, 95
 ring dikes, 281–298
 shapes of plutons, 283–284
 spatial pattern, 283
 structural control, 281–298
White Mountains, New Hampshire, 86

Williard Street fault, 136
Winnipesaukee pluton, 306, 308, 310
Winterville Formation, 245–246, 249, 251–254
 age, 245, 258
 lithology, 244–245, 250–254
Winthrop, 9
Wollaston syncline, 12
Woonsocket basin, 146, 174, 182–183
Worcester basin, 25, 26. *See also* Worcester Phyllite
Worcester coal, 183, 193
Worcester Formation. *See* Worcester Phyllite
Worcester, Massachusetts, 111, 182–183, 193
Worcester Phyllite, 24, 103, 108, 111, 122
 age, 39–40, 183
Wrentham, 186–187

Xenoliths, 82, 94, 119, 312
 Blue Hills igneous complex, 128, 132
 White Mountain Plutonic-Volcanic Series, 283, 286, 289–290, 292, 295–296, 298
X-ray data, 71, 84–86, 90, 96–97
X-ray diffraction analyses, 321, 343. *See also* Alkali feldspar

Zeolites, 254
Zircon, 94

WITHDRAWN